METHODS IN COMPUTATIONAL PHYSICS

Advances in Research and Applications

Volume 15

Vibrational Properties of Solids

Methods in Computational Physics

Advances in Research and Applications

1 STATISTICAL PHYSICS

2 QUANTUM MECHANICS

3 FUNDAMENTAL METHODS IN HYDRODYNAMICS

4 APPLICATIONS IN HYDRODYNAMICS

5 NUCLEAR PARTICLE KINEMATICS

6 NUCLEAR PHYSICS

7 ASTROPHYSICS

8 ENERGY BANDS OF SOLIDS

9 PLASMA PHYSICS

10 ATOMIC AND MOLECULAR SCATTERING

11* SEISMOLOGY: SURFACE WAVES AND EARTH OSCILLATIONS

12* SEISMOLOGY: BODY WAVES AND SOURCES

13* GEOPHYSICS

14 RADIO ASTRONOMY

15† VIBRATIONAL PROPERTIES OF SOLIDS

16‡ COMPUTER APPLICATIONS TO CONTROLLED FUSION RESEARCH

* Volume Editor: Bruce A. Bolt.
† Volume Editor: Gideon Gilat.
‡ In production.

METHODS IN COMPUTATIONAL PHYSICS

Advances in Research and Applications

Series Editors

BERNI ALDER

Lawrence Livermore Laboratory
Livermore, California

SIDNEY FERNBACH

Lawrence Livermore Laboratory
Livermore, California

MANUEL ROTENBERG

University of California
La Jolla, California

Volume 15

Vibrational Properties of Solids

Volume Editor

GIDEON GILAT

Department of Physics
Technion—Israel Institute of Technology
Haifa, Israel

1976

ACADEMIC PRESS New York San Francisco London

A Subsidiary of Harcourt Brace Jovanovich, Publishers

ACADEMIC PRESS, INC.
111 Fifth Avenue, New York, New York 10003

United Kingdom Edition published by
ACADEMIC PRESS, INC. (LONDON) LTD.
24/28 Oval Road, London NW1

LIBRARY OF CONGRESS CATALOG CARD NUMBER: 63-18406

ISBN 0-12-460815-9

PRINTED IN THE UNITED STATES OF AMERICA

Contents

LATTICE DYNAMICS AND RELATED PROPERTIES OF POINT DEFECTS

R. F. Wood

LATTICE DYNAMICS OF SURFACES OF SOLIDS

F. W. de Wette and G. P. Alldredge

VIBRATIONAL PROPERTIES OF AMORPHOUS SOLIDS

R. J. Bell

LATTICE DYNAMICS OF QUANTUM CRYSTALS

T. R. Koehler

METHODS OF BRILLOUIN ZONE INTEGRATION

G. Gilat

COMPUTER STUDIES OF TRANSPORT PROPERTIES IN SIMPLE MODELS OF SOLIDS

William M. Visscher

Contributors

Numbers in parentheses indicate the pages on which the authors' contributions begin.

G. P. ALLDREDGE,* *Department of Physics, The University of Texas, Austin, Texas* (163)

R. J. BELL, *Division of Quantum Metrology, National Physical Laboratory, Teddington, Middlesex, England* (215)

F. W. DE WETTE, *Department of Physics, The University of Texas, Austin, Texas* (163)

G. DOLLING,† *Chalk River Nuclear Laboratories, Atomic Energy of Canada, Limited, Chalk River, Ontario, Canada* (1)

G. GILAT, *Department of Physics, Technion—Israel Institute of Technology, Haifa, Israel* (317)

T. R. KOEHLER, *IBM Research Laboratory, San Jose, California* (277)

R. PYNN, *Physics Department, Brookhaven National Laboratory, Upton, New York* (41)

WILLIAM M. VISSCHER, *Theoretical Division, Los Alamos Scientific Laboratory, University of California, Los Alamos, New Mexico* (371)

JOHN L. WARREN, *Los Alamos Scientific Laboratory, University of California, Los Alamos, New Mexico* (77)

R. F. WOOD, *Solid State Division, Oak Ridge National Laboratory, Oak Ridge, Tennessee* (119)

THOMAS G. WORLTON, *Argonne National Laboratory, Argonne, Illinois* (77)

* Present address: Graduate Center for Materials Research, University of Missouri, Rolla, Missouri 65401.

† Temporary address: Institut Laue–Langevin, Grenoble, France.

Preface

THROUGH LARGE-SCALE COMPUTATIONS, it is becoming possible in many fields to examine theoretical models applied to complex systems. The solid state, being an aggregate of many interacting particles, is a good example of such a complex system. A previous volume was devoted to electronic band calculations, while this volume is devoted to another important aspect of the solid state, namely, its microscopic vibrational behavior. This field, known as lattice dynamics, has received considerable impetus recently by the very detailed information that has become available, principally through inelastic neutron scattering experiments. The analysis of such data inevitably requires computational methods. It is the purpose of this volume to examine these computational methods and to illustrate them for ordered lattices, quantum solids, impurity modes, surface modes, and amorphous solids.

The leading article is of a general nature, describing the basic theoretical models and their computational aspects for different solids of diverse chemical nature. The next article reviews both the methods of automation and computation in the highly sophisticated experiments in inelastic scattering of neutrons. The third article describes how group theoretical methods treated by computers can yield the proper symmetry assignments of phonon eigenvalues and eigenstates. The following group of four articles is concerned with different applications (as described by their titles) of traditional lattice dynamics, each having its own computational ramification. Since most of the properties of solids involve, in one way or another, integrations over the Brillouin zone that is obtained from lattice dynamics, the next article describes the computational methods used for that purpose, and a few applications. The last article concerns the dynamic or time-dependent aspect of lattice dynamics, namely, the calculation of thermal and electric conductivities in some models of solids. It is hoped that this fairly comprehensive survey will prove helpful in teaching and in advancing research in lattice dynamics.

G. GILAT*
B. J. ALDER
S. FERNBACH

* I wish to express my gratitude to the Department of Physics, Oregon State University, Corvallis, Oregon, for their kind hospitality during my sabbatical visit.

METHODS IN COMPUTATIONAL PHYSICS

Advances in Research and Applications

Volume 15

Vibrational Properties of Solids

The Calculation of Phonon Frequencies

G. DOLLING*

CHALK RIVER NUCLEAR LABORATORIES
ATOMIC ENERGY OF CANADA, LIMITED
CHALK RIVER, ONTARIO, CANADA

I. Introduction

A. Historical Background

THE STUDY OF THE vibrations of atoms and molecules in crystals can be said to have begun with the classic papers by Born and von Kármán, "On Vibrations in Space Lattices," and by Debye, "On the Theory of Specific Heat," both published in 1912. In the latter paper, the atomic vibrations were treated as if they were elastic waves in a continuous isotropic medium instead of a set of discrete atoms oscillating about their equilibrium positions. The Born–von Kármán treatment was very much more realistic in this respect and has formed the essential basis for most of the more recent work in this

* Temporary address: Institut Laue–Langevin, Grenoble, France.

field. In the early days, however, the Debye approach proved to be more tractable and thus enjoyed greater popularity for several decades. Even now, the concept of the Debye temperature θ_D is still widely used as a more or less accurate measure of the maximum phonon frequency in a material.

In the Born–von Kármán theory, the force on a given atom in a crystal depends on its position relative to the other atoms—in principle, to all other atoms, but in practice just to those surrounding atoms which are close enough to have a significant interaction with the given atom. The atomic motions are most readily described in terms of a superposition of traveling plane waves, the so-called lattice vibrations, each specified by a wave vector $\mathbf{q}$ ($= 2\pi$/wave length), a circular frequency ω, and a polarization index j. The relationship between these quantities

$$\omega = \omega(\mathbf{q}, j) \tag{1}$$

is called the phonon dispersion relation. A phonon is a quantum of vibration energy of the crystal. The precise form of this dispersion relation is directly related to the forces which exist between the atoms and/or molecules in the crystal. If we can measure or calculate the frequencies for all the normal modes of vibration of the crystal, then we can relate them back to the basic interatomic forces and calculate forward to obtain the many other properties, such as lattice heat capacity, that depend upon the atomic vibrations.

In this paper we shall discuss the various kinds of models for the interatomic forces that have been proposed for several different classes of crystalline solids. In every case, the models are specified by one or more parameters or force constants, which must be chosen in some way so that numerical calculations of the phonon frequencies can be made. In Section IV we consider some of the commonly used approaches to this problem, with their various difficulties and limitations. The starting point for all these investigations is the equations of motion for the atoms or molecules in the crystal. These equations are developed in detail in the comprehensive and authoritative treatise by Born and Huang (1954). The following abbreviated derivation is given so as to establish a convenient notation for the remainder of the article. (This notation is very similar to that employed in the article by Warren and Worlton, this volume; the main difference is the employment of $\mathbf{q}$ here, instead of $\mathbf{k}$, to represent the normal mode wave vector).

B. Basic Equations of Motion

We consider a perfect crystal composed of a very large number of identical unit cells, each cell specified by three non-coplanar vectors $\mathbf{a}(i)$, $i = 1, 2, 3$, and containing n atoms labeled κ, located at crystallographically distinct positions $\mathbf{r}^\kappa(\mathbf{l})$, where

$$\begin{aligned} \mathbf{r}^\kappa(l) &= \mathbf{x}^\kappa(l) + \mathbf{u}^\kappa(l) \\ \mathbf{x}^\kappa(l) &= \mathbf{l} + \mathbf{x}^\kappa \\ \mathbf{l} &= \sum_{i=1}^{3} l(i)a(i) \end{aligned} \tag{2}$$

Here, $\mathbf{l}$ is a set of integer numbers specifying the location of a unit cell origin, $\mathbf{x}^\kappa$ gives the equilibrium position of the κth atom with respect to that origin, and $\mathbf{u}^\kappa(l)$ is the displacement of the atom from its equilibrium position arising from thermal vibrations.

We assume that the potential energy Φ of the crystal can be expanded in a power series of the displacements $\mathbf{u}^\kappa(l)$:

$$\Phi = \Phi_0 + \Phi_1 + \Phi_2 + \Phi_3 + \cdots \tag{3}$$

where Φ_0 is the energy of the static lattice independent of $\mathbf{u}^\kappa(l)$, Φ_2 is quadratic in $\mathbf{u}^\kappa(l)$ and is called the *harmonic* term, while Φ_3 and the higher order terms represent the anharmonic contributions to the potential. In detail, we have

$$\Phi_1 = \sum_{l\kappa\alpha} \phi_\alpha{}^\kappa(1)\mathrm{u}_\alpha{}^\kappa(1) \tag{4}$$

$$\Phi_2 = \tfrac{1}{2}\sum_{l\kappa\alpha}\sum_{l'\kappa'\beta} \phi_{\alpha\beta}^{\kappa\kappa'}(ll')u_\alpha{}^\kappa(l)u_\beta{}^{\kappa'}(l') \tag{5}$$

$$\Phi_3 = \tfrac{1}{6}\sum_{l\kappa\alpha}\sum_{l'\kappa'\beta}\sum_{l''\kappa''\gamma} \phi_{\alpha\beta\gamma}^{\kappa\kappa'\kappa''}(ll'l'')u_\alpha{}^\kappa(l)u_\beta{}^{\kappa'}(l')u_\gamma{}^{\kappa''}(l'') \tag{6}$$

and so on, where the derivatives

$$\phi_\alpha{}^\kappa(l) = [\partial\Phi/\partial u_\alpha{}^\kappa(l)]|_0 \tag{7}$$

$$\phi_{\alpha\beta}^{\kappa\kappa'}(ll') = [\partial^2\Phi/\partial u_\alpha{}^\kappa(l)\,\partial u_\beta{}^{\kappa'}(l)]|_0 \tag{8}$$

are to be taken with all the atoms at their equilibrium positions. It follows from these definitions that the first-order contribution Φ_1 [Eq. (4)] must be identically zero. From the point of view of lattice vibrations, therefore, the most important term is Φ_2, with Φ_3, Φ_4 being relatively small perturbations for most materials, at least at moderately low temperatures. There are a few exceptional cases, such as solid He, which are so "anharmonic" that a treatment based on the harmonic term alone leads to absurd results (e.g., imaginary normal mode frequencies!). Methods of dealing with these cases are discussed by Koehler in this volume, to which the interested reader is referred. Here we confine the discussion to the harmonic approximation, in which only the second-order term Φ_2 is considered explicitly. The same discussion also applies to the so-called "quasi harmonic" approximation, in which the coefficients of Φ_2 are allowed to be temperature and/or pressure dependent, to compensate (so to speak) for the explicit omission of the higher order anharmonic terms which actually give rise to the temperature and pressure dependences of the normal mode frequencies.

The second-order coefficients, defined by (8), are the interatomic force constants; $-\phi_{\alpha\beta}^{\kappa\kappa'}(ll')$ is the force in the α direction on the κth atom in the lth unit cell when the atom $(l'\kappa')$ is displaced a unit distance (assumed small) along the β direction. Since a mixed partial derivative is independent of the order of differentiation, we have

$$\phi_{\alpha\beta}^{\kappa\kappa'}(ll') = \phi_{\beta\alpha}^{\kappa'\kappa}(l'l) \tag{9}$$

Further, since there can be no net force on an atom if the entire crystal undergoes a uniform translation,

$$\phi_{\alpha\beta}^{\kappa\kappa}(ll) = -\sum_{l'\kappa'}{}' \phi_{\alpha\beta}^{\kappa\kappa'}(ll') \tag{10}$$

where $\sum'$ denotes that the term $\kappa = \kappa', l = l'$ is omitted from the summation. It is important to note that $\phi_{\alpha\beta}^{\kappa\kappa}(ll)$ is defined by (10): it is *not* a second derivative of Φ. The potential energy of the crystal remains invariant against an infinitesimal rigid body rotation of the whole crystal, and also against all the allowed symmetry operations of the crystal. Application of these symmetry conditions can often effect a considerable reduction in the number of independent parameters $\phi_{\alpha\beta}^{\kappa\kappa'}(ll')$ required to specify the interatomic forces. One of the most general of these symmetry conditions, arising from the translational symmetry of the (infinite) perfect lattice, is that $\phi_{\alpha\beta}^{\kappa\kappa'}(l + l_1, l' + l_1) \equiv \phi_{\alpha\beta}^{\kappa\kappa'}(ll')$, where l_1 is any integer. Thus the interatomic force constants, if we ignore surface effects and other imperfections, depend on l and l' only through their difference $(l - l')$. The higher the crystal symmetry, the more conditions may be applied; extreme examples are provided by the face-centered or body-centered cubic structures in which many elements crystallize. The allowed independent force constants in these structures have been listed by Squires (1963). We shall give other examples of the simplifications arising from symmetry conditions in subsequent sections. Here we emphasize how important it is to make the maximum use of the crystal symmetry conditions to simplify the theoretical and numerical problems involved in the computation of normal mode frequencies.

The basic equations of motion are simply an elaboration of the well-known Hooke's law which governs the (small amplitude) extension of a spring: for the κth atom of mass m^κ we have

$$m^\kappa\, \partial^2 u_\alpha{}^\kappa(l)/\partial t^2 = -\sum_{l'\kappa'\beta} \phi_{\alpha\beta}^{\kappa\kappa'}(ll')u_\beta{}^{\kappa'}(l') \tag{11}$$

A simple solution to these equations is provided by plane waves of the form

$$u_\alpha{}^\kappa(l) = (Nm^\kappa)^{-1/2}\sum_{\mathbf{q}} B(\mathbf{q})e_\alpha{}^\kappa(\mathbf{q}) \exp i[\mathbf{q}\cdot\mathbf{x}^\kappa(l) - \omega(\mathbf{q})t] \tag{12}$$

where N is the number of unit cells in the crystal and $B(\mathbf{q})$ is a complex amplitude determined—in a classical picture—by the initial conditions. Substitution of (12) into (11) leads to a set of $3n$ equations in place of the $3nN$ equations (11):

$$\omega^2(\mathbf{q})e_\alpha{}^\kappa(\mathbf{q}) = \sum_{\kappa'\beta} D_{\alpha\beta}^{\kappa\kappa'}(\mathbf{q})e_\beta{}^{\kappa'}(\mathbf{q}) \tag{13}$$

where

$$D_{\alpha\beta}^{\kappa\kappa'}(\mathbf{q}) = (m^\kappa m^{\kappa'})^{-1/2}\sum_{l'} \phi_{\alpha\beta}^{\kappa\kappa'}(ll') \exp i\mathbf{q}\cdot[\mathbf{x}^{\kappa'}(l') - \mathbf{x}^\kappa(l)] \tag{14}$$

Slight variations of definition of the dynamical matrix $D_{\alpha\beta}^{\kappa\kappa'}(\mathbf{q})$ can be found in the literature, differing by a phase factor $\exp(i\mathbf{q}\cdot\mathbf{x}^\kappa)$. The expressions defined above are somewhat more convenient for numerical computations, particularly of the intensity of inelastic neutron scattering from crystals (see also Cochran and Cowley, 1967; Maradudin, 1974). Equation (13) can be written in matrix notation

$$\omega^2(\mathbf{q})\mathbf{e}(\mathbf{q}) = \mathsf{D}(\mathbf{q})\mathbf{e}(\mathbf{q}) \tag{15}$$

where $\mathbf{e}(\mathbf{q})$ is a column matrix:

$$\mathbf{e}(\mathbf{q}) = \begin{bmatrix} e_\alpha{}^1(\mathbf{q}) \\ e_\beta{}^1(\mathbf{q}) \\ e_\gamma{}^1(\mathbf{q}) \\ e_\alpha{}^2(\mathbf{q}) \\ \vdots \\ e_\gamma{}^n(\mathbf{q}) \end{bmatrix} \tag{16}$$

and $\mathsf{D}(\mathbf{q})$ is a $(3n \times 3n)$ square matrix:

$$\mathsf{D}(\mathbf{q}) = \begin{bmatrix} D_{\alpha\alpha}{}^{11} & D_{\alpha\beta}{}^{11} & D_{\alpha\gamma}{}^{11} & D_{\alpha\alpha}{}^{12} \cdots D_{\alpha\gamma}{}^{1n} \\ D_{\beta\alpha}{}^{11} & D_{\beta\beta}{}^{11} & D_{\beta\gamma}{}^{11} & D_{\beta\alpha}{}^{12} \cdots D_{\beta\gamma}{}^{1n} \\ \vdots & \vdots & \vdots & \vdots \qquad \vdots \\ D_{\gamma\alpha}{}^{n1} & D_{\gamma\beta}{}^{n1} & D_{\gamma\gamma}{}^{n1} & D_{\gamma\alpha}{}^{n2} \cdots D_{\gamma\gamma}{}^{nn} \end{bmatrix} \tag{17}$$

[We have omitted the argument $(\mathbf{q})$ after each element of $\mathsf{D}(\mathbf{q})$ to save space.] It is also helpful to define another matrix

$$M_{\alpha\beta}^{\kappa\kappa'}(\mathbf{q}) = (m^{\kappa}m^{\kappa'})^{1/2} D_{\alpha\beta}^{\kappa\kappa'}(\mathbf{q}) \tag{18}$$

and a related eigenvector $\mathbf{u}^{\kappa}(\mathbf{q}) = (m^{\kappa})^{-1/2}\mathbf{e}^{\kappa}(\mathbf{q})$, so that (14) can be expressed alternatively as

$$\omega^2(\mathbf{q})\mathsf{m}\mathbf{u}(\mathbf{q}) = \mathsf{M}(\mathbf{q})\mathbf{u}(\mathbf{q}) \tag{19}$$

where m is a diagonal matrix whose elements are the atomic masses. From Eq. (10), (14), and (18), we see that for $\kappa = \kappa'$

$$M_{\alpha\beta}^{\kappa\kappa'}(\mathbf{q}) = \sum_{l'}{}' \phi_{\alpha\beta}^{\kappa\kappa'}(ll') \exp i\mathbf{q} \cdot [\mathbf{x}^{\kappa'}(l') - \mathbf{x}^{\kappa}(l)] - \sum_{l'\kappa'}{}' \phi_{\alpha\beta}^{\kappa\kappa'}(ll') \tag{20}$$

where $\sum'$ indicates that the term $(l\kappa) = (l'\kappa')$ is to be omitted. If we now define

$$\bar{M}_{\alpha\beta}^{\kappa\kappa'}(\mathbf{q}) = \sum_{l'} \phi_{\alpha\beta}^{\kappa\kappa'}(ll') \exp i\mathbf{q} \cdot [\mathbf{x}^{\kappa'}(l') - \mathbf{x}^{\kappa}(l)] \tag{21}$$

then (20) can be written, for all κ, κ',

$$M_{\alpha\beta}^{\kappa\kappa'}(\mathbf{q}) = \bar{M}_{\alpha\beta}^{\kappa\kappa'}(\mathbf{q}) - \delta(\kappa, \kappa') \sum_{\kappa\kappa''} \bar{M}_{\alpha\beta}^{\kappa\kappa''}(0) \tag{22}$$

In this way, the condition of translational invariance [Eq. (10)] is automatically built into the elements $M_{\alpha\beta}^{\kappa\kappa'}(\mathbf{q})$, from which one easily constructs the dynamical matrix (17) with the help of (18).

The condition for the set of equations (15) to have a solution is that the determinant of the coefficients vanish:

$$|\mathsf{D}(q) - \omega^2(\mathbf{q})\, \delta(\alpha, \beta)\, \delta(\kappa, \kappa')| = 0 \tag{23}$$

The calculation of the normal mode frequencies thus reduces to the problem of finding the eigenvalues of the matrix $\mathsf{D}(\mathbf{q})$. In general, the elements of

$\mathsf{D}(\mathbf{q})$ will be complex numbers, obeying the hermitian condition

$$D_{\alpha\beta}^{\kappa\kappa'}(\mathbf{q}) = (D_{\beta\alpha}^{\kappa'\kappa}(\mathbf{q}))^* = (D_{\alpha\beta}^{\kappa\kappa'}(-\mathbf{q}))^* \qquad (24)$$

In the particularly simple case of a Bravais lattice ($n = 1$), $\mathsf{D}(\mathbf{q})$ becomes a real, symmetric (3×3) matrix which may be diagonalized without the aid of sophisticated modern computers. One may also dispense with the computer in certain instances where the symmetry of the wave vector $\mathbf{q}$ is high enough to permit some degree of block-diagonalization of $\mathsf{D}(\mathbf{q})$ by suitable choice of coordinates (we shall come back to this point in more detail later). However, it is in general too difficult and laborious to tackle the problem of solving Eq. (23) "by hand," one needs an adequate computer with a program library containing subroutines for diagonalizing (a) real symmetric and (b) complex hermitian matrices. It is further necessary that these subroutines should be able to deal with both degenerate and nondegenerate roots. Some of the more popular methods by which matrix diagonalization can be achieved are described in a book by Gourlay and Watson (1973). Computer programs based on these methods are now widely available at many laboratories, so that there is in principle no difficulty in obtaining the eigenvalues $\omega^2(\mathbf{q})$ and eigenvectors $\mathbf{e}(\mathbf{q})$ for any desired value of $\mathbf{q}$, given the crystal structure and a set of values for the basic force constants $\phi_{\alpha\beta}^{\kappa\kappa'}(ll')$. (There may of course be practical difficulties of storage space and computing time, if n is large or if the available computer is small!) For each $\mathbf{q}$ value, diagonalization of $\mathsf{D}(\mathbf{q})$ will generate $3n$ eigenvalues $\omega^2(\mathbf{q}, j)$, $j = 1 \ldots 3n$. Each eigenvalue will be associated with an eigenvector $\mathbf{e}(\mathbf{q}, j)$, having $3n$ components as listed in Eq. (16), which describe the type of atomic motion involved in the jth normal mode.

The above treatment of the crystal vibrations is completely classical: however, a rigorous quantum mechanical treatment leads to essentially the same equations, mainly because we are dealing with wave motion and simple harmonic oscillators. From a quantum-mechanical viewpoint we may speak of a crystal having $3n$ normal modes of vibration, each mode having a certain population of phonons—quanta of vibrational energy—determined by the equilibrium crystal temperature. The amplitudes of vibration of the atoms are given by these population values. These aspects of the subject are naturally of great importance for calculations of transition matrix elements in which some physical process, e.g., absorption of infrared radiation or inelastic scattering of thermal neutrons, depends upon the interaction with the phonons. But for the calculation of phonon energies or normal mode frequencies—we shall use these terms interchangeably—we only need to consider Eqs. (14) and (23).

C. Interatomic Potentials and Force Models

As we have seen in the previous section, the determination of the normal mode frequencies within the harmonic or quasiharmonic approximation requires at the outset a knowledge of the equilibrium atomic positions $\mathbf{x}^{\kappa}(l)$, i.e., the crystal structure, the atomic masses, and the interatomic force constants $\phi_{\alpha\beta}^{\kappa\kappa'}(ll')$. The second stage is to make any possible simplifications arising from symmetry conditions or other restrictions on the force constants that may be imposed, and finally the resulting expressions for the dynamical matrix elements are programmed ready for diagonalization by a suitable computer. The basic physics of the problem lies in the first stage, the establishment of the force constant values. In the original Born–von Kármán approach, the $\phi_{\alpha\beta}^{\kappa\kappa'}(ll')$ are regarded as disposable parameters, restricted only by the requirements of crystal symmetry. This continues to be extremely valuable as a framework within which to make calculations; the $\phi_{\alpha\beta}^{\kappa\kappa'}(ll')$ need not be treated as arbitrary parameters, but may be derived from some more fundamental theory of the interatomic forces. In certain circumstances it may be more appropriate to deal with their Fourier transformed values, i.e., directly with the elements $D_{\alpha\beta}^{\kappa\kappa'}(\mathbf{q})$. The basic framework of the calculation has, however, proved to be remarkably durable, as we shall see in the following sections.

It may be helpful, for visualizing more easily the significance of the $\phi_{\alpha\beta}^{\kappa\kappa'}(ll')$, to show how these are related to an assumed two-body potential function $V^{\kappa\kappa'}(r)$ which depends only on the distances $|\mathbf{r}|$ between pairs of atoms, κ, κ'. By definition (8),

$$\begin{aligned}\phi_{\alpha\beta}^{\kappa\kappa'}(ll') &= \left.\frac{\partial^2 V^{\kappa\kappa'}(r)}{\partial u_\alpha{}^\kappa(l)\,\partial u_\beta{}^{\kappa'}(l')}\right|_0 \\ &= \left.\frac{\partial^2 V^{\kappa\kappa'}(r)}{\partial r_\alpha{}^\kappa(l)\,\partial r_\beta{}^{\kappa'}(l')}\right|_{r=|\mathbf{x}^\kappa(l)-\mathbf{x}^{\kappa'}(l')|}\end{aligned} \tag{25}$$

If we consider any particular pair of atoms κ, κ', separated by $\mathbf{r}$, we can drop the superscripts on $V(r)$:

$$\begin{aligned}\phi_{\alpha\beta}^{\kappa\kappa'}(ll') &= [\partial^2 V(r)/\partial r_\alpha\,\partial r_\beta]|_{r=|\mathbf{x}^\kappa(l)-\mathbf{x}^{\kappa'}(l')|} \\ &= \partial/\partial r_\alpha[(x_\beta/x)(\partial V/\partial r)] \\ &= x_\alpha x_\beta/x^2[(\partial^2 V/\partial r^2) - (1/r)(\partial V/\partial r)] + \delta(\alpha,\beta)(1/r)(\partial V/\partial r) \\ &= x_\alpha x_\beta/x^2(A - B) + B\,\delta(\alpha,\beta)\end{aligned} \tag{26}$$

which serves to define the derivatives A and B of the potential $V(r)$ with respect to r. [This notation has been commonly used in studies of alkali

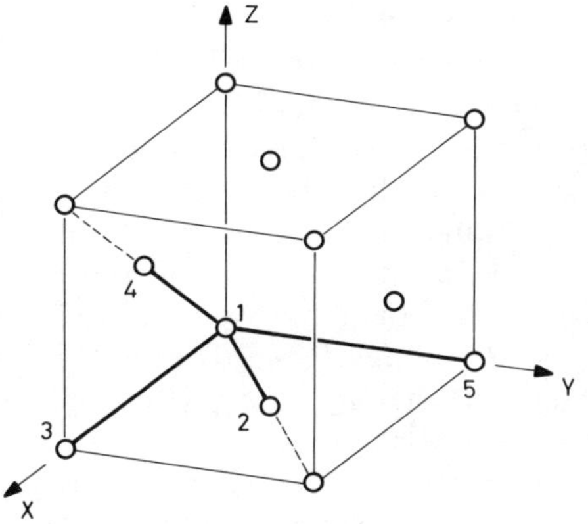

FIG. 1. Atoms on a face-centered cubic lattice. Axially symmetric interatomic force constants for the (1, 2) and (1, 3) pairs are given by Eq. (27). Force constants for (1, 4) are obviously related by symmetry to those for (1, 2), e.g. $\phi_{yy}{}^{14} = \phi_{zz}{}^{12}$. The interactions between atom 1 and its 12 nearest neighbors (only two of which are shown) can be readily derived in terms of those for (1, 2).

halides and other ionic crystals, for example, by Woods *et al.* (1963), Cowley (1964), and Dolling *et al.* (1965).] In a face-centered cubic crystal, as illustrated in Fig. 1, the force constant matrices for the pairs of atoms (1, 2) and (1, 3) are readily derived from Eq. (26) in terms of the A and B parameters for each interatomic distance:

(1, 2) Interaction, potential derivatives A, B:

$$\phi_{xx}{}^{12} = \phi_{yy}{}^{12} = (A + B)/2$$
$$\phi_{yz}{}^{12} = \phi_{zy}{}^{12} = \phi_{xz}{}^{12} = \phi_{zx}{}^{12} = 0$$
$$\phi_{xy}{}^{12} = \phi_{yx}{}^{12} = (A - B)/2$$
$$\phi_{zz}{}^{12} = B$$

(1, 3) Interaction, potential derivatives A', B':

$$\phi_{xx}{}^{13} = A'$$
$$\phi_{yy}{}^{13} = \phi_{zz}{}^{13} = B'$$

All other elements are zero. The two force constant matrices are, therefore,

$$\begin{bmatrix} (A + B)/2 & (A - B)/2 & 0 \\ (A - B)/2 & (A + B)/2 & 0 \\ 0 & 0 & B \end{bmatrix} \begin{bmatrix} A' & 0 & 0 \\ 0 & B' & 0 \\ 0 & 0 & B' \end{bmatrix}$$

It is easy to see how these are related to other matrices representing the (1, 4) and (1, 5) interactions, for example. It is clear that each "shell" of neighbors at a given distance from the origin atom 1 are represented by a set of closely related matrices (related by the symmetry operations required to transform from one pair of atoms to another), whose elements are simple functions of the appropriate A and B values. If we had not made the restrictive assumption that the (1, 2) interaction potential was a function only of the distance r between atoms 1 and 2, then the $\phi_{\alpha\beta}{}^{12}$ would have been restricted only by the crystal symmetry. By straightforward inspection of the symmetry of Fig. 1, one easily concludes that $\phi_{xx} = \phi_{yy} \neq \phi_{zz}$, $\phi_{xy} = \phi_{yx}$, and all other elements are still zero. There is, however, no relation between ϕ_{xx}, ϕ_{zz}, and ϕ_{xy}, so that the (1, 2) interaction is specified in this general case by three dependent parameters. A comprehensive discussion of the force constants for the face-centered cubic and body-centered cubic structures has been given by Squires (1963).

These force constant matrices and their interrelationships have been written down in some detail in the literature for a variety of other crystal structures. A selection of these references may serve as illustrative examples: alkali halides (Kellermann, 1940; Woods *et al.*, 1963); diamond (Smith, 1948; Herman, 1959); magnesium (Iyengar *et al.*, 1965); hexamethylenetetramine (Venkataraman and Sahni, 1970). The last-named example is of a molecular crystal in which the molecules have both translational and librational degrees of freedom. We shall discuss this problem in more detail in Section III.

In the last few years, a number of review papers have been written on the subject of force models for various kinds of crystal, in which the basic physics underlying the equations has been emphasized. The articles by Cochran (1971), Sinha (1973), and by Bilz *et al.* (1974) are especially valuable in this field. In the following sections we shall try not to duplicate these reviews too much, as far as the physics of the various problems is concerned; we emphasize instead the more computational aspects and give some hints on how the numerical calculations may be performed.

II. Interatomic and Interionic Force Models

A. RARE GAS SOLIDS

The rare gas solids are perhaps the simplest class of solids from a theoretical point of view, and the theoretical models which have been developed within the past few years are able to provide a very good description of most if not all of the observed vibrational properties of these materials. In the course of this development, it has proved necessary to go substantially beyond the ideas of the harmonic approximation and two-body potentials described in Section I. Earlier calculations of the thermodynamic properties of the rare gas solids (e.g., Horton and Leech, 1963; Grindlay and Howard, 1965) employed the familiar Lennard–Jones form for the two-body interaction:

$$V(r) = [\varepsilon/(m - 6)][6(\sigma/r)^m - m(\sigma/r)^6] \tag{27}$$

where m was taken to be 12 or 13. The parameters ε and σ were determined from the second virial coefficient of the appropriate rare gas at low temperatures. The effective Born–von Kármán force constants could then be found from Eq. (26), and anharmonic effects could be calculated from the higher order derivatives. In the case of He and Ne, these anharmonic effects are so large that standard perturbation theory does not work at all well, and even for the heavier rare gas solids, the perturbations became uncomfortably large at temperatures greater than one third of the melting point. The best solution to this problem thus far appears to be the self-consistent phonon theory (SCP), with modification to include the most important third-order terms in the expansion of the potential energy (Koehler, 1969). In the harmonic approximation, the force constants, or derivatives of the potential, are evaluated at the equilibrium positions of the atoms; in the SCP theory, on the other hand, the force constants are effectively averaged over a region defined by the mean square displacement, assumed Gaussian, of the atoms about their equilibrium positions. It turns out that this procedure is equivalent to a certain summation over all the even order anharmonic terms. Since the third-order term is at least as important as the fourth-order, the SCP theory is not satisfactory as it stands. It is usually necessary to make an additional correction to allow for the third-order anharmonic term.

Calculations based on this modified SCP theory, given the basic Lennard–Jones form (27) for the potential, were not in completely satisfactory agreement with experiment, however, and the need both to improve upon this two-body potential and also to include the leading three-body terms became apparent. The best available calculations would seem to be those of Barker and co-workers (Bobetic and Barker, 1970; Barker *et al.*, 1970), using a two-body potential of the form

$$V(r) = \varepsilon\left[\exp \alpha(1 - r_1) \sum_{i=0}^{L} A_i(r_1 - 1)^i - \sum_{i=0}^{2} \frac{C_{2i+6}}{\delta + r_1{}^{2i+6}}\right] \tag{28}$$

where $r_1 = r/r_m$, r_m being the interatomic spacing at the minimum of the potential. In addition to this $V(r)$, a three-body potential of the Axilrod–Teller triple–dipole form was incorporated:

$$V^3(r_{ijk}) = \frac{\nu(1 + 3\cos\theta_1 \cos\theta_2 \cos\theta_3)}{(r_1 r_2 r_3)^3} \tag{29}$$

where θ_i, r_i are the angles and sides of the triangle formed by the three atoms. This subject has recently been reviewed in detail by Klein (1975), and we shall not attempt to discuss here the intricate question of how all the parameters of (28) and (29) have been determined. It is sufficient to state that by suitable choices of parameters, followed by application of the modified SCP theory, it has been possible to compute rather accurate values not only of the phonon frequencies in the rare gas solids but also, with somewhat less precision, their temperature and pressure dependences.

It is nevertheless of interest to note that the phonon dispersion curves for these crystals, at least at low temperatures, are remarkably simple in shape (almost perfect sine curves), and it is extremely easy to fit them, to very high accuracy, with the aid of phenomenological models having Born–von Kármán force constants extending to one, two, or at most to three, nearest-neighbor atoms. The simple Lennard–Jones potential will also give an excellent description of the curves provided the parameters are allowed to vary somewhat from the values obtained from second virial coefficient data. These simple phenomenological models remain quite useful in practice, on account of the very much shorter computing times (several orders of magnitude!) they require as compared with the evaluation of (28) and (29) in the modified SCP theory. They can be used as a convenient shorthand form for representing large numbers of measured phonon frequencies, and they have even been used to represent the curves obtained by the more sophisticated theories in order to compute additional phonon frequencies (by interpolation) or other thermodynamic properties (H. Glyde, private communication, 1973). Since all the rare gas solids crystallize in the face-centered cubic structure (with the exception of two phases of He, body-centered cubic and hexagonal close-packed, respectively), it is possible to establish very accurate phenomenological models with the formulas conveniently listed by Squires (1963).

B. Ionic Crystals

If we attempt to describe the interionic forces in a typical ionic crystal by means of phenomenological Born–von Kármán force models, we find immediately that the short-range repulsive forces required are two orders of magnitude higher than those in the case of rare gas solids. It is also very difficult to justify the representation of the long-range Coulomb interactions by means of force constants extending from an origin ion to a small number of near-neighbor ions. However, a simple but fairly realistic model (Kellermann, 1940) can be constructed by considering separately the long-range Coulomb forces and the short-range repulsive forces arising from the overlap of electronic charge distributions when near-neighbor ions come too close to each other. The dynamical matrix (18) is split into a sum of two terms:

$$M_{\alpha\beta}^{\kappa\kappa'}(\mathbf{q}) = R_{\alpha\beta}^{\kappa\kappa'}(\mathbf{q}) + Z^{\kappa}Z^{\kappa'}C_{\alpha\beta}^{\kappa\kappa'}(\mathbf{q}) \tag{30}$$

where

$$R_{\alpha\beta}^{\kappa\kappa'}(\mathbf{q}) = \sum_{l'} {}^{R}\phi_{\alpha\beta}^{\kappa\kappa'}(ll') \exp i\mathbf{q} \cdot [\mathbf{x}^{\kappa'}(l') - \mathbf{x}^{\kappa}(l)] \tag{31}$$

$$C_{\alpha\beta}^{\kappa\kappa'}(\mathbf{q}) = \sum_{l'} {}^{C}\phi_{\alpha\beta}^{\kappa\kappa'}(ll') \exp i\mathbf{q} \cdot [\mathbf{x}^{\kappa'}(l') - \mathbf{x}^{\kappa}(l)] \tag{32}$$

and Z^{κ}, $Z^{\kappa'}$ are the electric charges on the ions κ, κ'.

The short-range force constants ${}^{R}\phi$ can be represented by empirical Born–von Kármán parameters or obtained from some assumed analytic form (e.g., Lennard–Jones) for the short-range potential. The "Coulomb" force constants ${}^{C}\phi$ are given by

$${}^{C}\phi_{\alpha\beta}^{\kappa\kappa'}(ll') = \left.\frac{\partial^2 V(r)}{\partial r_\alpha \, \partial r_\beta}\right|_{r=|\mathbf{x}^{\kappa}(l)-\mathbf{x}^{\kappa'}(l')|} \tag{33}$$

where

$$V(r) = |\mathbf{r}^{\kappa}(l) - \mathbf{r}^{\kappa'}(l')|^{-1}$$

It is easy to show that

$${}^{C}\phi_{\alpha\beta}^{\kappa\kappa'}(ll') = -[(3r_\alpha r_\beta/r^5) - (\delta(\alpha, \beta)/r^3)] \tag{34}$$

for $(l\kappa) \neq (l'\kappa')$, and that [as before Eq. (10)],

$${}^{C}\phi_{\alpha\beta}^{\kappa\kappa}(ll) = - \sum_{l'\kappa' \neq l\kappa} {}^{C}\phi_{\alpha\beta}^{\kappa\kappa'}(ll') \tag{35}$$

It is not feasible to evaluate the sum in (32) directly. However, the use of the theta transformation of Ewald (1921) enables the sum to be split into two parts, involving summations over limited regions of real space for the first part and of reciprocal space for the second part. We should also note that the sum (32) does not converge to a unique limit as $\mathbf{q} \to 0$, but that the limit depends upon the direction of $\mathbf{q}$ as $\mathbf{q}$ becomes very small. It would be inappropriate here to describe in detail the application of the theta transformation by Kellermann (1940) to derive rapidly convergent expressions for the sum (32). The result, in a form immediately suitable for numerical computation, is

$$C_{\alpha\beta}^{\kappa\kappa'}(\mathbf{q}) = -\frac{4\pi}{v}\sum_h \frac{(b_{h\alpha}+q_\alpha)(b_{h\beta}+q_\beta)}{(\mathbf{b}_h+\mathbf{q})^2}\exp -\frac{(\mathbf{b}_h+\mathbf{q})^2}{4K^2}\exp i\mathbf{b}_h\cdot(\mathbf{x}^\kappa - \mathbf{x}^{\kappa'})$$
$$+ K^2 \sum_l \left[\psi'(K|\mathbf{x}_l^{\kappa\kappa'}|)\frac{\delta(\alpha,\beta)}{|\mathbf{x}_l^{\kappa\kappa'}|} + \left[K\psi''(K|\mathbf{x}_l^{\kappa\kappa'}|)\right.\right.$$
$$\left.\left. - \frac{\psi'(K|\mathbf{x}_l^{\kappa\kappa'}|)}{|\mathbf{x}_l^{\kappa\kappa'}|}\right]\frac{x_{\alpha l}^{\kappa\kappa'}x_{\beta l}^{\kappa\kappa'}}{|\mathbf{x}_l^{\kappa\kappa'}|^2}\right]\exp -(i\mathbf{q}\cdot\mathbf{x}_l^{\kappa\kappa'}) \tag{36}$$

$$C_{\alpha\beta}{}^{\kappa\kappa}(\mathbf{q}) = -\frac{4\pi}{v}\sum_h \frac{(b_{h\alpha}+q_\alpha)(b_{h\beta}+q_\beta)}{(\mathbf{b}_h+\mathbf{q})^2}\exp -\frac{(\mathbf{b}_h+\mathbf{q})^2}{4K^2} + \frac{4K^3}{3\sqrt{\pi}}\delta(\alpha,\beta)$$
$$+ K^2 \sum_l \left[\psi'(K|\mathbf{l}|)\frac{\delta(\alpha,\beta)}{|\mathbf{l}|} + \left[K\psi''(K|\mathbf{l}|)\right.\right.$$
$$\left.\left. - \frac{\psi'(K|\mathbf{l}|)}{|\mathbf{l}|}\right]\frac{l_\alpha l_\beta}{|\mathbf{l}|^2}\right]\exp -(i\mathbf{q}\cdot\mathbf{l}) \tag{37}$$

where v is the volume of the unit cell, $\mathbf{x}_l^{\kappa\kappa'} = (\mathbf{x}_0{}^\kappa - \mathbf{x}_l{}^{\kappa'})$, $\mathbf{x}_l{}^{\kappa'} = (\mathbf{l} + \mathbf{x}^{\kappa'})$, and K is a parameter chosen so that each of the summations, over reciprocal lattice vectors $\mathbf{b}_h$ and over real space cells $\mathbf{l}$, converges rapidly. The function $\psi(y)$ and its derivatives are defined by

$$\psi(y) = (1 - G(y))/y$$
$$\psi'(y) = \partial\psi(y)/\partial y; \qquad \psi''(y) = \partial^2\psi(y)/\partial y^2$$

and

$$G(y) = \frac{2}{\sqrt{\pi}}\int_0^y \exp(-z^2)\,dz \tag{38}$$

These expressions (36), (37) are valid for all $\mathbf{q}$ except $\mathbf{q} = 0$, and will in general as noted above, give rise to different limiting values as $\mathbf{q} \to 0$ from different directions. The separation parameter K is normally of order 1; if a much smaller value is employed, the reciprocal space term converges extremely rapidly but the function ψ becomes large and many terms are required before the real space sum converges. Conversely, too large a value of K leads to very slow convergence of the reciprocal space sum. If we require an accuracy better than 0.1% in the matrix elements $C_{\alpha\beta}^{\kappa\kappa'}(\mathbf{q})$ then the summation should in a typical case extend over a radius of about 4 or 5 cells in each space, with K chosen in the range 1 to 3. Early calculations of these sums for simple crystal structures were done with the help of hand calculators only, but the use of modern high-speed computers is strongly recommended, particularly for more complicated crystals. Having written the computer program, one can try out various choices of limits on l and h, and of K values, for a small number of typical $\mathbf{q}$ values, and thereby discover by trial and error the optimum set of values for obtaining a given accuracy in the shortest time. Care should be taken to avoid unnecessary recomputation of various components of the calculation. For example, most of the terms in the real space sum are independent of $\mathbf{q}$, and thus need to be computed once and then stored for use at all $\mathbf{q}$ values. Typical calculation times for all the matrix elements in the case of solid carbon dioxide (cubic Pa3, 12 atoms per unit cell) are of order 5 seconds per $\mathbf{q}$ vector, for an accuracy of 0.05%, with the help of a CDC6600 computer. Much shorter times are possible if the Eqs. (36) and (37) are specialized as far as possible to the particular crystal structure involved. The special case of the alkali halide structure has been described in detail by Kellermann. The amount of analytical effort one should make to save on computer time is perhaps a matter of personal taste.

One final point should be made in connection with the translational invariance condition (10) or (35). By analogy with development leading to Eq. (22), we may automatically include this condition by writing

$$Z^{\kappa}Z^{\kappa'}C_{\alpha\beta}^{\kappa\kappa'}(\mathbf{q}) = Z^{\kappa}Z^{\kappa'}\bar{C}_{\alpha\beta}^{\kappa\kappa'}(\mathbf{q}) - \delta(\kappa,\kappa')Z^{\kappa}\sum_{\kappa''} Z^{\kappa''}\bar{C}_{\alpha\beta}^{\kappa\kappa''}(\mathbf{q}' \approx 0) \qquad (39)$$

For a given $\mathbf{q}$ value, we evaluate the coefficients $C_{\alpha\beta}^{\kappa\kappa'}$ not only for that $\mathbf{q}$ but also for a very small value of $\mathbf{q}'$ *in the same direction.* (We cannot put $\mathbf{q}'$ exactly to zero, as is possible in evaluating the short-range components of the matrix elements). The summation over κ'' for $\mathbf{q}' \approx 0$ can of course be utilized subsequently for all $\mathbf{q}$ values in the same direction, without recalculation. Equation (39), with $\mathbf{q}$ actually equal to zero in the second term, can be used to apply the translational invariance condition to the short-range force contributions (31). We notice that the matrix elements $R_{\alpha\beta}{}^{\kappa\kappa}(\mathbf{q})$, for $\mathbf{q} = 0$, contain no force constants of the type ${}^{R}\phi_{\alpha\beta}{}^{\kappa\kappa}(ll')$, since these are all cancelled out by translational invariance. This is only to be expected since all atoms or ions of a given sublattice κ move in phase with each other in all $\mathbf{q} = 0$ modes of vibration.

The simple force model for ionic crystals described above is called the rigid-ion model, since it assumes that the electronic charge on each ion moves rigidly with the ionic core; no allowance is made for the physical fact that the charge distributions are to some degree *polarizable*, that is, they may be distorted in various ways during the course of the ionic vibrations. The interionic forces may be modified substantially by such distortions, so that they should be taken into account if one wishes to calculate accurately the normal mode frequencies of an alkali halide, for example. During the past 15 years, several methods (all more or less closely related) have been developed to take the electronic polarizability into account in normal mode calculations, mainly for crystals with the alkali halide (NaCl), fluorite (CaF_2), and gallium arsenide (GaAs) structures. Reference should be made to the work of Woods *et al.* (1960), Cowley *et al.* (1963), Hardy (1962), Karo and Hardy (1963), Nüsslein and Schröder (1967), and Verma and Singh (1969). An excellent review of these variants has been given by Cochran (1971). The simplest way to visualize the basic ideas of these models is in terms of the shell model introduced originally by Dick and Overhauser (1958) in which the electronic distribution of an ion is split into two parts—an inner "core" consisting of the nucleus and the tightly bound inner electrons, surrounded by an outer "shell" of valence electrons, which may move and/or distort with respect to the core during the course of a lattice vibration. The resulting induced electric dipoles interact with each other and with the ionic displacements, leading to very significant changes in the overall force field and hence in the phonon frequencies. Force constants ${}^R\phi$, ${}^T\phi$, and ${}^S\phi$ are introduced to describe the ion–ion, ion–core, and shell–shell short-range interactions, respectively, Y^κ is the charge on the κth shell, and Δ^κ is the short-range interaction between the κth core and its own shell. The ionic displacements are, as before, $\mathbf{u}^\kappa$, and in addition we now have to consider the dipole moments $\mathbf{w}^\kappa$, assumed to be situated on the ions themselves. In effect, we are expanding the crystal potential energy Φ not only in powers of $\mathbf{u}^\kappa$ but also in powers of $\mathbf{w}^\kappa$ and their cross terms. Differentiation of Φ leads to the following equations of motion, in an obvious matrix notation:

$$\omega^2 \mathsf{m}\mathbf{u} = (\mathsf{R} + \mathbf{ZCZ})\mathbf{u} + (\mathsf{T} + \mathbf{ZCY})\mathbf{w} \tag{40}$$

$$0 = (\mathsf{T}^+ + \mathbf{YCZ})\mathbf{u} + (\mathsf{S}_1 + \mathbf{YCY})\mathbf{w} \tag{41}$$

Here $\mathbf{u}$ and $\mathbf{w}$ are column vectors [cf. Eq. (16)], R, T, and S are very similar looking matrices representing the short-range contributions to the dynamical matrix, $\mathbf{Y}$ and $\mathbf{Z}$ are diagonal matrices whose elements are the shell charges and ionic charges, respectively, and T^+ is the Hermitian adjoint (or transpose of the complex conjugate) of T. S_1 is related to S by

$$S_{1\alpha\beta}^{\kappa\kappa'}(\mathbf{q}) = S_{\alpha\beta}^{\kappa\kappa'}(\mathbf{q}) + \delta(\alpha, \beta)\, \delta(\kappa, \kappa')\, \Delta^{\kappa} \tag{42}$$

The trigonometric expressions for R, T, and S are essentially identical, aside from the appearance of force constants ${}^R\phi$, ${}^T\phi$ and ${}^S\phi$, respectively. Equation (40) is the equation of motion for the ions, and (41) for the electronic shells, or dipole moments. The zero on the left of (41) is thus, in effect, the adiabatic approximation which implies that the electron shells follow the ionic motion extremely rapidly, their configuration is always appropriate to the instantaneous configuration of the ion cores. Substitution of (41) into (40) to eliminate $\mathbf{w}$ leads to the well-known shell model equation

$$\omega^2 \mathsf{m}\mathbf{u} = \mathsf{M}\mathbf{u}$$

where

$$\mathsf{M} = \mathsf{R} + \mathbf{ZCZ} - (\mathsf{T} + \mathbf{ZCY})(\mathsf{S}_1 + \mathbf{YCY})^{-1}(\mathsf{T}^+ + \mathbf{YCZ}) \tag{43}$$

Thus we see that the numerical computation of phonon frequencies for ionic crystals on the basis of the shell model is substantially more time-consuming than for the simple rigid ion model. The calculation of the Coulomb coefficients C, and of the short-range contributions R (and by analogy T and S) is the same, but we must now invert the so-called S-matrix, multiply by two other matrices, and carry out a matrix subtraction, before finally calling a suitable subroutine to diagonalize the dynamical matrix. In addition, the force constants ${}^S\phi$ appear in the matrix elements in a nonlinear manner, which can cause complications in the procedure for obtaining their optimum values by the method of least-squares fitting to experimentally determined phonon frequencies.

The basic shell model, leading to Eq. (43), has been remarkably successful in describing the normal modes of vibration (both frequencies and eigenvectors) of a wide range of ionic, semi-ionic, and even covalent compounds, for which it is not especially appropriate from a physical point of view. It remains, however, a *dipole approximation* model, in which all the variations of electron charge distributions are described simply by point dipoles at the ion sites. It is not surprising, therefore, that shell model calculations suffer from certain kinds of difficulties in fitting the measured dispersion curves, and that several variants of the model have been developed to try to overcome these difficulties (see Cochran, 1971). The "breathing" shell model (Nüsslein and Schröder, 1967) and the charge transfer model (Verma and Singh, 1969) have progressively diminished these discrepancies between theory and experiment, with only small increases in complexity of the numerical calculations. The concept of a "double-shell" model has been successfully introduced (Weber, 1973) to describe the influence of d electrons on the phonon dispersion curves of certain transition metal carbides. These display very significant anomalies (Smith and Gläser, 1970) which are attributed to resonance effects in the d-electron polarizability. These electrons are represented by a second shell, parametrized by its own core–shell coupling constant and short-range force matrix. In each of these variants of the basic shell model, however, the actual equations one has to solve have essentially the same form as (43). Indeed, one moves steadily from a situation in which the equations of motion are relatively simple and easy to compute, but where the parameters are more or less phenomenological, toward a more fundamental type of approach, in which the equations of motion are rather complex, with few or zero adjustable parameters. In the former case, the ease of computation allows the establishment of the parameter values by the method of least-squares fitting, while in the latter, such a procedure is in principle inappropriate (as well as being extremely difficult) since the model parameters are derived from "first principles" rather than being adjusted to fit the data.

C. Metals

There have been a very large number of computations of phonon frequencies in metals based on the purely phenomenological approach involving Born–von Kármán force constants between various near-neighbor pairs of atoms, and this method is still widely used as a kind of preliminary analysis of experimental data. It is often very useful to know whether it is possible to fit the measured frequencies by means of such an empirical model, and if so, how many adjustable force constants are required to obtain a satisfactory fit. If a suitable model can be obtained, then it will serve, at the least, as an interpolation procedure for computing frequencies that have not actually been measured. At first sight, however, it is not at all clear why such simple interatomic force models should be able to describe successfully the phonon dispersion relation for a metal. The positive ionic cores are embedded in a more or less free electron gas, and it is evident that a very important (perhaps the most important) contribution to the interatomic forces must arise through the interaction between the ions and this electron gas, rather than directly between pairs of ions. The direct electrostatic interaction must be largely screened out by the electrons, which, as usual, move very rapidly indeed in response to ionic motion so as to minimize the total energy for every instantaneous configuration of the positive ion cores. A very fruitful approach to this problem of describing the ion–electron interactions in a metal is the so-called pseudopotential method, in which the very deep potential well about each ion core is replaced by a much weaker "effective" potential. This replacement is possible, in essence, because the wave functions of the conduction electrons must be orthogonal to those of the ionic cores. Detailed and authoritative discussions of this matter can be found in reviews by Sham and Ziman (1963) and by Harrison (1965). In order to introduce this concept of a screened pseudopotential, we must first recast our basic equations such as (20) by means of Fourier transformation.

The Fourier transform of a pair potential $V^{\kappa\kappa'}(r)$ is given by

$$V^{\kappa\kappa'}(Q) = \int V^{\kappa\kappa'}(r) \exp(i\mathbf{Q} \cdot \mathbf{r}) \, d^3r \tag{44}$$

and its inverse relation is

$$V^{\kappa\kappa'}(r) = (2\pi)^{-3} \int V^{\kappa\kappa'}(Q) \exp(-i\mathbf{Q} \cdot \mathbf{r}) \, d^3Q \tag{45}$$

Assuming that the force constants $\phi_{\alpha\beta}^{\kappa\kappa'}(ll')$ of Eq. (21) can indeed be derived from such a potential $V^{\kappa\kappa'}(r)$ depending only on the interatomic distance $r = |\mathbf{r}|$, we can combine (21), (25), (44), and (45) to obtain (Cochran and Cowley, 1967)

$$\begin{aligned} \bar{M}_{\alpha\beta}^{\kappa\kappa'}(\mathbf{q}) = (2\pi)^{-3} \sum_{l'} \int & Q_\alpha Q_\beta V^{\kappa\kappa'}(Q) \\ & \times \exp[i(\mathbf{Q} - \mathbf{q}) \cdot (\mathbf{x}^\kappa(l) - \mathbf{x}^{\kappa'}(l'))] \, d^3Q \end{aligned} \tag{46}$$

We reverse the order of summation and integration, and make use of the well-known relation

$$\sum_l \exp(-i\mathbf{Q} \cdot \mathbf{x}(l)) = \frac{(2\pi)^3}{v} \sum_h \delta(\mathbf{Q} - \mathbf{b}_h) \tag{47}$$

where v is the volume of the unit cell, to obtain finally

$$\begin{aligned} \bar{M}_{\alpha\beta}^{\kappa\kappa'}(\mathbf{q}) = \frac{1}{v} \sum_h & (\mathbf{b}_h + \mathbf{q})_\alpha (\mathbf{b}_h + \mathbf{q})_\beta V^{\kappa\kappa'}(|\mathbf{b}_h + \mathbf{q}|) \\ & \times \exp i\mathbf{b}_h \cdot (\mathbf{x}^\kappa - \mathbf{x}^{\kappa'}) \end{aligned} \tag{48}$$

The elements of the dynamical matrix (17) are constructed as before by use of Eqs. (18) and (22), so that the translational invariance condition is automatically taken into account. The form (48) is thus the reciprocal space representation analogous to that in real space [Eq. (14)]. The electrostatic interaction between point charges in an ionic crystal (Section II, B) is a special case of this kind of representation, in which $V^{\kappa\kappa'}(Q)$ is derived from the Coulomb potential $(Z^{\kappa}Z^{\kappa'}/r)$. In a metal, of course, we will have not only this Coulomb interaction between the bare ions, but also the screening effect of the conduction electrons. In covalent crystals and in semiconductors, the situation is rather more complicated, but the basic idea of a pseudopotential screened by some kind of dielectric function seems to be generally applicable, and the formulas for computing dynamical matrix elements for all these types of crystals turn out to be remarkably similar in form. A detailed discussion of the dielectric screening function method has been given in a recent review by Bilz *et al.* (1974).

For the simple case of a monatomic lattice (see, for example, potassium, Cowley *et al.*, 1966) the dynamical matrix [Eqs. (20) and (48)] reduces to

$$M_{\alpha\beta}(\mathbf{q}) = \frac{1}{v} \sum_{h} [(\mathbf{b}_h + \mathbf{q})_\alpha(\mathbf{b}_h + \mathbf{q})_\beta V(|\mathbf{b}_h + \mathbf{q}|) - b_{h\alpha}b_{h\beta}V(|\mathbf{b}_h|) \tag{49}$$

an equation first obtained by Pines (1963). It is possible to regard the total potential $V(Q)$ as a purely empirical function to be determined by fitting it to the measured phonon frequencies, in much the same spirit as one may derive the Born–von Kármán force constants of a real space potential. Such a potential has been called (Ziman, 1964) the potential between "neutral pseudo atoms." For the purposes of the fitting process, $V(Q)$, or more conveniently $Q^2V(Q)$, may be specified by means of a Fourier series in Q/Q_{max}, where Q_{max} is a limit beyond which $V(Q)$ is assumed to be zero. Care should be taken to ensure that $Q^2V(Q)$ tends to zero at $Q = 0$ with a curvature consistent with the slopes of the phonon dispersion curves near $q = 0$ (Buyers and Cowley, 1969). Alternatively, the function may be specified by a table of values at given intervals of Q; at intermediate Q values, $Q^2V(Q)$ is found by a suitable interpolation scheme (Cowley *et al.*, 1966).

However, a closer link to the microscopic theories of phonons in metals is obtained by dividing the dynamical matrix elements into three components:

$$M_{\alpha\beta}(\mathbf{q}) = M^R + M^C + M^E \tag{50}$$

These terms represent contributions arising from short-range overlap forces between ion cores, long-range Coulomb forces between the *bare* ions, and finally the interactions between the conduction electrons and the ions which screen out the direct Coulomb forces at large distances. The contribution M^R is often not very well known and it is normally represented by one or two Born–von Kármán force constants which are adjusted empirically to suit the data. For alkali metals it is believed to be quite small or zero. M^C is readily found with the help of the Ewald theta transformation given in Section II, B, while M^E is expressed in terms of a pseudopotential screened by a dielectric function $\varepsilon(Q)$. If the bare ion charge is Ze, this pseudopotential must be (Ze/r) at large r, but inside the ion core it varies smoothly to a finite value at $r = 0$. The calculation of M^E is straightforward, at least for free electronlike metals (see, for example, Cochran, 1971); the result is given by (49), with

$$V^E(Q) = -(4\pi e^2 Z^2(Q)/Q^2)(1 - \varepsilon^{-1}(Q)) \tag{51}$$

where $Z(Q)$ is the Fourier transform of an effective ionic charge distribution $Z(r)$. The dielectric screening function for a free electron gas, within the Hartree or self-consistent field approximation, is given by

$$\varepsilon(Q) = 1 + \frac{6\pi e^2 n}{vQ^2 E_F}\left[\frac{1}{2} + \frac{4k_F{}^2 - Q^2}{8k_F Q}\log\left|\frac{2k_F + Q}{2k_F - Q}\right|\right] \tag{52}$$

where n is the number of electrons in the volume v, and E_F, k_F are the Fermi energy and wave vector, respectively. The logarithmic singularity in $\varepsilon(Q)$ at $Q = 2k_F$ gives rise to discontinuities of slope of the phonon dispersion curves, which have been observed in several metals such as Pb and Mo. However, they appear to be extremely small in the free-electronlike metals (Na, K) for which this simple theory is expected to apply. The main problems in these calculations are the determination from first principles of the function $Z(Q)$, and, for the complex metals for which the free-electron approximation does not hold, the evaluation of a generalized dielectric matrix $\varepsilon(Q, Q')$. We shall not attempt to discuss these matters here. As regards the problem of deriving $Z(Q)$ from measured phonon dispersion curves, this may be done in a similar manner to that described above for the total neutral pseudoatom potential, by fitting $Q^2 V^E(Q)$ specified by some suitable analytic form or by a table of values. It is also easy to incorporate some of the more sophisticated theoretical expressions for $\varepsilon(Q)$ which have been developed by many authors, in attempts to go beyond Eqs. (51) and (52). These have been summarized by Brovman and Kagan (1974).

D. Covalent Crystals

In the case of many metal crystals, as mentioned previously, it has been possible to employ the Born–von Kármán theory to obtain a description, numerically satisfactory if not physically so, of the measured phonon dispersion curves. This approach has not met with much success in covalent crystals, as represented by the diamond structure semiconductors Ge and Si. An early paper by Herman (1959) demonstrated the long-range nature of the interatomic forces, and the inconveniently large number of force constants needed to achieve agreement with experiment. The alternative approach, using screened pseudopotentials, is also more difficult to apply to covalent materials than to metals, since the basic free-electron type of approximation and diagonal dielectric matrix is clearly much less appropriate. Nevertheless, considerable progress has been made in recent years toward satisfactory first-principles calculations (Martin, 1969; Sinha *et al.*, 1971). In an extensive survey of empirical and microscopic models for covalent crystals (including even the case of degenerate semiconductors), Sinha (1973) has discussed the various approximate methods which have been used to treat the awkward off-diagonal elements of the generalized dielectric screening matrix. The dynamical matrix he obtains is remarkably similar in form to the basic shell model equation (43). Indeed, it appears that a wide range of crystal types, such as free-electronlike metals, transition metals, ionic crystals, and semiconductors, can all be discussed within this framework; the original shell model equations of Cochran are simply a special case of this more general microscopic theory. This subject has also been reviewed recently by Sham (1974) and by Bilz *et al.* (1974).

Among the more empirical models that have been employed with more or less success to compute phonon frequencies in covalent crystals, we should mention are the shell model (SM) and the valence force field (VFF) model. The former model is inherently more suited to ionic crystals (Section II, B) since the idea of representing the electron distributions in covalent bonds of Ge by means of electron shells centered at the nuclear sites is not physically very attractive. There is, nevertheless, ample justification for its use (see, for example, Sham, 1974) and it has indeed been applied with reasonable success to several IV–IV and III–V semiconductors.

The existence of strongly directed covalent bonds in these structures leads quite naturally to the ideas of the valence force field model, in which the potential energy is quadratic in terms of the "valence coordinates." As applied by Musgrave and Pople (1962) to the case of diamond, these coordinates were taken to be (i) the nearest-neighbor C–C bond distance, and (ii) the angle between adjacent bonds on the same C atom. The "force constants" of the model are thus the C–C bond-stretching constant and the bond-angle change. McMurry *et al.* (1967) extended this model to include two bond-angle changes involving a common bond. Since there are obvious geometrical relations between atomic displacements and the bond-stretching and bond-angle change coordinates, it is easy to express the latter in terms of Born–von Kármán force constants, as described in detail by Musgrave and Pople. The VFF model of McMurry *et al.* is very satisfactory for diamond, but additional bond-angle change parameters are required in the case of silicon (Solbrig, 1971), as a result of the longer range forces in this material. It is evident that the VFF model is quite analogous to the standard Born–von Kármán model, except that certain classes of Born–von Kármán parameters which are of particular physical significance in covalent crystals are automatically grouped together in the VFF model. The mathematical computations are thus very similar to those of the Born–von Kármán model equations given in Section I.

Another type of empirical model for covalent crystals has been introduced more recently by Vasil'ev *et al.* (1971). In this model each atom carries rigidly with it a set of covalent bonds (in the case of diamond, four tetrahedrally oriented bonds produced by the sp^3 hybridization of the atomic s and p wave functions of the outermost electrons). As the atom plus its bonds is translated or rigidly rotated, the overlap of the bonds with those of neighboring atoms is changed, and this change of overlap gives rise to the interatomic forces. This "kinking" of the interatomic bonds is described empirically by a minimum of two parameters, one being analogous to the first-nearest-neighbor tangential force constant of conventional Born–von Kármán theory and the other denoting in some sense (not fully understood by this author) the position at which this force constant is applied. Two other parameters are also included to represent the radial force constants between an atom and its first- and second-nearest-neighbors (the second neighbor tangential force constant is assumed to be zero). With this basic minimum number of four adjustable parameters, Vasil'ev *et al.* succeed in reproducing the measured phonon dispersion curves and elastic constants (but not, of course, the dielectric constants) of diamond, Si, and Ge with remarkable accuracy. It would seem, however, that these calculations are not entirely free from programming faults (Kress, 1975; also private communication, 1975); further calculations are needed to check whether the model can retain its capacity to provide a good fit, not only for Si and Ge but also for III–V semiconductors, after correction of these computational errors.

III. Intermolecular Force Models

A. Intermolecular Force Constants

In the previous section we considered the vibrations of individual atoms in a crystal. In many solids, however, certain groups of atoms are more or less tightly bound to each other by strong covalent forces to form molecules (e.g., C_6H_6 molecules in solid benzene) or radicals (e.g., $NH_4{}^+$ in ammonium chloride, NO_2^- in sodium nitrite). These "molecules" are linked to each other or to other atoms by relatively weak "intermolecular forces." In this case it is often a good approximation to assume that the intermolecular motions or "external" modes can be described in terms of the translational and librational motions of rigid molecules. The classic example of such a molecular crystal is hexamethylenetetramine, $(CH_2)_6N_4$, which we abbreviate to HMT. The first calculations of molecular modes of vibration and libration in this crystal were made by Pawley (1962), Cochran and Pawley (1964), and Hahn and Biem (1963) and the frequency distribution of HMT was measured by Becka (1962); subsequently, neutron coherent inelastic scattering measurements and further theoretical computations were made (Dolling and Powell, 1970; Rafizadeh and Yip, 1970; Powell and Sandor, 1971; Dolling *et al.*, 1973) so that HMT is one of the most thoroughly studied molecular materials, at least from the viewpoint of its vibrational properties. It was also selected, in a very detailed and comprehensive review of intermolecular modes in crystals by Venkataraman and Sahni (1970), as an example to demonstrate how the intermolecular force constants are set up, simplified by means of symmetry arguments, and incorporated into the dynamical matrix. This review article is highly recommended for anyone planning to make calculations on molecular crystals. In this section, we shall mention only the main points pertinent to molecular normal mode calculations, based on a purely empirical approach very similar to the Born–von Kármán model described in Section I.

Some expansion of the previous notation is necessary because we must now consider both translational and librational displacements, $\mathbf{u}^\kappa(l)$ and $\boldsymbol{\theta}^\kappa(l)$ of the κth molecular group in the lth unit cell, and also define new categories of force constants which relate the force or the torque exerted on a molecule when a neighboring molecule is translated or rotated. In order to distinguish more clearly, for the moment, between these types of motion, we use axes (x, y, z) for the translations and (α, β, γ) for the librations. The moment of inertia of the κth molecule about the α axis (α being a principle axis of the molecule) is I_α^κ. The equations of motion for the molecular groups, corresponding to Eq. (11) are

$$\begin{aligned} m^\kappa\, \partial^2 u_x^\kappa(l)/\partial t^2 &= -\sum_{l'\kappa'}\left[\sum_y \phi_{xy}^{\kappa\kappa'}(ll')u_y^{\kappa'}(l') + \sum_\beta \phi_{x\beta}^{\kappa\kappa'}(ll')\theta_\beta^{\kappa'}(l')\right] \\ I_\alpha^\kappa\, \partial^2 \theta_\alpha^\kappa(l)/\partial t^2 &= -\sum_{l'\kappa'}\left[\sum_y \phi_{\alpha y}^{\kappa\kappa'}(ll')u_y^{\kappa'}(l') + \sum_\beta \phi_{\alpha\beta}^{\kappa\kappa'}(ll')\theta_\beta^{\kappa'}(l')\right] \end{aligned} \tag{53}$$

The force constants $\phi_{xy}^{\kappa\kappa'}(ll')$ are exactly analogous to the conventional interatomic Born–von Kármán force constants, while $-\phi_{\alpha y}^{\kappa\kappa'}(ll')$ represents the torque exerted about the α axis on the $(l\kappa)$ molecule when molecule $(l'\kappa')$ is translated a unit distance along y. The definitions of the other parameters such as $\phi_{\alpha\beta}^{\kappa\kappa'}(ll')$ are obvious by inspection. We note that the interaction between a given pair of molecules is now specified in general by a (6×6) force constant array, in place of the (3×3) array for interatomic interactions, and that ϕ_{xy}, $\phi_{\alpha\beta}$, and $\phi_{x\beta}$ (or $\phi_{\alpha y}$) all have different dimensions. The equations (53) may be solved in much the same way as outlined by Eqs. (12)–(14), and we obtain finally a dynamical matrix of dimensions $(6n \times 6n)$, where n is the number of molecules in one unit cell. If we know all the elements of the generalized force constant matrix for all the nonzero molecule–molecule interactions, we can set up the dynamical matrix elements in a computer and diagonalize to obtain the $6n$ normal mode frequencies for any desired wave vector $\mathbf{q}$. The molecular analog of Eq. (21) is easily written down by comparison with the "atomic" case, but the allowance for the conditions of translational and rotational invariance of the molecular crystal is not quite so straightforward. These so-called "self-terms" are the force and torque on a molecule when only that molecule is displaced or rotated. For example, an infinitesimal rotation of a molecule at the origin is equivalent to an opposite rotation of all the other molecules *plus* an appropriate set of translational displacements of those molecules (Cochran and Pawley, 1964). The expressions for the self-terms in the general case are quite cumbersome, as shown by Eqs. (IIA.16) to (IIA.19) of Venkataraman and Sahni (1970), and we need not reproduce them here. Application of these general expressions to the particular example of HMT is also given in detail by these authors.

It should be noted, however, that HMT is one of the very few crystals which can be dealt with in this manner (i.e., using empirical intermolecular force constants directly) since a very high degree of crystal symmetry is required to reduce the number of force constants to a manageable level. For the majority of molecular crystals, the analysis must be continued at least one stage further, to the stage of expressing the intermolecular force constants in terms of sums of interatomic force constants between atoms on different molecules (Section III, B). It is also of importance to assess the influence on the external modes of the fact that the molecules are not completely rigid units. It is often the case that the frequencies of the internal or intramolecular modes of vibration of a molecule are not very much higher than the external mode frequencies, so that the latter are significantly perturbed from their "rigid molecule" values (Section III, C). Whatever the force model employed, it is always most important to make the maximum use of crystal symmetry to reduce both the number of independent parameters and also the amount of computational effort. We shall return to this point again in Section IV, B, but it is worthwhile here to point out a significant difference in the behavior of intermolecular force constants under the operation of symmetry operations as compared with interatomic parameters. Consider two (6×6) force constant arrays $\Phi(ll')$ and $\Phi(ll'')$, representing the interaction between a given molecule l and two of its neighbors, l', l'', which are related by a symmetry operation S about the origin molecule l. S can be expressed as a (3×3) matrix whose determinant det S is $+1$ or -1 depending on whether the rotation is proper or improper. In both cases S is appropriate for transforming a polar vector (e.g., a *translation* vector) but for an axial vector (e.g., a *rotation* vector) we must choose $+S$ or $-S$ for proper or improper rotations, respectively. Thus the force constants may be related by

$$\Phi(ll'') = T\Phi(ll')\tilde{T} \tag{54}$$

where

$$T = \begin{bmatrix} S & 0 \\ 0 & S \det S \end{bmatrix} \tag{55}$$

Use of these symmetry relations in the case of HMT resulted in a reduction in the number of force constant elements for the 8 first-nearest-neighbor molecules from 288 ($= 8 \times 6 \times 6$) to 6. Very significant reductions can also be achieved even for relatively low symmetry crystals.

B. Expansion in Terms of Interatomic Forces

In an arbitrary displacement $\mathbf{u}^\kappa(l)$, $\boldsymbol{\theta}^\kappa(l)$ of the (l_κ) molecule, any given atom m at position $\mathbf{x}(m)$ relative to the molecular center of mass will suffer a total displacement

$$\mathbf{S}^\kappa(lm) = \mathbf{u}^\kappa(l) + \boldsymbol{\theta}^\kappa(l) \wedge \mathbf{x}(m) \tag{56}$$

The crystal potential is expanded as a Taylor series in powers of the displacements $\mathbf{S}^\kappa(lm)$, and within the harmonic approximation we retain only the quadratic term

$$V_2 = \tfrac{1}{2} \sum_{\alpha\beta} \sum_{l\kappa m} \sum_{l'\kappa'm'} \phi_{\alpha\beta}^{\kappa\kappa'}(ll'mm') S_\alpha{}^\kappa(lm) S_\beta{}^{\kappa'}(l'm') \tag{57}$$

where the interatomic force constants are to be evaluated at the equilibrium positions and orientations of the molecules:

$$\phi_{\alpha\beta}^{\kappa\kappa'}(ll'mm') = \partial^2 V / \partial S_\alpha{}^\kappa(lm)\, \partial S_\beta{}^{\kappa'}(l'm') \tag{58}$$

These are exactly analogous to the Born–von Kármán force constants of Section I. We may now rewrite the equations of motion (53) in terms of these parameters. In order to clarify as far as possible the relationship between intermolecular and interatomic forces, and to simplify the notation, we consider the case where the (x, y, z) and (α, β, γ) axes can be taken to be coincident, as is the case for HMT. We find that the equations of motion are as given in (53) with the following substitutions:

$$\phi_{xy}^{\kappa\kappa'}(ll') = \sum_{mm'} \phi_{\alpha\beta}^{\kappa\kappa'}(ll'mm') = \sum_{mm'} \phi_{\alpha\beta} \tag{59}$$

$$\phi_{x\beta}^{\kappa\kappa'}(ll') = \sum_{mm'} [x_{\beta+1}(m')\phi_{\alpha,\beta+2} - x_{\beta+2}(m')\phi_{\alpha,\beta+1}] \tag{60}$$

$$\phi_{\alpha y}^{\kappa\kappa'}(ll') = \sum_{mm'} [x_{\alpha+1}(m)\phi_{\alpha+2,\beta} - x_{\alpha+2}(m)\phi_{\alpha+1,\beta}] \tag{61}$$

$$\begin{aligned}\phi_{\alpha\beta}^{\kappa\kappa'}(ll') = \sum_{mm'} \{&x_{\alpha+1}(m)[x_{\beta+1}(m')\phi_{\alpha+2,\beta+2} - x_{\beta+2}(m')\phi_{\alpha+2,\beta+1}] \\ &- x_{\alpha+2}(m)[x_{\beta+1}(m')\phi_{\alpha+1,\beta+2} - x_{\beta+2}(m')\phi_{\alpha+1,\beta+1}]\}\end{aligned} \tag{62}$$

On the left-hand side we have all four types of intermolecular force constant, and on the right a series of summations over all possible pairs of atoms m, m', taken from different molecules $l\kappa$, $l'\kappa'$. [The dropping of superscripts and arguments shown in the right of (59) is continued in the other three equations]. At this stage, we do not appear to have made any significant simplification of the problem, since the new interatomic force constants are extremely numerous and, in most cases, not susceptible to calculation from first principles. Under such conditions, the next step would seem to be the expression of the interatomic force constants in terms of an assumed analytic form for the potential. The Lennard–Jones form

$$V(r) = -Ar^{-6} + Br^{-n} \tag{63}$$

and the Buckingham "6-exp" form

$$V(r) = -Ar^{-6} + B\exp(-\alpha r) \tag{64}$$

have been very commonly used in this context (Kitaigorodskii, 1966; Pawley, 1967; Lutz and Hälg, 1970; Pawley *et al.*, 1971; Reynolds *et al.*, 1972; Dolling *et al.*, 1973; Kjems and Dolling, 1975). Suitable parameters A, B, n or A, B and α are chosen for each of the types of atom pairs; in a molecular hydrocarbon, for example, such as benzene, one would need to specify V_{CC}, V_{CH}, and V_{HH} for the carbon–carbon, carbon–hydrogen, and hydrogen–hydrogen potentials, respectively. The interatomic force constans are then found by substituting (63) or (64) into (26) [note that the derivatives A, B in Eq. (26) should not be confused with the potential parameters of (63), (64)]. The problem of proliferation of adjustable parameters becomes more serious when more atomic species are present, for example, the nitrogen atoms in HMT or chlorine in β-paradichlorobenzene (Reynolds *et al.*, 1972). It is nevertheless to be hoped that the potentials V_{CC}, V_{NH}, etc., will be *transferable* from one molecule to another. We should be able to use the same potential between, say N and H, not only in HMT but also in a series of other nitrogenous organic molecules. Attempts have been made with variable success, to establish such transferable potential functions, but it is clear that more work remains to be done in this area. Recent calculations on solid $\alpha - N_2$ (Kjems and Dolling, 1975) have re-emphasized the fact that, insofar as the normal mode frequencies are concerned, the potential parameters are strongly correlated. The dynamical matrix elements, and hence the phonon frequencies, depend only on the first and second derivatives of the potential $V(r)$ at a relatively small number of discrete values of r. It turns out that a whole series of (A, B, and n or α) values can be found, which give essentially the same quality of fit to the measured phonon frequencies and eigenvectors. However, the potential should also be consistent with the known sublimation energy of the crystal, and the total free energy should be a minimum at the observed value of the lattice parameter. Insistence upon these conditions was found to be sufficient to select a well-defined set of parameter values from the above-mentioned series.

When calculations of normal mode frequencies are made with the help of analytic potential functions, it is, of course, important to take into account intermolecular bonds (normally refered to by chemists as "nonbonding contacts," as distinct from the bonds *within* each molecule) with sufficiently large r values to ensure adequate computational accuracy. In this connection, in order to obtain about 1% accuracy for a computed frequency in a typical crystal, one must consider all bonds up to at least 5.5 Å, and preferably up to 7 Å or more. This frequently involves a rather large number of distinct interatomic pairs. The rather slow convergence of the Van der Waals r^{-6} term is the obvious cause of this difficulty. One is often in the position of having to include, for computational reasons, a nonbonding contact between two atoms that are physically shielded from one another by an intervening third atom. The significance of such a bond is then not clear. The problem is more acute when one computes the total potential energy for comparison with the measured sublimation energy, since this depends directly on $V(r)$ rather than on its more rapidly convergent derivatives. Nonbonding contacts up to about 15 Å are needed to achieve 1% accuracy in the energy. It is feasible to perform the summations numerically up to some limiting $r = r_m$, and then continue analytically from r_m to $r = \infty$, assuming a continuous distribution of atoms having the correct density. One can also make use of the computational methods developed by Williams (1971) for summing series of the type r^{-n} for arbitrary n. The terms of the series are multiplied by a convergence function $F(r)$, chosen so that $F(0) = 1$ and $F(r)$ tends rapidly to zero as r increases. The Fourier transform of the remainder of the terms is then summed in reciprocal space, in much the same spirit as in the Ewald transformation described in Section II, B.

C. Molecular Distortion and Polarization

Calculations of the effect of internal molecular distortions on the external intermolecular modes of vibration have only been performed for a few molecular crystals (naphthalene: Pawley and Cyvin, 1970; HMT: Dolling *et al.*, 1973). These have demonstrated, perhaps surprisingly, that the percentage frequency shifts produced when the rigid molecule assumption is relaxed can be quite large, of order 10%, even when there is a substantial gap between the internal and external mode frequencies, In HMT, the lowest frequency internal mode has a frequency some 3.5 times higher than that of the highest external mode, as compared with a ratio of only 1.2 in naphthalene. The perturbations of the external modes as a result of molecular flexibility are nevertheless very comparable in the two materials, and are furthermore substantially larger than the experimental errors on the measured frequencies. It is evident, therefore, that these effects should be taken into account in the analysis of external mode dispersion curves in terms of potential functions (hopefully transferable from one molecular species to another) for the non-bonding contacts.

The procedure adopted in the above-mentioned calculations requires at the outset an analysis of the internal modes of vibration of the molecule in terms of internal force constants (Wilson *et al.*, 1955). These force constants, denoted by a matrix F, are based on a set of molecular coordinates S as opposed to the Cartesian coordinates C which are appropriate to the intermolecular motions in the crystal as a whole. In the case of HMT, one molecule of which contains 22 atoms, and possesses $3 \times (22 - 2) = 60$ degrees of internal freedom, F is a (60×60) matrix. The total force constant matrix, Φ, including both internal and external forces has dimensions (66×66), and it is found from the transformation $\Phi = \tilde{B}FB + \Phi_E$, where Φ_E are the external or nonbonding contributions and B is a matrix representing the transformation between coordinate systems C and S. In effect we are setting up a Born–von Kármán type of interatomic force model incorporating all the previous results for isolated molecules obtained from gas-phase experiments (Elvebredd and Cyvin, 1972). The diagonalization of the (66 × 66) Hermitian matrix for a general **q** value is a considerable but certainly feasible task for modern computers. In this particular case also, the high crystal symmetry can be used to simplify and hence speed up the computations for various high symmetry wave vectors.

The discussion of intermolecular force models up to this point has neglected what may well be a very significant physical effect in many molecular solids. As described in Section II, B, the electronic polarizability can play a vital role in the understanding of the normal modes of vibration of ionic crystals, so we should not be surprised to find similar effects in molecular crystals. An adequate mathematical treatment of the distortions of the electronic charge distributions around a molecule during vibrational or librational motion is less straightforward than in the case of alkali halides. Nevertheless, the basic shell model for alkali halides has recently been developed by Luty and Pawley (1974) to take account of this "molecular polarizability." The relative arrangement of the atomic cores within the molecule is assumed to be rigid, as is the relative arrangement of the "shells" representing the polarizable electrons, but this shell array can move with respect to the core array under the influence of an isotropic core–shell force constant. The equations of motion for the molecular displacements and dipole moments are very similar to Eqs. (40) and (41), except that there are twice as many on account of the translational *and* librational coordinates, as in Eq. (53). One of the difficulties of this kind of molecular shell model is the handling of the many disposable parameters generated by the model. It remains to be seen how useful this approach will be in practical cases.

IV. Discussion

A. Theoretical and Experimental Approaches to Model Building

In the previous sections we have described several types of force model, which may be classified very roughly into phenomenological and microscopic models, according to the extent to which the model parameters are determined by fitting to some experimental data or calculated from a more fundamental standpoint. The most satisfactory theoretical situation, which may perhaps be said to have been reached only for rare gas solids and for simple free-electron metals, is one in which the physical basis of each component of the interatomic force field is well understood, and where the values for the vital few parameters (if any) are determined quite independently of any kind of dynamical information for the solid, such as phonon frequencies, sound velocities, and so on. At the other extreme we have the Born–von Kármán model and also the valence force field model, in which the interatomic interactions are represented by purely empirical parameters, restricted only by conditions of crystal symmetry and translational and rotational invariance. The complexity of the analysis and calculation in the former case is often such that the possibility of making numerical refinements aimed at producing a good fit to the measured phonon frequencies is rather limited. In any event, the fundamental theorist prefers to do the best he can without distorting his model simply to achieve better agreement with experiment—such agreement should appear naturally from the model without the aid of (overtly) adjustable parameters.

The experimentalist, on the other hand, is normally confronted with a given theoretical model, which in the vast majority of cases fails to fit all his measurements within experimental error. The theory may, nevertheless, be a substantial improvement over earlier attempts, and it may have included an important physical idea, previously neglected, which dramatically improves the qualitative agreement with experiment. If the model has any adjustable parameters at all (and there are always one or two hidden in even the purest models!), it is tempting to try to choose new values for them which give better agreement. If this approach fails, one looks for small modifications or extensions to the basic model that are deliberately chosen to achieve better agreement with experiment where it is most needed. A classic example of this process is provided by the many variations of the shell model for ionic crystals. The original calculations for NaI (Woods *et al.*, 1960) were in themselves a great step forward over the earlier rigid ion model, but there remained a few stubborn points of disagreement with experimental phonon frequencies, especially at the L point $(\frac{1}{2}, \frac{1}{2}, \frac{1}{2})$, in the Brillouin zone. The modification developed by Schröder (1966), the "breathing" shell model, was introduced in such a way as to achieve better agreement at this particular wave vector, in full knowledge of the type of ionic motion involved there.

This approach, of taking an existing theoretical model and adjusting its parameters or extending it in some way to obtain a better fit to experiment, can be called the experimental approach to lattice dynamics. It is facilitated by the use of models that are relatively simple and direct, so that numerical methods of least-squares fitting, requiring many iterations or re-calculations of dynamical properties, are not too time consuming in the computer. The dielectric screening function methods for covalent crystals mentioned in Section II, D do not appear to be suitable for this kind of treatment; on the other hand, the purely empirical models almost demand it, since there is no other way to determine the appropriate parameter values. Although a detailed discussion of least-squares fitting theory would be inappropriate here, it may be worthwhile to describe very briefly the principles involved for the general case in which the phonon frequencies depend nonlinearly on the adjustable parameters or force constants.

Let us consider a force model, having force constants f_j, which we wish to adjust so as to give the best least-squares fit to a set of experimentally measured quantities P_i, with experimental errors σ_i (variance σ_i^2). These will typically consist of a set of squared phonon frequencies $P_i = \nu_i^2$, with $\sigma_i = 2\nu_i\,\Delta\nu_i$ where $\Delta\nu_i$ is the experimental error of ν_i, and a selection of elastic constants, dielectric and/or piezoelectric constants, appropriate to the material being studied. From the model, we can compute values ϕ_i of all these experimental quantities, and also their gradients $\partial\phi_i/\partial f_j$ with respect to each force constant f_j. This gradient is often most conveniently calculated by finite differences; one simply recalculates the ϕ_i for new sets of f_j in which each f_j in turn is changed by an amount Δf_j. These new $\phi_i = \phi_i'$ differ from the initial set by amounts $\Delta\phi_i$, and the appropriate gradients are then given by $\Delta\phi_i/\Delta f_j$. We define a set of weights $\rho_i = 1/\sigma_i^2 (= 1/4\nu_i^2\,\Delta\nu_i^2$ for phonon frequencies, for example), and we wish to minimize the function

$$S = \sum_i \rho_i(\phi_i - P_i)^2 \tag{65}$$

For each j, we require

$$\sum_i \rho_i(\phi_i - P_i)(\partial\phi_i/\partial f_j) = 0 \tag{66}$$

We expand the ϕ_i in a Taylor series about an initial set ϕ_i^0 obtained from force constants which are, hopefully, not too far removed from their eventual best-fit values:

$$\phi_i = \phi_i^0 + \sum_k \frac{\partial\phi_i^0}{\partial f_k}\Delta f_k + \cdots \tag{67}$$

Substitution into (66) gives, for each j,

$$\sum_i \rho_i\left(\phi_i^0 - P_i + \sum_k \frac{\partial\phi_i^0}{\partial f_k}\cdot\Delta f_k\right)\frac{\partial\phi_i^0}{\partial f_j} = 0 \tag{68}$$

We define a rectangular matrix L and a column vector M as follows:

$$\left.\begin{aligned} L_{ij} &= \rho_i^{1/2}(\partial\phi_i^0/\partial f_j) \\ M_i &= \rho_i^{1/2}(P_i - \phi_i^0) \end{aligned}\right\} \tag{69}$$

If X is a vector formed from the *changes* Δf in the force constants required to provide a better fit, then we can write (68) as

$$\sum_i M_i L_{ij} = \sum_i \left(\sum_k L_{ik} X_k \right) L_{ij}$$

or, in matrix notation

$$\mathsf{L}M = \mathsf{L}\tilde{\mathsf{L}}X \tag{70}$$

where $\tilde{\mathsf{L}}$ is the transpose of L. Multiplying both sides on the left by $(\mathsf{L}\tilde{\mathsf{L}})^{-1}$ gives finally the required changes in the force constants

$$X = (\mathsf{L}\tilde{\mathsf{L}})^{-1}\mathsf{L}M \tag{71}$$

If the problem is *linear*, that is, if all the ϕ_i depend linearly on the f_j, then the best least-squares fit is obtained with the first application of (71). If the problem is nonlinear, then this first iteration is a linearized approximation to the best fit, and several such iterations will be needed before the X_i become negligibly small. A typical least-squares fitting program will test the sizes of the X_i and M_i after each iteration: when all elements are less than a suitably chosen limit, the program automatically stops at the current iteration and outputs the results. The errors E_j on the derived force constant values can be obtained from the elements of the derivative matrix $A_{ij} = (L\tilde{L})_{ij}$

$$E_j^2 = \sum_i A_{ij}^2 \tag{72}$$

A second kind of error E'_j, which is approximately equivalent to what one might call the "experimental error" of each force constant, is given by

$$E'_j = E_j[S/(n - m)]^{1/2} \tag{73}$$

where S is given by (65), and n, m are the numbers of observables P_i and variable parameters f_j, respectively. The quantity in brackets on the right of (73) is often called the goodness-of-fit parameter χ:

$$\chi^2 = S/(n - m) \tag{74}$$

When a satisfactory fit has been obtained, that is, when the calculated ϕ_i agree with the measured P_i within experimental errors, then $\chi \approx 1$. Least-squares fitting programs for lattice dynamical problems often have a tendency to diverge, particularly if the initial set of f_j is far from a best-fit set. If χ^2 increases from one iteration to the next, it is often very useful to multiply all the computed X_i by a scale factor, of order 0.6 say, and to recalculate χ^2 at the current iteration. If χ^2 is still higher than at the previous iteration, the scale factor multiplication is again applied, until finally χ^2 does decrease. This procedure limits the divergence and often guides the program in a mathematically more reasonable direction. In the final stages of fitting, however, the scale factor must be put equal to 1 to ensure that χ^2 is at a genuine minimum in the m-dimensional parameter space.

In the theoretical approach to lattice dynamics described at the beginning of this section, the theorist usually tries to ensure that his model of the interatomic forces not only gives a good description of the phonon properties, but also is consistent with such basic quantities as the total energy of the crystal. This total energy should also be at its minimum value at the observed cell size and other structural parameters. It is often the case with the more phenomenological models that the model parameters cannot be used to specify a potential function, so that the crystal energy cannot be calculated from the model. If a model potential, such as (63) or (64), is available, however, the total energy can be expressed in terms of the adjustable parameters, and we can treat it as another observable, insisting that the model should fit the measured sublimation energy within its experimental error. Other fundamental conditions, such as the minimization of the energy at the observed lattice parameter, can also be treated in this manner. The resulting fitted force model is then on much firmer ground than the purely empirical Born–von Kármán or valence force field models which are fitted only to the phonon frequencies.

B. Application of Group Theory

This subject is dealt with at length by Warren and Worlton (this volume), so we shall confine ourselves here to a brief summary of the main points. The crystal symmetry enters directly in two ways in the establishment of force models and the calculation of phonon frequencies. First, the interatomic or intermolecular force constants should obey all the relations and conditions which follow from the operations of crystal symmetry. The determination of these relations is often very useful in reducing computing time, by avoiding unnecessary re-calculation of parameters already stored in the computer. In more complex crystals, it may be more efficient to compute each force constant as it is required without regard to the fact that it may already have been computed in a symmetry-related situation. The symmetry relations are still valuable, however, as a check on the correctness of the computer program. Second, and more interesting, the crystal symmetry imposes certain conditions on the form of the dynamical matrix for particular values of the phonon wave vector **q**. For such values, the dynamical matrix may be block-diagonalized with the aid of suitable matrices whose elements can be obtained by standard group-theoretical techniques. Practical applications of these techniques to particular cases have been described by Chen (1967), Montgomery (1969), Waeber (1969), Venkataraman and Sahni (1970), and Montgomery and Dolling (1972). One of the main purposes of the chapter by Warren and Worlton is to show how these matrices can be obtained from a computer program. The ability to reduce the dynamical matrix to block-diagonal form is of great importance in the process of least-squares fitting of models to experimental data. For example, the external or intermolecular modes of solid α-N_2 can be obtained by diagonalizing a (20×20) complex matrix (four molecules per unit cell, five external degrees of freedom per molecule). At the point R $(a\mathbf{q}/2\pi) = (\frac{1}{2}, \frac{1}{2}, \frac{1}{2})$, however, the space group and time-reversal symmetries produce a considerable simplification to five 4-fold degenerate modes having group-theoretical representations $R_{23}{}^{-}$, $R_{23}{}^{+}$, $R_1{}^{-}$, and $R_1{}^{+}$. (There are two $R_{23}{}^{-}$ modes.) The general (20×20) matrix can thus be block-diagonalized into four entirely separate matrices, three of which yield, upon further numerical diagonalization in the computer, only one distinct normal mode frequency each. The $R_{23}{}^{-}$ matrix yields the remaining two modes. There is no ambiguity as to which computed frequency belongs to which representation, or to which observed normal mode it refers. The process of refining an intermolecular force model for α-N_2 (Kjems and Dolling, 1975) is then relatively straightforward; there is much less risk of accidently mixing up the modes and fitting the $R_1{}^{+}$ computed frequency to the observed $R_1{}^{-}$, than would be the case if no block-diagonalization had been performed to separate out the representations.

Since the eigenvectors of modes within any given representation have a particular form, appropriate to that representation and orthogonal to all other representations, the sum of the scattering cross sections for all modes of the same representation is controlled by the crystal symmetry and can be obtained by application of group-theoretical methods. In the language of optical spectroscopy, we say that the crystal symmetry governs the *selection rules* for various light scattering or absorption processes. The same kind of arguments also apply to the inelastic scattering of thermal neutrons, and they can be very useful in the interpretation and analysis of such experiments (Casella and Trevino, 1972).

C. Limitations and Ambiguities: Utilization of Force Models

The relationship between the phonon frequencies and the interatomic forces is, as noted in Section I, a basic problem in the study of lattice dynamics, and it is very important to consider the limitations of this relationship. If we know all the phonon frequencies for a particular crystal, to what extent does this information restrict our choice of interatomic force model? Is it possible that several quite different models can reproduce any given set of phonon frequencies? Various aspects of this problem of uniqueness have been discussed, for example, by Foreman and Lomer (1957), by Herman (1959), and most recently by Leigh *et al.* (1971). The latter authors demonstrate that it is in principle impossible to determine uniquely the parameters of a Born–von Kármán force model (and by extension any other largely parametrized model) by comparison with a given set of phonon frequencies, irrespective of the accuracy with which these frequencies are known. It is possible to transform in a continuous manner an initial set of force constants, which fit the given frequencies, so as always to maintain that agreement. However (again in principle!) a determination of one or more phonon eigenvectors will suffice to pick out the "correct" force model from its "incorrect" brethren. The need to obtain eigenvector information and its importance in deciding on the correctness of a force model has long been recognized. One of the most significant achievements of the early shell model calculations for NaI (Woods *et al.*, 1960) was that the neutron scattering cross sections for certain specific modes were given correctly by the new shell model, whereas the rigid-ion model calculations were completely at variance with the observed scattered neutron intensities. The neutron scattering cross section for coherent one-phonon processes depends in a rather direct way on the phonon eigenvector (Waller and Fröman, 1952), so that the latter can in principle be determined from suitable intensity measurements (Brockhouse *et al.*, 1963). In practice it is rather difficult to achieve an accuracy of better than 10% in an eigenvector determination from scattered neutron intensities, but relative accuracies of 30 to 50% are quite feasible. Indeed, eigenvector determinations to this level of accuracy are very important for the correct identification of the phonons being observed during neutron scattering experiments. For all but the simplest crystal structures, these experiments proceed by a kind of iterative process, in which an initial educated guess as to the phonon assignments is made in order to establish values of the parameters of a suitably chosen force model. If a good fit is achieved, not only to the phonon frequencies but also to the observed scattered neutron intensities, then we can be fairly sure that the initial assignments were reasonable. Further experiments can then be made to test the model and further refine the parameter values. If inconsistencies between theory and experiment are found, the phonon label assignments can be revised and the fitting process repeated until consistency is achieved. The force model eventually established as a result of this procedure is then able to predict correctly a wide selection of phonon frequencies as well as their eigenvectors, albeit to relatively poor accuracy; in addition, for certain types of models, we may be able to demonstrate consistency with the crystal sublimation energy, the heat capacity, various equilibrium conditions, elastic constants, and other physical properties. It goes without saying that the more experimental data that can be successfully fitted by the model, the more reliance can be placed on its ability to predict the results of further experiments. On the other hand, we must always be on the lookout for other force models, or other minima in χ^2 for a given model, which may give an equally satisfactory description of the available experimental data. The possibility of multiple minima in χ^2 in nonlinear least-squares fitting is well known, and we can probably never be sure that we have found the absolutely and uniquely "correct" model in this process. One should not conclude, however, that all empirical models are worthless, rather one should consider each such model on its merits, keeping in mind its limitations and possible ambiguities and maintaining a critical but constructive attitude to its utilization for further calculations. There is no question but that the wide variety of more or less empirical force models for different materials has greatly stimulated further development in this field of solid state physics. Even the less adequate models play their part in drawing attention to the difficulties which must be overcome in future attempts to understand the basic physics underlying the phonon dispersion curves.

One of the most popular uses of empirical force models has been as an interpolation formula for computing frequencies of phonons throughout the Brillouin zone at **q** values where no neutron inelastic scattering measurements have been made. If a suitable network of **q** values is chosen, the frequency distribution function $g(\omega)$, for the phonons, and many other related properties can readily be calculated from any given force model (see chapter by Gilat). It is self-evident that any limitations in the force model, such as a failure to predict all phonon frequencies to within, say 5%, will be reflected in similar inaccuracies in the $g(\omega)$ derived from that model. A clear demonstration of this type of problem has been given by Stedman *et al.* (1967) for the cases of Al and Pb. In the latter case, the existence of large Kohn anomalies in the dispersion curves made it virtually impossible to achieve a good fit to the data with the aid of a purely empirical Born–von Kármán type model; the use of such a model to compute $g(\omega)$ was thus doomed to give rather unreliable results. For Al, however, the dispersion curves along several major symmetry directions appeared to be quite well fitted by a simple Born–von Kármán model. Nevertheless, this model predicted rather poorly (~10% accuracy) several other phonon frequencies which were subsequently measured. The computed $g(\omega)$ based on this model was therefore not so accurate as that derived by Stedman *et al.* by a Taylor series type of interpolation between their measurements made at uniformly spaced **q** values. This illustrates the point made above that the more data which can be fitted by a model, the more reliable that model will be. If we were to refine the Born–von Kármán model by fitting to all the data obtained by Stedman *et al.* and then use it to compute $g(\omega)$, we would naturally obtain a much more accurate and reliable distribution function than that based on a model fitted to a small selection of data points. Viewed strictly as an interpolation procedure, a well-fitted empirical force model has a considerable advantage over the limited Taylor series expansion method in that it does have the correct form and symmetry properties (branch crossings and anti-crossings, etc). There are several materials, however, for which no good force model exists at the present time, so that model-independent interpolation procedures must be used to obtain $g(\omega)$ or other related phonon properties. In any event, it would seem that methods for obtaining $g(\omega)$ by interpolation between phonon frequencies determined by coherent one-phonon inelastic neutron scattering experiments are superior to any other method now available.

ACKNOWLEDGMENT

A large part of this article was written at the Institut Laue-Langevin, Grenoble. I would like to express my appreciation of the hospitality extended to me by many members of the ILL staff.

REFERENCES

Barker, J. A., Klein, M. L., and Bobetic, M. V. (1970). *Phys. Rev. B* **2**, 4176.
Becka, L. N. (1962). *J. Chem. Phys.* **37**, 431.
Bilz, H., Gliss, B., and Hanke, W. (1974). *In* "Dynamical Properties of Solids" (G. K. Horton and A. A. Maradudin, eds.), Vol. 1, pp. 343–390. Amer. Elsevier, New York.
Bobetic, M. V., and Barker, J. A. (1970). *Phys. Rev. B* **2**, 4169.
Born, M., and Huang, K. (1954). "Dynamical Theory of Crystal Lattices." Oxford Univ. Press (Clarendon), London and New York.
Born, M., and von Kármán, T. (1912). *Phys. Z.* **13**, 297.
Brockhouse, B. N., Becka, L. N., Rao, K. R., and Woods, A. D. B. (1963). *In* "Inelastic Scattering of Neutron in Solids and Liquids," Vol. II, pp. 23–33. IAEA, Vienna.
Brovman, E. G., and Kagan, Yu. M. (1974). *In* "Dynamical Properties of Solids" (G. K. Horton and A. A. Maradudin, eds.), Vol. 1, pp. 191–300. Amer. Elsevier, New York.
Buyers, W. J. L., and Cowley, R. A. (1969). *Phys. Rev.* **180**, 755.
Casella, R. C., and Trevino, S. F. (1972). *Phys. Rev. B* **6**, 4533.
Chen, S. H. (1967). *Phys. Rev.* **163**, 532.
Cochran, W. (1971). *Crit. Rev. Solid State Sci.* **2**, 1–44.
Cochran, W., and Cowley, R. A. (1967). *In* "Handbuch der Physik" (S. Flügge, ed.), Vol. 25, Part IIa, pp. 59–156. Springer-Verlag, Berlin and New York.
Cochran, W., and Pawley, G. S. (1964). *Proc. Roy. Soc., Ser. A* **280**, 1.
Cowley, R. A. (1964). *Phys. Rev.* **134**, A981.

Cowley, R. A., Cochran, W., Brockhouse, B. N., and Woods, A. D. B. (1963). *Phys. Rev.* **131**, 1030.
Cowley, R. A., Woods, A. D. B., and Dolling, G. (1966). *Phys. Rev.* **150**, 487.
Debye, P. (1912). *Ann. Phys.* (*Leipzig*) [4] **39**, 789.
Dick, B. J., and Overhauser, A. W. (1958). *Phys. Rev.* **112**, 90.
Dolling, G., and Powell, B. M. (1970). *Proc. Roy. Soc., Ser. A* **319**, 209.
Dolling, G., Cowley, R. A., and Woods, A. D. B. (1965). *Can. J. Phys.* **43**, 1397.
Dolling, G., Pawley, G. S., and Powell, B. M. (1973). *Proc. Roy. Soc., Ser. A* **333**, 363.
Elvebredd, I., and Cyvin, S. J. (1972). *In* "Molecular Structures and Vibrations" (S. J. Cyvin, ed.), Chapter 17. Elsevier, Amsterdam.
Ewald, P. P. (1921). *Ann. Phys.* (*Leipzig*) [4] **64**, 253.
Foreman, A. J. E., and Lomer, W. M. (1957). *Proc. Phys. Soc., London* **70**, 1143.
Gourlay, A. R., and Watson, G. A. (1973). "Computational Methods for Matrix Eigenproblems." Wiley, New York.
Grindlay, J., and Howard, R. (1965). *In* "Lattice Dynamics" (R. F. Wallis, ed.), p. 129. Pergamon, Oxford.
Hahn, H., and Biem, W. (1963). *Phys. Status Solidi* **3**, 1911.
Hardy, J. R. (1962). *Phil. Mag.* [8] **7**, 315.
Harrison, W. (1965). "Pseudopotentials in the Theory of Metals." Benjamin, New York.
Herman, F. (1959). *Phys. Chem. Solids* **8**, 405.
Horton, G. K., and Leech, J. W. (1963). *Proc. Phys. Soc., London* **82**, 816.
Iyengar, P. K., Venkataraman, G., Vijayaraghavan, P. R., and Roy, A. P. (1965). *In* "Inelastic Scattering of Neutrons," Vol. 1, pp. 153–177. IAEA, Vienna.
Karo, A. M., and Hardy, J. R. (1963). *Phys. Rev.* **129**, 2024.
Kellermann, E. W. (1940). *Phil. Trans. Roy. Soc. London, Ser. A* **238**, 513.
Kitaigorodskii, A. J. (1966). *J. Chim. Phys.* **63**, 8.
Kjems, J. K., and Dolling, G. (1975). *Phys. Rev. B* **11**, 1639.
Klein, M. L. (1975). *In* "Rare Gas Solids." (to be published).
Koehler, T. R. (1969). *Phys. Rev. Lett.* **22**, 777.
Kress, W. (1975). To be published.
Leigh, R. S., Szigeti, B., and Tewary, V. K. (1971). *Proc. Roy. Soc., Ser. A* **320**, 505.
Luty, T., and Pawley, G. S. (1974). *Phys. Status Solidi B* **66**, 309.
Lutz, U. A., and Hälg, W. (1970). *Solid State Commun.* **8**, 165.
McMurry, H. L., Solbrig, A. W., Jr., Boytor, J. K., and Noble, C. (1967). *J. Phys. Chem. Solids* **28**, 2359.
Maradudin, A. A. (1974). *In* "Dynamical Properties of Solids" (G. K. Horton and A. A. Maradudin, eds.), Vol. 1, pp. 1–82. Amer. Elsevier, New York.
Martin, R. M. (1969). *Phys. Rev.* **186**, 871.
Montgomery, H. (1969). *Proc. Roy. Soc., Ser. A* **309**, 521.
Montgomery, H., and Dolling, G. (1972). *J. Phys. Chem. Solids* **33**, 1201.
Musgrave, M. J. P., and Pople, J. A. (1962). *Proc. Roy. Soc., Ser. A* **268**, 474.
Nüsslein, V., and Schröder, U. (1967). *Phys. Status Solidi* **21**, 309.
Pawley, G. S. (1962). Ph.D. Thesis, Cambridge University.
Pawley, G. S. (1967). *Phys. Status Solidi* **20**, 347.
Pawley, G. S., and Cyvin, S. J. (1970). *J. Chem. Phys.* **52**, 4073.
Pawley, G. S., Rinaldi, R. P., and Windsor, C. G. (1971). *In* "Phonons" (M. A. Nusimovici, ed.), pp. 223–227. Flammarion, Paris.
Pines, D. (1963). "Elementary Excitations in Solids." Benjamin, New York.
Powell, B. M., and Sandor, E. (1971). *J. Phys. C* **4**, 23.
Rafizadeh, H. A., and Yip, S. (1970). *J. Chem. Phys.* **53**, 315.

Reynolds, P. A., Kjems, J. K., and White, J. W. (1972). *In* "Neutron Inelastic Scattering," pp. 195–205. IAEA, Vienna.
Schröder, U. (1966). *Solid State Commun.* **4**, 347.
Sham, L. J. (1974). *In* "Dynamical Properties of Solids" (G. K. Horton and A. A. Maradudin, eds.), Vol. 1, pp. 301–342. Amer. Elsevier, New York.
Sham, L. J., and Ziman, J. M. (1963). *Solid State Phys.* **15**, 221–299.
Sinha, S. K. (1973). *Crit. Rev. Solid State Sci.* **3**, 273–334.
Sinha, S. K., Gupta, R. P., and Price, D. L. (1971). *Phys. Rev. Lett.* **26**, 1324.
Smith, H. G., and Gläser, W. (1970). *Phys. Rev. Lett.* **25**, 1611.
Smith, H. M. J. (1948). *Phil. Trans. Roy. Soc. London, Ser. A* **241**, 105.
Solbrig, A. W. (1971). *J. Phys. Chem. Solids* **32**, 1761.
Squires, G. L. (1963). *In* "Inelastic Scattering of Neutrons in Solids and Liquids," Vol. II, pp. 71–85. IAEA, Vienna.
Stedman, R., Almqvist, L., and Nilsson, G. (1967). *Phys. Rev.* **162**, 549.
Vasil'ev, L. N., Logachev, Yu. A., Moizhes, B. Ya., and Yurév, M. S. (1971). *Fiz. Tverd. Tela* **13**, 450; *Sov. Phys.—Solid State* **13**, 363 (1971).
Venkataraman, G., and Sahni, V. C. (1970). *Rev. Mod. Phys.* **42**, 409.
Verma, M. P., and Singh, R. K. (1969). *Phys. Status Solidi* **33**, 769.
Waeber, W. B. (1969). *J. Phys. C* [Solid State Physics] **2**, 882.
Waller, I., and Fröman, P. O. (1952). *Ark. Fys.* **4**, 183.
Weber, W. (1973). *Phys. Rev. B* **8**, 5082.
Williams, D. E. (1971). *Acta Crystallogr., Sect. A* **27**, 452.
Wilson, E. B., Decius, J. C., and Cross, P. C. (1955). "Molecular Vibrations." McGraw-Hill, New York.
Woods, A. D. B., Cochran, W., and Brockhouse, B. N. (1960). *Phys. Rev.* **119**, 980.
Woods, A. D. B., Brockhouse, B. N., Cowley, R. A., and Cochran, W. (1963). *Phys. Rev.* **131**, 1025.
Ziman, J. M. (1964). *Advan. Phys.* **13**, 89.

The Use of Computers in Scattering Experiments with Slow Neutrons*

R. PYNN

PHYSICS DEPARTMENT, BROOKHAVEN NATIONAL LABORATORY
UPTON, NEW YORK

I. Introduction

THE MAJORITY OF THE chapters of this volume are concerned with theoretical aspects of lattice vibrations in solids. These chapters describe theoretical models which may be used as bases for the calculation of various vibrational properties. While many of these theories are of interest in their own right, they are often rendered substantially less esoteric when compared with experimental data. In a large number of cases, the data with which calculations presented in this volume are to be compared have been obtained from

* Work performed under the auspices of the U.S. Atomic Energy Commission.

scattering experiments with slow neutrons. Thus, it is natural to include a chapter that attempts to provide the reader with some understanding of these experiments. In the present context "an understanding" comprises a summary of three aspects of neutron scattering experiments. First, one must know how the measured intensities of scattered neutrons are related to properties of the scattering sample that are customarily calculated. Second, it is of some interest to know how experiments are performed in practice and to what extent different experimental configurations are suitable for the measurement of particular physical quantities. Finally, the ability to assess the adequacy of a specific theoretical model is substantially enhanced if one understands the errors and experimental complications inherent in a relevant measurement.

In view of the foregoing remarks, subsequent parts of this article will be divided into three principle sections. In Section II a brief background of the neutron scattering method is presented and particular emphasis is placed on a definition of the vocabulary used in this field. An expression for the neutron scattering cross section is presented and is subsequently developed in order to display the equations which relate the results of neutron scattering experiments to the properties of the (nonmagnetic) solid sample under investigation. In Section III several of the most common types of spectrometer used in neutron scattering experiments are described. The use of small computers for the control of such spectrometers is considered, and the software required for one such control system is discussed in some detail. Section IV is concerned with the reliability of neutron scattering data and the computational techniques that are applied to raw data in order to minimize such systematic distorting effects as instrumental resolution and multiple scattering. It is hoped that this section will help readers of this book to judge to what extent they may blame experimentalists for any manifest disagreement between experiment and the theories which are presented in subsequent chapters. The final section (V) provides a bibliographic note which should enable the interested reader to pursue further the subject matter of this paper.

II. Vocabulary of Neutron Scattering

A. The Neutron Scattering Cross Section

In a typical neutron scattering experiment (Egelstaff, 1965) an ostensibly monochromatic beam of neutrons is allowed to impinge on a sample and the number of neutrons scattered through a particular angle is recorded as a function of the energy which the neutrons have lost to or gained from the sample. The scattering power of the sample is conveniently specified by the differential cross section $d^2\sigma/d\Omega\, d\varepsilon$ defined by

$$\frac{d^2\sigma}{d\Omega\, d\varepsilon} = \frac{1}{Nn_\mathrm{i}}\left\{\begin{matrix}\text{number of neutrons scattered per second into} \\ \text{the energy range } \varepsilon \rightarrow \varepsilon + d\varepsilon \text{ in the element of} \\ \text{solid angle } \Omega \rightarrow \Omega + d\Omega\end{matrix}\right\}\frac{1}{d\varepsilon\, d\Omega} \quad (1)$$

In Eq. (1) n_i is the number of neutrons incident on the sample per square centimeter per second, and N is the number of scattering nuclei in the system.

An expression for the cross section (Egelstaff, 1965; Marshall and Lovesey, 1971), for nuclear scattering of neutrons may be derived from the Born approximation (Messiah, 1961). To avoid muddle the details of this derivation are given in the Appendix; the result is

$$\frac{d^2\sigma}{d\Omega\, d\varepsilon} = \frac{1}{N}\frac{1}{2\pi\hbar}\frac{k_f}{k_\mathrm{i}}\sum_{ii'} a_i a_{i'}$$
$$\times \int_{-\infty}^{\infty} \langle \exp[-i\mathbf{Q}\cdot\mathbf{r}_i(0)]\exp[i\mathbf{Q}\cdot\mathbf{r}_{i'}(t)]\rangle_\mathrm{T} e^{-i\omega t}\, dt \quad (2)$$

where

$\mathbf{k}_\mathrm{i}$ is the wave vector of incident neutrons,
$\mathbf{k}_f$ is the wave vector of scattered neutrons,
$\mathbf{Q} = \mathbf{k}_\mathrm{i} - \mathbf{k}_f$ is the wave vector transfer to the sample,
$\hbar\omega = \hbar^2(k_\mathrm{i}^2 - k_f^2)/2m$ is the energy transfer to the sample,
m is the neutron mass,
a_i is the bound scattering length of the nucleus on the ith site of the sample,
$\mathbf{r}_i(t)$ is the Heisenberg operator for the position of the ith nucleus at time t,
and $\langle\cdots\rangle_\mathrm{T}$ denotes a thermal average over states of the sample.

Equation (2) is the alpha and the omega of neutron scattering; it is the fundamental equation that describes all experiments in which slow neutrons are scattered by the nuclei of a (nonmagnetic) sample. All other descriptions are derivable from Eq. (2) either as exact restatements or as approximations.

In spite of its fundamental nature, Eq. (2) does not easily lend itself either to intuitive understanding or to a simple statement which might begin "Neutron scattering measures the . . . ." To obtain the catch phrases of the neutron scattering fraternity some rearrangement of Eq. (2) is required. In the following four subsections some of these manipulations are developed and the required vocabulary is presented.

B. Coherent and Incoherent Scattering

Let us for simplicity consider a sample which contains only one type of atom. Suppose that several nuclear states exist as, for example, would be the case in a mixture of ortho- and para-hydrogen (Carneiro and Nielsen, 1974). The scattering lengths for these nuclear states are not equal but the states may be assumed to be randomly distributed among the atomic sites of the sample. If we define

$$P_{ii'}(\mathbf{Q}, \omega) = \int_{-\infty}^{\infty} \langle \exp[-i\mathbf{Q}\cdot\mathbf{r}_i(0)] \exp[i\mathbf{Q}\cdot\mathbf{r}_{i'}(t)]\rangle_{\mathrm{T}} e^{-i\omega t}\, dt \tag{3}$$

Eq. (2) may be written as

$$\frac{d^2\sigma}{d\Omega\, d\varepsilon} = \frac{1}{2\pi\hbar}\frac{1}{N}\frac{k_f}{k_i}\left\{\sum_i a_i{}^2 P_{ii} + \sum_{ii'}{}' a_i a_{i'} P_{ii'}\right\} \tag{4}$$

where $\sum'$ signifies that the term with $i = i'$ is to be omitted from the sum. The first term in braces is the sum of intensities scattered by each nucleus while the second term represents an interference between neutrons scattered from different atoms. In view of the situation proposed at the beginning of this subsection (nuclear species distributed randomly) Eq. (4) may be rewritten as

$$\frac{d^2\sigma}{d\Omega\, d\varepsilon} = \left\{(\overline{a^2} - \bar{a}^2)\sum_i P_{ii} + \bar{a}^2 \sum_{ii'} P_{ii'}\right\}\frac{1}{2\pi\hbar}\frac{1}{N}\frac{k_f}{k_i} \tag{5}$$

where the bar over a or a^2 denotes an average over the sites of the sample. The first term in braces in Eq. (5) represents *incoherent scattering* while the second is *coherent scattering*. Evidently incoherent scattering measures a sum of single-atom properties, while the coherent scattering gives some information about a sum of two-atom properties.

The spin incoherence (ortho- + para-hydrogen mixture) considered above is not the only cause of incoherent scattering (Lomer and Low, 1965). However, the origins of incoherence is hardly a problem since "The Barn Book" (Goldberg *et al*., 1966) gives for each atomic species a value for coherent and incoherent scattering lengths. Following the prescription of Eq. (5) these scattering lengths may be used to determine the coherent (coh) and incoherent (inc) scattering cross sections according to the relations

$$\left.\frac{d^2\sigma}{d\Omega\, d\varepsilon}\right)_{\mathrm{inc}} = \frac{1}{2\pi\hbar}\frac{1}{N}\frac{k_f}{k_i}\sum_i (a_i^{\mathrm{inc}})^2$$

$$\times \int_{-\infty}^{\infty} \langle \exp[-i\mathbf{Q}\cdot\mathbf{r}_i(0)] \exp[i\mathbf{Q}\cdot\mathbf{r}_i(t)]\rangle_{\mathrm{T}} e^{-i\omega t}\, dt \tag{6a}$$

and

$$\left.\frac{d^2\sigma}{d\Omega\, d\varepsilon}\right)_{\mathrm{coh}} = \frac{1}{2\pi\hbar}\frac{1}{N}\frac{k_f}{k_i}\sum_{ii'} a_i^{\mathrm{coh}} a_{i'}^{\mathrm{coh}}$$

$$\times \int_{-\infty}^{\infty} \langle \exp[-i\mathbf{Q}\cdot\mathbf{r}_i(0)] \exp[i\mathbf{Q}\cdot\mathbf{r}_{i'}(t)]\rangle_{\mathrm{T}} e^{-i\omega t}\, dt \tag{6b}$$

In this paper we shall be concerned almost exclusively with coherent scattering.

C. The Fluctuation Spectrum

One of the often-heard statements about neutron scattering may be obtained, for a system of identical particles, by making the definition

$$S(\mathbf{Q}, \omega) = \frac{1}{2\pi}\sum_{ii'} \int_{-\infty}^{\infty} \langle \exp[-i\mathbf{Q}\cdot\mathbf{r}_i(0)] \exp[i\mathbf{Q}\cdot\mathbf{r}_{i'}(t)]\rangle_{\mathrm{T}} e^{-i\omega t}\, dt \tag{7}$$

The quantity on the left of Eq. (7) is known as the *fluctuation spectrum* (Egelstaff, 1965; Marshall and Lovesey, 1971) or, to the peddlers of jargonese, simply as "S-of-Q-and-omega." In terms of this quantity the coherent scattering cross section becomes

$$\left.\frac{d^2\sigma}{d\Omega\, d\varepsilon}\right)_{\mathrm{coh}} = \frac{1}{\hbar}\frac{1}{N}\frac{k_f}{k_i}(a^{\mathrm{coh}})^2 S(\mathbf{Q}, \omega) \tag{8}$$

Actually the above rearrangement of the basic equation (2) is not as futile as it may at first sight appear. $S(\mathbf{Q}, \omega)$ can be related, via a piece of mathematics known as the fluctuation-dissipation theorem (Pines and Nozière, 1966), to a quantity often calculated. The latter, known as the generalized susceptibility, $\chi(\mathbf{Q}, \omega)$, relates the density fluctuations in a sample to an applied external potential which drives these fluctuations. If the space-and-time Fourier components of the density fluctuation are denoted by $\rho(\mathbf{Q}, \omega)$ and the components of the applied potential by $V(\mathbf{Q}, \omega)$ one has (Pines and Nozière, 1966)

$$\rho(\mathbf{Q}, \omega) = \chi(\mathbf{Q}, \omega) V(\mathbf{Q}, \omega) \tag{9}$$

It is often relatively straightforward to compute $\chi(\mathbf{Q}, \omega)$ from a particular model (Singwi, *et al.*, 1968, 1970). From $\chi(\mathbf{Q}, \omega)$ one may obtain $S(\mathbf{Q}, \omega)$ via the fluctuation-dissipation theorem and hence, from Eq. (8), one may find the neutron scattering cross section.

D. The Time-Dependent Correlation Function

A less abstract description of the neutron scattering cross section can be obtained from some analysis originally presented by van Hove (1954). Let us write

$$\sum_{ii'} \langle \exp[-i\mathbf{Q}\cdot\mathbf{r}_i(0)] \exp[i\mathbf{Q}\cdot\mathbf{r}_{i'}(t)] \rangle_{\mathrm{T}} = N \int_{-\infty}^{\infty} G(\mathbf{r}, t) \exp(i\mathbf{Q}\cdot\mathbf{r})\, d^3r \quad (10)$$

evidently Eqs. (6b) and (10) may be combined to give

$$\left.\frac{d^2\sigma}{d\Omega\, d\varepsilon}\right)_{\mathrm{coh}} = \frac{1}{2\pi\hbar} \frac{k_f}{k_{\mathrm{i}}} (a^{\mathrm{coh}})^2 \int d^3r \int dt\, G(\mathbf{r}, t) \exp[i(\mathbf{Q}\cdot\mathbf{r} - \omega t)] \quad (11)$$

for a system of identical atoms. To establish the meaning of Eq. (11) it is necessary to invert Eq. (10) to obtain an expression for $G(\mathbf{r}, t)$ of the form:

$$G(\mathbf{r}, t) = \frac{1}{N} \sum_{ii'} \left\langle \int d^3r'\, \delta(\mathbf{r} + \mathbf{r}_{i'}(0) - \mathbf{r}')\, \delta(\mathbf{r}' - \mathbf{r}_i(t)) \right\rangle_{\mathrm{T}} \quad (12)$$

In any real system the laws of quantum mechanics (Messiah, 1961) prohibit any further simplification of Eq. (12) because the position operators appearing in this equation do not commute. Nevertheless it is often useful to consider the result one would obtain from Eq. (12) for a system of classical particles. Evidently in such a case the position operators commute and the integral of Eq. (12) may be performed to give

$$G(\mathbf{r}, t) = \frac{1}{N} \sum_{ii'} \langle \delta(\mathbf{r} + \mathbf{r}_{i'}(0) - \mathbf{r}_i(t)) \rangle_{\mathrm{T}} \quad (13)$$

This equation says that $G(\mathbf{r}, t)$ is the probability that a particle is at the origin of a coordinate system at time zero *and* a particle is at position $\mathbf{r}$ at time t. Thus $G(\mathbf{r}, t)$ is a *time-dependent, pair correlation function.*

The above definition of $G(\mathbf{r}, t)$ may be used to provide a simple, verbal statement of the neutron scattering law: the neutron scattering cross section is proportional to the space and time Fourier transform of the time-dependent pair correlation function. As pointed out earlier, this correlation function is really defined by Eq. (12) for any system. However, the simplest mental picture is obtained by passing to the classical limit expressed by Eq. (13). As usual, the quantum mechanical description is not easily verbalized and defies the use of a mental picture not based on mathematical symbolism.

E. The Phonon Expansion

Suppose one applies the idea of Section II, D to neutron scattering from a crystalline solid. In this case ones knows that $G(\mathbf{r}, t)$ can be expressed in terms of the equilibrium coordinates of the atoms plus a sum of lattice waves or phonons (Ziman, 1960). Evidently if there is one running wave of wave vector $\mathbf{q}$ and frequency ω_q in the lattice, the double Fourier transform of Eq. (11) will give rise to a δ function peak in the neutron scattering cross section at $\mathbf{Q} = \mathbf{q}$ and $\omega = \omega_q$. As more running waves (or phonons) are added each will give rise to a new peak in $(d^2\sigma/d\Omega\ d\varepsilon)_{\text{coh}}$. In addition, however, a neutron may simultaneously excite several running waves providing frequency and wave vector conservation are satisfied. The expansion of $S(\mathbf{Q}, \omega)$ in terms of the creation (or annihilation) of one, two, three . . . phonons is known as the *phonon expansion* (Marshall and Lovesey, 1971) of the scattering cross section.

As mentioned above, one-phonon scattering gives rise to well-defined peaks in the coherent scattering cross section. The location of such a peak in the energy spectrum of neutrons scattered in a particular direction determines the frequency and wave vector of the phonon responsible for the scattering process. In general, scattering processes involving two or more phonons (conveniently assimilated in the term *multiphonon scattering*) do not yield well-defined peaks in the cross section and merely contribute to the background on which the single-phonon peaks ride.

In addition to their positions in $(\mathbf{Q}, \omega)$ space, the one-phonon peaks contain two additional pieces of information—intensity and width. Intensity is determined by three major factors (Ziman, 1960):

1. The polarization direction of the lattice waves (i.e., the directions of atomic motion in the wave) with respect to the direction of $\mathbf{Q}$. Evidently, for example, atomic motions perpendicular to $\mathbf{Q}$ cannot transfer momentum (wave vector) to a neutron and will not give rise to a peak in $S(\mathbf{Q}, \omega)$.
2. The amplitude of the lattice wave, which appears in the cross section as a Bose–Einstein population factor for each phonon state.
3. The probability that a given atom will be in its official equilibrium position when the lattice wave reaches it. This introduces the familiar Debye–Waller factor (Ziman, 1960), which is equal to the probability of the previous sentence.

These factors are all of interest to the experimentalist who, if he is to report phonon frequencies, must be able to record some scattered intensity! Thus the above factors frequently place severe constraints on "where one should look in $(\mathbf{Q}, \omega)$ space in order to see a particular phonon."

Finally, the width of the single-phonon peaks gives a measure of the anharmonicity of the scattering sample. If a lattice is anharmonic, phonons do not represent normal coordinates and each lattice wave will decay with time. This decay can, as usual, be expressed by giving each phonon frequency an imaginary part. The time Fourier transform of Eq. (11) then implies that the single-phonon peak in the scattering cross section will not be a δ function but will have a finite width.

F. Relation to Subsequent Chapters

In the previous five subsections some of the important vocabulary of neutron scattering has been introduced. The various concepts are of relevance to different chapters in this volume. Coherent neutron scattering has been used to investigate some of the phenomena described by Dolling, Wood, Bell, and Koehler in this volume. The phonon expansion is relevant to the articles by Dolling and Wood and, to a certain extent, to the quantum solids described by Koehler. Of more importance in Koehler's article is the use of $S(\mathbf{Q}, \omega)$, a function which is all important in the study of amorphous solids (Axe, 1974) described in the review by Bell.

III. Spectrometers

In the preceding section the relationship between the neutron scattering cross section and various sample properties has been discussed. We now turn to the second point mentioned in the introduction—a description of the methods used to obtain experimental information about the scattering cross section. From the definition of the cross section given by Eq. (1) it is evident that the latter quantity is completely determined if the energies and trajectories of incident and scattered neutrons are known. Thus the experimental determination of the cross section is reduced to a specification of these quantities. In the following paragraph two relevant methods are discussed.

A. Determination of Neutron Energies and Trajectories

One simple method of determining both the energy and trajectory of neutrons makes use of Bragg's law (Egelstaff, 1965)

$$|\boldsymbol{\tau}| = 2|\mathbf{k}| \sin \theta_B \tag{14}$$

where $\boldsymbol{\tau}$ is a lattice wave vector, $\mathbf{k}$ the neutron wave vector, and $2\theta_B$ the angle through which the neutron is Bragg reflected. Evidently, if a beam of neutrons

ciated with Eq. (2) it follows that only one point in ($\mathbf{Q}$, ω) space is examined during such a counting period. Thus, an extremely long and tedious experiment would be required if $S(\mathbf{Q}, \omega)$ [cf. Eq. (7)] were to be determined for all $\mathbf{Q}$ and ω with a triple-axis spectrometer. However, if only sharp features of $S(\mathbf{Q}, \omega)$ such as phonons are to be observed, the TAS provides a simple and direct way of making a measurement. Usually the TAS is operated (i.e., the values of $\mathbf{k}_\mathrm{I}$ and $\mathbf{k}_\mathrm{F}$ are chosen) so that counts at a sequence of points along a simple path in ($\mathbf{Q}$, ω) space are recorded. The two most common modes of operation, the *constant-Q* and *constant-ω* modes (Brockhouse, 1961), perform the scans their names suggest. In Fig. 4 we show a zone of the reciprocal lattice of a bcc material, a typical constant-Q scan (A–A′ of Fig. 4b) and the intensity profile that might be recorded during such a scan (Fig. 4c). We shall return to a discussion of such profiles in Section IV where instrumental resolution will be considered.

Consider the task of measuring the intensity of scattered neutrons at point P on the profile shown in Fig. 4c. Values of $\mathbf{Q}$ and ω are given and these determine $\mathbf{k}_\mathrm{I}$ and $\mathbf{k}_\mathrm{F}$ according to the energy and momentum conservation conditions [cf. Eq. (2)].

$$\mathbf{Q} = \mathbf{k}_\mathrm{I} - \mathbf{k}_\mathrm{F} \tag{15}$$

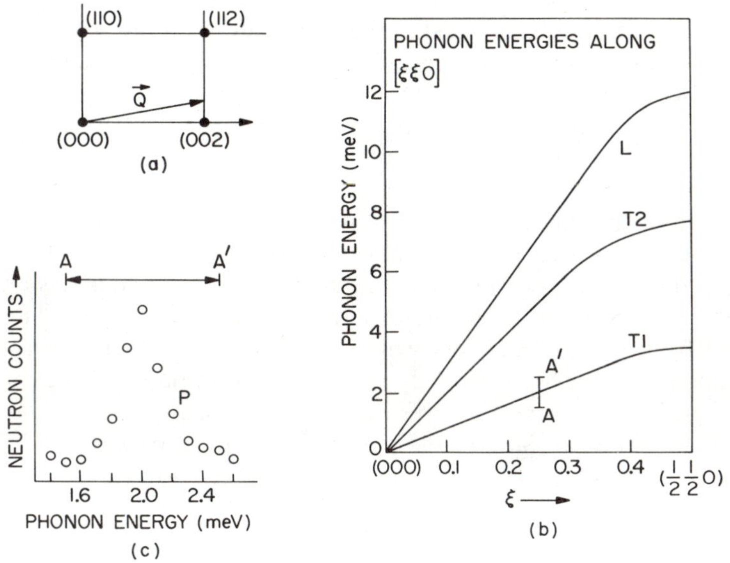

FIG. 4. (a) The [110] zone of the reciprocal lattice of a bcc structure. (b) A typical dispersion relation for phonons propagating in the [110] direction of a bcc structure. (c) A typical phonon profile for the scan denoted A–A′ in Fig. 4b at the $\mathbf{Q}$ point shown in Fig. 4a.

and

$$\omega = (\hbar/2m)(k_I^2 - k_F^2) \tag{16}$$

These values of $\mathbf{k}_I$ and $\mathbf{k}_F$ imply, in turn, certain values for the angles θ_M, θ_S, and θ_A shown in Fig. 3. The specification of the spectrometer configuration is completed by providing the orientation ϕ_S of the sample, a quantity that is related to the angle between $\mathbf{Q}$ and the sample reciprocal axes shown in Fig. 4a. Normally, measurements of profiles such as that of Fig. 4c are made with either $\mathbf{k}_I$ or $\mathbf{k}_F$ held constant and thus three independent angles, rather than the four described above, need be specified. The manipulation of these angles and of the counting mechanism are tasks which are ideally suited to computer control. In the following subsection such a control system is described in more detail.

2. *Computer Control of a Triple-Axis Spectrometer*

At this laboratory each triple-axis spectrometer is controlled by a PDP 11/40 computer which either operates in a relatively crude "stand alone" mode or in conjunction with a central PDP 11 machine (Pynn and Youngblood, 1975). Individual spectrometer-control computers have typewriter and paper-tape input facilities while the central computer has access to disk storage space and a number of magnetic tape stations. Under normal operating conditions (as opposed to the "stand alone" mode which provides a back-up facility should the central computer fail) programs are called to individual computers from the central disk and experimental data may be written either to disk, magnetic tape, or to the typewriter. The central computer is available for program-development work and for certain data analysis procedures.

In a scan of the type shown in Fig. 4c each of the angles θ_M, θ_S, θ_A, and ϕ_S shown in Fig. 3 is calculated by the spectrometer control computer from data provided by the experimenter. The computer system activates four synchronous motors which cause these angles to vary. Values of the angles are obtained from optical encoders which are attached to each rotation axis. When all angles have been correctly set, counting is initiated by the computer and continues until a preset number of neutrons have impinged on a monitor placed before the sample. At the end of a counting period the recorded intensity is printed out on the typewriter together with parameters which specify the spectrometer configuration. The computer then moves on to consider the next point of a scan.

The software part of the spectrometer control system may naturally be divided into two catagories: a set of basic routines for motor and detector manipulation and a series of high-level language programs which issue appropriate calls for the basic routines. The basic routines, which control directly such functions as the movement of a motor to a particular target angle, the initiation of a counting period, etc., are permanently resident in 8K of the core memory of each spectrometer control computer. These routines are called by FORTRAN programs which have been written as a series of overlays. This feature, which is made feasible by the availability of disk storage capacity, overcomes the limitations which might otherwise be imposed by the size of the core memory of the basic PDP 11 machine.

The FORTRAN programs consist of a supervisor routine which is permanently resident in 3K words of the core of individual computers, a number of permanently resident, often used subroutines (total of 3K words), and a series of overlays (about 2K words each), each of which controls a particular mode of spectrometer operation. Keyboard commands (which consist of two alphabetic characters) are accepted by the supervisor which either interacts directly with one of the basic motor or detector control routines or calls an appropriate overlay from disk. For example, during initial spectrometer adjustment, it is often necessary to move the motors that control the angles displayed in Fig. 3 and record the intensity of Bragg scattering from the sample. Such operations are controlled directly via the supervisor and the basic motor and counter routines. On the other hand, the control of a constant-Q scan such as that depicted in Fig. 4 is a somewhat more sophisticated operation that is controlled via the supervisor and a particular overlay program.

Data for each of the overlays are transmitted via common storage and may be manipulated (i.e., set or interrogated) via a universal READ package. The latter has been designed to overcome the limitations imposed by the usual FORTRAN input and output statements. In order to describe this part of the system it is useful to consider a typical sequence of spectrometer operations such as the performance of a series of constant-Q scans. Let us suppose that spectrometer adjustment has been completed and that an experimentalist has just brought to life the supervisor routine. This routine first inquires, via the typewriter, what type of spectrometer operation is required. Upon receiving the information that triple-axis scans are to be performed, the supervisor calls the overlay responsible for the control of such scans. This first call causes the overlay in question to define and to place in common storage the (two alphanumeric character) symbols that will be used to identify the variables that the experimenter must set in order for the overlay to function correctly. In the case of the triple-axis routine, for example, a scan identification number and the parameters ($\mathbf{Q}$, ω, etc.) which define a scan are required. When these identifiers have been set and added to the list of identifiers of general variables (such as sample lattice spacing, etc.) control is returned to the supervisor.

The experimentalist may now call the READ package (another overlay) which has access to the internally defined dictionary of variable identifiers. Values of variables may be set either by typing an equation of the form (IDENTIFIER) = (VALUE) or, if several variables are to be set simultaneously, by listing values in the order in which they are stored. Interrogation of the values of parameters proceeds in an analogous manner via a command of the form (IDENTIFIER): n. This command causes the value of the parameter to the left of the colon and the values of $(n - 1)$ variables stored in subsequent locations to be typed together with their variable identifiers.

Once all variables have been set control may be returned to the supervisor and thence to the triple-axis overlay. The latter checks that all necessary variables have been set and then performs a "dry run" of all scans loaded in the order which the user wishes to perform them. At this stage, no motors are moved but a series of error checks is performed to ascertain whether or not the scans that have been loaded are feasible. Thus, for example, suitable messages are typed out if intended values of $\mathbf{Q}$ or ω cannot be obtained for some reason. Once errors of this sort have been flagged the scan sequence is initiated.

The system described above has several important advantages. Since programs are written in a widely understood language (FORTRAN), modifications which permit esoteric types of spectrometer control (e.g., keeping the detector fixed in space during a scan) can easily be made. Such changes are greatly facilitated by the universal nature of the READ package, which receives its dictionary of variable identifiers directly from a spectrometer control overlay. Thus although, at the date of writing, the system described above has not been long operational it appears to be sufficiently versatile to handle most experimental situations that can be envisaged.

C. Time-of-Flight Spectrometers

1. *Instrumental Design*

There is very little variation in the basic functional specifications of TAS machines—the only latitude for choice lies in the size of the instrument, in the use of appropriate monochromator and analyzer crystals, and in the choice of collimators. Spectrometers that belong to the genus time of flight (TOF), however, display a far greater variation (Harris *et al.*, 1961; Kleb *et al.*, 1973; Otnes, 1975). The instrument (Harris *et al.*, 1961) that is perhaps the most straightforward is shown in Fig. 5. Here the energy and trajectory of incident neutrons are specified by a curved-slot rotor as described in Section II, A. Scattered neutrons are detected by one of a series of counters and their times of arrival are determined with respect to the electronic pulse

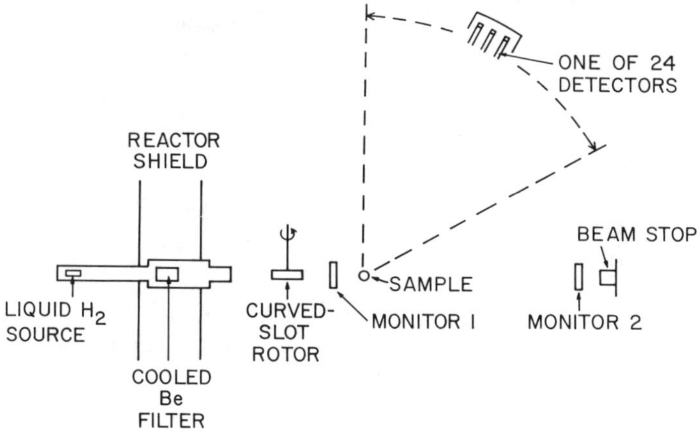

FIG. 5. Schematic side view of a typical TOF.

generated once per cycle by the rotor. The time of flight of the scattered neutrons determines their energy while the trajectory is specified by the counter position to an accuracy controlled by the aperture of the counters used. Since the TOF can readily be constructed with many detectors, this instrument provides an ideal method for studying $S(\mathbf{Q}, \omega)$ over large regions of $(\mathbf{Q}, \omega)$ space.

Variations in the design of the instrument shown in Fig. 5 include the use of evacuated flight tubes between sample and detectors (Kleb *et al.*, 1973), the insertion of a monochromator crystal before the rotor (Kleb *et al.*, 1973; Otnes, 1975), the use of several phased rotors (Brugger, 1965), and the deployment of more or fewer detectors than are shown in Fig. 5.

2. *Use of Computers in TOF Experiments*

From the point of view of computer control the TOF is simpler than the TAS described in the previous section since only one function, the rotor speed, need be monitored during a given run. However, in the case of the TOF, an on-line computer is an almost necessary tool for the collection and initial reduction of data. To illustrate this point we give here a brief description of a TOF system recently installed at the Argonne National Laboratory in the United States (Kleb *et al.*, 1973).

A schematic plan of the apparatus is shown in Fig. 6. This machine incorporates the double (or dog-leg) monochromator system that is used for many TAS machines (Stedman *et al.*, 1969) and which serves not only to monochromate the incident neutron beam but also to reduce background radiation levels at the sample. Other important design features include the enclosure of detectors and sample in an evacuated flight path (which ensures

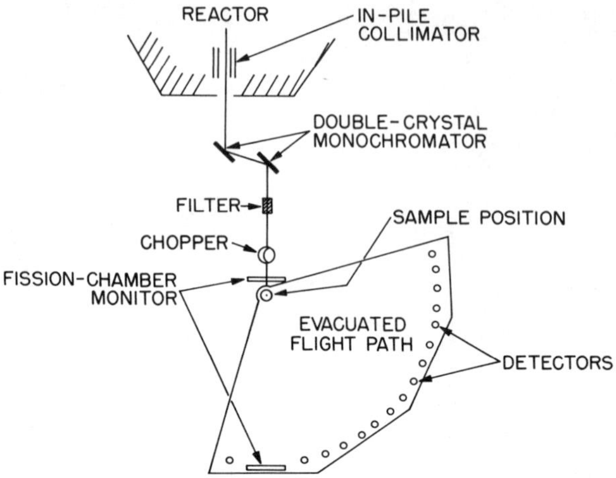

FIG. 6. Schematic plan of the Argonne time-of-flight machine.

that data will not be contaminated by gas scattering) and a constant spacing between sample and each of the detectors. This latter feature ensures that a given flight time of the scattered neutrons corresponds to the same energy transfer at the sample, independent of the detector considered. This simplifies the initial data reduction, some of which may be performed by an on-line computer.

At the end of a run with the TOF the memory of the on-line computer contains the TOF spectra for each detector. Each spectrum is in the form of a histogram, each step of which contains the aggregate of neutrons that arrived at the detector within a time interval known as the channel width. In practice this channel width varies between about 1 and 64 μsec depending on the length of the flight path and the experiment being performed. Evidently the "trick" of TOF spectrometry is to convert the raw TOF spectra to a plot of $S(\mathbf{Q}, \omega)$. To see how this is done we continue the discussion of the Argonne TOF machine (Copley *et al.*, 1973).

Each experiment with the TOF consists, in general, of four runs:

(a) A sample run.

(b) A run in which scattering from the sample container only is observed.

(c) A normalization run with a vanadium sample: this material is an incoherent scatterer which, because its cross section is well known, allows relative detector efficiencies and spectrometer energy-resolution to be determined.

(d) A timing run in which the minimum available channel width is used in order to obtain accurate positions for the monitor peaks. From this run k_I is determined.

Time-of-flight spectra from each of the above runs are punched on paper tapes which are then used as input to a complex program which eventually yields $S(\mathbf{Q}, \omega)$. The program first uses run (d) above to find the incident neutron energy and then computes, from runs (a) and (c), time-of-flight spectra that are corrected for variations in detector efficiency. Next, the normalized container counts [run (b)], corrected for attenuation by the sample material, are subtracted. In a subsequent stage of the program the effects of multiple scattering and of sample self-shielding (cf. Section IV, C) are removed and $S(\mathbf{Q}, \omega)$ is interpolated to a series of constant-Q plots. Finally, run (c) is used to determine the instrumental energy resolution (cf. Section IV, D) which is deconvolved from the data to yield the fluctuation spectrum $S(\mathbf{Q}, \omega)$ on an absolute scale.

The brevity of the above description of the analysis of TOF data may lead the reader to believe that this procedure is trivial. That this is not so is demonstrated by the fact that accurate, absolute values of $S(\mathbf{Q}, \omega)$ for disordered systems (Sköld, *et al.*, 1972; Copley and Rowe, 1974) have only become available during the past two or three years. The problems of computing multiple scattering and resolution corrections have required several man-years for their solution.

D. Correlation Choppers

In the preceding section the possibility of on-line reduction of time-of-flight data was mentioned. At the Argonne TOF such data reduction is customarily performed when the instrument is operated as a correlation-chopper spectrometer (CCS). Such an instrument belongs to the genus TOF but involves several features not yet discussed in this article. In the conventional TOF, each burst of neutrons transmitted by the rotor is allowed to decay before the subsequent burst is imposed. That is, neutrons arriving at each detector derive from a known incident burst. In the CCS this decay of the incident neutron burst is not required. Rather, the responses (time-of-flight spectra) from subsequent bursts overlap and the fluctuation spectrum of the sample is obtained by cross-correlating the detector signal with the incident burst pattern. Evidently this method has the advantage of increasing the duty cycle of the chopping device since more than one burst of neutrons is transmitted per rotor cycle.

Rather than give here a detailed discussion of the mathematics of the correlation-chopper method we refer the interested reader to the original papers (Price and Sköld, 1970; Sköld, 1968, and references cited by these authors) and merely quote a number of the important features of this method. It is possible to derive a criterion for the relative efficiencies of the CCS and conventional TOF methods in a given experiment. The CCS is found to be more efficient than the TOF if the background per channel of a TOF spectrum is more than twice the average signal per channel and if the signal in channels of interest is more than twice the average signal per channel. A list of specific experiments to which these conditions might apply is given by Price and Sköld (1970) and Sköld (1968).

In a conventional TOF experiment it is relatively easy to determine an optimum counting time by examining the TOF spectra at intervals during the experiment. Usually the observation of these spectra is implemented by hooking the memory of the on-line computer which controls the TOF machine to a television screen. This procedure allows the experimenter to keep track of the up-dated spectra registered by each detector. Since in the conventional TOF the spectra are directly related to $S(\mathbf{Q}, \omega)$, the statistical accuracy of the spectra and of the $S(\mathbf{Q}, \omega)$ derived from the experiment will be simply related. Therefore, an experiment may be terminated when sufficient statistical accuracy has been achieved.

With the CCS, $S(\mathbf{Q}, \omega)$ is determined, as mentioned above, by cross-correlating recorded TOF spectra with the burst pattern of the incident neutrons. Thus it is not possible, in this case, to determine the statistical accuracy that would be achieved for $S(\mathbf{Q}, \omega)$ merely by visual inspection of the relevant TOF spectra. To overcome this, the on-line computer at a CCS ought to be capable of performing the cross-correlation procedure and of displaying the result so that the length of an experiment may be optimized. This facility is available at the Argonne machine (Kleb *et al.*, 1973).

The mechanical chopper at the Argonne CCS is not the best that can be used for such instruments. This limitation arises because the mathematics on which the CCS method is based requires that the burst pattern of incident neutrons should be random. At Argonne a pseudorandom burst pattern is achieved by allowing neutrons to impinge close to the periphery of a rotating disk, segments of which are painted with a neutron absorbing material. Such a device transmits neutron bursts that are randomly spaced during each cycle of the disk but which repeat from one cycle to the next: hence the term "pseudorandom" is coined to describe this device.

A more truly random sequence of incident neutron bursts is available at the CCS at Oak Ridge National Laboratory (Mook *et al.*, 1974).* On this machine, pulsing of the incident beam is achieved by rapidly changing the direction of the magnetic moment of a ferrite crystal and by allowing neutrons to be magnetically Bragg reflected by this crystal. The principle of this system is based on the observation that the intensity of a magnetic Bragg reflection (Marshall and Lovesey, 1971) depends on the relative orientation of the

* In practice the principle advantage of the ORNL machine is that it provides a choice of both pulse duration and duty cycle.

atomic magnetic moments in the crystal and the neutron scattering vector **Q**. When the atomic moments are parallel to **Q** the intensity of magnetically scattered neutrons is zero, whereas, with the moments perpendicular to **Q**, the intensity is maximized. Thus the intensity of magnetic scattering can be varied between zero and some finite value by changing the orientation of the crystal magnetic moment with an external field. Pulsing of this field yields a pulsed beam of monochromatic neutrons and the duration and frequency of the neutron pulses can be varied by controlling the time dependence of the applied field. In practice, the Oak Ridge machine provides a risetime for each neutron pulse which is of order 1 μsec.

The Oak Ridge CCS is connected on-line to a PDP 15 30 computer which generates the pulse sequence for the incident neutron beam. In addition, the computer controls motors that vary the scattering angle from the ferrite monochromator, the detector positions, the placement of various radiation shields, and the sample orientation (cf. Section III, B, 2). A detailed discussion of the logic on which this control system is based is presented by Mook *et al.* (1974). In addition to its control function, the on-line computer is also responsible for the type of data analysis (cross-correlation) described earlier in this section.

IV. Analysis of Neutron Scattering Data

In the preceding two sections we have specified the relation between sample properties and scattering cross section and have described a number of devices that are used to measure all or part of this cross section. Like any other measurement, the experimental determination of a cross section is subject both to random and systematic errors. If experimental data are to be compared with theory, as they will be elsewhere in this volume, it is important to know how and to what extent these errors influence features of the experimentally measured cross section. In order to enable the reader to assess some of these factors, this section will provide a brief compendium of the corrections that are most frequently applied to neutron scattering data and a brief discussion of the adequacy of these corrections. In addition the possible magnitude of various insidious systematic errors will be discussed. Where corrections applied to raw data are dependent on the measuring instrument used, the discussion will concentrate on the TAS for reasons of brevity.

A. Random and Pseudorandom Errors

A neutron scattering experiment consists of recording the number of neutrons that arrive at one or more detectors for a given configuration of a particular spectrometer. The very nature of the experiment dictates that the number of neutrons counted will be subject to a statistical error: the counts generally being distributed according to the laws of Poisson statistics (Mack, 1967). For the purposes of assessing statistical errors, experimental results may be divided into two classes. In the cases in which $S(\mathbf{Q}, \omega)$ is plotted for large regions of $\mathbf{Q}$ and ω, statistical uncertainties are generally evident since most authors present plots of experimentally determined values of $S(\mathbf{Q}, \omega)$ (Sköld *et al.*, 1972; Copley and Rowe, 1974). Papers which report phonon frequencies, on the other hand, generally includes plots like Fig. 4c of a "typical" phonon profile whose ability to represent the quality of data depends, among other factors, upon the degree of astigmatism suffered by the author(s)! Usually statistical uncertainties are combined to give an error bar on a measured phonon frequency. However, it is not unusual for the experimenter to discover indications that such error bars are too small by a factor of about two! The additional random error, which is often discovered during attempts to fit phonon data to force-constant models (Dolling, page 1, this volume), arise primarily because the scattering angles of a TAS cannot be set or determined to much better than 0.01°. Such a misset of the (002) planes of a pyrolytic graphite monochromator set to reflect 14 meV neutrons yields, for example, a phonon energy that is shifted by 0.02 meV from its correct value. This shift is comparable, to within a factor of less than about three, to the statistical uncertainty in the energy of a phonon which might be measured with such a monochromator arrangement. During a particular scan (of the constant-Q type, for example) angular misset may introduce systematic error because motor angles are usually varied in the same sense throughout the scan. However, since many phonons with substantially different wave vectors are measured during a typical experiment, misset errors are usually distributed more or less at random among the phonon frequencies obtained in such an experiment.

B. Systematic Errors

In addition to the random factors described in the preceding subsection, a number of systematic effects plague the life of neutron spectroscopists. The most insidious of these errors arises from systematic misset of the spectrometer angles caused, for example, by poor calibration of monochromator and analyzer units or by faulty setting of the zero readings of the optical encoders which measure scattering angles. In a limited number of cases these errors can be located and corrected. For example, the results of neutron energy-loss and energy-gain experiments can be compared and discrepancies may often be attributed to errors of the type described above. A further consistency check is available if a given phonon frequency can be measured at different, but equivalent, points in the extended zone scheme of the reciprocal lattice. Naturally, the latter check is not available during the measurement of $S(\mathbf{Q}, \omega)$ of disordered systems.

In addition to the insidious errors just described there are certain important systematic effects for which corrections are customarily made. The most important of these are the errors that are introduced by multiple scattering processes, by sample self-shielding, and by the finite resolution of the neutron spectrometers. In the following sections these effects will be discussed in turn.

C. Multiple Scattering

In Section II an expression for the neutron scattering cross section was presented. This cross section refers explicitly to events in which a neutron is scattered only once by the sample before it is detected. Evidently, scattering processes which do not fall into this catagory cause contamination of measured values of the single scattering cross section and must be accounted for in the analysis of experimental data. Several sources of such error may be identified. As the incident neutron beam passes through a sample it is attenuated by scattering events and the intensity of the incident beam thus varies spatially within the sample. This sample self-shielding causes a depletion of the single scattering cross section. On the other hand, the latter cross section is augmented by events in which a neutron is multiply scattered within the sample before it is detected. Furthermore, a liquid, amorphous or polycrystalline sample is almost invariably surrounded by a container that contributes both single and multiple scattering events and which further attenuates the incident beam.

Corrections for all of the effects described above have been mentioned in Section III, C, 2 of this paper. In the current section attention will be focused on the calculation of multiple scattering corrections, a problem which has only recently been solved in a satisfactory manner.

While several programs (Honeck, 1964; Johnson, 1974) have been written to perform multiple scattering corrections, the most versatile and sophisticated code appears to be that generated by Copley (1974) as an extension of the original work of Bischoff (1970). This program uses an estimated scattering function to track the scattering events suffered by neutrons within the sample and/or container in a Monte Carlo fashion. Single and multiple scattering contributions to the total cross section are accumulated separately, a procedure which allows the multiple scattering contribution to be subsequently subtracted from an experimentally determined cross section. Since the details of the calculation and of the associated computer program have been presented elsewhere (Copley, 1974) only the salient features will be described here.

For the purposes of computation the scattering properties of a material are conveniently divided into three parts: elastic coherent scattering (Bragg scattering), which is characterized by δ functions in both $\mathbf{Q}$ and ω; elastic incoherent scattering which involves a δ function in ω; and inelastic scattering which, for the (isotropic) systems in which multiple scattering corrections are traditionally employed, involves no δ functions. In the first step of the Monte Carlo process a point of impact on the sample is chosen for a neutron. Next a scattering point is selected using a knowledge of the total absorption and scattering cross sections. At the scattering point, one of the three types of scattering process described above is allowed to occur and a new propagation wave vector for the neutron is determined. The probability that the neutron will now be able to escape from the sample–container system is then determined and the probability of observing a detector response is calculated. The neutron then proceeds to further scattering events similar to that just described and the detector response is calculated in each case. A cut-off is imposed in order to avoid tracking the history of a given neutron indefinitely.

The program described so briefly above has been written in FORTRAN IV and occupies 75K, 32 bit words of store when run on an IBM 370/195 machine. Of these 75K words, 26K words are used to store various arrays derived from the fluctuation spectrum $S(\mathbf{Q}, \omega)$. Typical running times for the complete program are, with this machine, between 1 and 10 minutes (Copley, 1974).

Correction for multiple scattering effects is one of the most complicated and sophisticated computational tasks involved in the reduction of neutron scattering data. However, not all neutron scattering experiments require corrections of this sort. In particular, when only sharp features of $S(\mathbf{Q}, \omega)$ such as phonons are to be measured, multiple and container scattering and sample self-shielding are usually negligible effects. Since, of the remaining articles in this volume, only those that deal with amorphous materials use neutron data that have to be corrected for these extra scattering effects, the description of the techniques used for these corrections has been substantially abbreviated. A review of this subject which provides references to many of the original papers may be found in the article of Copley and Lovesey (1975).

D. Resolution for a Triple-Axis Spectrometer

While measurements of sharp features of $S(\mathbf{Q}, \omega)$ such as phonons do not usually need to be corrected for multiple scattering effects, they are often severely influenced by the finite resolution of neutron spectrometers. In the following sections this problem will be examined in some detail. For reasons of brevity attention will be focused on a discussion of the computational aspects of the effects of the resolution of triple-axis spectrometers. Resolution effects for other types of spectrometer (Steinsvoll, 1973; Komura and Cooper, 1970) and other important aspects of the resolution problem, such as the calculation of integrated intensities (Dolling and Sears, 1973; Dorner, 1972; Chesser and Axe, 1973) have been considered in the literature but will not be discussed here.

1. *Analysis of the Resolution Problem*

Let us consider the case of a TAS set to observe energy and momentum transfers of $\hbar\omega_0$ and $\hbar\mathbf{Q}_0$ to a sample. The conservation conditions for these variables are nominally given by

$$\mathbf{k}_\mathrm{I} - \mathbf{k}_\mathrm{F} = \mathbf{Q}_0 \tag{17}$$

and

$$(\hbar/2m)(k_\mathrm{I}^2 - k_\mathrm{F}^2) = \omega_0 \tag{18}$$

However, neither the monochromator nor analyzer crystal is perfect and the collimators in the TAS system do not specify neutron trajectories absolutely. For these reasons scattered intensity is observed from points in $(\mathbf{Q}, \omega)$ space which are adjacent to $(\mathbf{Q}_0, \omega_0)$ as well as from this latter point. In order to account for this fact one may define a resolution function (Cooper and Nathans, 1967) $R(\mathbf{Q} - \mathbf{Q}_0; \omega - \omega_0)$ for the spectrometer as the probability of detecting a scattering event at $(\mathbf{Q}, \omega)$ when the spectrometer is set to observe such events at $(\mathbf{Q}_0, \omega_0)$. The number of counts recorded by the detector is then given by convoluting the function $R(\Delta\mathbf{Q}, \Delta\omega)$ with the scattering cross section of the sample.

In order to better understand the form of the resolution function one may examine each component of the TAS in turn. The scattering triangle (a geometrical expression of Bragg's law) for the monochromator is shown in Fig. 7. If the monochromator crystal and the collimator between monochromator and sample were perfect, the monochromator system would transfer neutrons from the center of the volume V_0 in momentum space to the center of the volume V_1. The imperfection of the monochromator may be taken into account by defining for this crystal a mosaic spread (I.L.L. Report, 1973). Thus the monochromator is regarded as an aggregate of mosaic grains that are individually perfect but whose orientations in space are distributed in a gaussian manner about a mean value. This has the effect of allowing an angular spread of the reciprocal lattice vectors $\mathbf{G}_\mathrm{M}$ of the monochromator. In turn this leads to a spread in the $\mathbf{\Delta}_1$ direction of the wave vectors of neutrons scattered by the monochromator. A spread in the $\mathbf{\Delta}_2$ direction is introduced by the finite extent of the collimation (in the plane of

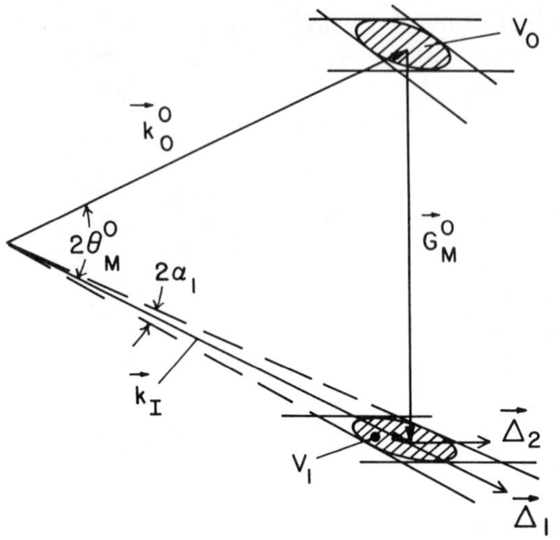

FIG. 7. Scattering triangle for a crystal monochromator. $\mathbf{k}_0{}^0$ and $\mathbf{k}_I$ are the most probable wave vectors of neutrons incident on and reflected from the monochromator while $\mathbf{G}_M{}^0$ is the most probable value of the monochromator reciprocal lattice vector $\mathbf{G}_M$. $2\alpha_1$ is the divergence of the collimator denoted C_1 in Fig. 3.

Fig. 7) of the scattered neutron beam. Finally, a spread in the direction $\mathbf{\Delta}_3$ which is perpendicular to the plane of Fig. 7 is introduced both by finite collimation and by the vertical mosaic spread of the monochromator. Thus, the overall effect of the monochromator system is to transfer neutrons from volume V_0 to volume V_1 and it is the neutrons in the latter volume of momentum space which impinge on the sample. It is worth noting at this stage that the volume V_1 is ellipsoidal and that the vectors $\mathbf{\Delta}_i$ may be chosen to be principal radii of the ellipsoid (Stedman, 1968). This implies that an expression, $P_M(\Delta)$, for the probability density of neutrons at any point within V_1 may be written as the product of three functions, each of which depends on only one of the vectors Δ_i. Symbolically one may express the number of neutrons of wave vector $\mathbf{k}_i$ incident on the sample per square centimeter per second as

$$\Phi(\mathbf{k}_i) = \Phi_0 P_M(\mathbf{\Delta}) \equiv \Phi_0 P_M^{(1)}(\mathbf{\Delta}_1) P_M^{(2)}(\mathbf{\Delta}_2) P_M^{(3)}(\mathbf{\Delta}_3) \tag{19}$$

where

$$\mathbf{k}_i = \mathbf{k}_I + \mathbf{\Delta}$$

and Φ_0 is the flux of neutrons incident on the monochromator.

Arguments similar to those given above apply to the analyzer system which which accepts neutrons of wave vectors $\mathbf{k}_f$ with a probability $P_A(\mathbf{\delta})$. Here,

$\boldsymbol{\delta}$ (= $\mathbf{k}_f - \mathbf{k}_F$) is a vector which describes analyzer resolution in the same way as Δ describes the monochromator properties.

Let us now define a sample cross section $d^3\sigma/dk_f{}^3$ which gives the probability that a scattering event in the sample changes the wave vector of a neutron from $\mathbf{k}_i$ to $\mathbf{k}_f$ and leaves the final wave vector in an element of $\mathbf{k}$ space of size d^3k_f. The total number of neutrons registered by the detector with the spectrometer set at $(\mathbf{Q}_0, \omega_0)$ is then given by

$$N = \Phi_0 \int d^3k_i \int d^3k_f P_M(\boldsymbol{\Delta}) \frac{d^3\sigma}{dk_f{}^3} P_A(\boldsymbol{\delta}) \tag{20}$$

One would like to rewrite Eq. (20) as an integral over $\Delta\mathbf{Q}$ (= $\mathbf{Q} - \mathbf{Q}_0$) and $\Delta\omega$ (= $\omega - \omega_0$) of the product of a scattering cross section and a resolution function $R(\Delta\mathbf{Q}, \Delta\omega)$. To see how this may be done one may examine the scattering triangle, shown in Fig. 8, for the sample. Figure 8 is to be interpreted as meaning that the sample scatters neutrons from volume V_1 into volume V_2. Suppose we consider the case in which $\Delta\mathbf{Q}$ and $\Delta\omega$ are held fixed. This implies the following constraints

$$\Delta\mathbf{Q} = \boldsymbol{\Delta} - \boldsymbol{\delta} \tag{21}$$

and

$$\Delta\omega = (\hbar/m)(\mathbf{k}_I \cdot \boldsymbol{\Delta} - \mathbf{k}_F \cdot \boldsymbol{\delta}) \tag{22}$$

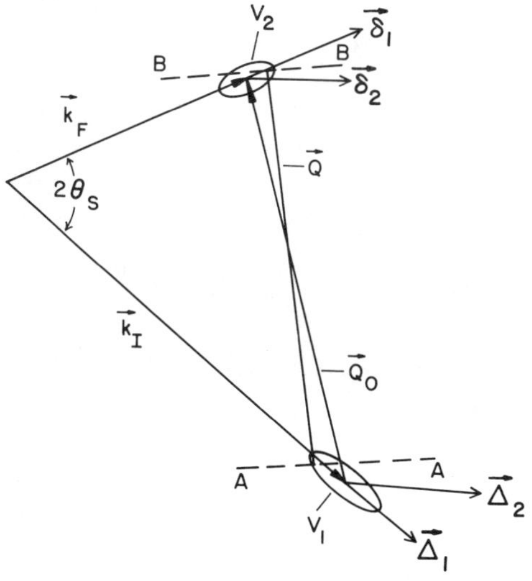

FIG. 8. Scattering triangle for the sample. The symbols are explained in the text.

For given $\Delta\mathbf{Q}$ and $\Delta\omega$ the tips of the vectors $\mathbf{k}_i$ and $\mathbf{k}_f$ are restricted to lie in planes A–A and B–B of Fig. 8, respectively. These planes are perpendicular to the plane of Fig. 8 and to the vector $\mathbf{Q}$ ($= \mathbf{k}_i - \mathbf{k}_f$) and there is a one-to-one correspondence between points (Δ_y, Δ_z) in plane A–A and points (δ_y, δ_z) in plane B–B. Thus the integrations of Eq. (20) may be written as integrations over $\Delta\mathbf{Q}$, $\Delta\omega$, Δ_y, and Δ_z. Changing to these variables and taking into account the Jacobian so introduced allows Eq. (20) to be restated as

$$N = \Phi_0 \int d^3(\Delta\mathbf{Q}) \int d(\Delta\omega) \frac{m}{\hbar Q} \frac{d^3\sigma}{dk_f{}^3} \int d\Delta_y \int d\Delta_z P_M(\boldsymbol{\Delta}) P_A(\boldsymbol{\delta}) \tag{23}$$

This equation now permits one to achieve the goal of expressing N as a convolution of a resolution function $R(\Delta\mathbf{Q}, \Delta\omega)$ and a cross section. Defining

$$R(\Delta\mathbf{Q}, \Delta\omega) = \int\int P_M(\boldsymbol{\Delta}) P_A(\boldsymbol{\delta})\, d\Delta_y\, d\Delta_z \tag{24}$$

one finds that Eq. (23) becomes

$$N = \Phi_0 \frac{m}{\hbar} \int\int \frac{1}{Q} \frac{d^3\sigma}{dk_f{}^3} R(\Delta\mathbf{Q}, \Delta\omega)\, d^3(\Delta\mathbf{Q})\, d(\Delta\omega) \tag{25}$$

It is clear from the foregoing discussion that the resolution function can be obtained from Eq. (24) in terms of collimator sizes and the crystal mosaics of monochromator and analyzer. The details of this calculation are lengthy and would be instructive only to lovers of algebraic manipulation. The result of the calculation is, however, simple. One finds that, if collimators are assumed to have gaussian transmission probabilities and that the mosaic distributions of the monochromator and analyzer are also gaussian, one may write (Cooper and Nathans, 1967; Werner and Pynn, 1971)

$$R = R_0 \exp\left(-\frac{1}{2} \sum_{ij} M_{ij} X_i X_j\right) \tag{26}$$

where $(X_1, X_2, X_3, X_4) \equiv (\Delta Q_x, \Delta Q_y, \Delta Q_z, \Delta\omega)$ and M_{ij} are matrix elements that can be determined from the spectrometer specifications. As a further development one may also include in the M_{ij} the effect of mosaic spread of the sample—this is discussed by Werner and Pynn (1971).

2. *Effects of Finite Resolution on Phonon Measurements*

Equation (26) allows one to achieve a mental picture of the effect of TAS resolution (Cooper and Nathans, 1967). The locus of points at which the resolution function has any fraction of its maximum value is an ellipsoid in (**Q**, ω) space. The size and orientation of this ellipsoid are determined by the spectrometer configuration and by the values of collimations and crystal mosaics. Restricting these considerations to a two-dimensional (Q, ω) plane yields a picture like that displayed in Fig. 9a, in which two possible orientations of a resolution ellipse are superimposed on a typical plot of a dispersion relation. The resolution ellipses are intended to represent contours on which R has some small fraction of its maximum value R_0. Neutrons which suffer scattering events within a resolution ellipse will be detected. As the spectrometer configuration is changed during, for example, a constant-Q scan, the orientation and size of the resolution ellipse is substantially unaltered, but the ellipse is translated parallel to the ω axis during the scan. Scanning of the ellipses gives rise to the profiles (plots of detector counts against neutron energy transfer) shown in Fig. 9b. In one case the phonon profile is well-defined or focused while in the other case this definition is lost. Evidently a spectrometer should be used in the focused mode whenever possible in order to minimize the uncertainty in the measured position of a phonon profile.

If the neutron scattering cross section does not vary substantially over that part of the phonon dispersion surface contained within the resolution ellipse and if the dispersion surface is essentially planar in this region, instrumental resolution merely broadens the observed phonon profile (Cooper and

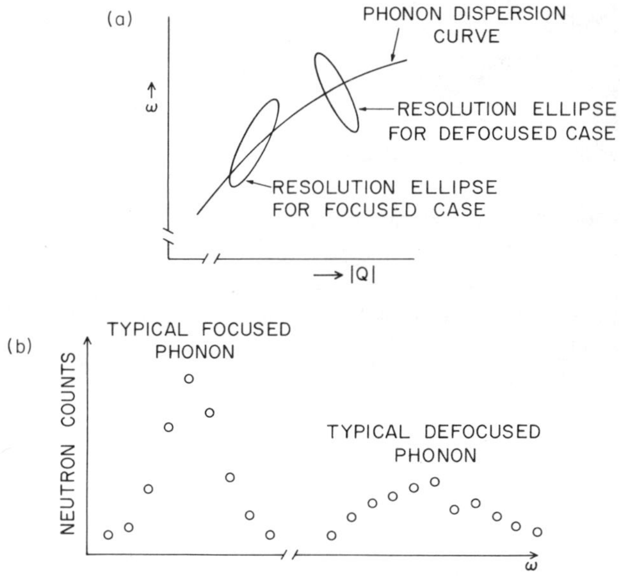

FIG. 9. Examples of resolution ellipses and phonon profiles for focused and defocused cases.

Nathans, 1967; Stedman, 1968). That is, the finite instrumental resolution introduces a width to the profile which, when combined with the natural linewidth resulting from sample anharmonity, yields the observed width of the profile. On the other hand, if either of the restrictions of the penultimate sentence is not satisfied, finite instrumental resolution may lead to the observation of a phonon profile whose mean (or peak) position does not correspond to the true phonon energy at wave vector $\mathbf{Q}_0$ (Werner and Pynn, 1971; Samuelsen *et al.*, 1970). That this is so should be evident from a brief consideration of plots like Fig. 9a.

Several common situations in which the mean of a recorded profile does not coincide with the phonon frequency at $\mathbf{Q}_0$ may be identified.

1. In many phonon measurements the scattering plane (the plane containing $\mathbf{k}_\mathrm{I}$ and $\mathbf{k}_\mathrm{F}$) is chosen to be a mirror plane of the sample. Thus the phonon frequencies plotted along a line perpendicular to the scattering plane achieve, in general, maximum or minimum values in the scattering plane. This variation of frequency is sufficient to allow a systematic displacement of a phonon profile from the true phonon energy by an amount which is principally determined by the degree of collimation perpendicular to the scattering plane (vertical collimation). In addition, variation of the phonon polarization vectors as $\mathbf{Q}$ moves out of the scattering plane cause a variation of scattering cross section and a concomitant discrepancy between observed and actual phonon frequency.

2. If two or more phonons are degenerate in the scattering plane and if the degeneracy is lifted as $\mathbf{Q}$ moves out of this plane, anomalously broad and/or shifted profiles may be observed. Examples of this effect have been noted and correctly explained by several authors (Werner and Pynn, 1971; Raunio *et al.*, 1969; Cowley and Pant, 1970; Copley, 1971).

3. A particularly important case of the systematic errors introduced by spectrometer resolution occurs in the measurement of acoustic phonons of small reduced wave vector. At small wave vector the phonon dispersion surface has considerable curvature, both in and perpendicular to the scattering plane. In addition, in this region the scattering cross section varies strongly (approximately as $1/\omega^2$) over the dispersion surface. Both of these effects lead to systematic displacements of the center of a measured profile from the true phonon frequency.

3. *Corrections for Resolution Effects*

Approximate analytic methods of dealing with *some* of the effects of finite resolution have been available for some years (Stedman, 1968, 1973; Nielsen and Bjerrum Møller, 1969; Hayward, 1971). However, only recently have numerical procedures (Werner and Pynn, 1971; Samuelsen *et al.*, 1970) been developed which include all such effects via an evaluation of the integral given by Eq. (25). In order to illustrate how such a procedure works in practice we present in the following paragraphs the history of two phonons in argon (Fujii *et al.*, 1974), measured at this laboratory, as they pass through the data analysis procedure.

In Fig. 10 two constant-Q phonon profiles measured in ^{36}Ar at 82°K are shown. One of them (Fig. 10a) has an apparently clean, symmetric profile riding on an essentially constant background while the other displays two overlapping peaks. In each case the first stage in our data analysis involves fitting the observed profile to the sum of one or two gaussian functions plus a linear, possibly sloping, background. The fits so obtained are shown by the solid lines in Fig. 10. From the parameters of the fitted gaussian curves the observed phonon frequency and the error in this quantity due to counting statistics may immediately be deduced. In the argon experiment these quantities were obtained for all measured phonons, and the phonon frequencies were fitted to a force-constant model (Svensson *et al.*, 1967). This model allowed the scattering cross section to be computed for the entire phonon dispersion surface and hence permitted, in principle, the evaluation of the scattering cross section at any $\mathbf{Q}$ and ω.

In order to use this cross section to evaluate the resolution correction given by Eq. (25) a program written by the author (Pynn and Werner, 1970) was used. The program first constructs, in reciprocal space, a rectangular prism which just contains that resolution ellipsoid on which the resolution function $R(\Delta\mathbf{Q}, \Delta\omega)$ has some small, predetermined fraction of its maximum value. This fraction is usually chosen to be e^{-4}. The rectangular prism is next divided into an array of approximately cubic cells whose size guarantees the adequacy, within a given cell, of an expansion of the one-phonon scattering cross section as a second-order Taylor series about the center of that cell.

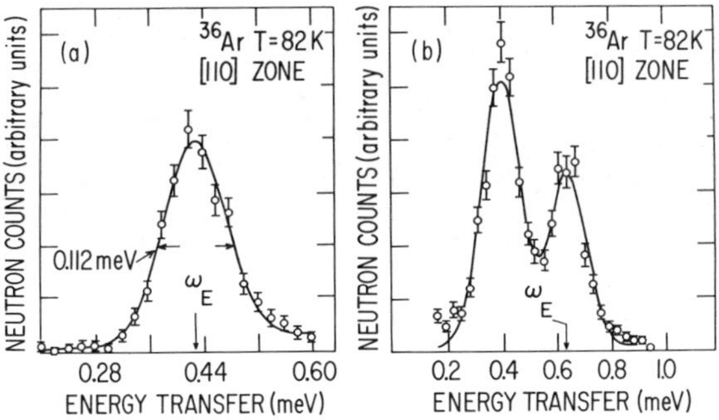

FIG. 10. Measured profiles for two phonons in ^{36}Ar obtained with $k_I \simeq 1.5$ Å^{-1}. For Fig. 10a, $\mathbf{Q}_0 \equiv (0.93, 1.07, 1.07) \times 2\pi/a$, $\omega_E = 0.430$ meV and for Fig. 10b, $\mathbf{Q}_0 \equiv (-1.0, 1.04, 1.04) \times 2\pi/a$ and $\omega_E = 0.639$ meV, where a is the lattice constant.

If a similar polynomial expansion of the resolution function $R(\Delta\mathbf{Q}, \Delta\omega)$ can be obtained, the integrals of Eq. (25) can evidently be evaluated in closed form. However, since $R(\Delta\mathbf{Q}, \Delta\omega)$ may vary rapidly within a particular cell (R is, after all, a gaussian function of $\Delta\mathbf{Q}$ and $\Delta\omega$), the expansion of this quantity is carried out within a set of subcells which fill the original cell. The size of subcells is chosen within the program in a manner which guarantees that the error made in expanding $R(\Delta\mathbf{Q}, \Delta\omega)$ within a subcell as a polynomial series is commensurate with the error introduced by ignoring contributions to the scattering that arise from points outside the rectangular prism that contains the chosen resolution ellipsoid. Naturally, many of the subcells constructed by the program yield negligible contributions to the integral of Eq. (25). Thus, the program contains a series of successfully more restrictive criteria that isolate those subcells over which the integration needs to be performed.

The program in its present form is relatively fast and takes about $\frac{1}{4}$ second on a CDC 6600 machine to evaluate the intensity N, which should be recorded at each point of a triple-axis scan. A listing of the program which occupies about 28,000 words of storage is available on request.

Let us now return to consideration of the ^{36}Ar phonons of Fig. 10. From the force-constant model described earlier, the anticipated shapes of the phonon profiles were computed using the resolution program discussed above. These profiles are shown in Fig. 11. Each open circle on this figure has been obtained from the resolution program. The computer-generated profiles have been fitted to gaussian functions in the manner described for the observed profiles and the fitted curves are displayed as solid lines in Fig. 11. From the gaussian parameters the mean phonon frequency ω_C of the computer-generated profile can be determined. However, ω_C will, in general, differ from the phonon frequency ω_0, which the force-constant model

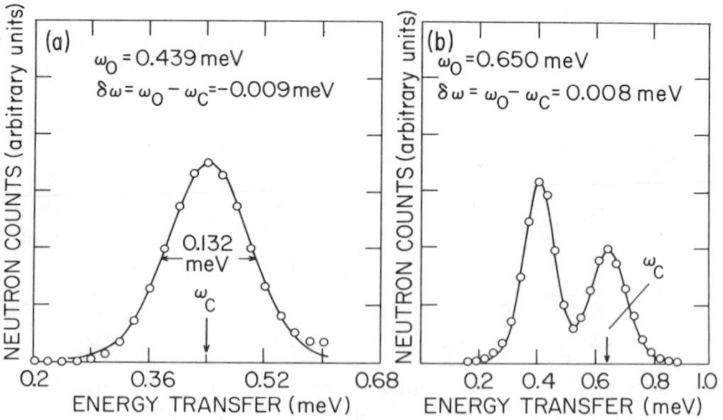

FIG. 11. Computer-generated profiles for the phonons of Fig. 10.

TABLE I

VALUES OF ω_E, ω_0, ω_C, AND $\delta\omega$ FOR THE PHONON PROFILES DISPLAYED IN FIGS. 10 AND 11

Phonon	ω_E(meV)	ω_0(meV)	ω_C(meV)	$\delta\omega$(meV)
Figs. 10a, 11a	0.430 ± 0.002	0.439	0.441	−0.009
Figs. 10b, 11b	0.639 ± 0.007	0.650	0.642	+0.008

predicts at the wave vector setting of the spectrometer. The quantity $\delta\omega = \omega_0 - \omega_C$ is the shift of the phonon frequency introduced by the finite spectrometer resolution. In order to find a good approximation to the true phonon frequency in ^{36}Ar, the shift $\delta\omega$ must be added to the observed phonon frequency ω_E obtained from Fig. 10. In Table I the quantities ω_E, ω_0, ω_C, and $\delta\omega$ are displayed for each of the phonon profiles of the latter figure. Evidently in all cases the correction $\delta\omega$ ($= \omega_0 - \omega_C$) is significant in comparison to the (statistical) error quoted for ω_E. Further iterations of the procedure described above are possible but do not generally yield significant values of the correction $\delta\omega$. However, it is clear that the first correction is significant. Indeed, in the argon experiment considered here, part of the intention was to measure zero-sound elastic constants and these were changed by between 4 and 5% when resolution corrections were included. Since the difference between first and zero-sound elastic constants is typically 10% of the elastic constant (Cowley *et al.*, 1968) the correction is essential if these two types of elastic constant are to be compared with any pretence of reliability.

One further point concerning Figs. 10a and 11a is worthy of note. The computed width (0.132 meV) of the profile shown in Fig. 11a is greater than that (0.112 meV) of the observed profile of the same phonon displayed in Fig. 10a. This error arises because the instrumental parameters (crystal mosaics and collimations) used in the resolution program do not adequately represent those of the spectrometer. In fact, it is extremely difficult to determine such instrumental parameters with any degree of reliability (Werner and Pynn, 1971). This problem arises partly because the transmission probabilities of spectrometer components may not, as is usually assumed, be gaussian. In addition *in situ* measurements of some parameters are difficult to perform—the mosaic spread of a crystalline sample, for example, may not be the same for phonon scattering and for the Bragg scattering process which is usually used to measure this mosaic. Furthermore, in many cases, incident and scattered neutron beams are not truly collimated (Cocking and Webb, 1965) but have only a controlled divergence. Such a situation, which often pertains to the "vertical collimations" of triple-axis spectrometers, cannot readily be included in the analytic formulation of the resolution problem that has been presented here. In the light of these remarks it should be clear that it is an almost hopeless task to measure the anharmonic broadening of a phonon by deconvolving from the observed phonon profile a calculated instrumental linewidth. Indeed, the only reasonable procedure would seem to be to compare observed phonon profiles at a sequence of temperatures and to identify the width of the profile at low temperatures with the instrumental linewidth. Even this procedure is not particularly accurate and it is the author's opinion that measurements of phonon lifetimes should, at present, be regarded with healthy suspicion.

V. A Note on the Bibliography

Some effort has been made to avoid cluttering this article with an excessive number of references. Readers who wish to pursue any of the topics considered in the foregoing pages will find an excellent source of references in the report edited by Larose and Vanderwal (1973).

Appendix: Neutron Scattering Cross Section

In this simplified derivation of the neutron scattering cross section, we use a stationary-state method and build in energy conservation as an additional postulate. First, let us consider a monochromatic beam of neutrons of wave vector $\mathbf{k}_i$ incident on a scattering system. At a point $\mathbf{r}$ far from the scatterer, the neutron wave function will have the form (Messiah, 1961)

$$\exp(i\mathbf{k}_i \cdot \mathbf{r}) + (1/r)f_j(\theta, \phi) \exp(i\mathbf{k}_f \cdot \mathbf{r}) \tag{A1}$$

Here $\mathbf{k}_f$ is the wave vector of a scattered neutron, θ and ϕ are the polar and azimuthal angles which separate $\mathbf{k}_i$ and $\mathbf{k}_f$, and $f_j(\theta, \phi)$ is an amplitude factor which depends on the final state $|j\rangle$ of the scatterer. In terms of $f_j(\theta, \phi)$ one has

$$(d\sigma/d\Omega)_j = (k_f/k_i)|f_j(\theta, \phi)|^2 \tag{A2}$$

where $(d\sigma/d\Omega)_j$ is the scattering cross section when the scatterer changes from initial state $|j_0\rangle$ to final state $|j\rangle$.

In the first Born approximation $f_j(\theta, \phi)$ is given by the usual expression (Messiah, 1961)

$$f_j(\theta, \phi) = \frac{m}{2\pi\hbar^2} \int \exp(i\mathbf{Q} \cdot \mathbf{r}') \langle j|V(\mathbf{r}')|j_0\rangle \, d^3r' \tag{A3}$$

where

$$\mathbf{Q} = \mathbf{k}_i - \mathbf{k}_f, \tag{A4}$$

m is the neutron mass, and $V(\mathbf{r})$ is the interaction potential of the neutron and the scatterer. If we denote the initial and final energies of the scatterer by ε_{j0} and ε_j and the corresponding energies of the neutron by ε_i and ε_f we obtain from (A2) and (A3)

$$\left(\frac{d^2\sigma}{d\Omega\, d\varepsilon}\right)_j = \frac{1}{N}\left(\frac{m}{2\pi\hbar^2}\right)^2 \frac{k_f}{k_i}\left|\int \exp(i\mathbf{Q}\cdot\mathbf{r})\,\langle j|V(\mathbf{r})|j_0\rangle\, d^3r\right|^2 \times \delta(\varepsilon_{j0} + \varepsilon_i - \varepsilon_j - \varepsilon_f) \tag{A5}$$

In writing (A5) explicit use has been made of the energy conservation condition.

In order to find $d^2\sigma/d\Omega\, d\varepsilon$ from Eq. (A5), this equation must be summed over the final states $|j\rangle$ and averaged over a distribution of the initial states $|j_0\rangle$. To perform the sum over the $|j\rangle$ we note that

$$\delta(\varepsilon_{j0} + \varepsilon_i - \varepsilon_j - \varepsilon_f) = \frac{1}{2\pi\hbar}\int_{-\infty}^{\infty} \exp[(\varepsilon_{j0} - \varepsilon_j)t/i\hbar - i\omega t]\, dt \tag{A6}$$

where

$$\hbar\omega = \varepsilon_i - \varepsilon_f$$

Further, we make use of the fact that, for neutron scattering, $V(\mathbf{r})$ is the sum of Fermi pseudopotentials (Fermi, 1936) located at the sites of nuclei in the scattering system. Thus

$$V(\mathbf{r}) = \frac{2\pi\hbar^2}{m}\sum_i a_i\delta(\mathbf{r} - \mathbf{r}_i) \tag{A7}$$

where a_i is the bound scattering length of the ith nucleus.

Combining (A5), (A6), and (A7) gives after a little algebra

$$\frac{d^2\sigma}{d\Omega\, d\varepsilon} = \frac{1}{N}\frac{1}{2\pi\hbar}\frac{k_f}{k_i}\sum_{ii'} a_i a_{i'} \times \int_{-\infty}^{\infty} \langle \exp[-i\mathbf{Q}\cdot\mathbf{r}_i(0)]\exp[i\mathbf{Q}\cdot\mathbf{r}_{i'}(t)]\rangle_T e^{-i\omega t}\, dt \tag{A8}$$

where $\langle \cdots \rangle_T$ denotes a thermal average over the scattering system and $\mathbf{r}_i(t)$ is the Heisenberg position operator for the ith nucleus.

Equation (A8) is the desired result for the neutron scattering cross section.

References

Axe, J. D. (1974). Preprint No. 19179. Brookhaven Nat. Lab., Upton, New York.

Bischoff, F. G. (1970). Ph.D. Thesis, Rensselaer Polytechnic Institute, Troy, New York (available from University Microfilms, Ann Arbor, Michigan; Catalogue No. 70-19931).

Brockhouse, B. N. (1961). *In* "Inelastic Scattering of Neutrons in Solids and Liquids," Vol. 1, p. 113. IAEA, Vienna.

Brugger, R. M. (1965). In Egelstaff (1965), p. 53.

Carneiro, K., and Nielsen, M. (1974). "Anharmonic Lattices, Structural Transitions and Melting" (N.A.T.O. School). Nordhoff, Leiden.

Chesser, N. J., and Axe, J. D. (1973). *Acta Crystallogr. Sect. A* **29**, 160.

Cocking, S. J., and Webb, F. J. (1965). In Egelstaff (1965).

Cooper, M. J., and Nathans, R. (1967). *Acta Crystallogr.* **23**, 357.

Copley, J. R. D., Price, D. L., and Rowe, J. M. (1973). *Nucl. Instrum. & Methods* **107**, 501.

Copley, J. R. D. (1971). *Solid State Commun.* **9**, 531.

Copley, J. R. D. (1974). *Comput. Phys. Commun.* **7**, 289.

Copley, J. R. D., and Lovesey, S. W. (1975). *Rep. Progr. Phys.* **38**, 461.

Copley, J. R. D., and Rowe, J. M. (1974). *Phys. Rev. A* **9**, 1656.

Cowley, E. R., and Pant, A. K. (1970). *Acta Crystallogr., Sect. A* **26**, 439.

Cowley, R. A., Buyers, W. J. L., and Svensson, E. C. (1968). *In* "Neutron Inelastic Scattering," Vol. 1, p. 281. IAEA, Vienna.

Dolling, G., and Sears, V. F. (1973). *Nucl. Instrum. & Methods* **106**, 419.

Dorner, B. (1972). *Acta Crystallogr., Sect. A* **28**, 319.

Egelstaff, P. A. (1965). "Thermal Neutron Scattering." Academic Press, New York.

Fermi, E. (1936). *Ric. Sci.* **7**, 13.

Fujii, Y., Lurie, N. A., Pynn, R., and Shirane, G. (1974). *Phys. Rev. B* **10**, 3647 (1974).

Goldberg, M. D. *et al.* (1966). "Neutron Cross Sections," Rep. BNL 325. Brookhaven Nat. Lab., Upton, New York.

Harris, D., Cocking, S. J., Egelstaff, P. A., and Webb, F. J. (1961). *In* "Inelastic Scattering of Neutrons in Solids and Liquids," Vol. I, p. 107. IAEA, Vienna.

Hayward, B. C. (1971). *Acta Crystallogr., Sect. A* **27**, 408.

Honeck, H. C. (1964). General Atomic Report GA-5968.

I.L.L. Report. (1973). "Neutron Monochromators," Rep. No. 74F42S. Institut Laue-Langevin, Grenoble, France.

Johnson, M. W. (1974). *UKAEA* Rep. AERE-R7682.

Kleb, R., Ostrowski, G. E., Price, D. L., and Rowe, J. M. (1973). *Nucl. Instrum. & Methods* **106**, 221.

Komura, S., and Cooper, M. J. (1970). *J. Appl. Phys.* **9**, 866.

Larose, A., and Vanderwal, J. (1973). "Bibliography of Papers Relevant to the Scattering of Thermal Neutrons." Available from B. N. Brockhouse, Institute for Materials Research, McMaster University, Hamilton, Ontario, Canada for the modest fee of $15.00.

Lomer, W. M., and Low, G. G. (1965). In Egelstaff (1965).

Mack, C. (1967). "Essentials of Statistics." Plenum, New York.

Marshall, W., and Lovesey, S. W. (1971). "Theory of Thermal Neutron Scattering." Oxford Univ. Press, London and New York.

Messiah, A. (1961). "Quantum Mechanics." Wiley (Interscience), New York.
Mook, H. A., Snodgrass, F. W., and Bates, D. D. (1974). *Nucl. Instrum. & Methods* **116**, 205.
Nielsen, M., and Bjerrum Møller, H. (1969). *Acta Crystallogr., Sect. A* **25**, 574.
Otnes, K. (1975). To be published; also see Steinsvoll (1973).
Pines, D., and Nozière P. (1966). "The Theory of Quantum Liquids." Benjamin, New York.
Price, D. L., and Sköld, K. (1970). *Nucl. Instrum. & Methods* **82**, 208.
Pynn, R., and Werner, S. A. (1970). Report No. AE-FF-112. A. B. Atomenergi, Studsvik, Sweden.
Pynn, R., and Youngblood, R. (1975). Technical Report No. 20195. Brookhaven Nat. Lab., Upton, New York.
Raunio, G., Almqvist, L., and Stedman, R. (1969). *Phys. Rev.* **178**, 1496.
Samuelsen, E. J., Hutchings, M. T., and Shirane, G. (1970). *Physica (Utrecht)* **48**, 13.
Sköld, K. (1968). *Nucl. Instrum. & Methods* **63**, 114.
Sköld, K., Rowe, J. M., Ostrowski, G., and Randolph, P. D. (1972). *Phys. Rev. A* **6**, 1107.
Singwi, K. S., Sköld, K., and Tosi, M. (1968). *Phys. Rev. Lett.* **21**, 881.
Singwi, K. S., Sköld, K., and Tosi, M. (1970). *Phys. Rev. A* **1**, 454.
Stedman, R. (1968). *Rev. Sci. Instrum.* **39**, 878.
Stedman, R. (1973). Report 1973-11-22. A. B. Atomenergi, Studsvik, Sweden.
Stedman, R., Almqvist, L., Raunio, G., and Nilsson, G. (1969). *Rev. Sci. Instrum.* **40**, 249.
Steinsvoll, O. (1973). *Nucl. Instrum. & Methods* **106**, 453.
Svensson, E. C., Brockhouse, B. N., and Rowe, J. M. (1967). *Phys. Rev.* **155**, 619.
Van Hove, L. (1954). *Phys. Rev.* **95**, 249.
Werner, S. A., and Pynn, R. (1971). *J. Appl. Phys.* **42**, 4736.
Ziman, J. M. (1960). "Electrons and Phonons." Oxford Univ. Press, London and New York.

Group Theory of Lattice Dynamics by Computer*

JOHN L. WARREN

LOS ALAMOS SCIENTIFIC LABORATORY, UNIVERSITY OF CALIFORNIA
LOS ALAMOS, NEW MEXICO

AND

THOMAS G. WORLTON

ARGONNE NATIONAL LABORATORY, ARGONNE, ILLINOIS

* Work supported by the U.S. Atomic Energy Commission.

I. Introduction

DETERMINATION OF THE NORMAL mode eigenvalues and eigenvectors is of primary interest in lattice dynamics. Group theory can be used to determine the number of eigenvalues of each symmetry type and to determine which modes will be degenerate. It can also be used to find the form of the eigenvectors and the dynamical matrix and can be used to bring the dynamical matrix into block diagonal form.

The application of group theory to lattice dynamics was first given by Yanagawa (1953), however, the method did not receive much attention until it was extended by Chen (1964) and applied to the analysis of β-tin. A more complete theory was presented by Maradudin and Vosko (1968) with extensions and applications by Warren (1968). The first applications of the theory were for simple crystals with one or two atoms per unit cell, but it soon became evident that crystals with much larger unit cells were of interest. In this case the group theory becomes very tedious because of the manipulation of large matrices required. Errors become very likely. The tendency then was to do only the easiest portions of the group theory. This seemed an unsatisfactory situation and prompted Worlton and Warren (1972) to write a program to do the group theory on the computer. This program was rather ambitious in its aim of trying to obtain all the information about the symmetry of the lattice vibrations of crystals while keeping the amount of input that must be supplied by the user to a minimum. The bulk of the input to this program consisted of tables of irreducible multiplier representations (IMR's), but later the need for this part of the input was eliminated by a subroutine to calculate the IMR's (Worlton, 1973).

The program is limited by dimension and format statements to no more than 20 atoms per unit cell (60 degrees of freedom). Actually the volume of output goes up very rapidly with the number of atoms per unit cell and a complete analysis of spinel, which is a cubic crystal with 14 atoms per unit cell resulted in approximately 100 pages of output. This could be considered a practical limit to the size of the problems to be treated with a complete analysis. There are two methods of approach for more complex crystals. One technique, which has been reviewed by Venkataraman and Sahni (1970), is to treat tightly bound clusters of atoms as rigid bodies and only analyze the external modes of vibration. The modes internal to the molecule will be very similar to the modes of the free molecule and will show very little dispersion (Bhagavantam, 1941). A modification which allows the program to treat external modes of molecular crystals was added by Worlton (1972).

The other approach to more complex crystals is to do only part of the group theory. The part of the analysis requiring the most storage space is the symmetry reduction and block diagonalization of the dynamical matrix. Moreover, the information will probably not be used since it is unlikely that anyone will fit a Born–von Kármán model to a crystal of great complexity. Therefore, for those interested in crystals with more than 60 degrees of freedom, a trimmed-down version of the program might be useful. A program to calculate the symmetry coordinates for more complex crystals has been written by Boyer (1974). Boyer's program has the advantage of being able to treat crystals with very large unit cells, but has the disadvantages of requiring more complicated input and requiring two passes through the computer. The form of the output is very compact.

The theory of Maradudin and Vosko (1968) was incomplete in that it did not include the effect of time reversal symmetry on the eigenvectors in a way that could be adapted to the computer. This addition to the theory has recently been discussed by Warren (1974), and a new version of the program was written by Warren and Worlton (1974) to include this improvement and the other two revisions mentioned above.

Most of this article will be concerned with the most recent version of our program which will be referred to hereafter as GROUP2. GROUP2 has been used to analyze the symmetry properties of the lattice vibrations of a large number of crystals and the results of 23 analyses have been published (Warren and Worlton, 1973).

Section II of this article discusses the equations of motion and the effect of space–time translational symmetry. The time reversal point group of the wave vector and its representations are discussed in Section III. This section also includes a discussion of the method of generating IMR's on the computer. Sections IV, V, and VI discuss the symmetry reduction of the dynamical matrix, the determination of the symmetry coordinates, and the block diagonalization of the dynamical matrix. Optical selection rules for first-order Raman scattering and infrared absorption are discussed in Section VII. Section VIII suggests other uses of the symmetry coordinates.

II. The Equations of Motion and the Space–Time Translation Group

A General Approach and Assumptions

The purpose of Section II is to formulate the lattice dynamics and bring it into a mathematical form which can be treated by the computer. The procedure for doing this is (1) define the coordinates of the atoms in the crystal, (2) express these coordinates in terms of the translational and rotational fields, (3) write down the Lagrangian for the external modes of the crystal in the harmonic approximation, (4) derive the equations of motion (EOM) from the Lagrangian, and (5) use the properties of the space–time translation group to transform the EOM into an eigenvalue problem in wave vector space. Because the main emphasis of this article is on computer applications, we can only give a sketchy treatment of the procedure outlined above. It must be assumed that the reader is familiar with the standard formulation of lattice dynamics in the harmonic approximation as given, for example, by Maradudin *et al.* (1971). Familiarity with the external mode formulation as reviewed by Venkataraman and Sahni (1970) would also be useful.

B. Atomic Coordinates and the Displacement Fields

1. *Equilibrium Positions*

The crystal is divided into unit cells. The location of the corner of a unit cell is specified by a direct lattice vector

$$\mathbf{l} = \sum_{j=1}^{3} l(j)\mathbf{a}(j) \tag{1a}$$

where the $\mathbf{a}(j)$'s are the primitive translation vectors and the integers $l(j)$ range over the crystal. For definiteness, let

$$-L < l(j) \leqslant L \tag{1b}$$

where L is a large integer. Any summation over the lattice will assume this range. The choice of the $\mathbf{a}(j)$'s is not unique and has been a source of confusion for people using GROUP2. The choice of the cartesian components of the primitive vectors we prefer for the 14 Bravais lattices is given in Table I. This choice is consistent with the IRE Standards (1949).

There are two ways to describe the positions of the atoms in the unit cell. The usual way is to assign a vector $\mathbf{x}(\kappa)$, $\kappa = 1, \ldots, N$, going from the origin of the cell to the κth atom. For molecular crystals, it is more usual to group the atoms into clusters containing one or more atoms. For example in $CaWO_4$, which has two formula units per unit cell, we can define four clusters: (1) Ca, (2) Ca, (3) WO_4, and (4) WO_4. The location of the center of mass of a cluster with respect to the origin of the unit cell can be called $\mathbf{r}(\mathrm{K})$, $\mathrm{K} = 1, \ldots, C$, where C is the number of clusters. The location of the atoms in the cluster with respect to the center of mass of the cluster we call $\mathbf{s}(\mathrm{K}, \nu)$,

TABLE I
CARTESIAN COMPONENTS OF PRIMITIVE LATTICE VECTORS[a]

Lattice type	$\mathbf{a}(1)$	$\mathbf{a}(2)$	$\mathbf{a}(3)$
(P) Simple	$(aS, 0, aC)$	$b(d, e, f)$	$(0, 0, c)$
(I) Bodycentered	$\frac{1}{2}(-a, b, c)$	$\frac{1}{2}(a, -b, c)$	$\frac{1}{2}(a, b, -c)$
(F) Facecentered	$\frac{1}{2}(0, b, c)$	$\frac{1}{2}(a, 0, c)$	$\frac{1}{2}(a, b, 0)$
(B) Basecentered	$\frac{1}{2}(aS, 0, aC + c)$	$(0, b, 0)$	$\frac{1}{2}(-aS, 0, -aC + c)$
(R) Trigonal	$\frac{1}{2}(a, a/\sqrt{3}, 2c/3)$	$\frac{1}{2}(-a, a/\sqrt{3}, 2c/3)$	$(0, a/\sqrt{3}, c/3)$
Cubic *P, I, F*	$b = a = c, \quad \alpha = \beta = \gamma = 90°,$		$d = f = 0, \quad e = 1$
Tetragonal *P, I*	$b = a \neq c, \quad \alpha = \beta = \gamma = 90°,$		$d = f = 0, \quad e = 1$
Orthorhombic *P, I, F, B*	$b > a > c, \quad \alpha = \beta = \gamma = 90°,$		$d = f = 0, \quad e = 1$
Monoclinic *P, B*	$a > c, \quad \alpha = \beta = 90°, \quad \gamma > 90°,$		$d = f = 0, \quad e = 1$
Triclinic *P*	$b > a > c, \quad \alpha > 90°, \quad \beta > 90°$		
Hexagonal *P*	$b = a, \quad \alpha = \beta = 90°, \quad \gamma = 120°,$		$d = -\frac{1}{2}, \quad e = \sqrt{3}/2, \quad f = 0$

[a] Key to symbols: $S = \sin\beta$, $C = \cos\beta$, $d = (\cos\gamma - \cos\alpha\cos\beta)/\sin\beta$, $f = \cos\alpha$, $e = (1 - d^2 - f^2)^{1/2}$. α is the angle between $\mathbf{a}(2)$ and $\mathbf{a}(3)$; β is the angle between $\mathbf{a}(1)$ and $\mathbf{a}(3)$; and γ is the angle between $\mathbf{a}(1)$ and $\mathbf{a}(2)$.

$\nu = 1, \ldots, n(\mathrm{K})$, where $n(\mathrm{K})$ is the number of atoms in the Kth cluster. In $CaWO_4$, $n(1) = n(2) = 1$ and $n(3) = n(4) = 5$. Finally, we have the relations

$$\mathbf{x}(\kappa) = \mathbf{r}(\mathrm{K}) + \mathbf{s}(\mathrm{K}, \nu) \tag{2a}$$

$$\kappa = \sum_{\iota=1}^{\mathrm{K}-1} n(\iota) + \nu \tag{2b}$$

Figure 1 illustrates these atomic coordinates.

The input to GROUP2 is the $\mathbf{x}(\kappa)$'s. If the point group of the cluster is smaller than the point group of the lattice it is important to put in positions of the other atoms in the clusters to get the correct space group symmetry of the crystal.

Three warnings must be issued here. Careful attention must be paid to the coordinate system. The coordinate system used to describe the atoms must be the same as that used for the primitive vectors. Figuring out cartesian components of atoms can be nontrivial because of the lack of uniformity in the literature. The special positions in the unit cell as defined in the *International Tables for X-Ray Crystallography*, Vol. I (Henry and Lonsdale, 1965) are useful for standardization. The second caution is that the special positions must be interpreted by the rule

$$xyz \rightarrow x\mathbf{a}(1) + y\mathbf{a}(2) + z\mathbf{a}(3). \tag{3}$$

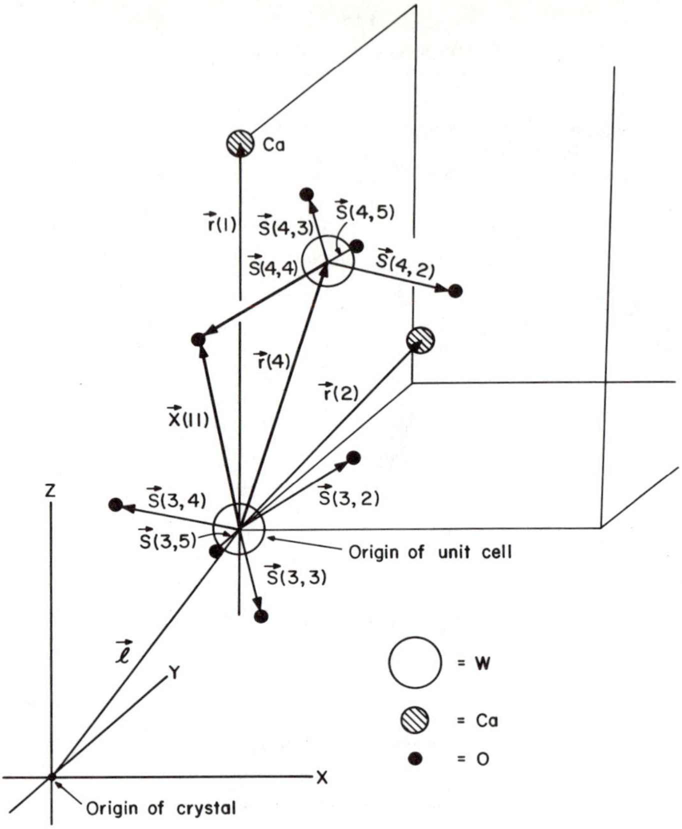

FIG. 1. Atomic coordinates for $CaWO_4$ in the external mode formulation. Note that $\mathbf{r}(3) = \mathbf{0}$ because the first WO_4 unit is centered on the origin of the unit cell. Also $\mathbf{s}(3, 1) = \mathbf{s}(4, 1) = \mathbf{0}$ because the W atoms are at the center of mass of the cluster. The vector $\mathbf{x}(11)$ gives a typical atom position in ordinary atomic coordinates.

The third warning is that GROUP2 assumes that some atom in the cluster will be located at the center of mass of the cluster. That atom is tagged in the input and used as the position of the cluster. In some cases this will not be true. As an example the radical CN in KCN might be thought of as a cluster. The position of the carbon atom could be used as a pseudo center of mass and the nitrogen position added as an extra atom position to ensure proper crystal symmetry. In essence, it is as if the carbon atom were given an infinite mass.

2. *Time-Dependent Positions*

So far we have been discussing the equilibrium positions of the atoms. There are two displacement fields which must be included in the time-dependent position vectors. These correspond to the translation of the center of mass of the cluster, $\mathbf{v}(\mathbf{l}, \mathrm{K}, t)$ and the displacement of the atoms about the center of mass, $\mathbf{w}(\mathbf{l}, \mathrm{K}, \nu, t)$. The displacement fields contain $\mathbf{l}$ as a variable because the displacements of the atoms in different unit cells are in general different. In summary, the time-dependent position vector for the νth atom in the Kth cluster in the lth unit cell is

$$\mathbf{x}(\mathbf{l}, \mathrm{K}, \nu, t) = \mathbf{l} + [\mathbf{r}(\mathrm{K}) + \mathbf{v}(\mathbf{l}, \mathrm{K}, t)] + [\mathbf{s}(\mathrm{K}, \nu) + \mathbf{w}(\mathbf{l}, \mathrm{K}, \nu, t)] \quad (4)$$

In the external mode approximation, it is assumed that the displacement field $\mathbf{w}(\mathbf{l}, \mathrm{K}, \nu, t)$ is due only to rotation of the atoms in the cluster about the center of mass. This means that the displacement is describable by a rotation field $\boldsymbol{\theta}(\mathbf{l}, \mathrm{K}, t)$.

$$\mathbf{w}(\mathbf{l}, \mathrm{K}, \nu, t) = \boldsymbol{\theta}(\mathbf{l}, \mathrm{K}, t) \times \mathbf{s}(\mathrm{K}, \nu) \quad (5)$$

Note that $\boldsymbol{\theta}(\mathbf{l}, \mathrm{K}, t)$ is independent of the atom index ν.

C. The Lagrangian and the Equations of Motion

The Lagrangian is

$$L = T - V, \quad (6)$$

where T is the total kinetic energy and V is the total potential energy. After much algebra and the use of the center-of-mass condition

$$\sum_{\nu=1}^{n(\mathrm{K})} M(\mathrm{K}, \nu)\mathbf{s}(\mathrm{K}, \nu) = 0 \quad (7)$$

where $M(\mathrm{K}, \nu)$ is the mass of the νth atom in the Kth cluster, one can express the kinetic energy as

$$T = \frac{1}{2} \sum_{\mathbf{l}} \sum_{\nu=1}^{C} [\dot{\mathbf{v}}(\mathbf{l}, \mathrm{K}, t) \cdot \mathbf{M}(\mathrm{K}) \cdot \dot{\mathbf{v}}(\mathbf{l}, \mathrm{K}, t) + \dot{\boldsymbol{\theta}}(\mathbf{l}, \mathrm{K}, t) \cdot \mathbf{I}(\mathrm{K}) \cdot \dot{\boldsymbol{\theta}}(\mathbf{l}, \mathrm{K}, t)] \quad (8a)$$

where

$$\mathbf{M}(\mathrm{K}) = \sum_{\nu=1}^{n(\mathrm{K})} M(\mathrm{K}, \nu)\, \mathbf{E} \quad (8b)$$

and

$$\mathbf{I}(\mathrm{K}) = \sum_{\nu=1}^{n(\mathrm{K})} M(\mathrm{K}, \nu)[\mathbf{s}(\mathrm{K}, \nu) \cdot \mathbf{s}(\mathrm{K}, \nu)\mathbf{E} - \mathbf{s}(\mathrm{K}, \nu)\mathbf{s}(\mathrm{K}, \nu)] \tag{8c}$$

are the mass and moment of inertia dyadics of the cluster. The unit dyadic is called **E**.

The form of T can be further simplified by introducing a generalized displacement field with two components.

$$\mathbf{u}^1(\mathbf{l}, \mathrm{K}, t) = \mathbf{v}(\mathbf{l}, \mathrm{K}, t) \tag{9a}$$

$$\mathbf{u}^2(\mathbf{l}, \mathrm{K}, t) = \boldsymbol{\theta}(\mathbf{l}, \mathrm{K}, t) \tag{9b}$$

In order to save storage in the computer, it is useful to recognize that some of the clusters may contain only one atom. One atom clusters do not have rotational degrees of freedom. This means that the superscript index ζ on the generalized field $\mathbf{u}^\zeta(\mathbf{l}, \mathrm{K}, t)$ may take on two values for multiatom clusters but only one value for single atom clusters. We therefore make the upper limit of ζ variable,

$$\zeta = 1 \quad \text{to} \quad f(\mathrm{K})$$

where $f(\mathrm{K}) = 1$ if K is a single atom cluster and $f(\mathrm{K}) = 2$ if K is a multiatom cluster.

In a similar way, the mass and moment of inertia dyadics can be written

$$\mathbf{J}^1(\mathrm{K}) = \mathbf{M}(\mathrm{K}) \tag{9c}$$

$$\mathbf{J}^2(\mathrm{K}) = \mathbf{I}(\mathrm{K}) \tag{9d}$$

In this notation, the kinetic energy becomes

$$T = \frac{1}{2} \sum_{\mathbf{l}} \sum_{\mathrm{K}=1}^{C} \sum_{\zeta=1}^{f(\mathrm{K})} \dot{\mathbf{u}}^\zeta(\mathbf{l}, \mathrm{K}, t) \cdot \mathbf{J}^\zeta(\mathrm{K}) \cdot \dot{\mathbf{u}}^\zeta(\mathbf{l}, \mathrm{K}, t) \tag{10}$$

It is assumed that the potential energy can also be expressed in terms of the generalized displacement field. In the harmonic approximation V is expanded to second order in $\mathbf{u}^\zeta(\mathbf{l}, \mathrm{K}, t)$'s and one obtains in the usual way

$$V = V(0) - \frac{1}{2} \sum_{\mathbf{l},\mathbf{m}} \sum_{\mathrm{K},\Lambda=1}^{C} \sum_{\zeta=1}^{f(\mathrm{K})} \sum_{\eta=1}^{f(\Lambda)} \mathbf{u}^\zeta(\mathbf{l}, \mathrm{K}, t) \cdot \mathbf{K}^{\zeta\eta}(\mathbf{l} - \mathbf{m}, \mathrm{K}, \Lambda) \cdot \mathbf{u}^\eta(\mathbf{m}, \Lambda, t). \tag{11}$$

Until now we have worked in dyadic notation. There are some advantages for the computer code if we work in tensor notation. This simply means replacing the vectors and dyadics with their cartesian components. The Einstein summation convention will be assumed for the Greek subscripts, i.e., repeated Greek subscripts are automatically summed from 1 to 3. In this notation the Lagrangian is

$$L = \frac{1}{2} \sum_{\mathbf{l}} \sum_{K=1}^{C} \sum_{\zeta=1}^{f(K)} \Bigg\{ J_{\alpha\beta}{}^{\zeta}(K)\dot{u}_{\alpha}{}^{\zeta}(\mathbf{l}, K, t)\dot{u}_{\beta}{}^{\zeta}(\mathbf{l}, K, t) + \sum_{\mathbf{m}} \sum_{\Lambda=1}^{C} \sum_{\eta=1}^{f(K)} K_{\alpha\beta}^{\zeta\eta}(\mathbf{l} - \mathbf{m}, K, \Lambda)u_{\alpha}{}^{\zeta}(\mathbf{l}, K, t)u_{\beta}{}^{\eta}(\mathbf{m}, \Lambda, t) \Bigg\} \tag{12}$$

The Lagrangian form of the EOM are

$$\frac{\partial}{\partial t}\left(\frac{L}{\partial \dot{u}_{\alpha}{}^{\zeta}(\mathbf{l}, K, t)}\right) - \frac{\partial L}{\partial u_{\alpha}{}^{\zeta}(\mathbf{m}, K, t)} = 0 \tag{13}$$

This gives the result

$$J_{\alpha\beta}{}^{\zeta}(K)\ddot{u}_{\beta}{}^{\zeta}(\mathbf{l}, K, t) - \sum_{\mathbf{m}} \sum_{\Lambda=1}^{C} \sum_{\eta=1}^{f(\Lambda)} K_{\alpha\beta}^{\zeta\eta}(\mathbf{l} - \mathbf{m}, K, \Lambda)u_{\beta}{}^{\eta}(\mathbf{m}, \Lambda, t) = 0 \tag{14}$$

The three Greek indices α, K, ζ can be combined into one latin index called œ. This avoids multisubscripted variables in the computer program. The combined index becomes

$$\text{œ} = 3\{[\zeta - 1] + [k(K) - 1]\} + \alpha \tag{15a}$$

where

$$k(1) = 1 \tag{15b}$$

and

$$k(K) = k(K - 1) + f(K - 1). \tag{15c}$$

In GROUP2 the index œ is called II; K becomes K; ζ is KI; $k(K)$ is called $KX(K)$; $C = NA$; and $f(K)$ becomes $[JM(K) + 1]$. For symmetry we write

$$J_{\alpha\beta}{}^{\zeta}\ddot{u}_{\beta}{}^{\zeta}(\mathbf{l}, \mathrm{K}, t) = \sum_{\Lambda=1}^{C} \sum_{\eta=1}^{f(\Lambda)} J_{\alpha\beta}{}^{\eta}\delta(\mathrm{K}, \Lambda)\delta(\zeta, \eta), \ddot{u}_{\beta}{}^{\eta}(\mathbf{l}, \Lambda, t)$$
$$= \sum_{\mathfrak{b}=1}^{p} J_{\mathfrak{a}\mathfrak{b}}\ddot{u}_{\mathfrak{b}}(\mathbf{l}, t) \tag{16}$$

where

$$p = \sum_{\mathrm{K}=1}^{C} 3f(\mathrm{K}) \tag{17}$$

and $J_{\mathfrak{a}\mathfrak{b}}$ is defined in an obvious way. If every cluster contains only one atom, then $p = 3N$. Many arrays in the program have dimensions which depend on p. In terms of the combined indices the EOM become

$$\sum_{\mathfrak{b}=1}^{p} \left[J_{\mathfrak{a}\mathfrak{b}}\ddot{u}_{\mathfrak{b}}(\mathbf{l}, t) - \sum_{m} K_{\mathfrak{a}\mathfrak{b}}(\mathbf{l} - \mathbf{m})u_{\mathfrak{b}}(\mathbf{m}, t) \right] = 0 \tag{18}$$

D. The Space–Time Group

Equation (18) stands for a large set of equations. For each lattice vector $\mathbf{l}$, there are p equations. This gives one set of $p(2L)^3$ coupled equations. In Section II, E below we will use group theory to reduce these equations to $(2L)^3$ sets of p coupled equations. Before doing this, it is necessary to define the symmetry group of Eqs. (18). This group consists of operators of the form

$$\mathscr{G}(i, \mathfrak{h}|\mathbf{l}, t) \equiv (\mathscr{R}(i), \mathscr{T}(\mathfrak{h})|\mathbf{l} + \mathbf{v}(i), t) \tag{19a}$$
$$i = 1, \ldots, g; \quad \mathfrak{h} = 1, 2 \tag{19b}$$

where $\mathscr{R}(i)$'s are rotation and rotoinversion operators in the point group of the crystal, g is the order of the group, $\mathscr{T}(1)$ is the identity element for operators acting on the time variable, $\mathscr{T}(2)$ is time inversion, and $\mathbf{v}(i)$ is a fractional direct lattice translation associated with $\mathscr{R}(i)$ when there is a screw axis or glide plane in the space group of the crystal. The effect of $\mathscr{G}(i, \mathfrak{h}|\mathbf{l}, t')$ on a space–time point $(\mathbf{x}, t)$ is given by

$$\mathscr{G}(\mathrm{i}, \mathfrak{h}|\mathbf{l}, t')(\mathbf{x}, t) = (\mathbf{R}(i) \cdot \mathbf{x} + \mathbf{l} + \mathbf{v}(i), T(\mathfrak{h})t + t') \tag{20}$$

where $\mathbf{R}(i)$ is a 3×3 matrix which represents $\mathscr{R}(i)$ in euclidian space, $T(1) = 1$, and $T(2) = -1$. The effect of these operations on the components of a 3-dimensional vector function $\mathbf{f}(\mathbf{x}, t)$ is

$$\mathscr{G}(i, \mathfrak{h}|\mathbf{l}, t')f_\alpha(\mathbf{x}, t) = \mathbf{R}_{\alpha\beta}(i)f_\beta(\mathscr{G}^{-1}(i, \mathfrak{h}|\mathbf{l}, t')(\mathbf{x}, t)) \tag{21}$$

where $\mathscr{G}^{-1}$ denotes the inverse of $\mathscr{G}$. From these properties of the group elements we can deduce the effect of applying them to the p-dimensional generalized displacement field, namely,

$$\mathscr{G}(i, \mathfrak{h}|\mathbf{n}, t'')u_{\mathfrak{a}}(\mathbf{l}, t) = T_{\mathfrak{a}\mathfrak{b}}(i)u_{\mathfrak{b}}(\mathbf{m}, t') \tag{22}$$

where the matrix elements $T_{\mathfrak{a}\mathfrak{b}}(i)$ are given by

$$T_{\mathfrak{a}\mathfrak{b}}(i) = T^{\zeta\eta}_{\alpha\beta}(i, \mathrm{K}, \Lambda) = \delta(\zeta, \eta)\, \delta(\mathrm{K}, F_0(\Lambda, i))C^{\zeta}(i)R_{\alpha\beta}(i). \tag{23}$$

The operator $\mathscr{G}^{-1}(i, 1|\mathbf{n}, 0)$ acting on cluster coordinate vector $[\mathbf{l} + \mathbf{r}(\mathrm{K})]$ gives another cluster point $[\mathbf{m} + \mathbf{r}(\Xi)]$ where

$$\Xi = F_0(\mathrm{K}, i^{-1}). \tag{24}$$

One can show that the space–time point $(\mathbf{m}, t')$ in Eq. (23) is given by

$$(\mathbf{m}, t') = (\mathbf{R}^{-1}(i)[\mathbf{l} + \mathbf{r}(\mathrm{K}) - \mathbf{n} - \mathbf{v}(i)] - \mathbf{r}(\Lambda), \qquad T(\mathfrak{h})t - t'') \tag{25}$$

This will prove useful later on. The factor $C^{\zeta}(i)$ in Eq. (23) is given by

$$C^1(i) = 1 \tag{26a}$$

$$C^2(i) = \det(\mathbf{R}(i)) \equiv |\mathbf{R}(i)|. \tag{26b}$$

This factor comes in because the angular displacement part of $u_{\mathfrak{a}}(\mathbf{l}, t)$ behaves like a 3-dimensional axial vector. Similar but more complicated expressions can be written down for the effects of $\mathscr{G}(i, \mathfrak{h}|\mathbf{n}, t'')$ on tensors such as the force constants.

E. Generation of the Space Group by Computer Methods

This is an appropriate place to explain how GROUP2 determines the operations in the space group from the input information, which consists of the primitive lattice vectors and the atom positions. The general procedure is as follows: (1) generate the $\mathbf{R}(i)$'s, (2) determine the crystal system by determining which $\mathbf{R}(i)$'s leave the lattice invariant, (3) apply the operations of the point group of the lattice to the atomic positions and determine which operations leave the atomic arrangement unchanged, (4) concurrently with the last step, determine the fractional translations $\mathbf{v}(i)$ and the atom transformation table, $F_0(\mathrm{K}, i)$.

In a preliminary version of the program, the 48 matrices of the cubic point group or the 24 matrices of the hexagonal group were part of the program input. This was not very elegant. The approach taken in GROUP2 is to generate these matrices in a subroutine called ROT(N) starting from a few generating elements. For the cubic group all rotation matrices are generated from $\mathbf{R}(9) = C_3^{xyz}$, $\mathbf{R}(19) = C_4{}^x$ and $\mathbf{R}(25) = I$, while for the hexagonal crystals, they are generated by $\mathbf{R}(2) = C_6$, $\mathbf{R}(7) = C_2^{(3)}$ and $\mathbf{R}(13) = I$. One should not confuse $\mathbf{R}(i)$'s for the cubic group with those of the hexagonal group, because only one set of matrices is used at a time. Subroutine ROT(N) is called from subroutine PGL (Point Group of the Lattice). If the calling parameter is $N = 24$, the hexagonal matrices are generated. If $N = 48$, the cubic matrices are obtained. The matrices are stored in an array called $R(I, J, K)$, $I, J = 1, 2, 3$ and $K = 1, \ldots, 48$. The first call from PGL is with $N = 24$. The 24 hexagonal matrices are multiplied with the lattice vectors. The operation $\mathbf{R}(i)$ will be a member of the point group of the lattice if

$$\mathbf{l}' = \mathbf{R}(i)\mathbf{a}(j), \qquad j = 1, 2, 3 \tag{27}$$

is a lattice vector, i.e., if the coefficients $l'(j)$ in the decomposition of $\mathbf{l}'$ into primitive lattice vectors,

$$\mathbf{l}' = \sum_{j=1}^{3} l'(j)\mathbf{a}(j), \tag{28}$$

are integers. To make this test, subroutine RLV (Remove Lattice Vectors), is called to remove the part of $\mathbf{l}'$ with integer coefficients. If the remaining part of $\mathbf{l}'$ is zero, then $\mathbf{R}(i)$ is in the group. If the number of operations, NC, obtained for the point group of the lattice is 24, then the crystal if hexagonal. If $NC = 12$, the crystal is trigonal. If $NC < 12$, the crystal is neither hexagonal nor trigonal. In this case, PGL calls ROT(N) with $N = 48$ and the cubic matrices are obtained. PGL again applies the rotations to the lattice vectors and subtracts off integral multiples of the $\mathbf{a}(j)$'s. As in the hexagonal case, the order of the point group of the lattice determines the crystal system. Table II summarizes how the crystal systems can be determined from the order of the point group of the lattice.

Having this information, it is easy to proceed to the next step and eliminate operations that do not leave the atomic arrangement invariant. This is done in subroutine ATFTMT (Atom Transformations, Fractional Translations, and Multiplication Table), which also determines the function $F_0(\mathrm{K}, i)$ and obtains the multiplication table for the operations in the point group of the crystal.

TABLE II

CRYSTAL SYSTEMS AS DETERMINED BY THE ORDER OF THE POINT GROUP OF THE LATTICE

System	Order, NC	System number
Hexagonal	24	7
Trigonal	12	6
Cubic	>16	5
Tetragonal	16	4
Orthorhombic	>4 but <16	3
Monoclinic	4	2
Triclinic	<4	1

It will be recalled from Eq. (20) that the effect of a space group operation on an atomic position vector $\mathbf{x}(\mathbf{l}, \kappa) = \mathbf{l} + \mathbf{x}(\kappa)$ should be

$$(\mathscr{R}(i)|\mathbf{m} + \mathbf{v}(i))\mathbf{x}(\mathbf{l}, \kappa) = \mathbf{R}(i) \cdot \mathbf{x}(\mathbf{l}, \kappa) + \mathbf{m} + \mathbf{v}(i) = \mathbf{x}(\mathbf{n}, \lambda) \quad (29)$$

where the index of the transformed atomic position is given by the function $\lambda = F_0(\kappa, i)$ of Maradudin and Vosko (1968). The operations must carry atoms of a given type, e.g., Na's in NaCl, into atoms of the same type. This is why, in the input, each atom position has associated with it a type number TY(I).

The procedure for eliminating the extra $\mathbf{R}(i)$'s, finding the $\mathbf{v}(i)$'s and constructing $F_0(\kappa, i)$ relies heavily on the computer's ability to try many cases rapidly. An $\mathbf{R}(i)$ in the point group of the lattice is applied to an atomic position $\mathbf{x}(\kappa)$ and $[\mathbf{x}(\kappa) - \mathbf{x}(\lambda)]$ is calculated for some λ. The subroutine RLV is used to subtract off any lattice vector $\mathbf{n}$ which might have come in. The resultant vector $[\mathbf{x}(\kappa) - \mathbf{x}(\lambda) - \mathbf{n}]$ should be a fractional translation vector $\mathbf{v}(i)$. The operation $(\mathscr{R}(i)|\mathbf{v}(i))$ is then applied to all other atomic positions to see if they are taken into positions occupied by atoms of the same type (equivalent positions). If so, the operation is a member of the space group. If not, a new value of λ is assumed and the process is repeated until a $\mathbf{v}(i)$ is found which works for all atoms. If none can be found, then $\mathbf{R}(i)$ is eliminated from the group. In the process, a table of $F_0(\kappa, i)$'s is obtained for later use.

F. THE TRANSLATION GROUP AND THE DYNAMICAL MATRIX

One can show that the EOM [Eq. (18)], are invariant under the operations in the space–time group. For reasons of simplicity, one divides the applications of group theory to the EOM into two parts. One first examines the effect of translations and then later treats the rotations and rotoinversions.

There is a very general theorem from group theory which is applicable to our problem. The proof of the theorem will not be given here. It can be constructed by analogy to the proof given by Vilenkin (1968) for continuous groups.

Theorem: Let $G = \{\mathscr{G}_1, \mathscr{G}_2, \ldots, \mathscr{G}_g\}$ be a group. Suppose that G has r irreducible unitary representations.

$$\Gamma(\rho) = \{\Gamma(\rho; \mathscr{G}_1), \Gamma(\rho; \mathscr{G}_2), \ldots, \Gamma(\rho; \mathscr{G}_g)\}, \qquad \rho = 1, \ldots, r \tag{30}$$

of dimensions $d(\rho)$. Let the matrix elements of the representations be $\gamma_{\delta\varepsilon}(\rho; \mathscr{G}_k)$, where $1 \leqslant \delta, \varepsilon \leqslant d(\rho)$. The functions $(d(\rho))^{1/2}\gamma_{\delta\varepsilon}(\rho; \mathscr{G}_k)$ form a complete orthonormal system on G, and therefore any function on the group $f(\mathscr{G}_k)$ can be expanded in terms of these functions,

$$f(\mathscr{G}_k) = \sum_{\rho=1}^{r} \sum_{\delta=1}^{d(\rho)} \sum_{\varepsilon=1}^{d(\rho)} b_{\delta,\varepsilon}(\rho)(d(\rho))^{1/2}\gamma_{\delta\varepsilon}(\rho; \mathscr{G}_k) \tag{31}$$

This theorem is used the following way. The functions $u_\alpha(\mathbf{l}, t)$ are in fact functions on the group whose elements are of the form $(\mathscr{E}, \mathscr{T}(1)|\mathbf{l}, t)$ where $\mathscr{E}$ is the identity rotation, i.e.,

$$(\mathbf{l}, t) = (\mathscr{E}, \mathscr{T}(1)|\mathbf{l}, t)(\mathbf{0}, 0) \tag{32}$$

where $(\mathbf{0}, 0)$ is the space–time origin point. This is an abelian group and therefore has only 1-dimensional irreducible representations which can be indexed by a discrete set of vectors $\mathbf{k}$ and a continuous index ω. It can be shown that the matrix elements of the representations are

$$\gamma_{1,1}(\mathbf{k}, \omega; (\mathscr{E}, \mathscr{T}(1)|\mathbf{l}, t)) = \exp(i[\mathbf{k} \cdot \mathbf{l} - \omega t]) \tag{33}$$

where the representation index $\mathbf{k}$ ranges over the Brillouin zone (BZ) of wave vector space and ω varies from $-\infty$ to $+\infty$. For completeness we include a definition of the $\mathbf{k}$ vectors and the BZ. We have defined the *direct* lattice in Eq. (1). The wave vector lattice is defined by the equations,

$$\mathbf{h} = \sum_{j=1}^{3} h(j)\mathbf{b}(j) \tag{34a}$$

where the $h(j)$'s are integers and

$$\mathbf{a}(i) \cdot \mathbf{b}(j) = 2\pi\, \delta(i, j) \tag{34b}$$

The wave vector **k** is given by

$$\mathbf{k} = (\tfrac{1}{2L}) \sum_{j=1}^{3} k(j)\mathbf{b}(j) \tag{35a}$$

where the integers $k(j)$ are in the range,

$$-L < k(j) \leqslant L. \tag{35b}$$

Wave vectors on the boundary of the BZ satisfy the equation

$$2\mathbf{k} \cdot \mathbf{h} = \mathbf{h} \cdot \mathbf{h} \tag{36}$$

where the **h** vectors are *small* wave vector lattice vectors. The word *small* is somewhat vague and this causes us some extra effort in our computer program. GROUP2 has a subroutine called BZB which determines the intersections of the BZ boundary with the three axes of the cartesian coordinate system. What we have done is calculate 125 cases given by the conditions $-2 \leqslant h(j) \leqslant 2, j = 1, 2, 3$. For a given **h**, the magnitude of the intersections with the cartesian axes are

$$y_x = |\mathbf{h}|^2/(2|h_x|) \tag{37a}$$

$$y_y = |\mathbf{h}|^2/(2|h_y|) \tag{37b}$$

$$y_z = |\mathbf{h}|^2/(2|h_z|) \tag{37c}$$

If h_x, h_y, or h_z are zero, then we ignore the corresponding y_x, y_y, or y_z because the plane defined by Eq. (36) does not intersect the corresponding axis. By finding the smallest value of y_x from the set of 125 intersections, we have found the intersection of the BZ with the x axis. The same is true for the set of y_y's and y_z's. These intersections, which in the program are called QM(1), QM(2), and QM(3), are useful for scaling wave vectors. This will be discussed in Section III, F.

Let us return now to the application of the theorem. The expansion formula, Eq. (31) translates to

$$u_{a}(\mathbf{l}, t) = \sum_{\mathbf{k}}^{\mathrm{BZ}} \int_{-\infty}^{\infty} d\omega\, u_{a}(\mathbf{k}, \omega) \exp(i[\mathbf{k} \cdot \mathbf{l} - \omega t]). \tag{38}$$

When this expression is substituted into the EOM [Eq. (18)] the result is

$$\sum_{\mathbf{k}}^{\mathrm{BZ}} \int_{-\infty}^{\infty} d\omega \{[B_{ab}(\mathbf{k}) - \omega^2 J_{ab}] u_b(\mathbf{k}, \omega) \exp(i[\mathbf{k} \cdot \mathbf{l} - \omega t])\} = 0 \tag{39}$$

where the Einstein summation convention holds on subscripts and

$$B\ (\mathbf{k}) = \sum_{\mathbf{m}} K_{ab}(\mathbf{l} - \mathbf{m}) \exp(-i\mathbf{k} \cdot [\mathbf{l} - \mathbf{m}]) \tag{40}$$

which is independent of $\mathbf{l}$. We will call $\mathbf{B}(\mathbf{k})$ the generalized dynamical matrix (GDM). There are no solutions of Eq. (39) unless ω is a function of $\mathbf{k}$; one representation index is a function of the other. The only allowed values of ω are the eigenvalues of the set of equations

$$[B_{ab}(\mathbf{k}) - \omega^2 J_{ab}]e_b(\mathbf{k}) = 0, \qquad a = 1, \ldots, p \tag{41}$$

where the p-dimensional eigenvectors $\mathbf{e}(\mathbf{k})$ are sometimes called polarization vectors. It is well known that there are at most p distinct eigenvalues, which we shall call $\omega(\mathbf{k}, j)$, $j = 1$ to p. Actually, since ω^2 appears in the eigenvalue equations, the solutions can be $\pm\omega(\mathbf{k}, j)$. For all other values of ω, the coefficients $u_b(\mathbf{k}, \omega)$ must vanish. Let

$$u_b(\mathbf{k}, \pm\, \omega(\mathbf{k}, j)) = A(\mathbf{k}, \pm j)\, e_b(\mathbf{k}, j). \tag{42}$$

The displacement field [Eq. (38)] becomes

$$\begin{aligned} u_a(\mathbf{l}, t) = \sum_{\mathbf{k}}^{\mathrm{BZ}} \sum_{j=1} e_a(\mathbf{k}, j) \exp(i\mathbf{k} \cdot \mathbf{l})[A(\mathbf{k}, +j) \exp(-i\omega(\mathbf{k}, j)t) \\ + A(\mathbf{k}, -j) \exp(i\omega(\mathbf{k}, j)t)] \end{aligned} \tag{43}$$

The arbitrary coefficients $A(\mathbf{k}, \pm j)$ must be determined from the boundary conditions and initial conditions of the problem.

The detailed symmetry properties of the GDM will be treated in Section IV, but it is worthwhile recording some general properties of $\mathbf{B}(\mathbf{k})$, $\omega(\mathbf{k}, j)$, and $\mathbf{e}(\mathbf{k}, j)$ which follow from their definitions,

$$\mathbf{B}^{\dagger}(\mathbf{k}) = \mathbf{B}(\mathbf{k}) \quad \text{(hermitian)} \tag{44a}$$

$$\mathbf{B}^{*}(\mathbf{k}) = \mathbf{B}(-\mathbf{k}) \tag{44b}$$

$$\omega(-\mathbf{k}, j) = \omega(\mathbf{k}, j) \tag{44c}$$

$$\mathbf{e}^{*}(\mathbf{k}, j) = \mathbf{e}(-\mathbf{k}, j). \tag{44d}$$

Proof of these formulas can be found in Maradudin and Vosko (1968).

III. The Time Reversal Point Group of the Wave Vector

A. Time Reversal and Complex Conjugation

In the previous section we have seen how the irreducible representations of the space–time translation group can be used to simplify the EOM to the point that **l** and **t** no longer appear as variables. Note that the GDM still depends on the positions of atoms or clusters in the unit cell through the index K. We can use the other operations of the full space–time group to further simplify the problem. The set of operations of the form $(\mathscr{R}(i), \mathscr{T}(\mathfrak{h})|\mathbf{0}, 0)$ do not in general form a subgroup of the space–time group because $\mathscr{R}(i)$ still involves fractional translations. The product of two operations of this form might have a nonzero lattice translational part and hence would not be of the original form. If we define multiplications of group elements as being multiplication modulo the lattice translational part, then the operations $(\mathscr{R}(i), \mathscr{T}(\mathfrak{h})|\mathbf{0}, 0)$ do form a group isomorphic with the time reversal point group of the crystal.

At this point in the discussion, the space–time approach breaks down. The EOM [Eq. (41)] are independent of time. Time reversal symmetry has already been included in the problem because we have allowed both positive and negative values of the eigenvalues $\omega(\mathbf{k}, j)$. The following heuristic argument can be made to show a relation between time reversal and complex conjugation symmetry. Consider a typical plane wave solution, say, $\cos(\mathbf{k} \cdot \mathbf{x} - \omega t)$ which describes a wave propagating in the direction **k**. Changing t to $-t$ results in a wave propagating in the $-\mathbf{k}$ direction. Changing **k** to $-\mathbf{k}$ in the transformed EOM is the same as complex conjugating the equations, because of Eq. (44b). In this indirect sense, time reversal symmetry of the original EOM is equivalent to complex conjugation symmetry of the transformed EOM. Let us therefore replace $\mathscr{T}(\mathfrak{h})$ by $\mathscr{K}(\mathfrak{h})$, the complex conjugation operation; $\mathscr{K}(1)$ is the identity operation and $\mathscr{K}(2)$ applied to a function, complex conjugates it.

B. Transformation of the Eigenvectors and the T Matrices

We wish to find the invariance group of Eq. (41). To do this we need a representation of the operations $(\mathscr{R}(i), \mathscr{K}(\mathfrak{h})|\mathbf{0}, 0)$. This is obtained by applying these operations to the components of the eigenvectors. Consider one of the partial solutions of the EOM,

$$u_{\alpha}(\mathbf{k}, j; \mathbf{l}, t) = e_{\alpha}(\mathbf{k}, j) \exp(\mathfrak{i}[\mathbf{k} \cdot \mathbf{l} - \omega(\mathbf{k}, j)t]) \tag{45}$$

From Eq. (22) we know how this quantity transforms. It can be shown that

$$\mathscr{G}(i, \mathfrak{h}|\mathbf{0}, 0)e_{\mathfrak{a}}(\mathbf{k}, j) = T_{\mathfrak{a}\mathfrak{b}}(\mathbf{k}; i, \mathfrak{h})e_{\mathfrak{b}}(\mathbf{k}, j). \tag{46}$$

The new **k**-dependent transformation matrix is given by

$$\begin{aligned} T_{\mathfrak{a}\mathfrak{b}}(\mathbf{k}; i, \mathfrak{h}) = T^{\zeta\eta}_{\alpha\beta}(\mathbf{k}; i, \mathfrak{h}, K, \Lambda) &= \delta(\zeta, \eta)\delta(K, F_0(\Lambda, i))C^{\zeta}(i)R_{\alpha\beta}(i) \\ &\times \exp\{(-1)^{3-h}\mathfrak{i}\mathscr{R}(i) \cdot \mathbf{k} \cdot [\mathbf{r}(K) - \mathscr{R}(i) \cdot \mathbf{r}(\Lambda)]\}\mathscr{K}(\mathfrak{h}) \end{aligned} \tag{47}$$

It should be noted that the conjugation operator $\mathscr{K}(\mathfrak{h})$ has been brought through to the right of the exponential function and the factor $(-1)^{3-\mathfrak{h}}$ has been placed before $\mathfrak{i} = \sqrt{-1}$ in the exponential function.

C. The Invariance Group of the Transformed EOM

We know intuitively that when a plane wave propagating in the direction **k** is subjected to a rotation $\mathbf{R}(i)$ the resulting wave will be propagating in the direction $\mathbf{R}(i) \cdot \mathbf{k}$. Therefore the polarization vector $\mathbf{e}(\mathbf{k}, j)$ should likewise transform into a possible polarization vector for the mode propagating in the direction $\mathbf{R}(i) \cdot \mathbf{k}$, i.e.,

$$\mathscr{G}(i, \mathfrak{h}|\mathbf{0}, 0)\mathbf{e}(\mathbf{k}, j) = \mathbf{e}(\mathbf{R}(i) \cdot \mathbf{k}, j') \tag{48}$$

Another way of saying this is that if $\mathscr{G}(i, \mathfrak{h}|\mathbf{0}, 0)$ is applied to Eq. (41), which describes a mode propagating in direction **k**, then the result will be the EOM for a mode propagating in the direction $\mathbf{R}(i) \cdot \mathbf{k}$. What we really want to do is find the set of operations which leave Eq. (41) invariant. Clearly this is the set of operations which leave $\mathbf{B}(\mathbf{k})$ unchanged. It is easily shown that this set consists of two parts,

$$\mathscr{G}(i, 1|\mathbf{0}, 0) \qquad \text{such that} \quad \mathbf{R}(i) \cdot \mathbf{k} = \mathbf{k} + \mathbf{h}(i) \tag{49a}$$

and

$$\mathscr{G}(i', 2|\mathbf{0}, 0) \qquad \text{such that} \quad \mathbf{R}(i) \cdot \mathbf{k} = -\mathbf{k} + \mathbf{h}(i') \tag{49b}$$

where $\mathbf{h}(i)$ and $\mathbf{h}(i')$ are vectors of the wave vector lattice. We have to worry about these lattice vectors because $\mathbf{B}(\mathbf{k})$ is periodic in wave vector space, i.e.,

$$\mathbf{B}(\mathbf{k} + \mathbf{h}) = B(\mathbf{k}) \tag{50}$$

The set of operations described above is a group when group multiplication is defined modulo lattice vectors. This group is called the time reversal point group of the wave vector (TRPGWV). The corresponding set of quantities $\{\mathbf{T}(\mathbf{k}; i, 1), \mathbf{T}(\mathbf{k}; i', 2)\}$ defines a multiplier operator representation (MOR) of this group, i.e.,

$$\mathbf{T}(\mathbf{k}, i, \mathfrak{h}) \cdot \mathbf{T}(\mathbf{k}, j, \mathfrak{l}) = \phi(\mathbf{k}, i, j)\mathbf{T}(\mathbf{k}, k, \mathfrak{m}) \tag{51a}$$

where

$$\mathscr{G}(i, \mathfrak{h}|\mathbf{0}, 0)\mathscr{G}(j, \mathfrak{l}|\mathbf{0}, 0) \bmod(\mathbf{l}) = \mathscr{G}(k, \mathfrak{m}|\mathbf{0}, 0) \tag{51b}$$

and $\mathbf{l}$ is any lattice vector which might result from the multiplication. The multiplier is given by (Maradudin and Vosko, 1968)

$$\phi(\mathbf{k}, i, j) = \exp(i[\mathbf{k} - \mathbf{R}^{-1}(i) \cdot \mathbf{k}] \cdot \mathbf{v}(j)) \tag{51c}$$

This is called an operator representation because the $\mathbf{T}(\mathbf{k}; i, \mathfrak{h})$'s contain the operator $\mathscr{K}(\mathfrak{h})$. We note that the multipliers are unity except for special wave vectors on the BZ boundary in nonsymmorphic crystals. The use of this MOR to symmetry-reduce the dynamical matrix is discussed in the next section.

In an early version of our code we stored the matrices of the MOR in a large array. This required auxiliary storage such as Extended Core Storage on the CDC 6600. This is obviously inefficient because most of the array was filled with zeros because of the δ functions. At present we only store the 3×3 matrices $\mathbf{R}(i)$, the atom transformation table $F_0(K, i)$ and the factors $C^{\zeta}(i)$. For the evaluation of matrix products such as $\mathbf{T}(\mathbf{k}; i, \mathfrak{h})\mathbf{B}(\mathbf{k})\mathbf{T}^{-1}(\mathbf{k}; i, \mathfrak{h})$, we have carried out all the algebra by hand to the point where the results can be expressed in terms of the 3×3 matrix multiplications by $\mathbf{R}(i)$'s, multiplications by scaler functions such as $C^{\zeta}(i)$ and summations over the proper terms as dictated by the atom transformation table. The slight amount of algebra was well worth the savings in storage.

D. Irreducible Multiplier Operator Representations and Decomposition

The block-diagonalization of the GDM requires a knowledge of the irreducible multiplier operator representations (IMOR's) of the TRPGWV. We will now discuss how these are handled on the computer. First of all, let $\tau_{\delta\varepsilon}(\mathbf{k}, \rho; i, \mathfrak{h})$, $1 \leqslant \delta, \varepsilon \leqslant d(\rho)$ be the matrix elements of the ρth IMOR that has dimension $d(\rho) \times d(\rho)$. It has been shown by Warren (1974) that

$$\tau_{\delta\varepsilon}(\mathbf{k}, \rho; i, \mathfrak{h}) = \theta_{\delta\varepsilon}(\mathbf{k}, \rho; i)\mathscr{K}(\mathfrak{h}) \tag{52}$$

where the set $\{\boldsymbol{\theta}(\mathbf{k}, \rho; i)\}$ is called an irreducible multiplier corepresentation (IMC). The IMC's can in turn be expressed in terms of the irreducible multiplier representations (IMR's) of the ordinary point group of the wave vector (PGWV), which is defined by Eq. (49a). The formulas for expressing the IMC's in terms of the IMR's will be discussed below when we address the question of time reversal degeneracy.

One immediate use of the IMR's is the decomposition of the representation of the PGWV into its irreducible parts. This is important because the unitary transformation which decomposes, i.e., block-diagonalizes, the representation of the PGWV is very closely related to the transformation which block-diagonalizes the GDM. The representation of the PGWV is derived from the MOR of the TRPGWV. It is given by Eq. (47) with $\mathfrak{h} = 1$ and i restricted to operations satisfying Eq. (49a). Let us call that representation $\mathbf{T}(\mathbf{k}; i)$. The decomposition formula, which calculates the number of times that each IMR occurs in the decomposition, can be derived using orthogonality of the characters of the IMR's. The formula is

$$m(\rho) = \sum_{i=1}^{g(k)} \chi^*(\mathbf{k}, \rho; i)\chi(\mathbf{k}; i) \tag{53a}$$

where

$$\chi(\mathbf{k}; i) = \sum_{\alpha=1}^{3} \sum_{\mathrm{K}=1}^{C} \sum_{\zeta=1}^{f(\mathrm{K})} T_{\alpha\alpha}^{\zeta\zeta}(\mathbf{k}; i, \mathrm{K}, \mathrm{K}) \tag{53b}$$

and

$$\chi(\mathbf{k}, \rho; i) = \sum_{\delta=1}^{d(\rho)} \tau_{\delta\delta}(\mathbf{k}, \rho, i) \tag{53c}$$

are the characters of the representation of the PGWV and the ρth IMR, respectively. The interpretation of this formula is that in the block-diagonalized form of $\mathbf{T}(\mathbf{k}; i)$, there will appear $m(\rho)$ identical blocks of the form $\boldsymbol{\tau}(\mathbf{k}, \rho; i)$ having dimension $d(\rho)$. For the block-diagonalized form of $\mathbf{B}(\mathbf{k})$, however, there will appear $d(\rho)$ blocks of dimension $m(\rho)$.

E. Generation of the IMR's by Computer

1. *Induction Methods*

A number of authors have suggested methods of constructing IMR's of the group of the wave vector $G(\mathbf{k})$, where $G(\mathbf{k})$ is the group of rotational symmetry operations which leave the wave vector unchanged or changed by a lattice vector in wave vector space. One author (Neto, 1973) has suggested a method which could be used in a computer calculation, but he did not publish a program based on the method. Miller and Love (1967) have published tables of IMR's for all space groups, which were generated by computer, but they have not discussed the program used to generate the tables.

Subroutine GIRACO of GROUP2 follows the method of Sahni and Venkataraman (1970), which starts by constructing the IMR's of an invariant cyclic subgroup H_1 of $G(\mathbf{k})$. An invariant subgroup is a subgroup which contains elements in complete classes so that if $\mathcal{G}_1$ is an element of H_1 ($\mathcal{G}_1 \in H_1$), then $\mathcal{G}'_j = \mathcal{G}_j^{-1}\mathcal{G}_i\,\mathcal{G}_j \in H_1$ for all elements $\mathcal{G}_j \in G(\mathbf{k})$. It is necessary to start with an invariant subgroup because we plan to augment H_1 by elements of $G(\mathbf{k})$ in a systematic manner to build up the IMR's of $G(\mathbf{k})$. In the case of cubic groups of order 24 or 48, we must start with an invariant cyclic subgroup of order 2. We start with the group $C_2{}^x$. Once an invariant cyclic subgroup has been chosen we choose an augmenting element $\mathcal{G}_2 \notin H_1$ and form the augmented group $H_2 = H_1 + \mathcal{G}_2 H_1 + \cdots + \mathcal{G}_2{}^{(n-1)}H_1$, where n is the order of the element $\mathcal{G}_2$, i.e., $\mathcal{G}_2{}^n = e$, the identity element. In the case of all groups except O_h, O, T_h, and T_d, the largest cyclic subgroup may be selected as the starting point. If H_2 is smaller than $G(\mathbf{k})$, we choose another augmenter and form a new augmented group. This process is repeated until the augmented group equals $G(\mathbf{k})$. This induction method can be written in the general form

$$H_i = H_{i-1} + \mathcal{G}_i H_{i-1} + \cdots + \mathcal{G}_i^{(n-1)} H_{i-1}. \tag{54}$$

If the order of the subgroup H_i is g_i, then the number g_i/g_{i-1} is called the index of H_{i-1} in H_i. The theory requires this index to be a prime number. For the point groups it turns out that the index is either 2 or 3 and is equal to n.

2. *Choosing an Invariant Cyclic Subgroup*

Since all crystallographic point groups contain an invariant cyclic subgroup, it is always possible to construct the IMR's using this method. We have mentioned the method we have used to select the subgroup H_1. This is not the only choice that can be made, but in choosing a starting subgroup there are two pitfalls of which one must be aware: (1) not all subgroups are invariant, and (2) not all invariant subgroups have prime index. A table showing the genealogy of the 32 crystallographic point groups is given by Bradley and Cracknell (1972). This shows that there are several possible sequences of induction for most point groups. In GROUP2 H_1 is chosen as indicated above and the augmenters are chosen to be the next element in numerical sequence of elements not in H_{i-1}, but in $G(\mathbf{k})$. The numerical sequence which we use follows Kovalev (1964). Worlton and Warren (1972) may be consulted for other notations.

3. *Defining the IMR's of a Cyclic Group*

The reason for starting with a cyclic subgroup is that the IMR's of a cyclic group are all 1-dimensional and may be easily determined. For ordinary vector representations, the representations of the generating element C_n will just be the n roots of unity, and the representation of the other elements will be powers of the representations of C_n. For multiplier representations, the representations of the generating element must also obey the relation

$$[\tau(C_n)]^n = \phi(C_n, C_n)\phi(C_n^{\,2}, C_n)\cdots\phi(C_n^{n-1}, C_n)\tau(E) \qquad (55)$$

The n roots of Eq. (55) give the n IMR's of C_n and the IMR's of the other elements may then be generated from the multiplication rules of the group elements.

4. *Orbit Classification*

To find the IMR's of the augmented group from those of a subgroup, we must first classify the IMR's of the subgroup into orbits with respect to the augmenting element $\mathcal{G}_i$. This classification may be accomplished by forming the conjugate representations whose characters are given by

$$\chi(\rho'; j) = \chi(\rho; j')[\phi(j, i)/\phi(i, j')] \qquad (56a)$$

where $\mathcal{G}_i$ is the augmenting element and $\mathcal{G}_{j'}$, is the element obtained from

$$\mathcal{G}_{j'} = \mathcal{G}_i^{\,-1}\mathcal{G}_j\mathcal{G}_i. \qquad (56b)$$

In Eqs. (56) $\chi(\rho'; j)$ is the character of element $\mathcal{G}_j$ in the representation ρ'. The conjugate representation will either be the same as the original representation ρ, or will be equal to one of the other representations. Representations which are taken into each other by conjugation are members of the same orbit. Self-conjugate representations are said to have an orbit of order 1. Each IMR ρ which is not self-conjugate will belong to an orbit of order n, where $\mathcal{G}_i^n = e$. Only one member of each orbit will be used in building up the IMR's of $G(\mathbf{k})$. The IMR's for orbits of order 1 and of order n must be derived separately.

5. *Deriving IMR's from an Orbit of Order n*

In the case of an orbit of order n the IMR's of the augmented group are given by

$$\tau(\rho_i; j) = \begin{bmatrix} \gamma_{1,1}(\rho_i; j) & \cdots & \gamma_{1,n}(\rho_i; j) \\ & \cdot & \\ & \cdot & \\ & \cdot & \\ \gamma_{n,1}(\rho_i; j) & \cdots & \gamma_{n,n}(\rho_i; j) \end{bmatrix} . \tag{57a}$$

where $\gamma_{ab}(\rho_i; j)$ is a submatrix of the same dimension as the IMR ρ_{i-1} from which it was derived and is given by

$$\gamma_{ab}(\rho_i; j) = \tau_{ab}(\rho_{i-1}; j')[\phi(j, i^b)/\phi(i^a, j')]. \tag{57b}$$

In Eq. (57b) i^a and i^b are the indices of the ath and bth power of the augmenting element $\mathscr{G}_i$, e.g., $\mathscr{G}_{i^a} = (\mathscr{G}_i)^a$. The index j' is determined from the equation

$$\mathscr{G}_{j'} = (\mathscr{G}_i)^{-a}\mathscr{G}_j(\mathscr{G}_i)^b \tag{57c}$$

if this product gives an element in H_{i-1}. If the product is not in H_{i-1} then the corresponding matrix $\gamma_{ab}(\rho_i; j)$ is zero.

6. *Deriving the IMR's from Orbits of Order 1*

For each IMR of H_{i-1} which is self-conjugate, we can derive n IMR's of H_i. In order to do this we must find n representations of the augmenter which satisfy the multiplication rules of the group. These representations must satisfy the relations

$$\tau(\rho_i; j)\tau(\rho_i; i) = \tau(\rho_i, i)\tau(\rho_i; j')[\phi(j, i)/\phi(i, j')], \tag{58}$$

where $\tau(\rho_i; j) \equiv \tau(\rho_{i-1}; j)$, j indexes all the elements of H_{i-1}, and j' is given by Eq. (56b). The new matrices $\tau(\rho_i, i)$ must also obey an equation like Eq. (55). A matrix which satisfies Eq. (58) can be derived from an arbitrary matrix if we multiply Eq. (58) from the left by $\tau(\rho_i; j^{-1})$ and sum over all the elements in the subgroup H_{i-1}. This gives us a self-consistent equation for $\tau(\rho_i; i)$,

$$\tau(\rho_i; i) = (1/g_{i-1}) \sum_{j=1}^{g_{i-1}} \frac{\tau(\rho_i; j^{-1})\tau(\rho_i; i)\tau(\rho_i; j')\phi(j, i)}{\phi(j^{-1}, j)\phi(i, j')} \tag{59}$$

Equation (59) will yield a matrix which obeys Eq. (58) if an arbitrary matrix is substituted for $\tau(\rho_i; i)$ on the right. Since numerical methods on a computer are faster and simpler than algebraic methods, we substitute a matrix of random numbers for $\tau(\rho_i; i)$ and apply Eq. (59). This gives one representation of $\mathscr{G}_i$ which satisfies the group multiplication table. This matrix is only uniquely defined to within a phase factor. Appropriate phase factors for determining n other IMR's of $\mathscr{G}_i$ can be obtained from taking the n roots of Eq. (55) where $\mathscr{G}_i$ is substituted for C_n.

Once n representations of $\mathscr{G}_i$ have been obtained, the n IMR's of all elements may be obtained from the following rules.

$$\tau(\rho_i; j) = \tau(\rho_{i-1}; j) \qquad \text{for} \quad \mathscr{G}_j \in H_{i-1} \tag{60a}$$

$$\tau(\rho_i; i^2) = [\tau(\rho_i; i)]^2/\phi(i, i) \tag{60b}$$

$$\tau(\rho_i; i \cdot j) = \tau(\rho_i; i)\tau(\rho_i; j)/\phi(i, j) \tag{60c}$$

$$\tau(\rho_i; i^2 \cdot j) = \tau(\rho_i; i^2)\tau(\rho_i; j)/\phi(i^2, j). \tag{60d}$$

Although we have used a single representation index, ρ_i, in Eqs. (60), the substitution of the n values of $\tau(\rho_i; i)$ will yield n IMR's for each old index ρ_{i-1} and the numbering of ρ_i must reflect that.

It is noteworthy that this method is not dependent on a particular factor system so it could be used to obtain the vector representations where the factors are all unity or it could be used with a modified multiplication table for use in finding the double group representations useful in electron band theory. To do this one should remove much of the program for lattice dynamics. How much could be removed depends on what information the user desired to supply to the subroutine. In GROUP2 the space group operations are determined from the lattice vectors and atomic positions by other subroutines. This reduces the amount of input to a minimum, but does increase the length of the program quite a bit.

F. Wave Vectors, Brillouin Zones, and Labeling

As GROUP2 presently stands, one must read in a title card and a scaled wave vector for every special symmetry point, line, or plane in the BZ. As mentioned in Section II, F, we use scaling factors for constructing wave vectors from the input. As a simple example of what we mean by scaling, consider a wave vector on the BZ boundary of a simple cubic lattice. Take in particular the plane perpendicular to the positive x axis. A typical wave vector on that plane might be $\mathbf{k} = (2\pi/a)\ (1, 0.2, 0.3)$ where a is the cubic lattice constant. The factor $(2\pi/a)$ is obviously a scale factor for each of the cartesian components. The input would be $k/k_{\max} = (1, 0.2, 0.3)$. The scale factors QS(1) = QS(2) = QS(3) = $2\pi/a$ would normally be computed by the program from the array QM(I) mentioned in Section II, F. The scale factors can also be read in for those cases where the intersections of the BZ with the cartesian axes are not the most convenient scale factors.

Once the wave vector has been constructed, the program determines the PGWV by using the criteria in Eq. (49a) and proceeds to simplify the GDM as discussed in Section IV.

It would be nice if the program itself determined which wave vectors of the BZ have special symmetry and provided standard labeling for these vectors. This is a nontrivial task. The choice of symmetry vectors and their labeling is a matter of some confusion. There are many books and articles on BZ's and group representations which purport to set up a uniform labeling, but all have flaws, inconsistencies, or omissions. The most complete book of this type to date is by Bradley and Cracknell (1972). The only minor flaw we have found so far is the omission of labels for planes of symmetry in the BZ. If anyone were to attempt to get automatic wave vector selection into a computer program, it would be worth trying to use their book as a standard. Another interesting idea, which could be pursued if automatic wave vectors were available, is the creation of compatibility relations between the lines of symmetry and their endpoints. The IMR's of the group of the wave vector at the endpoints of a line must be compatible with those of the line itself (Tinkham, 1964).

The labeling of the IMOR's is also of some concern because the eigenvectors and the eigenvalues can be labeled with the names of the IMOR's. Consider an eigenvalue $\omega(\mathbf{k}, j)$ which is degenerate. By this we mean that there are several eigenvectors $\mathbf{e}(\mathbf{k}, j, \delta)$, $\delta = 1, \ldots, d(j)$, all associated with the given eigenvalue. The degeneracy of transverse modes in a homogeneous solid is an example. It can be shown that the eigenvectors belong to the IMOR's, i.e., they have the property that

$$\mathbf{T}(\mathbf{k}; i, \mathfrak{h})\mathbf{e}(\mathbf{k}, j, \delta) = \sum_{\varepsilon=1}^{d(\rho)} \tau_{\varepsilon\delta}(\mathbf{k}, \rho; i, \mathfrak{h})\mathbf{e}(\mathbf{k}, j, \varepsilon) \tag{61}$$

for some IMOR ρ of the TRPGWV. This means that the $\mathbf{e}(\mathbf{k}, j, \delta)$'s can be labeled with the name of the IMOR to which they belong and similarly for the eigenvalues. There may be more than one eigenvalue associated with the IMOR ρ. We introduce a multiplicity index $\mu = 1, \ldots, m(\rho)$, where $m(\rho)$ is the number of eigenvalues belonging to the IMOR ρ. The index j is replaced by the double index (ρ, μ) to give eigenvalues $\omega(\mathbf{k}, \rho, \mu)$ and eigenvectors $\mathbf{e}(\mathbf{k}, \rho, \mu, \delta)$.

In GROUP2 the IMR's (not the IMOR's) are numbered sequentially as they are generated. Since the order in which they are generated is somewhat arbitrary, there is no correspondence between our labels and those found in any published table. The correspondence can be made by comparing character vectors. The character vector of a representation,

$$\chi(\mathbf{k}, \rho) = \{\chi(\mathbf{k}, \rho, 1), \chi(\mathbf{k}, \rho, 2), \ldots, \chi(\mathbf{k}, \rho, g)\} \tag{62a}$$

where

$$\chi(\mathbf{k}, \rho, i) = \sum_{\delta=1}^{d(\rho)} \tau_{\delta\delta}(\mathbf{k}, \rho, i) \tag{62b}$$

and g is the order of the PGWV, is a unique identifier of a representation. GROUP2 prints out these character vectors. If there were a standard set of tables that almost everyone agreed to, the computer could be made to label the IMOR's in a standard way by a comparison of character vectors. This is another possible frill for the future.

G. Time Reversal Degeneracy

There is one last topic that must be discussed in connection with labeling of eigenvalues and eigenvectors. That is the topic of time reversal degeneracy. Enlarging the symmetry group of an eigenvalue equation often increases the degeneracy of the eigenvalues. A trivial example is the fact that all three acoustic modes become degenerate at $\mathbf{k} = 0$ as a consequence of the fact that the group of the zero wave vector is larger than the group of a finite wave vector. The extra degeneracy resulting from the time reversal operations is manifest in the relations between the IMOR's and the IMR's. The IMR's are representations of PGWV,

$$G(\mathbf{k}) = \{\mathbf{R}(i); \qquad i = 1, \ldots, g\} \tag{63}$$

where g is the order of the group. The IMOR's are representations of the TRPGWV,

$$\bar{G}(\mathbf{k}) = \{\mathbf{R}(\bar{i}); \qquad \bar{i} = 1, \ldots, \bar{g}\} \tag{64a}$$

where $\bar{g}$ is usually twice as large as g. The group $\bar{G}(\mathbf{k})$ can be divided into two parts,

$$\bar{G}(\mathbf{k}) = \{G(\mathbf{k}), G(\mathbf{k})\,\mathbf{R}(1-)\} \tag{64b}$$

where $\mathbf{R}(1-)$ is any rotoinversion operation which changes $\mathbf{k}$ to $-\mathbf{k}$ modulo, a lattice vector in wave vector space. The elements of the coset $G(\mathbf{k})\mathbf{R}(1-)$ will be denoted

$$\mathbf{R}(i-) = \mathbf{R}(i) \cdot \mathbf{R}(1-), \qquad i = 1, \ldots, g \tag{65}$$

In the process of inducing the IMOR's of $\bar{G}(\mathbf{k})$ from the IMR's of $G(\mathbf{k})$, one of three things can happen, depending on the matrix which represents the operation $\mathbf{R}(1-)$ in the extension. If there is no additional degeneracy we find that $\mathbf{R}(1-)$ is represented by a matrix $\boldsymbol{\beta}$ which has the same dimension as the other matrices in the IMR ρ. The induced IMC, which is closely related to the IMOR [see Eq. (52)], becomes

$$\boldsymbol{\theta}(\mathbf{k}, \rho, i) = \boldsymbol{\tau}(\mathbf{k}, \rho, i) \tag{66a}$$

$$\boldsymbol{\theta}(\mathbf{k}, \rho, 1-) = \boldsymbol{\beta} \tag{66b}$$

$$\boldsymbol{\theta}(\mathbf{k}, \rho, i-) = \phi(\mathbf{k}, i, 1-)\boldsymbol{\tau}(\mathbf{k}, \rho, i)\boldsymbol{\beta} \tag{66c}$$

where $\boldsymbol{\tau}(\mathbf{k}, \rho, i)$'s are matrices of the IMR of $G(\mathbf{k})$. The multiplier $\phi(\mathbf{k}, i, 1-)$ is still given by Eq. (51c).

The second possibility is that there is a degeneracy between two eigenvalues both belonging to the same representation, i.e.,

$$\omega(\mathbf{k}, \rho, \mu') = \omega(\mathbf{k}, \rho, \mu) \tag{67}$$

for $\mu' \neq \mu$. In this case the operation $\mathbf{R}(1-)$ is represented by a matrix with twice the dimension of the matrices in the IMR ρ. The resulting IMC is given by

$$\boldsymbol{\theta}(\mathbf{k}, \rho, i) = \begin{pmatrix} \boldsymbol{\tau}(\mathbf{k}, \rho, i) & \mathbf{0} \\ \mathbf{0} & \boldsymbol{\tau}(\mathbf{k}, \rho, i) \end{pmatrix} \tag{68a}$$

$$\boldsymbol{\theta}(\mathbf{k}, \rho, 1-) = \begin{pmatrix} 0 & -\boldsymbol{\beta} \\ \boldsymbol{\beta} & 0 \end{pmatrix} \tag{68b}$$

$$\boldsymbol{\theta}(\mathbf{k}, \rho, i-) = \phi(\mathbf{k}, i, 1-)\boldsymbol{\theta}(\mathbf{k}, \rho, i)\boldsymbol{\theta}(\mathbf{k}, \rho, 1-) \tag{68c}$$

The third possibility is that

$$\omega(\mathbf{k}, \rho, \mu) = \omega(\mathbf{k}, \rho', \mu) \tag{69}$$

for $\rho' \neq \rho$. Once again $\mathbf{R}(1-)$ must be represented by a matrix with twice the dimension of the IMR ρ. The dimension of ρ' must equal that of ρ for this degeneracy to occur. The IMC is

$$\boldsymbol{\theta}(\mathbf{k}, \rho, i) = \begin{pmatrix} \boldsymbol{\tau}(\mathbf{k}, \rho, i) & 0 \\ 0 & \boldsymbol{\tau}(\mathbf{k}, \rho', i) \end{pmatrix} \tag{70a}$$

$$\boldsymbol{\theta}(\mathbf{k}, \rho, 1-) = \begin{pmatrix} 0 & \boldsymbol{\beta} \\ \boldsymbol{\beta} & 0 \end{pmatrix} \tag{70b}$$

$$\boldsymbol{\theta}(\mathbf{k}, \rho, i-) = \phi(\mathbf{k}, i, 1-)\boldsymbol{\theta}(\mathbf{k}, \rho, i)\boldsymbol{\theta}(\mathbf{k}, \rho, 1-). \tag{70c}$$

One can now see a small problem in the labeling of IMOR's. We have done the following thing in GROUP2. There is a subroutine called TRDEG which tests the IMR's to determine if time reversal degeneracy exists. If there is a degeneracy, then we say that there is a time reversal equivalence between the representations. An equivalence table is set up as shown for example in Table III. The first column gives the labels of the IMR's. The second column shows the equivalence of IMR's 2 and 3, the time reversal degeneracy of 4 with itself and the absence of extra degeneracy for IMR 1. Column 3 tells how the IMC's are constructed. Because of the equivalence of IMR 2 and IMR 3, there are only three IMC's and the label $\rho = 3$ is dropped. This equivalence table is used in other subroutines to determine the labeling of the symmetry coordinate vectors and to determine the decomposition of the reducible representation $\mathbf{T}(\mathbf{k})$ into its irreducible parts.

The generation of the matrix $\boldsymbol{\beta}$ can be done in two ways. One way is to start with a random matrix and restrict the matrix elements by the conditions that the induced IMC must satisfy the group multiplication table for $\bar{G}(\mathbf{k})$. This method was used in the generation of the IMR's themselves. Augmenters were generated this way [see Eq. (59)]. One can also use the analytical formula (Wigner, 1959, p. 80). The formula, which still contains an arbitrary matrix $\mathbf{Y}$, is

$$\boldsymbol{\beta} = \sum_{i=1}^{g} \boldsymbol{\tau}(\mathbf{k}, \rho', i) \cdot \mathbf{Y} \cdot \boldsymbol{\tau}'(\mathbf{k}, \rho, i)^{\dagger} \tag{71a}$$

TABLE III

TIME REVERSAL EQUIVALENCE

ρ	IET(ρ)	$\boldsymbol{\theta}(\mathbf{k}, \rho)$
1	1	Eqs. (66)
2	3	Eqs. (70)
3	2	Dropped
4	−4	Eqs. (68)

where

$$\tau'(\mathbf{k}, \rho, i) = \phi(\mathbf{k}, i, 1-)\phi(\mathbf{k}, 1-, i')\boldsymbol{\tau}(\mathbf{k}, \rho, i') \tag{71b}$$

and

$$\mathbf{R}(i') = \mathbf{R}(1-)^{-1}\mathbf{R}(i) \cdot \mathbf{R}(1-). \tag{71c}$$

The usual way of choosing **Y** is to let it be a $d(\rho) \times d(\rho)$ matrix having only one nonzero element, which is determined by the normalization condition

$$\boldsymbol{\beta\beta}^{\dagger} = 1. \tag{71d}$$

With this choice of **Y**, the summation in Eq. (71a) will either give zero or a proper form of $\boldsymbol{\beta}$, depending on which element of **Y** is made nonzero. Once $\boldsymbol{\beta}$ is determined, there is a criterion to determine which type of degeneracy exists. It is straightforward to code the conditions,

$$\text{If} \quad \boldsymbol{\tau}'(\mathbf{k}, \rho) = \boldsymbol{\tau}(\mathbf{k}, \rho) \quad \text{and}$$
$$\boldsymbol{\beta\beta}^* = \phi(\mathbf{k}, 1-, 1-)\boldsymbol{\tau}(\mathbf{k}, \rho, 1- \cdot 1-) \rightarrow \text{Eqs. (66)} \tag{72a}$$

$$\text{If} \quad \boldsymbol{\tau}'(\mathbf{k}, \rho) = \boldsymbol{\tau}(\mathbf{k}, \rho) \quad \text{and}$$
$$\boldsymbol{\beta\beta}^* = -\phi(\mathbf{k}, 1-, 1-)\boldsymbol{\tau}(\mathbf{k}, \rho, 1- \cdot 1-) \rightarrow \text{Eqs. (68)} \tag{72b}$$

or

$$\text{If} \quad \boldsymbol{\tau}'(\mathbf{k}, \rho) = \boldsymbol{\tau}(\mathbf{k}, \rho') \quad \text{and} \quad \rho' \neq \rho \rightarrow \text{Eqs. (70)}. \tag{72c}$$

It should be noted that there exists a formula [Maradudin and Vosko, 1968, Eq. (5.60)], involving only the character of a given IMR, which tells if there is time reversal degeneracy associated with that IMR. The formula does not, however, tell which ρ' is equivalent to which ρ in the third type of degeneracy and that is why we have used Eqs. (72) instead.

IV. Symmetry Reduction of the Dynamical Matrix

A. The Self-Consistent Equation for Symmetry Reduction

Symmetry requires that **B(k)** be invariant under the operations in the TRPGWV, i.e.,

$$\mathbf{T}(\mathbf{k}, \bar{i})^{-1} \cdot \mathbf{B}(\mathbf{k}) \cdot \mathbf{T}(\mathbf{k}, \bar{i}) = \mathbf{B}(\mathbf{k}) \tag{73}$$

for $\bar{i}$ in $\bar{G}(\mathbf{k})$. This results in some matrix elements being zero and others being linearly related. For example in Mg at the point H, where $k/k_{max} =$ [101] elements B_{12}, B_{23}, B_{14}, B_{24}, B_{15}, B_{25}, B_{36}, B_{46}, and B_{56} are zero. Furthermore B_{12} is pure imaginary (Warren and Worlton, 1973, p. 261). This simplification of $\mathbf{B}(\mathbf{k})$ is referred to as symmetry reduction and is necessary before $\mathbf{B}(\mathbf{k})$ can be brought into block-diagonal form.

Warren (1974) has shown that the effects of time reversal together with invariance under the space group symmetry imposes the condition

$$\mathbf{B}(\mathbf{k}) = 1/(2g) \sum_{i=1}^{g} \mathbf{T}(\mathbf{k}, i)[\mathbf{B}(\mathbf{k}) + \mathbf{\Theta}(\mathbf{k}, 1-)\mathbf{B}^*(\mathbf{k})\mathbf{\Theta}(\mathbf{k}, 1-)]\mathbf{T}^\dagger(\mathbf{k}, i), \quad (74)$$

where g is the order of the PGVW. The matrix $\mathbf{\Theta}(\mathbf{k}, 1-)$ is a multiplier corepresentation of a typical operation which carries $\mathbf{k}$ into $-\mathbf{k}$. Equation (74) is a self-consistent equation for the dynamical matrix. Substitution of an arbitrary matrix on the right-hand side of Eq. (74) will give a matrix in symmetry-reduced form on the left. This summation method is more convenient than Eq. (73) on the computer because it is only necessary for the computer to make decisions concerning the relationships between elements once, instead of making these decisions every time $\mathbf{B}(\mathbf{k})$ is transformed by one of the $\mathbf{T}(\mathbf{k}, i)$'s.

Before the symmetry reduction is started, subroutine PGWV determines whether the wave vector is on the BZ boundary and whether $\mathbf{k} = 1/2\mathbf{h}$. If the latter is true, than $\mathbf{B}(\mathbf{k})$ must be real. A message is printed out if either of these conditions is satisfied. The time reversal invariance condition contained in the square brackets of Eq. (74) is carried out in subroutine TRINV.

B. Use of Random Numbers

The symmetry reduction of the dynamical matrix is the most difficult part of the analysis to do on the computer because of the algebra involved. One method of handling the unknown $\mathbf{B}(\mathbf{k})$ is to represent it by a matrix of random numbers. In GROUP2 the FORTRAN library function RANF(0) is used to fill an array $G(I)$ with random numbers, where $I = 1$ to $p(p + 1)/2$. This array is then multiplied by 10^5 and loaded into the upper triangle of a square array $P(I, J)$. The diagonal elements are made real and the lower triangle is defined as the complex conjugate of the upper triangle to make $P(I, J)$ hermitian. Multiplication by 10^5 brings all the random numbers to values which should be greater than unity so that they may be converted to integers later. GROUP2 is dimensioned to handle 60×60 matrices. The maximum number of random numbers needed is 3660. Precautions are taken that the random numbers entering $G(I)$ have real and imaginary parts which are greater than 10^{-5} and that no two numbers in the array are closer together than 10^{-4}. The matrix of random numbers is then symmetry reduced using Eq. (74).

C. Conversion of Random Numbers to Symbols for Printout

After the random hermitian matrix has been simplified using Eq. (74), we are left with a matrix of complex numbers and zeros. For ease of comparison, we find the magnitudes of the complex numbers and store them in an amplitude matrix DA(I, J). The phase angles, determined from the relation $\tan^{-1}(\mathrm{Re}(B_{\alpha b})/\mathrm{Im}(B_{\alpha b}))$, are stored in the array DP (I, J). We know that some elements are related to others by phase factors and therefore we compare only the magnitude of the elements with each other. If two or more numbers in DA(I, J) are equal to within a small uncertainty, we set the numbers equal to the same integer. This comparison proceeds from left to right and from top to bottom. The integers assigned to the amplitudes are incremented from 1 to NSY, where NSY is the number of independent amplitudes in the symmetry-reduced $\mathbf{B}(\mathbf{k})$. Thus DA(1, 1) should always be 1 unless $B_{1,1}$ is required by symmetry to be zero. This conversion of amplitudes to a sequence of integers makes conversion to alphabetic characters for printout possible. The printout contains a different letter of the alphabet for each different amplitude. If there are more than 26 independent amplitudes, two letters are used for each amplitude, e.g., AA, AB, AC, etc.

This printout format makes it easy for the user to find elements that are related. Following the printout of the matrix of symbols representing the amplitudes, we print the matrix of phase angles. The phase is printed as an angle in degrees. The user must determine whether the phase is specified by symmetry or is arbitrary. For instance, in the Mg structure along the line U (Warren and Worlton, 1973), where $\mathbf{k}/k_{\max} = [01\zeta]$, with $\zeta = 0.2$, the phases are all determined by symmetry and involve combinations of $\pi/2$, π, and $\zeta\pi/2$. On the other hand, the phase of the off-diagonal elements along the line R in Mg are not determined by symmetry. We have not found a good way to replace the phases by symbols or to recombine the amplitude and phase information.

In most cases the nonzero elements that are related can all be expressed as a phase factor times one other element. Of the crystals which we have studied (Warren and Worlton, 1973), only those with the D_3 point group cannot be expressed this way. For D_3, and presumably C_3, the elements are related by linear equations with more than one term and GROUP2 is unable to determine these relationships. The complete symmetry reduction on the computer for this group of crystals will probably require algebraic methods.

D. Future Developments

It is now possible by using computer languages such as REDUCE or ALTRAN to do algebraic manipulation of matrices. This opens the way for doing the symmetry reduction analytically on the computer. The problem of recombining the amplitude and phase information could be solved in this way. We have not attempted to use these algebraic methods yet, but have approached persons familiar with the languages. It is generally their opinion that symmetry reduction using algebra will be rather time consuming and that it might not be practical to do any but the simplest cases, such as 6×6 GDM's. It remains for some venturesome person to see if this is true.

There is some concern that accidental equalities will occur because of the random numbers used for $\mathbf{B}(\mathbf{k})$. We tried to find a formula which would tell how many independent elements could be expected in the symmetry-reduced GDM. One formula we tried is based on the rule of decomposition of the representation $\mathbf{T}(\mathbf{k})$ into IMOR's. The decomposition formula [Eq. (53)] tells the structure of the block-diagonalized GDM. Each $m(\rho) \times m(\rho)$ block will be hermitian and hence will have $[m(\rho) + 1]m(\rho)/2$ independent elements. This formula says that the block-diagonalized GDM will contain

$$N_{\text{IE}} = \sum_{\rho=1}^{r} [m(\rho) + 1]m(\rho)/2 \tag{75}$$

independent elements. Since the symmetry-reduced GDM differs by a unitary transformation from the block-diagonalized GDM, it was hoped that Eq. (75) would also predict the number of independent elements in the GDM before block diagonalization. It does not. We have found several simple cases that could be checked by hand, but which gave either more or less independent elements than predicted by Eq. (75). We have put a message in GROUP2 which tells the user when the number NSY is more or less than N_{IE}. Anyone who wants to work on finding the correct predictive criterion can compare the large amount of data in our report (Warren and Worlton, 1973).*

* *Note added in proof*: The problem has been solved by R. Cassela (1975), see *Phys. Rev B* **11**, 4795.

V. Projection Operators and Symmetry Coordinates

A. DEFINITIONS

The basis functions of an IMR form a complete set so that any function belonging to that IMR can be expanded in terms of the basis functions. In particular the normal modes or phonon eigenvectors can be expanded in terms of a set of crystal symmetry coordinate vectors, $\{\mathbf{f}(\mathbf{k}, \rho, \mu, \delta)\}$ which transform according to the IMR $\boldsymbol{\tau}(\mathbf{k}, \rho)$. The $\mathbf{f}(\mathbf{k}, \rho, \mu, \delta)$'s satisfy Eq. (61). The eigenvectors can be written

$$\mathbf{e}(\mathbf{k}, \rho, \mu, \delta) = \sum_{\nu=1}^{m(\rho)} \xi(\mu, \nu)\mathbf{f}(\mathbf{k}, \rho, \nu, \delta) \tag{76}$$

The expansion coefficients $\xi(\mu, \nu)$ depend on the dynamics of the problem and cannot be determined from symmetry alone. The p-dimensional symmetry coordinate vectors can be obtained by projection from an arbitrary p-dimensional vector $\boldsymbol{\psi}$,

$$\mathbf{P}(\mathbf{k}, \rho, \delta) \cdot \boldsymbol{\psi} = \sum_{\mu=1}^{m(\rho)} c(\mu)\mathbf{f}(\mathbf{k}, \rho, \mu, \delta) \tag{77a}$$

where

$$\mathbf{P}(\mathbf{k}, \rho, \delta) = [d(\rho)/\bar{g}] \sum_{i=1}^{\bar{g}} \theta_{\delta\varepsilon}{}^*(\mathbf{k}, \rho; \bar{i})\mathbf{T}(\mathbf{k}, \bar{i}) \tag{77b}$$

The coefficients $c(\mu)$ have in the past been determined intuitively by looking at the form of $\mathbf{P}(\mathbf{k}, \rho, \delta) \cdot \boldsymbol{\psi}$. In order to do the job on the computer, a better method was found. It should be mentioned that one can also define character projection operators $\mathbf{P}(\mathbf{k}, \rho)$ which are obtained from Eq. (77b) by setting $\varepsilon = \delta$ and summing over δ. The use of the character projection operator is less desirable than the full projection operator because Eq. (77a) becomes

$$\mathbf{P}(\mathbf{k}, \rho) \cdot \boldsymbol{\psi} = \sum_{\mu=1}^{m(\rho)} \sum_{\delta=1}^{d(\rho)} c(\mu, \delta)\mathbf{f}(\mathbf{k}, \rho, \mu, \delta) \tag{77c}$$

and there are $d(\rho)$ times as many coefficients to find. The only advantage of $\mathbf{P}(\mathbf{k}, \rho)$ is that it requires only the characters of the IMR's, which are simpler and more readily available in books.

B. Examples

Perhaps the best way to understand how projection operators can be used to generate symmetry coordinates is to do an example. Consider $\mathbf{P}(\Lambda, 1, 1)$ for $CaWO_4$ where Λ stands for the wave vector $\mathbf{k}/k_{max} = \zeta[111]$ and the representation $\rho = 1$ is a 1-dimensional representation which assigns ones to all operations in the PGWV. We shall ignore time reversal symmetry in this example for simplicity. If we operate on an arbitrary 18-dimensional vector $\boldsymbol{\psi}$, we obtain, in the notation of Venkataraman and Sahni (1970), the following

$$\mathbf{P}(\Lambda, 1, 1)\boldsymbol{\psi} = \begin{bmatrix} \mathbf{A} & u\mathbf{A} & 0 & 0 & 0 & 0 \\ u^*\mathbf{A} & \mathbf{A} & 0 & 0 & 0 & 0 \\ 0 & 0 & \mathbf{A} & 0 & u\mathbf{A} & 0 \\ 0 & 0 & 0 & \mathbf{A} & 0 & u\mathbf{A} \\ 0 & 0 & u^*\mathbf{A} & 0 & \mathbf{A} & 0 \\ 0 & 0 & 0 & u^*\mathbf{A} & 0 & \mathbf{A} \end{bmatrix} \begin{bmatrix} \boldsymbol{\psi}_1 \\ \boldsymbol{\psi}_2 \\ \boldsymbol{\psi}_3 \\ \boldsymbol{\psi}_4 \\ \boldsymbol{\psi}_5 \\ \boldsymbol{\psi}_6 \end{bmatrix} \tag{78a}$$

where

$$\mathbf{A} = \begin{pmatrix} 0 & 0 & 0 \\ 0 & 0 & 0 \\ 0 & 0 & 2 \end{pmatrix} \tag{78b}$$

and $u^*u = 1$. In the compact notation we are using, each $\boldsymbol{\psi}_i$ has three components. After carrying out the multiplication indicated we obtain

$$\boldsymbol{\psi}(\Lambda_1) = \begin{bmatrix} A\boldsymbol{\psi}_1 + uA\boldsymbol{\psi}_2 \\ u^*A\boldsymbol{\psi}_1 + A\boldsymbol{\psi}_2 \\ A\boldsymbol{\psi}_3 + uA\boldsymbol{\psi}_5 \\ A\boldsymbol{\psi}_4 + uA\boldsymbol{\psi}_6 \\ u^*A\boldsymbol{\psi}_3 + A\boldsymbol{\psi}_5 \\ u^*A\boldsymbol{\psi}_4 + A\boldsymbol{\psi}_6 \end{bmatrix} \tag{78c}$$

We see immediately that $\boldsymbol{\psi}(\Lambda_1)$ consists of a sum of at least three different symmetry coordinates because of the three different combinations of $\boldsymbol{\psi}_i$'s.

The decomposition formula tells us that there are just three symmetry coordinates to be obtained for this IMR. The first symmetry coordinate is obtained by letting $\boldsymbol{\psi}_3 = \boldsymbol{\psi}_4 = \boldsymbol{\psi}_5 = \boldsymbol{\psi}_6 = 0$, factoring out $(\mathbf{A}\psi_1 + u\mathbf{A}\psi_2)$ and normalizing,

$$\mathbf{f}(\Lambda, 1, 1, 1) = \frac{1}{\sqrt{2}}\begin{bmatrix} \mathbf{1} \\ u^*\mathbf{1} \\ \mathbf{0} \\ \mathbf{0} \\ \mathbf{0} \\ \mathbf{0} \end{bmatrix}, \qquad \text{where} \quad \mathbf{1} = \begin{pmatrix} 0 \\ 0 \\ 1 \end{pmatrix} \tag{79}$$

However it is not necessary to go through this procedure. Normalization of the first nonzero column of $\mathbf{P}(\Lambda, 1, 1)$ yields just the same vector. Once the projection operator matrix has been generated on the computer, the symmetry coordinate vectors can be obtained from the independent columns of $\mathbf{P}(\mathbf{k}, \rho, \delta)$ by a Gram–Schmidt orthonormalization procedure, which is carried out in subroutine GSOP. The orthonormalization procedure is really not necessary in the example given above, but generally it is required. The basic idea behind GSOP is illustrated by the following. Let the columns of $\mathbf{P}(\mathbf{k}, \rho, \delta)$ be designated $\mathbf{p}_i$ and let the symmetry coordinates be labeled $\mathbf{f}_i$. Suppose the first nonzero column of $\mathbf{P}(\mathbf{k}, \rho, \delta)$ is $\mathbf{p}_1$. We set

$$\mathbf{f}_1 = \mathbf{p}_1/|\mathbf{p}_1| \tag{80}$$

Suppose $\mathbf{p}_3$ is the next nonzero column of $\mathbf{P}(\mathbf{k}, \rho, \delta)$. Assume that

$$b\mathbf{f}_2 = \mathbf{p}_3 - a\mathbf{f}_1 \tag{81}$$

If $\mathbf{f}_1$ and $\mathbf{f}_2$ are to be orthonormal, then

$$b(\mathbf{f}_1, \mathbf{f}_2) = (\mathbf{f}_1, \mathbf{p}_3) - a(\mathbf{f}_1, \mathbf{f}_1) = 0 \tag{82a}$$

or

$$a = (\mathbf{f}_1, \mathbf{p}_3) \tag{82b}$$

and b is chosen so as to normalize $\mathbf{f}_2$. This requires

$$|b|^2 = (\mathbf{p}_3, \mathbf{p}_3) - |a|^2. \tag{83}$$

The phase of b is not important and therefore b is taken to be real. Obviously if $\mathbf{p}_3$ is parallel to $\mathbf{f}_1$, then $\mathbf{f}_2$ is zero and we must choose another column of $\mathbf{P}(\mathbf{k}, \rho, \delta)$. The procedure is easily extended to give us $m(\rho)$ vectors $\mathbf{f}_i$.

C. Complications Caused by Time Reversal Invariance

Using time reversal invariant projection operators causes several complications in the computer code. The only apparent advantage is that the coefficients $\xi(\mu, v)$ can be taken to be real numbers. As pointed out in Section III, the form of the IMC's $\boldsymbol{\theta}(\mathbf{k}, \rho; \bar{i})$ in terms of the IMR's depends on the time reversal degeneracy which occurs. There are three cases. Accordingly, the projection operator will have three different forms.

Furthermore since $\mathbf{T}(\mathbf{k}, \bar{i})$ is an operator, the projection operator consists of two parts

$$\mathbf{P} = \mathbf{P}^{(1)} + \mathbf{P}^{(2)}\mathscr{K} \tag{84}$$

where $\mathscr{K}$ is the complex conjugation operator and $\mathbf{P}^{(2)}$ is the contribution to $\mathbf{P}$ from time reversal invariance (TRI). In the first version of the code without TRI, we simply used the columns of $\mathbf{P}^{(1)}$ as trial symmetry coordinate vectors. In the present version we have had to use a more complicated procedure. The columns of $\mathbf{P}^{(1)}$ are multiplied by $ST = (1 + \sqrt{-1})$ and the columns of $\mathbf{P}^{(2)}$ are multiplied by ST^*. The two matrices are then added to obtain the matrix whose columns become the trial symmetry coordinates. For some wave vectors this choice of ST accidently results in too few independent column vectors being available. A second choice of $ST = (1 + \frac{1}{2}\sqrt{-1})$ is then tried. If this choice of ST also gives too few symmetry coordinates, the program stops. This has not happened in any crystals analyzed thus far. This reason for using ST is illustrated by the effect of $\mathbf{P}$ on an arbitrary vector $\boldsymbol{\psi}$,

$$\mathbf{P} \cdot \boldsymbol{\psi} = \mathbf{P}^{(1)} \cdot \boldsymbol{\psi} + \mathbf{P}^{(2)} \cdot \boldsymbol{\psi}^*. \tag{85}$$

For the sake of easy computation, we have chosen a vector $\boldsymbol{\psi}$ which is not quite arbitrary. It is a real vector times the quantity ST. This is why it is sometimes necessary to make a second choice in order to get a complete set.

As we have mentioned, the symmetry coordinates form a square matrix $\mathbf{U}$ which will block-diagonalize the dynamical matrix. In order to do this, the columns of $\mathbf{U}$ must be in the proper order. The columns are given by the vectors $\mathbf{f}(\mathbf{k}, \rho, \delta, \mu)$ where the index on the right varies most rapidly (Worlton and Warren, 1972). Here μ is the multiplicity index, δ is the degeneracy index, and ρ is the IMR index. With the addition of TRI we find that if IMR ρ is degenerate with IMR ρ', it is necessary to reorder the columns of **U** and this is done in subroutine RCOD.

Symmetry coordinates are in general complex numbers so the symmetry coordinate printout includes both the real and imaginary parts. Each symmetry coordinate printed out has three labels printed at the top: IR, D, and M, which correspond to ρ, δ, and μ, respectively. We find that many of the components of the symmetry coordinates are zero. One could save additional space in the printout by only printing the nonzero components. This would be convenient when treating crystals with a large number of atoms per unit cell. Boyer (1974) has such a compact printout format for symmetry coordinates.

VI. Block Diagonalization of the Dynamical Matrix

A. Advantages

The block diagonalization is carried out in subroutine BDODM using the formula

$$(B^{\mathrm{BD}})_{\mathscr{a}\mathscr{b}} = \sum_{l,m=1}^{p} (U^{-1})_{\mathscr{a}l}(B^{\mathrm{SR}})_{lm}U_{m\mathscr{b}}, \tag{86}$$

where the block structure of the block-diagonalized matrix can be determined from the decomposition formula [Eq. (53)]. For example in Mg along the line R, we start with the symmetry-reduced matrix

$$\mathbf{B}^{\mathrm{SR}} = \begin{bmatrix} A & \cdot & \cdot & \cdot & \cdot & C \\ \cdot & B & \cdot & \cdot & C & \cdot \\ \cdot & \cdot & D & \cdot & \cdot & \cdot \\ \cdot & \cdot & \cdot & A & \cdot & \cdot \\ \cdot & \cdot & C^* & \cdot & B & \cdot \\ \cdot & C^* & \cdot & \cdot & \cdot & D \end{bmatrix} \tag{87}$$

The reducible representation $\mathbf{T}(\mathbf{k})$ decomposes into the IMR's $2\boldsymbol{\tau}(\mathbf{k}, 1) + \boldsymbol{\tau}(\mathbf{k}, 2) + \boldsymbol{\tau}(\mathbf{k}, 3) + 2\boldsymbol{\tau}(\mathbf{k}, 4)$ with $\boldsymbol{\tau}(\mathbf{k}, 1)$ and $\boldsymbol{\tau}(\mathbf{k}, 4)$ time reversal degenerate and $\boldsymbol{\tau}(\mathbf{k}, 2)$ and $\boldsymbol{\tau}(\mathbf{k}, 3)$ also time reversal degenerate. Transforming this B with U according to Eq. (86) gives

$$B^{\mathrm{BD}} = \begin{bmatrix} B & C & \cdot & \cdot & \cdot & \cdot \\ C^* & D & \cdot & \cdot & \cdot & \cdot \\ \cdot & \cdot & A & \cdot & \cdot & \cdot \\ \cdot & \cdot & \cdot & A & \cdot & \cdot \\ \cdot & \cdot & \cdot & \cdot & D & C \\ \cdot & \cdot & \cdot & \cdot & C^* & B \end{bmatrix} \tag{88}$$

We have achieved a simplification that is very useful for model fitting. Fitting experimental data to a 2×2 matrix is much easier than fitting to a 6×6 matrix. Besides reducing the time needed for a least squares fit, it prevents correlation of the fitting parameters for modes belonging to different IMR's. Of course one must be able to identify the experimental modes with the IMR's. This is usually possible through selection rules and compatibility relations. Selection rules for neutron scattering have been discussed by Casella and Trevino (1972). They have also written a computer program based on this theory (S. F. Trevino and R. C. Casella, unpublished, 1973).

B. Procedure

We would like to express the printout of the block diagonalized dynamical matrix in terms of the same symbols used for the symmetry reduction. Each matrix element of $\mathbf{B}^{\mathrm{BD}}$ may be a linear combination of at most NSY symbols. The coefficient of the first symbol will be stored in G(1), the coefficient of the second symbol in G(2) and so forth. In general we can write

$$\mathrm{G}(J) = \sum_{L,M=1}^{p} U^*(L, A)U(M, B)\delta[J - \mathrm{DA}(L, M)] \exp[i\mathrm{DP}(L, M)] \tag{89}$$

where DA(L, M) and DP(L, M) are the amplitude and phase matrices constructed during the symmetry reduction. It will be recalled that the elements of DA are a sequence of integers identifying the nonzero, nonequivalent elements of $\mathbf{B}^{\mathrm{SR}}$. Equation (89) gives the coefficients of the symbols for the A, Bth element of the block diagonalized matrix. If the coefficient is nonzero then the coefficient and the associated symbol is printed out. For example, the printout for a typical matrix element might be

$$B(1, 2) = (0.707, 0.000)D + (0.707, 0.000)E \tag{90}$$

where "D" and "E" are symbols and the complex coefficients are G(4) and G(5), respectively.

Since the block-diagonalized dynamical matrix will still be hermitian, we need only print out the matrix elements for $1 \leqslant A \leqslant m(\rho)$ and $A \leqslant B \leqslant m(\rho)$. Following each block, a message is printed out which identifies the IMR and tells how many times the block occurs on the diagonal $d(\rho)$. Since the block diagonalization depends on the symmetry reduction it will be incomplete if the symmetry reduction is incomplete. In the case of crystals with the D_3 point group, more symbols than necessary will be printed out, but the elements of $\mathbf{B}^{BD}$ will still have the correct relationship to the symbols printed for $\mathbf{B}^{SR}$.

VII. Optical Selection Rules and Acoustic Mode Identification

A. Optical Selection Rules

Much of the experimental work being done in lattice dynamics involves the optical techniques of infrared absorption and Raman scattering. Only modes of certain symmetry types are observable and the selection rules for these processes may be determined by group theory. We have included in GROUP2 an analysis of the selection rules for these processes when $\mathbf{k} = \mathbf{0}$. Higher order effects may be observed at other symmetry points but we have not taken these into account.

In order for a transition to take place symmetry requires that the irreducible representation of the ground state be in the same subspace as the reducible representation of the physical process involved. In terms of characters, the condition is

$$\sum_{i=1}^{g} \chi(i)\chi(\rho; i) > 0, \tag{91}$$

where $\chi(i)$ is the character of the reducible representation for the physical process and $\chi(\rho, i)$ is the character of the ρth IMR.

Infrared absorption involves electric dipole transitions. The electric dipole moment operator $e\mathbf{r}$ is a polar vector which means that its components transform as the coordinates x, y, and z. These transformations can be represented by the 3×3 rotation matrices that are generated by GROUP2. Thus $\chi(i)$ in Eq. (91) is taken as the trace of the matrices $\mathbf{R}(i)$.

Raman scattering involves the optical polarizability which transforms as the bilinear components x^2, y^2, z^2, xz, and yz. It was shown by Bhagavantam (1941) that the character of the reducible representations of these transforms can be written as

$$\chi(i) = \pm 2 \cos \phi_i(1 + 2 \cos \phi_i), \tag{92a}$$

where ϕ_i is the rotation angle, and the plus sign is for rotations, while the minus sign is for rotoinversions. Rotations are numbered 1 through 12 for hexagonal subgroups and 1 through 24 for cubic subgroups in Kovalev's scheme (1964). The other operations are rotoinversions and have determinant -1 instead of $+1$.

Equation (92a) can be written in terms of the trace of the 3×3 rotation matrices as

$$\chi(i) = [\mp 1 + \mathrm{Tr}(\mathbf{R}(i))]\, \mathrm{Tr}(\mathbf{R}(i)) \tag{92b}$$

Substitution of Eq. (92b) into Eq. (91) gives the selection rules for Raman scattering.

B. Acoustic Mode Identification

At the center of the BZ the frequencies of three modes of vibration always go to zero. These are the acoustic modes and they involve all the atoms moving together in phase. If the IMR to which an acoustic mode belongs only occurs once, the symmetry coordinate vector will be an eigenvector. GROUP2 can easily identify the mode as acoustic. When the IMR occurs more than once, there will not necessarily be a symmetry coordinate corresponding to the acoustic eigenvector and GROUP2 will not print out an identifying message. In this case the user must inspect the symmetry coordinates to see which ones could combine to give an acoustic mode. It would be possible to modify the subroutine that creates the symmetry coordinates so that it always gives three acoustic modes when $\mathbf{k} = \mathbf{0}$, but this has been left as a future improvement.

C. A Problem of Optic Modes at the BZ Center

It should be mentioned that our program will not predict the splitting of optical modes at the BZ center in ionic crystals. For instance, it is well known in NaCl that the longitudinal optic mode is quite a bit higher in frequency than the degenerate transverse optical modes. The group theory prediction is that all three optical modes belong to the same 3-dimensional IMR and should therefore be degenerate. The optical frequencies are actually measured at a finite, though very very small, wave vector where group theory would predict a splitting into one 1-dimensional IMR and one 2-dimensional IMR. The optical selection rules at $\mathbf{k} = \mathbf{0}$ are adequate for ionic crystals in the sense that some, but not all, of the modes belonging to a given IMR may be infrared or Raman active. The situation in ionic crystals belonging to the hexagonal system is more confusing, because the degeneracy of the transverse modes is dependent on the direction of the small wave vector associated with the optical measurement.

VIII. Some Uses of Symmetry Coordinates

Besides using the symmetry coordinates to provide the unitary transformation which block-diagonalizes the GDM, there are several other uses of these vectors. The most obvious is to give the physicist an intuitive idea of how the atoms are moving in a given mode. Let us take an example from the motions of CaF_2 at the zone center. There are three atoms per unit cell or nine degrees of freedom. The decomposition is $\mathbf{T}(\mathbf{0}) = \tau(\mathbf{0}, 7) + 2\tau(\mathbf{0}, 10)$. Both $\tau(\mathbf{0}, 7)$ and $\tau(\mathbf{0}, 10)$ are 3-dimensional representations. Warren and Worlton (1973, p. 493) show that the symmetry coordinates tell us that in the $\tau(\mathbf{0}, 7)$ modes the Ca ions are standing still and the F ions are vibrating against one another in any one of three orthogonal directions. The modes belonging to $\tau(\mathbf{0}, 10)$ are mixtures of acoustic and optic modes. We shall not go into this any further.

The symmetry coordinates could also be used as input to a program which calculates the one-phonon neutron scattering cross section, but apparently S. F. Trevino and R. C. Casella (unpublished, 1973) do not require symmetry coordinates as input to their program.

The final use that we want to mention is in the calculation of selection rules for two-phonon-assisted optical transitions. The theory is given by Maradudin and Vosko (1968, pp. 35–36). All the group-theoretical machinery exists in our program to evaluate these selection rules. It remains for someone with sufficient interest to make the application.

References

Bhagavantam, S. (1941). *Proc. Indian Acad. Sci., Sect. A* **13**, 543.
Boyer, L. L. (1974). *J. Comput. Phys.* **16**, 167.
Bradley, D. J., and Cracknell, A. P. (1972). "The Mathematical Theory of Symmetry in Solids." Oxford Univ. Press (Clarendon), London and New York.
Casella, R. C., and Trevino, S. F. (1972). *Phys. Rev. B* **6**, 4533.
Chen, S. H. (1964). Ph.D. Thesis, Physics Dept., McMaster University, Hamilton, Canada (unpublished).
Henry, N. F., and Lonsdale, K. (1965). "International Tables for X-Ray Crystallography," Vol. I. Kynock Press, Birmingham, England.
IRE Standards. (1949). *Proc. IRE* **37**, 1378.
Kovalev, O. V. (1964). "Irreducible Representations of the Space Groups." Gordon & Breach, New York.
Maradudin, A. A., and Vosko, S. H. (1968). *Rev. Mod. Phys.* **40**, 1.
Maradudin, A. A., Ipatova, I. P., Montroll, E. W., and Weiss, G. H. (1971). "Theory of Lattice

Dynamics in the Harmonic Approximation," 2nd ed., Solid State Phys., Suppl. 3. Academic Press, New York.
Miller, S. C., and Love, W. F. (1967). "Tables of the Irreducible Representations of Space Groups and Corepresentations of Magnetic Groups." Pruett Press, Boulder, Colorado.
Neto, N. (1973). *Acta Crystallogr., Sect. A* **29**, 464.
Sahni, V. C., and Venkataraman, G. (1970). *Phys. Kondens. Mater.* **11**, 199.
Tinkham, M. (1964). "Group Theory and Quantum Mechanics." McGraw-Hill, New York.
Venkataraman, G., and Sahni, V. C. (1970). *Rev. Mod. Phys.* **42**, 409.
Vilenkin, N. J. (1968). "Special Functions and the Theory of Group Representations." Amer. Math. Soc., Providence, Rhode Island.
Warren, J. L. (1968). *Rev. Mod. Phys.* **40**, 38.
Warren, J. L. (1974). *Phys. Rev. B* **9**, 3603.
Warren, J. L., and Worlton, T. G. (1973). "Symmetry Properties of the Lattice Dynamics of Twenty-Three Crystals," USAEC Rep., ANL-8053 and LA-5465-MS. U.S. At. Energy Comm., Argonne, Illinois.
Warren, J. L., and Worlton, T. G. (1974). *Comp. Phys. Comm.* **8**, 71.
Wigner, E. P. (1959). "Group Theory and its Application to the Quantum Mechanics of Atomic Spectra." Academic Press, New York.
Worlton, T. G. (1972). *Comput. Phys. Commun.* **4**, 249.
Worlton, T. G. (1973). *Comput. Phys. Commun.* **6**, 149.
Worlton, T. G., and Warren, J. L. (1972). *Comput. Phys. Commun.* **3**, 88; errata in **4**, 382.
Yanagawa, S. (1953). *Progr. Theor. Phys.* **10**, 83.

Lattice Dynamics and Related Properties of Point Defects*

R. F. Wood

SOLID STATE DIVISION, OAK RIDGE NATIONAL LABORATORY
OAK RIDGE, TENNESSEE

I. Introduction

A. Introductory Comments

The presence in materials of point defects such as substitutional and interstitial impurities, self-interstitials, vacancies, and color centers can strongly

* Research sponsored by the U.S. Atomic Energy Commission under contract with Union Carbide Corporation.

modify the physical properties of the materials. The influence of such defects is particularly obvious in insulating crystals where they frequently give rise to a characteristic coloration. The control of the electrical properties of semiconductors by doping is another classical example. The effects observed in these two examples are associated primarily with changes of the electronic structure of the crystals. Modifications of the vibrational properties are less obvious but easily observed by a variety of physical measurements. They are important on a microscopic level because they provide detailed information about the interatomic forces in the crystal and their dependence on the electronic structure. They can also strongly effect the electronic properties of the materials by way of the electron–phonon coupling as, for example, in superconductors. The theory of the lattice dynamics of crystals containing point defects has been well developed in recent years. The application of the theory to certain classes of problems is now possible because of the rapidly accumulating store of information about phonons in perfect crystals obtained from the inelastic scattering of thermal neutrons and the easy accessibility to large, fast computers. The purpose of this article is to demonstrate how applications of the theory can utilize perfect crystal data and adequate computational facilities to lead, in many cases, to quite good agreement with the results of various types of experiments.

In principle, neutron scattering is the most direct experimental technique for studying the lattice dynamics of both perfect and imperfect crystals. However, the sensitivity is such that even with the highest flux reactors and the best spectrometers presently available, such as those at Oak Ridge, Brookhaven, and Grenoble, defect concentrations of 1 to 2 atomic percent are required to induce observable effects. Although this allows a wide range of problems to be investigated by neutron scattering it prevents the method from being applied to crystals with the very low concentrations of vacancies, interstitials, and impurity ions that are of interest in many solid state phenomena. In insulating crystals, the effects of defect concentrations as low as 10^{-5} atomic percent are easily measured by optical techniques and by spin resonance methods when unpaired electron spins are present. The defects frequently give rise to absorption and emission bands in or near the visible region which are not present in the perfect crystal. Careful measurements of band moments and their temperature dependence provide an indirect probe of the perturbed phonons in the vicinity of the defect. If the defect–host interaction is strong, a large number of phonons will accompany the electronic transition and in such cases only average quantities (e.g., frequencies) of the phonon field can be obtained. If, on the other hand, the interaction is weak the bands may show vibronic structure from which effective, perturbed one-phonon densities of states of various symmetries can be extracted by a deconvolution procedure. This method is often useful but it is rather clumsy and indirect. Infrared absorption and Raman scattering give much more direct information about the perturbed phonons. With the commercial availability of lasers, improvements in detection systems and automation of data gathering systems via small on-line digital computers, infrared and Raman techniques have now become highly refined. They give detailed information about one phonon spectra in perturbed crystals and they tend to complement one another both experimentally and theoretically. In this article, results from infrared, Raman, and neutron scattering experiments will be used in the study of the lattice dynamics of point defects in solids.

B. Localized Perturbation Theory and the Coherent Potential Approximation

The introduction of a defect into a crystal destroys the translational invariance of the perfect lattice. Bloch's theorem is therefore no longer applicable and the group theoretical factorization of the electronic and vibrational secular equations is determined by the point symmetry group of the defect rather than by the full space group of the crystal. Without further simplification, the block-diagonalized secular determinants would be so large that it would be virtually impossible to calculate the physical properties of crystals containing point defects. A theoretical formulation is needed which retains the simplifications associated with translational invariance insofar as possible. This suggests a type of perturbation theory with the perfect crystal as the unperturbed system. Lifshitz (1956) developed such a theory for lattice dynamical problems and, independently, Koster and Slater (1954) gave a roughly equivalent development for electronic problems. The feasibility of applying these theoretical formulations to real problems depends on the condition that the perturbation of the crystal by the defect be well localized in a region around the defect. Hence, the whole theory is often referred to as localized perturbation theory. Many people have contributed to the development and refinement of this theory and references to some of the work will be given below. However, since this article is not intended to be a review of the theory, some readers may want to consult review articles such as those by Izyumov (1965) and Maradudin *et al.* (1971).

Closely related to the question of localization is that of concentration. The original formulation of localized perturbation theory assumed the concentration of defects to be so low that each defect could be treated as completely isolated in the crystal. The necessary conditions of localization and concentration are probably most often met when studying the optical, infrared, and Raman properties of ionic crystals containing substitutional impurities of very nearly the same charge as the ion replaced. In a metal, however, the lattice dynamics have to be studied by inelastic neutron scattering and, as already mentioned, this will require a concentration of 1–2% under the most favorable conditions. At such concentrations, even if the perturbation is highly localized, it must be assumed that the probability of the impurities interacting is nonnegligible. The conditions for the original form of localized perturbation theory are therefore not met and modifications suitable for dilute alloys must be sought.

Attempts to extend the basic ideas of localized perturbation theory to finite concentrations led to a major step forward in the field of alloy theory. The ideas and approximations involved in this extension were given simultaneously by Taylor (1967) for phonons and Soven (1967) for electrons and are now generally referred to as the coherent potential approximation (CPA). The basic idea of the CPA is to replace the real alloy by an effective crystal chosen so that scattering of phonons or electrons from all constituents of the alloy is zero on the average. This leads to a self-consistency requirement that makes the equations more complicated than those for standard localized perturbation theory, but not intractable. The CPA is a correct theory of alloys as far as it goes, and a number of lattice dynamical calculations within its framework have been published. The basic approach has so many elements in common with localized perturbation theory that many of the numerical problems discussed in this article are also applicable to the CPA calculations. Hence, a brief summary of the CPA and some of its applications will be included here.

C. Scope of the Article

This article is not a review of developments in the theory of the lattice dynamics of crystals containing point defects. Rather, its purpose is to set forth and discuss a set of computational procedures and numerical techniques that can be utilized in the application of the theory to certain types of real systems. The discussion is based almost entirely on the experiences gathered by myself and my colleagues in the Theory Section of the Solid State Division at Oak Ridge National Laboratory. We have carried out extensive calculations on the infrared, Raman, and neutron scattering properties of crystals containing point defects; we have also been actively engaged in the calculation of generalized electronic and magnetic susceptibilities by somewhat similar techniques. Of course, the development of our methods have followed fairly closely trends in the literature so that the restriction of the discussion to our own work is not a severe one and, in fact, it is almost a necessity in view of the heavy emphasis on computational procedures.

Our point defect work in the area of lattice dynamics has been mostly limited to substitutional impurities in cubic crystals and the illustrative examples used here will be confined to such cases. However, much of the general discussion and many of the computational approaches are applicable to a much larger class of problems. The scope of the paper is intended to be wide enough to give a clear indication of how to proceed with a calculation, yet not so narrow and detailed as to yield a cookbook recipe. Indications of computer time and core storage will be given toward the end of the article, but since these are not critical for present day computers, the discussion will be brief. Comparisons between theory and experiment will also be given so the reader can see how well the calculations do in selected cases.

In the next section, the bare essentials of the equations needed for the examples to be discussed will be presented. Apart from giving a very sketchy motivation for the equations, the main point of this section is to indicate that the physical quantities all involve the imaginary part of the perturbed Green's function or closely related quantities. The number of such physical quantities is much greater than the four listed. In Section III, a sketch of localized perturbation theory, formulated within the symmetry coordinate representation, will be given. Some remarks on the application of the coherent potential approximation and its relation to the isolated defect case will be made. Computational considerations will be discussed in Section IV with particular emphasis on how a typical calculation might proceed. In Section V, specific examples will be discussed and comparisons between theory and experiment given.

II. Measurable Quantities in Terms of Single-Particle Green's Functions

A. Far Infrared Absorption

The first-order, resonance absorption of infrared radiation by lattice vibrations in a diatomic, cubic crystal such as an alkali halide leads to a single absorption band at the *reststrahl* frequency. In the harmonic approximation, the absorption consists of a δ-function spike, but in a real crystal it is broadened by a number of effects. When an impurity destroys the translational invariance of the host lattice, the so-called $q = 0$ selection rule no longer holds and absorption can occur throughout the entire range of phonon frequencies of the crystal. The absorption may show in-band resonances of varying degrees of sharpness and true local modes both above the continuum and in the gap (if one exists) between the acoustical and optical regions. Absorption peaks due to local mode and sharp resonances may be considerably broadened by the anharmonic interaction between the impurity and host crystal. If the local mode is well removed from the continuum of band modes and if the anharmonic interaction between the impurity and host is strong, one-phonon side bands may be observed; these will be considered in the next subsection.

Klein (1968) has derived the following expression for the impurity-induced infrared absorption coefficient of an insulator:

$$\alpha(\omega) = NK(\omega)\,\mathrm{Im}\langle T0, 0|t(\omega^2 + io^+)|T0, 0\rangle, \tag{1}$$

with

$$K(\omega) \propto \omega/[n(\omega)(\omega_0{}^2 - \omega^2)^2]. \tag{2}$$

N is the defect concentration per unit volume; $n(\omega)$ is the real part of the complex refractive index which can be assumed to have a simple form such as

$$n^2(\omega) = \varepsilon_\infty + (\varepsilon_0 - \varepsilon_\infty)[1 - \omega^2/\omega_0{}^2]^{-1}; \tag{3}$$

ε_0 and ε_∞ are the static and high frequency dielectric constants, respectively; ω_0 is the frequency of the transverse optical mode at $q = 0$ and $|T0, 0\rangle$ denotes the corresponding eigenvector. $t(\omega^2 + i\varepsilon)$ is the t matrix within the defect subspace (defined later) of an isolated impurity and the notation in Eq. (1) indicates that the imaginary part is to be found from the limit as ω approaches the real axis from above. In the formal theory of scattering (Messiah, 1962), the t matrix is related to the unperturbed Green's function G^0 by

$$T = \Delta(1 + G^0\Delta)^{-1} \tag{4}$$

where Δ is the perturbation matrix. The perturbed G is given by

$$G = G^0 - G^0 T G^0. \tag{5}$$

In the calculation of the total absorption it is the imaginary part of the perturbed Green's function which enters, but in the impurity-induced part the expression reduces to Eq. (1).

B. Anharmonic Sidebands of Local Modes

Substitutional H^- ions in ionic crystals are the most thoroughly studied systems that have a localized vibrational mode accompanied by anharmonic sidebands. These defects, also known as U centers, have both ultraviolet bands due to electronic transitions and infrared bands due to vibrational transitions associated with them. The vibrational properties of U centers have proved to be especially interesting. The H^- ion oscillates at a frequency well above that of the highest in-band mode. Sidebands containing pronounced structure accompany the local mode absorption. Since the H^- ions destroy the translational invariance of the host crystal, absorption also occurs in the in-band region of the spectrum. Because of the light mass of H^-, the U center was quickly recognized as an interesting system on which to test the theory of localized perturbations. The first calculations of the local mode frequency treated the H^- ion in the mass defect approximation (MDA). Later calculations showed that force constant changes must also be considered in order to explain both the local mode frequency and the structure in the sidebands.

Timusk and Klein (1966) have derived an expression for the absorption coefficient in the one-phonon sideband region of the U-center local mode. They point out that the lowest order anharmonic interaction that will produce the bands gives a contribution to the Hamiltonian of the form

$$H_{\text{int}} = BQ^2X/2, \tag{6}$$

where Q is the local mode dynamical coordinate of T_{1u} symmetry and X is a configuration coordinate which must have even parity. Because of the extremely light mass of the H^- ion, it is sufficient to limit Q to any one of the cartesian displacements of that ion. Also, since the anharmonic coupling occurs almost entirely through the short range forces, it is a good approximation to include only first-nearest-neighbor (1nn) ions in the coordinate X. One then finds that Q couples only to A_{1g} and E_g lattice modes. B in Eq. (6) is the anharmonic coupling coefficient, which can be absorbed in an overall normalization factor. The expression obtained by Timusk and Klein can be written as

$$l^+(\omega) = (hB^2/4\pi M^2\omega^2\Omega^2)\,\text{Im}\langle X|G(\omega^2 + io^+)|X\rangle, \tag{7}$$

where M is the mass of the hydrogen atom and Ω is the local mode frequency with respect to which ω is given. As will be shown later, G can be expressed in terms of the unperturbed Green's function G^0 as

$$G = (1 + G^0\Delta)^{-1}G^0. \tag{8}$$

Δ is the change in the dynamical matrix resulting from the introduction of the impurity; its form will be considered in more detail in the next section.

C. Raman Scattering

The theory of Raman scattering is basically more complicated than the theory of infrared absorption and even a sketch of the theory can quickly become quite involved. Some indication of the complexity can be gained from an equation for the intensity of the scattered light which can be derived, for example, from the discussion of Raman scattering given in Born and Huang (1954).

The intensity of the scattered light of frequency ω_s is given by

$$\mathscr{I}(\omega_s) = \frac{\omega_i^{\,4}}{2\pi c^3} \sum_{\alpha\beta\gamma\lambda} \hat{\eta}_\alpha \hat{\eta}_\beta I_{\alpha\gamma,\beta\lambda}(\omega) E_\gamma(\omega_i) E_\lambda^*(\omega_i) \tag{9}$$

with

$$I_{\alpha\gamma,\beta\lambda}(\omega) = \mathrm{Av}_m \sum_n \langle m|P_{\beta\lambda}|n\rangle\langle n|P_{\alpha\gamma}^*|m\rangle\, \delta(\omega - \omega_{mn}). \tag{10}$$

The various symbols in these two expressions are defined as follows: ω_i is the frequency of the incident light; ω is the frequency difference between incident and scattered light; $\hat{\boldsymbol{\eta}}$ is a unit vector parallel to the electric field of the *scattered* light; $\mathbf{E}(\omega_i)$ is the Fourier component of the *incident* electric field; $|m\rangle, |n\rangle$ are vibrational wavefunctions whose difference in frequency is given by ω_{mn}; $\mathrm{Av}_m \sum_n$ means that an average over initial and a sum over final vibrational states is to be taken; $P_{\alpha\gamma}$ is the transition polarizability tensor given by

$$P_{\alpha\gamma}(\mathbf{u}) = \sum_{\mu\neq 0} [\hbar\omega_{\mu 0}(\mathbf{u})]^{-1}\{\langle 0|M_\alpha(\mathbf{u})|\mu\rangle\langle\mu|M_\gamma(\mathbf{u})|0\rangle + \text{c.c.}\}, \tag{11}$$

in which $|0\rangle$ and $|\mu\rangle$ are electronic wavefunctions and $\hbar\omega_{\mu 0}$ is an electronic transition energy. Both $\omega_{\mu 0}(\mathbf{u})$ and $\langle 0|M_\alpha(\mathbf{u})|\mu\rangle$ are functions of nuclear positions denoted collectively by $\mathbf{u}$. It is usually a good approximation to ignore the $\mathbf{u}$ dependence of the matrix element (the so-called Condon approximation) but the dependence of $\omega_{\mu 0}$ on $\mathbf{u}$ cannot be neglected and will give rise to most of the scattering. It should also be noted that $P_{\alpha\gamma}(\mathbf{u})$ is both real and symmetric.

A thorough understanding of the physical significance of Eqs. (9), (10), and (11) requires thoughtful study and here only one or two aspects of the scattering will be pointed out. Since ω is the difference between the scattered and incident light, i.e., $\omega = \omega_s - \omega_i$, a contribution from the δ function in Eq. (10) occurs whenever $\omega_s = \omega_i - \omega_{mn} \cdot \omega_{mn}$ is the difference between two phonon frequencies so that the scattering considered here is from the lattice vibrations. On the other hand, from Eq. (11) it can be seen that the electronic states are also involved in the process so that one can also say that the scattering goes by way of virtual transitions to excited electronic states (resonance Raman effects are neglected). Another way of looking at the process is the following. The polarizability of the crystal can be described in terms of virtual electronic excitations whose energies and dipole matrix elements depend on the positions of the ions in the crystals, that is, on $\mathbf{u}$. The expansion of $P_{\alpha\gamma}(\mathbf{u})$ in a power series in $\mathbf{u}$ will lead to a classification of scattering processes according to the number of phonons involved. Expansions of this type are basic to the calculation of all physical properties involving phonons and it may be useful to sketch the procedure here.

$P_{\alpha\gamma}(\mathbf{u})$ can be expanded as

$$P_{\alpha\gamma}(\mathbf{u}) = P_{\alpha\gamma}(0) + \sum_{\nu l} \left.\frac{\partial P_{\alpha\gamma}}{\partial u_\nu(l)}\right|_0 u_\nu(l) + \cdots. \tag{12}$$

$u_\nu(l)$ is the νth cartesian component of the displacement of the ion at the lth site. $P_{\alpha\gamma}(\mathbf{u})$ can be written in terms of the normal coordinates by using the transformation

$$u_\nu(l) = \sum_j \varepsilon_\nu(l;j)v_j \tag{13}$$

in which v_j is the jth normal coordinate (not necessarily of the Bloch form) and $\varepsilon_\nu(l;j)$ is an element of the transformation matrix. Equation (12) becomes

$$P_{\alpha\gamma}(\mathbf{u}) = P_{\alpha\gamma}(0) + \sum_{\nu lj} \left.\frac{\partial P_{\alpha\gamma}}{\partial u_\nu(l)}\right|_0 \varepsilon_\nu(l,j)v_j + \cdots. \tag{14}$$

The proliferation of subscripts in the above equation indicates how complicated the Raman scattering can become. Fortunately, the impurity-induced scattering comes already from the first-order terms in v_j. The zeroth-order term will not induce any phonon transitions since it does not depend on v; it gives rise to elastic or Rayleigh scattering. The next step is to consider the matrix elements of the first order term, i.e.,

$$\langle n|P_{\alpha\gamma}|m\rangle = \sum_{\nu lj} \left.\frac{\partial P_{\alpha\gamma}}{\partial u_\nu(l)}\right|_0 \varepsilon_\nu(l;j)\langle n1, n2, \cdots |v_j|m1, m2 \cdots\rangle \tag{15}$$

In this equation, the total vibrational wavefunction has been written as a product of one-particle functions with a one-to-one correspondence between n and the set $\{n1, n2 \cdots\}$ implied. For simplicity it will be assumed that the scattering is measured at 0°K so that the average over m in Eq. (10) is unnecessary. Hence, the matrix element of v_j will be zero unless the final state is one in which the jth oscillator has been raised to its first excited state. Equation (15) then reduces to

$$\langle n|P_{\alpha\gamma}|m\rangle = \sum_{\nu l} \frac{\partial P_{\alpha\gamma}}{\partial u_\nu(l)} \varepsilon_\nu(l;j)(\hbar/2\omega_j)^{1/2} \tag{16}$$

and in the δ function of Eq. (10), $\omega_{mn} = \omega_j$. The expression for $I_{\alpha\gamma,\beta\lambda}$ becomes

$$I_{\alpha\gamma,\beta\lambda} = \sum_{\nu l}\sum_{\mu l'} \frac{\partial P_{\alpha\gamma}}{\partial u_\nu(l)} \frac{\partial P_{\beta\lambda}}{\partial u_\mu(l')} \sum_j \frac{\varepsilon_\nu(l;j)\varepsilon_\mu(l';j)\hbar\,\delta(\omega - \omega_j)}{2\omega_j}. \tag{17}$$

The factor involving the sum over j is just the imaginary part of a matrix element of the Green's function and so

$$I_{\alpha\gamma,\beta\lambda} = \sum_{\nu l}\sum_{\mu l'} \frac{\partial P_{\alpha\gamma}}{\partial u_\nu(l)} \frac{\partial P_{\beta\lambda}}{\partial u_\mu(l')} \operatorname{Im}\langle \nu l|G(\omega^2 + io^+)|\mu l'\rangle. \tag{18}$$

Thus, the first-order Raman scattering involves the one-phonon Green's functions in much the same way as the expressions for the infrared absorption.

D. Neutron Scattering

The lattice dynamics of many nominally pure crystals have now been studied by inelastic neutron scattering. The phonon dispersion curves thus obtained can be used in a least-squares procedure to determine the parameters in various lattice dynamical models. Once these parameters are known, the eigenfrequencies and eigenvectors at general points in the Brillouin zone (BZ) can be obtained. It is this information which is needed in the application of the theoretical expressions given in the preceding three subsections. On the other hand, neutron scattering itself is the most direct method for obtaining information about the perturbed as well as the unperturbed lattice if the concentration of the defects is sufficiently high. The theory of neutron scattering from crystals containing defects has been discussed by several authors, e.g., Elliott and Taylor (1964, 1967) and Lakatos and Krumhansl (1968, 1969).

The one-phonon contribution to the cross section for the inelastic scattering of thermal neutrons from the crystal into the solid angle $d\Omega$ is given by

$$\frac{d^2\sigma}{d\Omega\, d\omega} = \frac{1}{2\pi}\left|\frac{\mathbf{k}_2}{\mathbf{k}_1}\right| S(\mathbf{k}, \omega). \tag{19}$$

$\mathbf{k}_1$ is the initial wavevector of the neutron; $\mathbf{k}_2$ is the wavevector after the scattering event and, therefore, $\mathbf{k} = \mathbf{k}_2 - \mathbf{k}_1$ is the change in wavevector; the energy transfer is $\hbar\omega = \hbar({k_2}^2 - {k_1}^2)/2m$, where m is the mass of the neutron. The general form of the scattering function $S(\mathbf{k}, \omega)$ was introduced by van Hove (1954). In order to simplify the equations somewhat, it will again be assumed that the scattering occurs at 0°K so that only phonon excitation processes can occur. Then $S(\mathbf{k}, \omega)$ can be written as

$$S(\mathbf{k}, \omega) = \frac{1}{N} \sum_{ll'} A_l A_{l'} \sum_{\alpha\beta} k_\alpha k_\beta \operatorname{Im} G_{\alpha\beta}(ll'; \omega) \times \exp\{i\mathbf{k}\cdot[\mathbf{R}(l) - \mathbf{R}(l')]\} \tag{20}$$

with

$$\operatorname{Im} G_{\alpha\beta}(ll'; \omega) \equiv \operatorname{Im}\langle \alpha l | G(\omega^2 + io^+) | \beta l' \rangle$$

A_l is an effective, temperature-dependent scattering length given by

$$A_l = a_l \exp[-\tfrac{1}{2}\langle \mathbf{k}\cdot\mathbf{u}(l)\rangle^2] \tag{21}$$

in which a_l is the actual scattering length of the atom at site l. a_l for the impurity atoms will differ from that of the atoms of the host crystal and this will lead to coherent and incoherent (angle-independent) impurity-induced scattering. However, both contributions involve the calculation of $\operatorname{Im} G_{\alpha\beta}(ll'; \omega)$ and for the purposes of this article it is not necessary to discuss them separately.

III. Localized Perturbations

A. Outline of Localized Perturbation Theory

The equation of motion for the ions in a crystal can be written in matrix form in the harmonic approximation as

$$[M\omega^2 - K]\mathbf{u} = 0 \tag{22}$$

in which the elements of the force constant matrix K are given by

$$K_{\alpha\beta}(l\kappa; l'\kappa') = [\partial^2\Phi/\partial u_\alpha(l\kappa)\, \partial u_\beta(l'\kappa')]_0. \tag{23}$$

α labels the cartesian component of the displacement, $\mathbf{u}(l\kappa)$, of the κth ion in the lth cell and the derivatives of the total potential energy Φ are evaluated at the equilibrium sites of the perfect crystal. Due to translational invariance the solutions to Eq. (22) for the perfect lattice yield normal coordinates that can be written in the Bloch form.

After a single substitutional impurity has been introduced into the otherwise perfect lattice, the new matrix equation of motion can be written as

$$[(M_0 + \delta M)\omega^2 - (K_0 + \delta K)]\mathbf{u} = 0 \tag{24}$$

δM and δK are matrices representing the mass and force constant changes due to the presence of the impurity. The equation can be rewritten as

$$(D_0 + \Delta)\mathbf{u} = 0 \tag{25}$$

with the definitions

$$D_0 \equiv M_0\omega^2 - K_0; \qquad \Delta = \delta M\omega^2 - \delta K. \tag{26}$$

Δ will usually be referred to as the defect matrix. After multiplying through from the left by the inverse of D_0, one obtains

$$(1 + D_0{}^{-1}\Delta)\mathbf{u} = 0. \tag{27}$$

The inverse of the dynamical matrix is the Green's function matrix G so that for the perfect crystal

$$G^0 = [M_0\omega^2 - K_0]^{-1}. \tag{28}$$

The Green's function matrix for the perturbed crystal can then be written in the Dyson form as

$$\begin{aligned} G &= [(M_0\omega^2 - K_0) + \Delta]^{-1} = [(G^0)^{-1} + \Delta]^{-1} \\ &= [(G^0)^{-1}(1 + G^0\Delta)]^{-1} = (1 + G^0\Delta)^{-1}G^0. \end{aligned} \tag{29}$$

An expression for the t matrix was given in Eq. (4). By formally expanding $(1 + G^0\Delta)^{-1}$ and regrouping terms, G can be related to the t matrix as in Eq. (5). The form of Eq. (29) is especially important for localized perturbation theory because it relates the perturbed and unperturbed Green's function matrices through the defect matrix Δ in a relatively simple way. The calculation of G would still be a hopeless task if it were not for the fact that there are many cases of physical interest in which the perturbation is so well localized that the defect matrix Δ is of small dimension. In the preceding section it was seen that all physical properties of interest here are directly related to G. That this should be the case is easily understood because the emphasis has simply been shifted from the dynamical matrix to the Green's function matrix.

B. Defect Space and Symmetry Coordinates

It is often useful to think of the defect problem in terms of a "defect space." This is a subspace of that space formed by the displacements of *all* ions in the crystal; it consists of the displacements of those ions whose interactions with the defect and with one another are assumed to be altered by the presence of the defect. Hence the defect matrix Δ has nonvanishing elements only within the defect space. As a specific example of what is meant by the defect space, we may consider a substitutional H^- ion and its six neighboring K^+ ions in KCl; this case is illustrated in Fig. 1. In the mass defect approximation it is assumed that there are no force constant changes whatsoever. However,

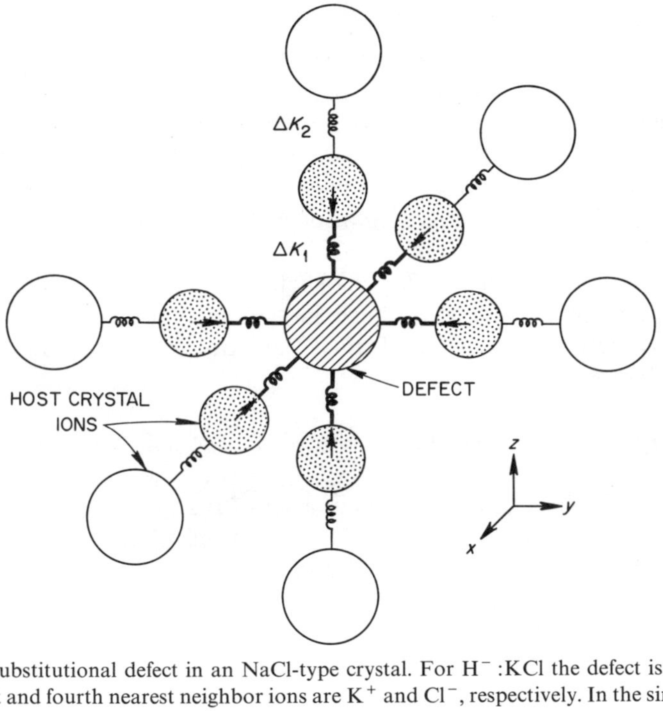

FIG. 1. Substitutional defect in an NaCl-type crystal. For H^-:KCl the defect is an H^- ion and the first and fourth nearest neighbor ions are K^+ and Cl^-, respectively. In the simple model of Section III, B, $\Delta K_2 = 0$ but in the relaxation model of Section V, $\Delta K_2 \neq 0$.

the change of the mass alters the self-energy at the defect site itself so that the defect space consists solely of the three cartesian displacements of the H^- ion. If the force constants between the defect site and the six 1nn ions are assumed to be altered, the defect space will have a dimension of 21. Although present day computers can easily handle 21×21 matrices, the repetitive nature of the calculations called for by the theory makes the full use of symmetry properties to reduce the sizes of the matrices almost mandatory. Furthermore, it is the calculation of the matrix elements themselves as well as the manipulation of the matrices that is time consuming.

Since an impurity destroys the translational invariance of the lattice, the spatially periodic phonons of the perfect crystal are no longer eigenfunctions of the problem. It seems natural therefore to fall back on the point symmetry properties of the defect and to introduce combinations of cartesian coordinates that are adapted to that symmetry. These orthonormalized, linear combinations of cartesian displacements will be referred to as symmetry coordinates and written as

$$Q_n(\Gamma, p) = \sum_{l\alpha} c(n\Gamma p; l\alpha) u_\alpha(l), \tag{30}$$

where Γ gives the irreducible representation according to which the symmetry coordinate transforms, n is an index which refers to shells of equivalent ions in the crystal, p labels the orthogonal components which can be constructed for each n and Γ, l is a site, and α a cartesian index. The notation is not complete because some shells may give rise to more than one appearance of some of the irreducible representations; for simplicity this possibility will be ignored here. Continuing with KCl:H^- as an example, we can set $n = 0$ for the defect site and $n = 1$ for the shell of six 1nn ions. The defect site, (000), alone yields a simple result since the symmetry coordinates are identical to the cartesian coordinates and must therefore transform like T_{1u} of the 0_h group. Thus,

$$Q_0(T_{1u}, 1) = u_x(000); \quad Q_0(T_{1u}, 2) = u_y(000); \quad Q_0(T_{1u}, 3) = u_z(000). \tag{31}$$

Inclusion of the six 1nn ions results in a set of symmetry coordinates which can be obtained by group theoretical calculations or from the appropriate literature. The general case to which the present example belongs has been treated often and it is well known that the subspace of the 1nn displacements decomposes as

$$\Gamma = A_{1g} + E_g + T_{2g} + T_{1g} + T_{2u} + 2T_{1u}. \tag{32}$$

It should be noted that the threefold degenerate T_{1u} irreducible representation appears twice, though, as already mentioned, our simplified notation for the Q's does not provide for it.

It would be too cumbersome to write down all the symmetry coordinates in the manner of Eq. (30). Therefore, it is useful to give them in a more compact form such as that shown in Table I for the even modes. In this table, the σ_j give the directions and relative magnitudes of the displacement of each of the ions in a particular Q; $j = 1$–6 labels the ions at (100), (010), (001), ($\bar{1}$00), (0$\bar{1}$0), and (00$\bar{1}$), respectively; N is a normalization constant; ε and θ label the components of the twofold degenerate E_g modes. As an example, from Table I the symmetry coordinate for the θ component of the E_g modes can be written as

$$Q_1(E_g, \theta) = \tfrac{1}{2}(u_x(1) - u_y(2) - u_x(4) + u_y(5)). \tag{33}$$

We have been assuming that the concentration of defects is so low that it is a good approximation to treat each one as if it were completely isolated in an otherwise perfect crystal. The normal coordinates of the system then transform according to the irreducible representations of the point group of the defect. It will seldom be true that a single symmetry coordinate will also be a normal coordinate, although it is nearly the case for the H^- ion in many crystals. The H^- ion vibrates in such a highly localized mode that its motion can be described rather well in the harmonic approximation by the coordinates given in Eq. (31). Symmetry coordinates constructed for all shells of ions in the crystal form a complete set just as do the cartesian coordinates and all quantities of interest can be expanded in terms of them. For example, the normal coordinates of the system can be expressed as linear combinations

TABLE I
EVEN MODES OF THE (100) SHELL IN NACL CRYSTALS

Γ_p	N	σ_1	σ_2	σ_3	σ_4	σ_5	σ_6
A_{1g}	$1/\sqrt{6}$	$\hat{x}$	$\hat{y}$	$\hat{z}$	$-\hat{x}$	$-\hat{y}$	$-\hat{z}$
$E_g\varepsilon$	$1/2\sqrt{3}$	$-\hat{x}$	$-\hat{y}$	$2\hat{z}$	$\hat{x}$	$\hat{y}$	$-2\hat{z}$
$E_g\theta$	$\frac{1}{2}$	$\hat{x}$	$-\hat{y}$	0	$-\hat{x}$	$\hat{y}$	0
$T_{2g}x$	$\frac{1}{2}$	0	$\hat{z}$	$\hat{y}$	0	$-\hat{z}$	$-\hat{y}$
$T_{2g}y$	$\frac{1}{2}$	$\hat{z}$	0	$\hat{x}$	$-\hat{z}$	0	$-\hat{x}$
$T_{2g}z$	$\frac{1}{2}$	$\hat{y}$	$\hat{x}$	0	$-\hat{y}$	$-\hat{x}$	0
$T_{1g}x$	$\frac{1}{2}$	0	$\hat{z}$	$-\hat{y}$	0	$-\hat{z}$	$\hat{y}$
$T_{1g}y$	$\frac{1}{2}$	$-\hat{z}$	0	$\hat{x}$	$\hat{z}$	0	$-\hat{x}$
$T_{1g}z$	$\frac{1}{2}$	$\hat{y}$	$-\hat{x}$	0	$-\hat{y}$	$\hat{x}$	0

[a] See text for notation.

of the symmetry coordinates. The potential energy can be written as

$$U = U_0 + \tfrac{1}{2}\sum_{nm}\sum_{\Gamma p} \mathscr{K}_{nm}(\Gamma p)Q_n(\Gamma p)Q_m(\Gamma p) \tag{34}$$

with

$$\mathscr{K}_{nm}(\Gamma p) \equiv [\partial^2 U/\partial Q_n(\Gamma p)\,\partial Q_m(\Gamma p)]_0. \tag{35}$$

It should be noted that there is a double sum over the shells of ions but a single sum over the irreducible representations and their components. In the symmetry coordinate representation there are no cross terms between Γ, Γ' and p, p'. In the perfect crystal, a complete set of symmetry coordinates centered about the site at which the substitution will occur can also be defined. Denoting the corresponding force constant by $\mathscr{K}_{nm}{}^0(\Gamma p)$, the force constant change can be written as

$$\Delta\mathscr{K}_{nm}(\Gamma p) = \mathscr{K}_{nm}(\Gamma p) - \mathscr{K}_{nm}{}^0(\Gamma p). \tag{36}$$

It is possible to evaluate the second derivatives with respect to the symmetry coordinates numerically on the computer if a suitable expression for the potential energy is known (Mostoller and Wood, 1973). Nevertheless, it is more common to have the force constants expressed in terms of ion–pair interactions as in the Born–von Kármán and shell models of lattice dynamics. The force constants K of Eq. (23) and $\mathscr{K}$ of Eq. (35) are related by the transformation (with $l\kappa \to l$ for convenience)

$$u_\alpha(l) = \sum_n \sum_{\Gamma p} \langle u_\alpha(l)|Q_n(\Gamma p)\rangle Q_n(\Gamma p) \tag{37}$$

and we find that

$$\mathscr{K}_{nm}(\Gamma p) = \sum_{\alpha l}\sum_{\beta l'} \langle Q_n(\Gamma p)|u_\alpha(l)\rangle\, K_{\alpha\beta}(l, l')\, \langle u_\beta(l')|Q_m(\Gamma p)\rangle. \tag{38}$$

The same transformation obviously holds for the defect matrix.

C. Green's Functions, Projection Operators, and Overlap Integrals

An element of the Green's function matrix of the perfect crystal can be expressed in the symmetry coordinate representation. We use the notation

$$G_{nm}{}^0(\omega^2, \Gamma p) \equiv \langle Q_n(\Gamma p)|G^0|Q_m(\Gamma p)\rangle \tag{39}$$

and note that the Green's function matrix like the dynamical matrix is diagonal in Γ and p. G^0 is, of course, also diagonal in the Bloch function representation, and it is convenient to utilize this property whenever possible. The phonon eigenvectors of the perfect crystal with wavevector $\mathbf{q}$ and branch index j can be written as

$$v(\mathbf{q}j) = N^{-1/2} \sum_{\alpha\kappa l} (M_\kappa)^{1/2} e_\alpha(\kappa; \mathbf{q}j) \exp[i\mathbf{q} \cdot \mathbf{R}(l)] u_\alpha(\kappa l), \tag{40}$$

where $\mathbf{e}$ is the polarization vector for which α labels the cartesian components and κ the atomic species; $\mathbf{R}(l)$ is a lattice vector. The $v(\mathbf{q}j)$ form a complete set and can therefore be introduced into the expression for $G_{nm}{}^0(\omega^2, \Gamma p)$ by way of the closure relation. Thus,

$$\begin{aligned} G_{nm}{}^0(\omega^2, \Gamma p) &= \sum_{\mathbf{q}j} \sum_{\mathbf{q}'j'} \langle Q_n(\Gamma p)|v(\mathbf{q}j)\rangle \langle v(\mathbf{q}j)|G^0|v(\mathbf{q}'j')\rangle \langle v(\mathbf{q}'j')|Q_m(\Gamma p)\rangle \\ &= \sum_{\mathbf{q}j} P_{nm}(\mathbf{q}j; \Gamma p)[\omega^2 - \omega^2(\mathbf{q}j)]^{-1} \end{aligned} \tag{41}$$

with

$$P_{nm}(\mathbf{q}j; \Gamma p) \equiv \langle Q_n(\Gamma p)|v(\mathbf{q}j)\rangle \langle v(\mathbf{q}j)|Q_m(\Gamma p)\rangle \tag{42}$$

The second line of Eq. (41) follows from the first because G^0 is diagonal in the $v(\mathbf{q}j)$ basis with eigenvalues $[\omega^2 - \omega^2(\mathbf{q}j)]^{-1}$. The discussion here will usually be carried out in terms of Green's functions expressed directly in the symmetry coordinate basis. Many people prefer, and in some cases it is clearly advisable, to express the Green's function in the cartesian representation. In terms of the same symbols appearing in Eq. (40), we can write

$$G_{\alpha\beta}{}^0(\kappa l, \lambda l'; \omega^2) = \sum_{\mathbf{q}j} \langle u_\alpha(\kappa l)|v(\mathbf{q}j)\rangle [\omega^2 - \omega^2(\mathbf{q}j)]^{-1} \langle v(\mathbf{q}j)|u_\beta(\lambda l')\rangle. \tag{43}$$

The transformations between G^0 in the two representations can be carried out readily by use of Eq. (37) and its inverse.

Regardless of the representation one uses, the problem of dealing with the singularities in Eqs. (41) and (42) must be resolved. A straightforward method for doing this consists of adding a small imaginary part $i\varepsilon$ to the frequency and using the relationship

$$\lim_{\varepsilon \to +0} (x + i\varepsilon)^{-1} = \mathrm{PP}(x^{-1}) - i\pi\, \delta(x). \tag{44}$$

PP denotes the principle part and $\varepsilon \to +0$ means that ε should approach zero from above the real axis. One thus finds that

$$\operatorname{Im} G_{nm}{}^{0}(\omega^2, \Gamma p) = \pi \sum_{\mathbf{q}j} P_{nm}(\mathbf{q}j; \Gamma p)\, \delta(\omega^2 - \omega^2(\mathbf{q}j)). \tag{45}$$

It is generally more convenient to work with ω than with ω^2, so the quantity that is usually computed is

$$\operatorname{Im} G_{nm}{}^{0}(\omega, \Gamma p) = \frac{\pi}{2\omega} \sum_{\mathbf{q}j} P_{nm}(\mathbf{q}j; \Gamma p)\, \delta(\omega - \omega(\mathbf{q}j)). \tag{46}$$

Once the imaginary parts of the elements of the Green's function matrix have been found by numerical techniques, the real parts can be found by the Kramers–Kronig dispersion relation in conjuction with a scheme for evaluating the principal part of the integral. Briefly the dispersion calculation runs as follows. From Eqs. (41) and (43), we have

$$\begin{aligned}
\operatorname{Re} G_{nm}{}^{0}(\omega, \Gamma p) &= \mathrm{PP} \sum_{\mathbf{q}j} P_{nm}(\mathbf{q}j; \Gamma p)[\omega^2 - \omega^2(\mathbf{q}j)]^{-1} \\
&= \mathrm{PP} \sum_{\mathbf{q}j} \int \frac{P_{nm}(\mathbf{q}j; \Gamma p)}{\omega^2 - \omega'^2}\, \delta[\omega'^2 - \omega^2(\mathbf{q}j)]\, d(\omega'^2) \\
&= \mathrm{PP} \int \frac{\operatorname{Im} G_{nm}{}^{0}(\omega', \Gamma p)}{\omega^2 - \omega'^2}\, 2\omega'\, d\omega'
\end{aligned} \tag{47}$$

This integral is not difficult to evaluate numerically by a method described in Timusk and Klein (1966). This and other methods are likely to be covered in the article by Gilat (this volume, p. 317) and it will not be discussed here.

Let us return now to Eq. (41) and consider its content in more detail. The quantities $\langle v(\mathbf{q}j)|Q_n(\Gamma p)\rangle$ which are just the inner products of the symmetrized coordinates and the phonon normal coordinates of the perfect crystal will be referred to as "overlap integrals." The overlap integrals and their products [Eq. (42)] are the basic quantities in the numerical calculation of $\operatorname{Im} G_{nm}{}^{0}$. In terms of the $v(\mathbf{q}j)$ and Eq. (30)

$$\begin{aligned}
\langle v(\mathbf{q}j)|Q_n(\Gamma p)\rangle &= N^{-1/2} \sum_{\alpha\kappa l} (M_\kappa)^{1/2} e_\alpha^*(\kappa; \mathbf{q}j) \exp[-i\mathbf{q} \cdot \mathbf{R}(l)] \\
&\quad \times \sum_{l'\beta} c(n\Gamma p; l'\beta) \langle u_\alpha(l)|u_\beta(l')\rangle
\end{aligned}$$

$$= N^{-1/2} \sum_{\alpha\kappa l} (M_\kappa)^{1/2} c(n\Gamma p, l\alpha) e_\alpha^*(\kappa; \mathbf{q}) \times \exp[-i\mathbf{q} \cdot \mathbf{R}(l)]. \tag{48}$$

Next we note that the quantity

$$P(\mathbf{q}j) = |v(\mathbf{q}j)\rangle\langle v(\mathbf{q}j)| \tag{49}$$

is a projection operator ($P^2 = P$) which projects out that component of $|v(\mathbf{q}j)\rangle$ which is contained within $|Q_m(\Gamma p)\rangle$. Thus $P_{nm}(\mathbf{q}j;\ \Gamma p)$ gives the product of the projections of $v(\mathbf{q}j)$ onto $Q_n(\Gamma p)$ and $Q_m(\Gamma p)$. The total density of states of the perfect crystal is given by

$$\rho^0(\omega) = \frac{\pi}{2\omega} \sum_{\mathbf{q}j} \delta[\omega - \omega(\mathbf{q}j)] \tag{50}$$

so that the imaginary parts of ${G_{nm}}^0$ are just projected densities of states. There is a sum rule for Im G^0 which is very useful as a check of numerical work. Let us rewrite Eq. (46) slightly and integrate both sides over ω^2 to get

$$\begin{aligned} 2\pi^{-1} \int \text{Im}\, {G_{nm}}^0(\omega, \Gamma p)\omega\, d\omega &= \langle Q_n(\Gamma p)| \int \sum_{\mathbf{q}j} |v(\mathbf{q}j)\rangle \\ &\quad \times \langle v(\mathbf{q}j)|\delta[\omega - \omega(\mathbf{q}j)]\, d\omega |Q_m(\Gamma p)\rangle \\ &= \sum_{\mathbf{q}j} P_{nm}(\mathbf{q}j; \Gamma p) = \delta_{nm} \end{aligned} \tag{51}$$

The final line follows because (1) the $|v(\mathbf{q}j)\rangle$ do not depend explicitly on ω, (2) $\sum |v(\mathbf{q}j)\rangle\ \langle v(\mathbf{q}j)| = 1$ and, (3) the $Q_n(\Gamma p)$ are orthonormal with respect to the subscript n. Another way of thinking about the role of the δ function is the following. For each value of ω in the integration, the sum over $\mathbf{q}j$ insures that the δ function will pick out each and every $|v(\mathbf{q}j)\rangle$ going with that frequency and none other. Thus the combined effect of the integration and the summation leads directly to the use of the closure relation on the $|v(\mathbf{q}j)\rangle$. The equation can be used by first calculating the integrals of the diagonal elements of Im G^0 and using them as normalization factors to force the condition on the diagonal elements. Then if the off-diagonal elements do not integrate to zero something has gone amiss, most likely in the construction of the symmetrized coordinates or in the calculation of the overlap integrals. Before proceeding with further discussion of the numerical work, it is useful to consider the transformation properties of the overlap integrals and their products since these are necessary for efficient numerical BZ integration.

D. Transformation Properties of Overlap Integrals and Green's Functions

Sums over the wave vectors in the BZ must be carried out when constructing the Green's function matrices. This numerical integration can be the step in an entire calculation that consumes the most computer time. Therefore, it is desirable to reduce the time by utilizing the symmetry properties inherent in the problem. For example, for a cubic system, the irreducible sector of the BZ is only $\frac{1}{48}$ of the whole zone and the summations over $\mathbf{q}$ can be restricted to this sector since all other points can be reached by one or more operations of the point group on $\mathbf{q}$. Since we are working here primarily with symmetry coordinates, we want to examine the transformation properties of the Green's functions in this representation.

Let us first study how an overlap integral at one point in the BZ is related to the overlap integrals at another point generated by operating on $\mathbf{q}$ with an element s of the point group. In orther words, how is $\langle Q_n(\Gamma p)|v(s^{-1}\mathbf{q}j)\rangle$ related to $\langle Q(\Gamma p)|v(\mathbf{q}j)\rangle$? To simplify the notation as much as possible, the branch index j and the atomic specie index κ in $v(\mathbf{q}j)$ and the shell index n in $Q_n(\Gamma p)$, which are unimportant in this discussion, will be suppressed. Then, as a result of the well-known relationship,

$$\exp[is\mathbf{q}\cdot\mathbf{R}(l)] = \exp[i\mathbf{q}\cdot s^{-1}\mathbf{R}(l)], \tag{52}$$

$$|v(s^{-1}\mathbf{q})\rangle = (M/N)^{1/2}\sum_{\alpha l} e_\alpha(s^{-1}\mathbf{q})\exp[i\mathbf{q}\cdot s\mathbf{R}(l)]|u_\alpha(l)\rangle. \tag{53}$$

For a symmorphic lattice, under the operations of the point group acting on $\mathbf{q}$, the polarization vectors transform (Maradudin and Vosko, 1968) according to

$$e_\alpha(s\mathbf{q}) = \sum_\beta s_{\alpha\beta}e_\beta(\mathbf{q}); \tag{54}$$

since $s_{\alpha\beta} = \tilde{s}_{\beta\alpha} = (s^{-1})_{\beta\alpha}$, Eq. (52) can be written

$$|v(s^{-1}\mathbf{q})\rangle = (M/N)^{1/2}\sum_{\beta l} e_\beta(\mathbf{q})\exp[i\mathbf{q}\cdot s\mathbf{R}(l)]\Big|\sum_\alpha s_{\beta\alpha}u_\alpha(l)\Big\rangle \tag{55}$$

Now s operating on the lattice vector $\mathbf{R}(l)$ takes it into another lattice vector $\mathbf{R}'(L)$ and takes the displacement $u_\alpha(l)$ into $u_\beta(L)$ given by

$$u_\beta(L) = \sum_\alpha s_{\beta\alpha} u_\alpha(l).$$

Thus,

$$|v(s^{-1}\mathbf{q})\rangle = (M/N)^{1/2} \sum_{\beta L} e_\beta(\mathbf{q}) \exp[i\mathbf{q} \cdot \mathbf{R}(L)] |u_\beta(L)\rangle = s|v(\mathbf{q})\rangle \quad (56)$$

where the last line means that s is to act only on real-space vectors. Hence, the operation of s^{-1} on $\mathbf{q}$ in $|v(\mathbf{q})\rangle$ gives the same result as s operating on the real-space vectors in $|v(\mathbf{q})\rangle$. For an overlap integral then,

$$\langle Q(\Gamma p)|v(s^{-1}\mathbf{q})\rangle = \langle Q(\Gamma p)|s|v(\mathbf{q})\rangle = \langle v(\mathbf{q})|s^{-1}|Q(\Gamma p)\rangle. \quad (57)$$

The point group transformation properties of the symmetrized coordinates yield the relation

$$s^{-1}|Q(\Gamma p)\rangle = \sum_{p'} A^\Gamma_{pp'}(s^{-1})|Q(\Gamma p')\rangle \quad (58)$$

in which p' runs over all the components of the irreducible representation Γ. $A^\Gamma_{pp'}(s^{-1})$ is an element of an appropriate matrix realization of Γ going with element s^{-1} of the point group; the rows and columns of the A's form orthonormal vectors. Using Eq. (30), with the shell index suppressed, gives

$$s^{-1}|Q(\Gamma p)\rangle = \sum_{p'} A^\Gamma_{pp'}(s^{-1}) \sum_{l\beta} c(\Gamma p; l\beta)|u_\beta(l)\rangle \quad (59)$$

and

$$\begin{aligned}
\langle Q(\Gamma p)|v(s^{-1}\mathbf{q})\rangle &= \langle v(\mathbf{q})|s^{-1}|Q(\Gamma p)\rangle \\
&= (M/N)^{1/2} \sum_{l'\alpha p'} A^\Gamma_{pp'}(s^{-1}) e_\alpha(\mathbf{q}) \exp[i\mathbf{q} \cdot \mathbf{R}(l')] \\
&\quad \times \sum_{l\beta} c(\Gamma p'; l\beta) \langle u_\alpha(l')|u_\beta(l)\rangle \\
&= (M/N)^{1/2} \sum_{l'\alpha p'} A^\Gamma_{pp'}(s^{-1}) c(\Gamma p'; l'\alpha) e_\alpha(\mathbf{q}) \exp[i\mathbf{q} \cdot \mathbf{R}(l')] \\
&= \sum_{p'} A^\Gamma_{pp'}(s^{-1}) \langle v(\mathbf{q})|Q(\Gamma p')\rangle. \quad (60)
\end{aligned}$$

This last line follows from Eq. (48) and defines the transformation properties of the overlap integrals. Thus the overlap integrals between one component of $Q(\Gamma)$ and a phonon at point $s\mathbf{q}$ in the BZ are linear combinations of overlap integrals of all components of $Q(\Gamma)$ with a phonon at BZ point $\mathbf{q}$. This is not yet a particularly convenient result, so we now examine the product of the overlap integrals which appear in Im G^0 and find

$$P(s\mathbf{q}j;\Gamma p) = \sum_{p',p''} [A^{\Gamma}_{pp'}(s)]^* A^{\Gamma}_{pp''}(s) \langle Q(\Gamma p')|v(\mathbf{q})\rangle \langle v(\mathbf{q})|Q(\Gamma p'')\rangle. \quad (61)$$

As it stands, this equation does not look very promising, but let us sum over p on both sides. Then

$$\sum_p P(s\mathbf{q}j;\Gamma p) = \sum_p P(\mathbf{q}j;\Gamma p) \quad (62)$$

because of the orthonormality properties of A. This is a simple and useful result; it says that the sum over p of products of overlap integrals is invariant under the operation of s on $\mathbf{q}$. From symmetry, the real-space Green's functions formed by summation over the entire BZ cannot depend on which component of Γ is used. Therefore Eq. (62) allows the sum over $\mathbf{q}$ to be restricted to the irreducible sector provided one works with the sum over p rather than just a single value of p when calculating Im $G_{nm}(\Gamma p)$ from Eq. (46).

E. The Coherent Potential Approximation

Thus far in this section it has been assumed that the concentration of defects is so low that each defect can be treated as completely isolated in the crystal. As the concentration increases, this approximation, which will be referred to in this subsection as the isolated defect approximation (IDA), breaks down and first the dilute alloy region and then yet more general situations are encountered. The coherent potential approximation (CPA), already discussed briefly in the Introduction, has proved to be quite successful in handling certain aspects of the lattice dynamics of alloys. The CPA has many elements in common with the IDA but it also differs in certain fundamental aspects which make it inherently more complicated to apply. Here, an extremely abbreviated and formal sketch of the derivation of the CPA will be given.

Equation (29) can be rewritten as

$$G = G^0 + G^0\,\Delta G. \quad (63)$$

In the IDA, the matrix Δ originates from a single defect and is of small dimension. In the CPA, Δ must include the contributions of many defects whose perturbations of the lattice may overlap in patterns which vary with the configurations of the random alloy. The basic approach of the CPA succeeds in reducing this apparently very complex problem to one quite similar in form to the IDA but with the real crystal replaced by an effective crystal determined in such a manner that the scattering of phonons from any site (host or defect) vanishes on the average. Let us use Eq. (63) to define the Green's function operator $\hat{G}$ for the effective crystal, i.e.,

$$\hat{G} = G^0 + G^0 E \hat{G}, \tag{64}$$

where E is a complex, periodic, self-energy function which is to be determined by minimizing the difference between $\hat{G}$ and the configuration average of G, denoted by $\langle\langle G \rangle\rangle$, at some level of approximation. The equation for the true perturbed Green's function before averaging can be written in terms of $\hat{G}$ as

$$G = \hat{G} + \hat{G}(\Delta - E)G. \tag{65}$$

Even though the matrix E is a crystal periodic function, it can be written as a sum of local, self-energy matrices A associated with each site in the crystal. Thus, with i running over lattice sites,

$$E = \sum_i A_i. \tag{66}$$

The mass and force constant change matrix Δ can also be written in an approximate form as a sum over sites of local functions, e.g.,

$$\Delta = \sum_i D_i(\delta_i) \tag{67}$$

in which δ_i specifies the type of atom at the ith site. Although this is a restrictive assumption, Kaplan and Mostoller (1974b) point out that it is valid for any system in which the force constants superimpose linearly and also when the overlap of perturbations due to individual defects can be neglected as in the IDA; it should therefore hold for both very small and very large concentrations. The difference between Eqs. (67) and (66) is

$$\Delta - E = \sum_i [A_i - D_i(\delta_i)] \equiv \sum_i V_i(\delta_i) \tag{68}$$

and Eq. (65) becomes

$$G = \hat{G} + \hat{G} \sum_{i} V_i(\delta_i)G. \tag{69}$$

The t matrix [see Eq. (4)] for scattering from an individual site with scattering potential $V_i(\delta_i)$ is just

$$t_i(\delta_i) = V_i(\delta_i)[1 - \hat{G}V_i(\delta_i)]^{-1} \tag{70}$$

After making certain approximations concerning the configurational average of Eq. (69), it is possible to define the average t matrix at site i and set it equal to zero, i.e.,

$$\sum_{\delta_i} c(\delta_i)t_i(\delta_i) = 0, \tag{71}$$

where $c(\delta_i)$ is the concentration of atoms of type δ at site i. The effect of this condition is to make the configuration averaged function $\langle\langle G \rangle\rangle$ equal to the CPA Green's function $\hat{G}$ to a good approximation.

Equation (71), when applied to the case where only one type of defect may be present at only one site in the unit cell, leads to a matrix equation of the form

$$A - cD + (A - D)\hat{G}A = 0. \tag{72}$$

In this equation, both A and $\hat{G}$ are unknown and so the equation must be solved self-consistently. A procedure for solving it might consist of the the following steps:

1. Guess a starting A matrix.
2. Construct E from A by way of Eq. (66).
3. Solve for $\hat{G}$ via Eq. (64).
4. Check to see if Eq. (72) is satisfied; if it is not return to 1 and continue iterating until it is.

The procedure for making successive estimates of A can involve some interesting numerical methods (keep in mind, for example, that E and A are complex quantities) but time and space preclude consideration of them here. Instead, it is important for the purposes of this article to observe the similarities with the IDA calculations described in the preceding subsections. It should be noted particularly, that the whole problem retains the point symmetry characteristics of the IDA and therefore the construction of symmetry coordinates, overlap integrals, projection operators, and Green's functions follow the procedures outlined above.

IV. Computational Considerations

A. Outline of a Typical Calculation

The essential equations for a calculation of the lattice dynamics of a crystal containing point impurities have now been given and we can proceed to a consideration of some of the computational details. A typical calculation can be broken down into six steps:

1. Generation of frequencies and eigenvectors of the perfect crystal.
2. Construction of symmetry coordinates, overlap integrals, and projection operators.
3. Calculation of imaginary parts of Green's functions by BZ integration and real parts by dispersion relations.
4. Construction of the defect matrix.
5. Calculation of real and imaginary parts of perturbed Green's functions, t matrix, etc.
6. Calculation of the desired physical quantity.

In the first step, the perfect crystal data must be generated at points on a mesh in the irreducible sector of the BZ for use in steps 2 and 3. Computer programs for lattice dynamical calculations on ionic crystals based on the shell model and various of its modifications are now common. A variety of approaches based on Born–von Kármán and pseudopotential models have been employed in calculations on metals. Other models originating from the microscopic theory of lattice dynamics and applicable to differing types of crystals are either already in existence or under development. For our purposes here, the simple shell model for insulators and Born–von Kármán models for metals are quite adequate; for further details of these and other models the interested reader can consult the article by G. Dolling in this volume (p. 1).

The completion of the second step requires an appropriate set of symmetry coordinates. These are not difficult to construct for the 1nn ions of a substitutional defect in crystals with the NaCl structure; they appear frequently in the literature and are given here in Table I. For the 1nn shells in fcc and bcc metals, the construction is more tedious. The coordinates found in the literature sometimes contain errors or are not orthonormalized and due caution should be exercised. Note that the second and third shells in an alkali halide include the 1nn shell for both fcc and bcc metals. For more complex crystals, interstitial defects, extended defects, etc., the construction of symmetry coordinates can become a real chore and it is quite useful to have the work carried out by a computer if possible. Techniques for doing this are described by Warren and Worlton in their contribution to this volume. Once the symmetry coordinates are available, the derivation of the algebraic expressions for the overlap integrals is straightforward but may become tiresome as the defect space is extended and the number of coordinates increases; any help from the computer in this area would also be welcome. At the risk of being overly explicit, the overlap integral between $Q_1(E_g\theta)$ of Eq. (33) and the phonons of the KCl:H^- problem will be written out as an example. The nearest neighbors of the H^- ions are K^+ ions so first the integral can be simplified to

$$\langle Q_1(\mathrm{E_g}\theta)|v(\mathbf{q}j)\rangle = N^{-1/2}M_+^{1/2}\sum_{\alpha l} c(1\mathrm{E}\ \theta)e_\alpha(+;\mathbf{q}j)\exp[i\mathbf{q}\cdot\mathbf{R}(l)]. \tag{73}$$

$\mathbf{R}(l)$ is nonvanishing for ions 1, 2, 4, and 5 of Fig. 1 with $\mathbf{R}(1) = -\mathbf{R}(4) = a\mathbf{i}_x$ and $R(2) = -R(5) = a\mathbf{i}_y$. Then

$$\begin{aligned}\langle Q_1(\mathrm{E_g}\theta)|v(\mathbf{q}j)\rangle &= N^{-1/2}M_+^{1/2}\tfrac{1}{2}\{e_x(+;\mathbf{q}j)(e^{iq_xa} - e^{-iq_xa})\\ &\quad -e_y(+;\mathbf{q}j)(e^{iq_ya} - e^{-iq_ya})\}\\ &= N^{-1/2}M_+^{1/2}ie_x(+;\mathbf{q}j)\sin q_xa - e_y(+;\mathbf{q}j)\sin q_ya\}.\end{aligned} \tag{74}$$

Attention is called to the mass factor in this equation. Most lattice dynamical programs assume mass reduced displacement, e.g., $w_+(l) = M_+^{1/2}u_+(l)$, and the polarization vectors reflect this fact. One should therefore check the consistency of the definitions of phonon normal coordinates, polarization vectors, and symmetrized coordinates. Here, for example, it might have been better to express both the phonon and symmetry coordinates in mass reduced displacements so that the mass factor would not enter in the equations for the overlap integrals.

The third step of the calculation consists of the construction of the imaginary parts of the Green's functions by numerical integration over the irreducible sector of the Brillouin zone. In Section IV, C below we discuss this problem briefly. Once the imaginary parts are available, the construction of the real parts of G^0 by way of Eq. (47) is very rapid.

Steps 4, 5, and 6 are straightforward although one or two questions are worth considering. The defect matrix must be constructed in the symmetry coordinate representation by way of Eqs. (36)–(38). In an ionic crystal, the fact that the ions are charged means that there will be long-range coulomb forces involved and these can have a strong influence on the perturbation matrix. They, together with second neighbor repulsive interactions, will lead to force constant changes which are symmetry dependent (Mostoller and Wood, 1973) and this can be an important effect. In fact, the whole problem of constructing the defect matrix can become rather complex when examined carefully. There is not enough time to do this here, but the major point to be made is that one may be quite justified in choosing different force constants for different symmetry modes. Finally, it should perhaps be pointed out that although the multiplication and inversion of complex matrices is probably most conveniently handled by prepackaged subroutines written in complex arithmetic, this may not represent the most efficient use of the computer. If computer time is a serious problem, this is an area in which some savings may be realized.

B. Extent of the Defect Space

In applications of localized perturbation theory, important factors influencing computer time are directly related to the extent of the perturbation. Obviously, a realistic model of the physical problem under consideration is needed but at the same time the demands on computer time cannot be ignored. The defect space dictates the number of imaginary Green's functions that must be constructed by BZ integration and determines the dimensions of the complex matrices that must be manipulated to find the perturbed Green's functions. It should also be kept in mind that the force constant changes in the defect matrix are usually not known and must be found by trial and error or by some least squares procedure. Since the number of force constant changes increases rapidly with the size of the defect space, meaningful variations of these adjustable parameters can present a problem. For this reason, even in simple cases one is usually forced to make compromises.

Although here the Green's functions have been constructed directly in the symmetry coordinate representation, it is not clear that this is the most efficient approach in all applications. Situations occur where it would appear to be preferable to calculate the cartesian elements of the Green's functions first and then transform to the symmetry coordinate representation. Attention to this aspect of the problem may result in increased efficiency of computer usage as the following discussion illustrates. Early applications of localized perturbation theory to calculations of infrared and Raman properties of alkali halide crystals showed that fairly good agreement with experiments could be obtained with a defect space consisting of the cartesian displacements of the substitutional impurity and its six 1nn ions. The local mode of the H^- ion and the induced far infrared absorption associated with it will involve only the T_{1u} coordinate of the central site and the two T_{1u} coordinates of the 1nn shell. The matrices are therefore only 3×3 and one must calculate six Green's functions in the symmetry coordinate representation. Calculation of the anharmonic sidebands under the assumptions leading to Eq. (6) is even simpler since only the A_{1g} and E_g modes are needed and each of these requires only one Green's function. Raman scattering involves the T_{2g} mode in addition to the A_{1g} and E_g modes; there is no coupling to the T_{1g} mode. Hence, a calculation of all the impurity-induced infrared and Raman properties would require nine Green's functions. Coherent neutron scattering at a general point in the BZ would require the computation of all 11 Green's functions in the symmetry coordinate representation. But in the cartesian coordinate representation there are only eight independent Green's functions within the entire defect space. Therefore, for calculations of neutron scattering cross sections and general infrared and Raman properties construction of the elements of Im G^0 in the cartesian representation may save time. But in the more likely case that a property involving symmetries, e.g., the sidebands, is desired it probably will be better to work with symmetrized coordinates.

To further emphasize the point of this discussion let us consider briefly calculations in fcc and bcc metals. In an fcc crystal, the defect ion and the 12 1nn ions give a defect space of dimension 39. Again the x, y, z displacements of the central ion transform according to T_{1u} and the space of the 1nn ions decomposes into

$$\Gamma_{1nn} \rightarrow A_{1g} + A_{2g} + 2E_g + 2T_{2g} + 2T_{1g} + A_{2u} + E_u + 2T_{2u} + 3T_{1u}.$$

Thus, it is easy to see that problems in which only a small number of irreducible representations occur may best be handled in the symmetry coordinate representation. However, only 13 independent Green's functions enter into a general calculation in the cartesian coordinate representation, and so it is not obvious which representation is better for problems such as neutron scattering in which many symmetry modes must be considered. In a bcc metal the first and second neighbors are at nearly the same distance, and it seems advisable to include both in the defect space. One then finds that there are 15 independent Green's functions in the cartesian coordinate representation whereas the decomposition of Γ_{1nn} and Γ_{2nn} gives

$$\Gamma_{1nn} \rightarrow A_{1g} + E_g + 2T_{2g} + T_{1g} + A_{2u} + E_u + T_{2u} + 2T_{1u}$$

$$\Gamma_{2nn} \rightarrow A_{1g} + E_g + T_{2g} + T_{1g} + T_{2u} + 2T_{1u}.$$

Obviously, some forethought in the choice of basis is advisable.

C. Brillouin Zone Integration

Calculations of many electronic, magnetic, and vibrational properties of solids require numerical integration over the Brillouin zone of highly singular functions. The lattice dynamical Green's functions discussed above represent only one class of such integrals. The development of fast, large-core, digital computers has made the accurate evaluation of BZ integrals feasible and the last few years has seen the development of a number of integration schemes. The article by Gilat in this volume should cover the subject in detail; here only a few comments about our experience in the Solid State Division at ORNL will be given.

At the time we began to study BZ integration we were interested in the calculation of generalized electronic and magnetic susceptibilities of metals as well as those properties related to lattice dynamics discussed in this article. There are, of course, many differences between the two types of calculations but one is particularly significant so far as computational time is concerned. Susceptibility calculations require the electronic energies and wavefunctions at mesh points in the irreducible sector of the BZ. Regardless of the type of band program (KKR, APW, OPW, etc.) one elects to use, the generation of this "first principles" data is a time consuming process. In lattice dynamical problems, on the other hand, the various models yield computer programs that are so fast that generation of the mesh data usually does not represent a significant factor in computer time. However, it should also be kept in mind that, thus far, lattice dynamical calculations have been mainly limited to crystals with one and two ions per unit cell so that at most there are only six dispersion curves. Calculations of electronic and magnetic properties often require many more energy bands that may have complex structure just in the important region near the Fermi surface. A difficulty with such calculations is to find an interpolation scheme that cuts down on the number of "first principles" points that must be calculated. These considerations led us to compare BZ integration methods based on linear (Gilat and Raubenheimer, 1966), quadratic (Mueller *et al.*, 1971), and combined linear-quadratic (Cooke and Wood, 1972; Janak, 1969, 1971) interpolation for points in the zone not lying on the mesh. Extensive studies of interpolation schemes (Slater and Koster, 1954; Hodges *et al.*, 1966) based on the tight-binding method of energy band calculations were also made.

As a result of our experience with these various techniques, we have fallen back on simple modifications of the original Gilat–Raubenheimer (GR) linear scheme in our lattice dynamical calculations. There are several reasons for this, only two of which will be discussed here. First, our tests indicated that under comparable conditions the original GR scheme applied to phonon problems is only slightly slower than the combined linear-quadratic scheme of Cooke and Wood (1972), which is the fastest we have tested. Assuming one has an IBM 360/91 or roughly equivalent computer available, the computation time involved in a phonon calculation is so short that the differences are not very significant. The second reason arises from a difficulty with quadratic schemes which Gilat will undoubtedly discuss in detail (see p. 317) but which we want to emphasize here also. Quadratic schemes used thus far require that the dispersion surfaces be ordered in energy, whereas the GR scheme does not. This ordering produces cusps and ridges in the energy surfaces in regions where bands cross or become nearly degenerate. Quadratic interpolation smoothes out these cusps and ridges and in so doing tends to introduce spurious peaks or spikes in the projected densities of states. The spikes usually do not contain many states but with experimental techniques as sensitive as infrared and Raman they could be misleading in comparisons between theory and experiment. When these two reasons are weighed together they make a rather good argument for using the GR scheme for lattice dynamical calculations. This is not necessarily the case for electronic problems but they will not be discussed here.

Thus far, nothing has been said about our treatment of the interpolation of "matrix elements," i.e., the $P_{nm}(\mathbf{q}j;\,\Gamma p)$ of Eq. (42), in the GR scheme. In our programs, we have simply given these quantities their mesh-point values throughout each GR cell. With the mesh we normally use the cell size is sufficiently small that we have yet to find any evidence that this is not an entirely adequate approximation. Should such evidence crop up in a calculation it can be tested very easily by varying the mesh.

D. Comments on Computer Time and Storage

Some qualitative remarks about factors influencing the computational time have already been made; here a few comments of a more quantitative nature will be given. Most of our calculations have been carried out on the IBM 360/91 at ORNL. With a machine of this or equivalent capacity no problems with computer time or storage for calculations within the isolated defect approximation have been encountered. The CPA calculations when force constant changes are included require orders of magnitude more time. However, since the CPA results discussed here are among the first of their type, it may be too soon to obtain a realistic assessment of the computer demands and none will be attempted. An example from work now underway will serve as a rough indication of the computer requirements for calculations within the isolated defect approximation.

The example concerns the calculation of the neutron scattering cross section from a copper crystal containing substitutional impurities. The problem requires the construction of six elements of the Green's function matrix by way of BZ integration and dispersion relations. It involves the manipulation of two 2×2 complex matrices to find the t matrices and the perturbed Green's functions and then the looping over six scattering vectors to obtain the coherent scattering cross section. The calculations are carried out with a mesh parameter in the Gilat–Raubenheimer scheme such that there are 2030 cells in the irreducible sector of the BZ. Although the GR scheme is not a histogram method in principle, in practice histograms are constructed and the number of bins will influence the running time. In the present example, the histogram bins have a width of 0.02 Terrahertz (THz) and the histogram arrays are dimensioned at 1000 so that a frequency range of 20 THz could be accommodated. This would allow a search for very high frequency local modes or a further reduction in the bin width when desired; in most applications the arrays need not be so large. In any case, the storage requirements for the program are approximately 250K and the running time is approximately 25 seconds of CPU time for the GO step. In another calculation of local modes in copper, it was necessary to construct ten elements of the Green's function matrix and to manipulate 4×4 matrices. The running time was still less than one minute for several sets of force constant changes. Clearly the program is very fast for this type of application and we have not bothered to run time checks on the individual steps.

In other applications such as infrared and Raman calculations in alkali halide crystals, the time can run up to three or four minutes per run and the storage up to 300–400K depending on the complexity of the problem, the degree of accuracy desired, and the number of sets of force constant changes considered. Since the force constant changes are usually treated as adjustable parameters, one might want to start off with a fairly coarse mesh and progress to finer meshes as the force constant changes giving good agreement with experiment are approached. Before leaving this discussion, it should be emphasized that the Green's functions are calculated each time the program is put on the computer; there is no storage of these quantities on tapes or disks. At installations with smaller computers than the 360/91 it may be desirable to store the Green's functions but this hardly seems worthwhile for larger machines.

If the foregoing discussion seems somewhat abbreviated and off-hand, it is simply because computer demands in these lattice dynamical calculations have never been much of a problem and therefore do not seem to warrant an extended treatment until the CPA is further developed.

V. Comparison between Theory and Experiment

A. Infrared Studies of the H^- Ion in Alkali Halides

The infrared absorption of substitutional H^--ions in alkali halides was discovered by Schaefer (1960), and the theoretical interpretation of its origin followed immediately (Rosenstock and Klick, 1960; Wallis and Maradudin,

1960). Shortly thereafter, Fritz and co-workers (Fritz, 1964; Fritz *et al.*, 1965) investigated the temperature dependence of both the main band and the side-bands. They concluded that the sidebands arise from the anharmonic interaction between the local mode and in-band lattice modes. Timusk and Klein (1966) studied the sideband region in KBr and developed a model which accounted for most of its features. Discrepancies between experimental and theoretical results remained, however, and led Gethins *et al.* (1967) and MacPherson and Timusk (1970) to extend the measurements to other crystals and to improve the Timusk–Klein model. It was recognized that force constant changes between the defect and 1nn sites almost certainly imply a displacement or relaxation of the 1nn ions. Because the relaxation is one in which the 1nn ions move in the [100] directions, the 1nn–4nn interactions are expected to be most effected. It is not difficult to show that even a small change in the distance between ions interacting by way of Born–Mayer repulsive potentials will produce large changes in the force constants. The so-called "relaxation model" then consists in extending the defect space to include the displacements of the 4nn ions as well as those of the 1nn ions and the defect itself. MacPherson and Timusk (1970) also studied the modifications of the in-band modes due to the presence of the H^- ions. The absorption in the in-band region is difficult to measure with high accuracy and the agreement between theory and experiment is not entirely satisfactory. Basically, however, the overall features of the absorption induced throughout the infrared region by the H^- ion is now understood for several crystals.

As already mentioned, one would normally expect to treat the force constant changes appearing in the defect matrix as adjustable parameters but the H^- ion is sufficiently simple that fairly complete, quantum mechanical calculation of its electronic structure in the alkali halides have been carried out by Wood and Gilbert (1967) and Wood and Opik (1967). Wood and Ganguly (1973) have calculated force constant changes in KCl, KBr, and KI based on this quantum-mechanical model. These computed force constants corresponded closely to the parameterized values and were used by Wood and Ganguly for calculations of the infrared properties of the H^- ion with the relaxation model. The theoretical results discussed here are taken from that work.

Figures 2 and 3 show typical results for the sidebands in KCl and KBr. The frequency of the T_{1u} local mode of the H^- ion itself is at $\nu = 0$. The dashed curves were extracted from the experimental work of Gethins *et al.* (1967). The overall agreement is quite good and tends to indicate that the relaxation model is basically correct. The very low frequency structure in the experimental curves is thought to arise from defect pairing and should be ignored here. In both crystals, the large broad peak which dominates the curves can be obtained from a model in which 4nn ion displacements are not

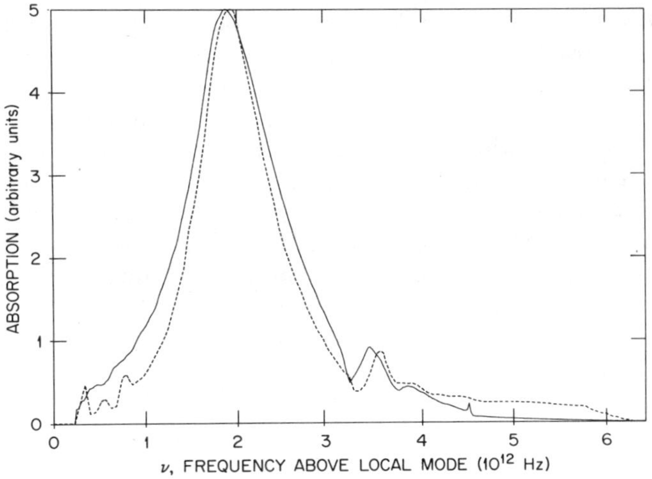

FIG. 2. Anharmonic sideband of the local-mode frequency of H^- ions in KCl. Dashed curve, experiment; solid curve, calculations.

included in the defect space. Comparison of the figures with unperturbed, projected A_{1g} and E_g densities of states shows that these peaks are due to a near resonance in the E_g acoustical modes. To obtain the higher frequency structure in KBr correctly the full relaxation model is needed but it is relatively unimportant in KCl. The force constant changes shown in Figs. 2

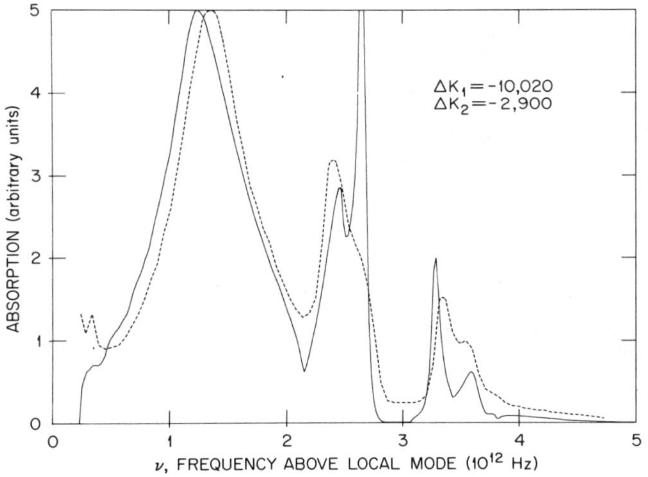

FIG. 3. Anharmonic sideband of the local-mode frequency of H^- ions in KBr. Dashed curve, experiment; solid curve, calculations.

and 3 represent roughly a 50% decrease in the nearest neighbor force constant when H^- is substituted for Cl^- and Br^- The quantum mechanical calculations indicate that the 1nn relax inwardly by approximately 1% of the nearest neighbor distance. This results in a reduction of the 1nn–4nn force constants (ΔK_2 in the figures) which, in the case of KBr, is roughly 30% as large as the defect 1nn force constants ΔK_1. The very sharp peak in H^-:KBr at ~2.6 THz which goes off scale for the theoretical calculations is due to a sharp E_g resonance which has just moved out of the gap between the acoustical and optical modes. This peak will be broadened and reduced in intensity by anharmonic interactions, an effect which the calculations do not include.

The agreement between theory and experiment in the case of the induced in-band absorption is less impressive than in the case of the sidebands. Nevertheless, quantitatively similar features do exist between the calculated and measured absorption spectra. The absorption spectrum measured by MacPherson and Timusk (1970) in KCl:H^- shows a broad band peaking at about 2.97 THz and a much narrower peak at 3.27 THz followed by sharply rising absorption in the restrahl region. Later work by Ward and Timusk (1972) showed that there is a peak at about 3.27 THz, a sharp minimum at 3.30 THz, and a broad shoulder and sometimes an actual peak at about 3.51 THz. The calculated curve in Fig. 4 gives the broad peak at 2.97 THz very well, shows a fairly sharp minimum in the region of 3.30 THz and indicates a pronounced shoulder at 3.50 THz; there is very little evidence of a separate peak just before the sharp 3.30 minimum. Ward and Timusk suggested that the 2.97 peak is a resonance or near resonance peculiar to

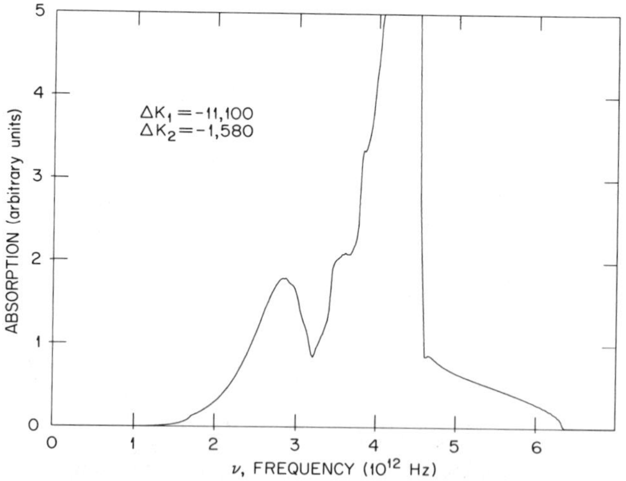

FIG. 4. Induced infrared absorption in KCl:H^-.

H^- in KCl and that the 3.27 peak is the same one exhibited by other impurities in KCl. If it is real, the 3.27 peak may be due to a van Hove singularity in the unperturbed density of states but it is unaccounted for by the calculations. Assuming the experimental results to be correct, there are two reasonable explanations for the discrepancy: (a) the shell-model phonon frequencies and eigenvectors for KCl are not accurate in the region around 3.27 THz and (b) the BZ interpolation scheme may encounter difficulties in regions where the eigenvectors and frequencies are changing rapidly. Wood and Ganguly point out that there is an interesting feature of the KCl calculations which does not appear in the experimental curves published thus far but which should show up with better resolution. This is the type M_2 van Hove singularity which occurs at 1.72 THz in both the infrared absorption and the perfect crystal density of states. The agreement between the calculated and measured absorption curves for KBr:H^- is roughly comparable to that for KCl:H^- but the results for KI:H^- are more difficult to assess. Space precludes a consideration of the experimental–theoretical discrepancies here and, moreover, it would be premature to attribute them entirely to the calculations.

B. Infrared and Raman Studies of NaCl:Ag^+

Substitutional silver in sodium chloride has been extensively investigated by thermal conductivity, infrared, and Raman techniques. A recent study of the infrared and Raman properties of NaCl:Ag^+ by Montgomery *et al.* (1972), hereafter referred to as MKGW, demonstrates the excellent agreement which can now be obtained between experiment and theory in such systems. A brief review of the work preceding that of MKGW will show how rapidly this agreement has improved in a relatively short time.

Caldwell and Klein (1967) observed a dip near 40°K in the thermal conductivity of NaCl crystals doped with Ag^+ and attributed it to phonon scattering from a resonant mode at 53 cm^{-1} previously observed in the infrared (Weber, 1964). They used a shell model, with parameters obtained before neutron scattering data were available, to calculate the Green's functions. The perturbation was given by a change in the mass and in the central force constant between the impurity and the first nearest neighbor Cl^- ions. Thermal conductivity data were reasonably well explained by this model if the central force constant was decreased by about 60% of the corresponding perfect crystal value. Subsequently Macdonald *et al.* (1969) studied the far infrared absorption of NaCl doped with various impurities, including Ag^+. They found that the model used by Caldwell and Klein could explain neither the strength of the 53 cm^{-1} resonance mode nor the defect-induced one-phonon peaks observed above 100 cm^{-1}. Agreement with experiment was improved with a model in which the noncentral force constant between the impurity and its neighbors was decreased by 10–15% and the central force constant slightly increased. This apparent failure of the central force constant model is now felt to be an artifact due primarily to the use of inaccurate "pre-neutron" shell model parameters. Möller and associates (Kaiser and Möller, 1969; Möller *et al.*, 1970) observed a strong 85 cm^{-1} E_g Raman peak at room temperature as well as several peaks at higher frequencies, the most prominent being a T_{2g} peak at 171 cm^{-1}. They interpreted their data with the help of earlier calculations by Benedek and Nardelli (1968) which were able to predict correctly the 53 cm^{-1} infrared resonance and the 85 cm^{-1} E_g Raman resonance for a 50% decrease in the central force constant. These calculations did not adequately describe the strong high-frequency T_{2g} peak or the high-frequency infrared structure and they failed to account for the finer details of the low temperature Raman spectra. In fact, until the recent work to be discussed here, no single model has been able to explain all the experimental results on NaCl:Ag^+. The force constant changes that were used to fit a given set of data seemed to vary strongly with the model used to calculate the host-crystal phonons. Since the previous calculations were done before the phonon dispersion curves for NaCl were experimentally determined, it was difficult to decide whether some of the discrepancies were due to inadequacies in the force constant models used for the impurity or in the model host-crystal phonons. Low temperature neutron scattering data are now available for NaCl and these have resulted in a better shell model. The primary purpose of the work of MKGW was to determine if this model used with a more elaborate treatment of the defect space (relaxation model) and more accurate computational procedures, could yield better agreement with both the infrared and Raman spectra for NaCl:Ag^+.

In Fig. 5, the experimental and calculated results for the far infrared absorption are compared, and in Figs. 6 and 7 similar comparisons are made for the E_g and T_{2g} Raman results. The calculated result for the 53 cm^{-1} peak in Fig. 5 can be brought into excellent agreement with experiment when $\Delta K_1 = -5000$ dyn/cm and $\Delta K_2 = -1000$ dyn/cm. The lack of better agreement between theory and experiment for the higher frequency peaks can probably be attributed to shortcomings of the shell model fit to the neutron data as suggested by the following considerations. The 120.5 cm^{-1} experimental infrared peak occurs at 115 cm^{-1} in the calculated curve. The source of this peak appears to be the acoustical modes in the region around the point $W(1, 0.5, 0)$ in the fcc Brillouin zone. Group theory predicts that W_3 will contribute to a T_{1u} defect activated spectrum. The neutron scattering data give a value of 120 cm^{-1} for this phonon frequency, whereas the shell model calculations give 115 cm^{-1}. Although the experimental peak at 131 cm^{-1} is not readily associated with phonons at any high symmetry points

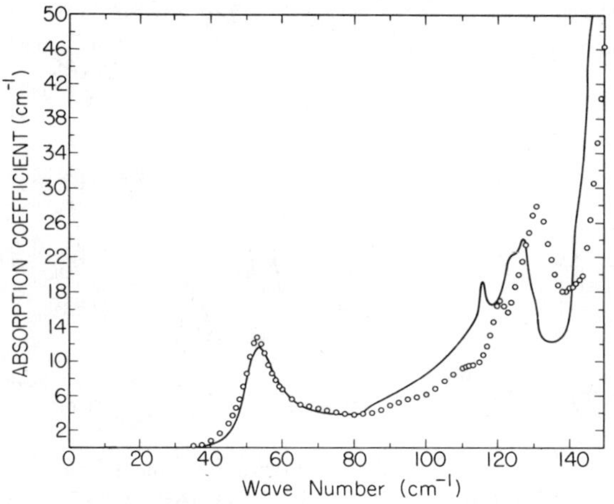

FIG. 5. The calculated (——) and experimental (○) infrared absorption in NaCl:Ag^+. The experimental data, collected at 7°K, are taken from Macdonald *et al.* (1969).

in **q** space, similar difficulties with the shell model fit may be responsible for the mismatch in this region.

Figures 6 and 7 show that the calculations of MKGW account for most of the features observed experimentally in the E_g and T_{2g} Raman spectra. The peaks at ~37 cm^{-1} in the E_g and T_{2g} spectra and 60 cm^{-1} in the E_g spectrum are not given by the calculations. These peaks are somewhat analogous to those appearing in the very low frequency region of the infrared sidebands in KCl:H^- and may be due to pairing effects with other impurities in the crystals. There is also a difficulty with a consistent mismatch between

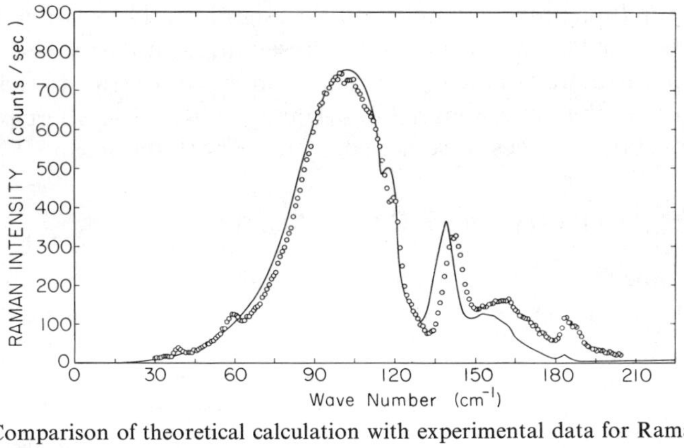

FIG. 6. Comparison of theoretical calculation with experimental data for Raman scattering from NaCl:Ag^+ at 7°K for the E_g mode.

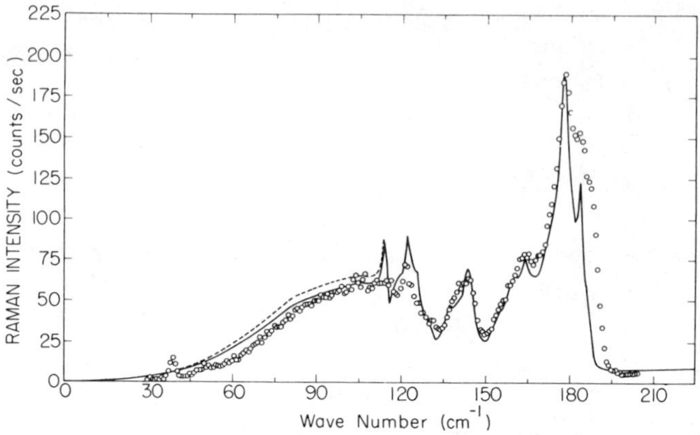

FIG. 7. Comparison of theoretical calculation with experimental data for Raman scattering from NaCl:Ag^+ at 7°K for the T_{2g} mode. Solid line, $\Delta K_1 = +750$ dyn/cm; dashed line, $\Delta K_1 = 0$.

the experimental and theoretical intensities in the high frequency regions; it is particularly noticeable in the E_g spectrum but may also occur in the T_{2g} spectra. The intensities of the experimental and theoretical curves were normalized to one another to give overall good agreement, but the difficulty in the high frequency region is not related to this. The origin of this mismatch was studied to a limited extent by MKGW but no satisfactory explanation was found.

The observed T_{2g} spectrum is in quite good agreement with the relevant unperturbed Green's function. This implies very little change in the tangential force constant between the Ag^+ and Cl^- ions relative to that of Na^+ and Cl^- in the perfect crystal. In contrast, for the E_g spectrum it is necessary to decrease the corresponding central force constants significantly in order to obtain the strong peak at 96 cm^{-1} which is not present in the unperturbed spectrum. This large perturbation of the E_g spectrum is in sharp contrast to the situation found for Tl^+ in the potassium halides (Harley *et al.*, 1971) where practically no force constant changes were necessary to fit both the E_g and T_{2g} spectra.

C. NEUTRON SCATTERING FROM DILUTE ALLOYS

The examples used in this subsection are taken entirely from the work of Kaplan and Mostoller in which they applied the CPA to one-dimensional alloys and to alloys of aluminum in copper and NH_4^+ in KCl. As stated in the Introduction, this reliance on work carried out at Oak Ridge is a matter of convenience and is not a slight of other workers in the field. The literature must be studied for a well-rounded view of recent developments in the theory of the lattice dynamics of random alloys.

Before considering the neutron scattering results for real crystals, let us consider a few features of the CPA that make it superior to the IDA. Figure 8 (Kaplan and Mostoller, 1974a) shows the comparison between CPA results and essentially exact results for a binary linear alloy with mass changes only. C gives the concentration, $\Delta = 0$ means there are no force constant changes, and $\varepsilon = 0.50$ is the ratio of the masses. In Fig. 8a, an impurity band has split off from the frequencies characteristic primarily of the host crystal. The CPA results are shown as a smooth curve and the "exact" results as histograms. The CPA gets the width of the impurity band approximately correct but smoothes over all the structure; it also counts the number of

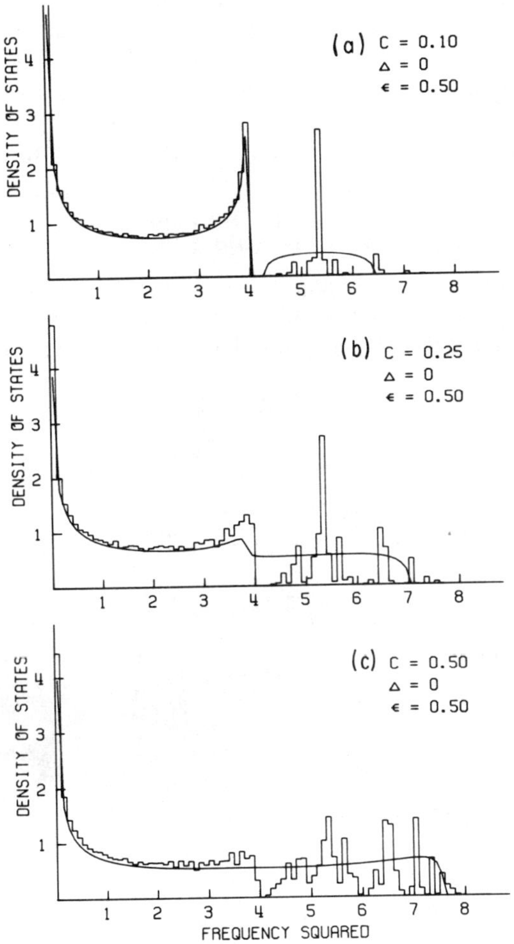

FIG. 8. CPA and exact densities of states for a monatomic alloy with mass changes only.

modes in each region correctly. As the concentration is increased (Fig. 8b and c) the impurity band merges with the host band in both the CPA and exact calculations. In contrast, at low concentration the IDA gives a δ-function spike at approximately the position of the highest peak in the exact results and becomes increasingly inadequate as the concentration is raised.

A comparison of the calculated and experimental neutron scattering cross section at the point $\mathbf{q} = (0.5, 0.5, 0.5)(2\pi/a)$ for a $Al_{0.1}Cu_{0.9}$ alloy is shown on Fig. 9 (Kaplan and Mostoller, 1974b). The rapid rise in the experimental values below the LA in-band peak at approximately 7 THz is due to high-order diffraction processes and should be discounted. The calculations were done in the mass defect approximation. It can be seen that the CPA gives the local mode position rather well but not the intensity relative to the in-band peak. This could indicate a force constant effect. There may be structure in the high-frequency region of the local mode peak that could reflect the existence of pair correlations not treated in the single-site CPA calculations discussed here. An effect should be mentioned here which has become apparent from attempts to calculate the shift of the in-band modes of metal alloys. In these calculations, both the IDA and the CPA give small but strongly frequency-dependent shifts of the in-band modes at low ($\sim 10\%$) concentrations. In contrast, the experiments clearly give much larger shifts which seem to suggest that an overall volume effect on the force constants is present. Nicklow *et al.* (1968) accounted for this in the aluminum–copper

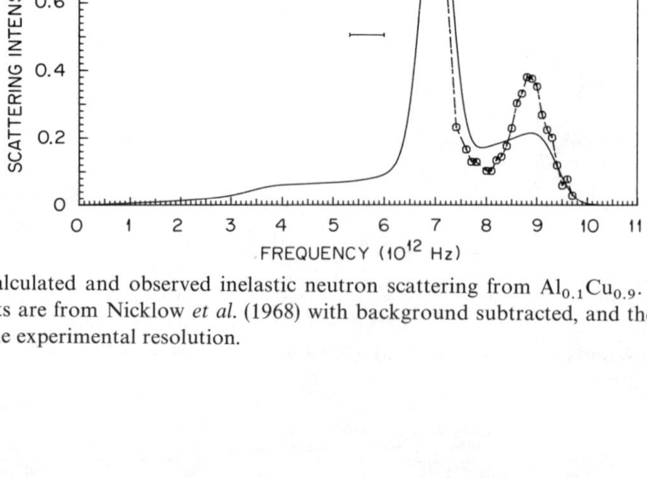

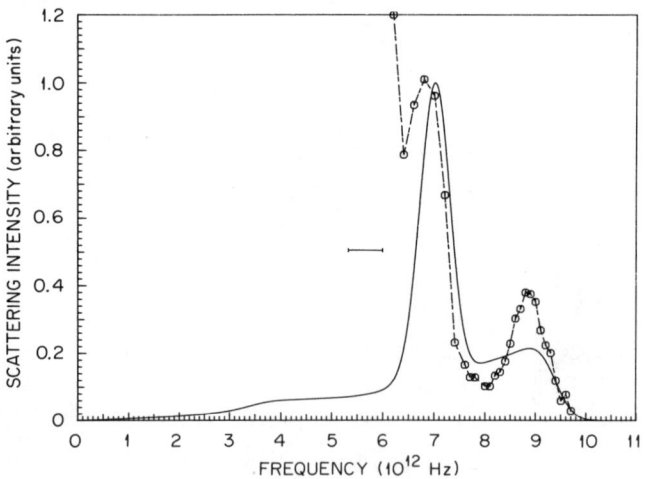

FIG. 9. Calculated and observed inelastic neutron scattering from $Al_{0.1}Cu_{0.9}$. The experimental points are from Nicklow *et al.* (1968) with background subtracted, and the horizontal bar shows the experimental resolution.

system by a simple scaling of the frequencies with a single-mode Gruneisin parameter. Bruno and Taylor (1971) attempted to do a more thorough job on $Au_{0.10}Cu_{0.90}$ but with mixed success. It should be recognized that bulk effects of this type are not easily treated within the framework of localized perturbation theory.

As a final example of the neutron scattering results, we can consider some results of the CPA calculations of Kaplan and Mostoller (1974c) on the system $(NH_4)_{0.07}K_{0.93}Cl$ in which both mass and force constant changes were included. The neutron scattering experiments (Smith *et al.*, 1972) revealed a local mode peak just above the top of the continuum of host-crystal modes. Unpublished calculations of the present author within the IDA showed that the mass defect approximation gave a sharp resonance just within the continuum and that an increase of 1nn central force constants of about 20% was necessary to shift the local mode to the observed position. Of course, these calculations gave no width to the local mode band. Figure 10 shows how nicely the CPA calculations do in the local mode region. Figure 10a gives the overall fit with experiment and Fig. 10b shows that even before the instrumental resolution function is folded in, the CPA gives a substantial width to the local mode band, as indicated by the dashed curve. The force constant change required to place the local mode in the right position is the same, within the accuracy of the calculations and experiments, as that found in the above-mentioned IDA calculations. This suggests that such calculations may be useful in providing estimates of force constant changes in CPA calculations. A cursory experimental study of the shifts of a few in-band phonons in the $(NH_4{}^+)$ KCl system was also made by Smith *et al.*, but the

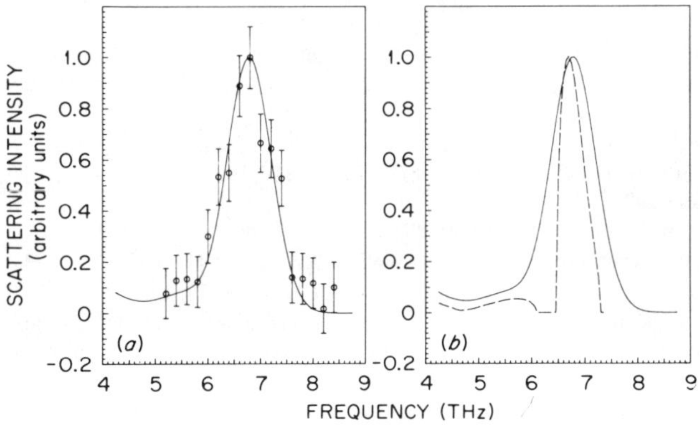

FIG. 10. (a) The inelastic neutron scattering cross section of $(NH_4)_{0.07}K_{0.93}Cl$ for the CPA with instrumental resolution included (solid line) and for the experiment with background subtracted (circles with error bars). (b) The CPA inelastic scattering cross section with (solid line) and without (dotted line) instrumental resolution.

associated error bars are quite large. Nevertheless, the rough comparison between calculated and measured shifts made by Kaplan and Mostoller suggest that the bulk effects discussed above may not be as large in insulators as they are in metals.

Acknowledgments

I wish to thank my colleagues and co-workers in the Solid State Division of ORNL, particularly J. F. Cooke, M. Mostoller, and T. Kaplan, for many useful discussions.

References

Benedek, G., and Nardelli, G. F. (1968). *Phys. Rev.* **167**, 837.
Born, M., and Huang, K. (1954). *In* "Dynamical Theory of Crystal Lattices," Chapter VII, p. 368. Oxford Univ. Press, London and New York.
Bruno, R., and Taylor, D. W. (1971). *Can. J. Phys.* **49**, 2496.
Caldwell, R. F., and Klein, M. V. (1967). *Phys. Rev.* **158**, 851.
Cooke, J. F., and Wood, R. F. (1972). *Phys. Rev. B* **4**, 1276.
Elliott, R. J., and Taylor, D. W. (1964). *Proc. Phys. Soc., London* **83**, 189.
Elliott, R. J., and Taylor, D. W. (1967). *Proc. Roy. Soc., Ser. A* **296**, 161.
Fritz, B. (1964). *In* "Proceedings of the 1963 International Conference on Lattice Dynamics." Pergamon, Oxford.
Fritz, B., Gross, U., and Bäuerle, D. (1965). *Phys. Status Solidi* **11**, 231.
Gethins, T., Timusk, T., and Woll, E. J., Jr. (1967). *Phys. Rev.* **157**, 744.
Gilat, G., and Raubenheimer, L. J. (1966). *Phys. Rev.* **144**, 390.
Harley, R. T., and Page, J. B., and Walker, C. T. (1971). *Phys. Rev. B* **3**, 1365.
Hodges, L., Ehrenreich, H., and Lang, N. D. (1966). *Phys. Rev.* **152**, 505.
Izyumov, Y. A. (1965). *Advan. Phys.* **14**, 569.
Janak, J. F. (1969). *Phys. Lett. A* **28**, 570.
Janak, J. F. (1971). *In* "Computational Methods in Band Theory" (P. M. Marcus, J. F. Janak, and A. R. Williams, eds.), p. 323. Plenum, New York.
Kaiser, R., and Möller, W. (1969). *Phys. Lett. A* **28**, 619.
Kaplan, T., and Mostoller, M. (1974a). *Phys. Rev. B* **9**, 353.
Kaplan, T., and Mostoller, M. (1974b). *Phys. Rev. B* **9**, 1783.
Kaplan, T., and Mostoller, M. (1974c). *Phys. Rev. B* **10**, 3610.
Klein, M. V. (1968). *In* "Physics of Color Centers" (W. Beall Fowler, ed.), Chapter 7, p. 429. Academic Press, New York.
Koster, G. F., and Slater, J. C. (1954). *Phys. Rev.* **96**, 1208.
Lakatos, K., and Krumhansl, J. A. (1968). *Phys. Rev.* **175**, 841.
Lakatos, K., and Krumhansl, J. A. (1969). *Phys. Rev.* **180**, 729.
Lifshitz, I. M. (1956). *Nuovo Cimento* **3**, Suppl. 4, 716. (This reference contains references to Lifschitz earlier work on the subject published in the Russian literature beginning around 1947.)
Macdonald, H. F., Klein, M. V., and Martin, T. P. (1969). *Phys. Rev.* **177**, 1292.
MacPherson, R. W., and Timusk, T. (1970). *Can. J. Phys.* **48**, 2917.
Maradudin, A. A., and Vosko, S. H. (1968). *Rev. Mod. Phys.* **40**, 1.
Maradudin, A. A., Montroll, E. W., Weiss, G. H., and Ipatova, I. P. (1971). *Solid State Phys., Suppl.* **3**.

Messiah, A. (1962). *In* "Quantum Mechanics," Vol. II, Chapter XIX, p. 830. North–Holland Publ., Amsterdam.

Möller, W., Kaiser, R., and Bilz, H. (1970). *Phys. Lett. A* **32**, 171.

Montgomery, G. P., Jr., Klein, M. V., Ganguly, B. N., and Wood, R. F. (1972). *Phys. Rev. B* **6**, 4047.

Mostoller, M., and Wood, R. F. (1973). *Phys. Rev. B* **7**, 3953.

Mueller, F. M., Garland, J. W., Cohen, M. H., and Bennemann, K. H. (1971). *Ann. Phys. (N.Y.)* **67**, 19.

Nicklow, R. M., Vijayaraghavan, P. R., Smith, H. G., Dolling, G., and Wilkinson, M. K. (1968). *In* "Neutron Inelastic Scattering," Vol. I, p. 43. IAEA, Vienna.

Rosenstock, H., and Klick, C. (1960). *Phys. Rev.* **119**, 1198.

Schaefer, G. (1960). *J. Phys. Chem. Solids* **2**, 233.

Slater, J. C., and Koster, G. F. (1954). *Phys. Rev.* **94**, 1498.

Smith, H. G., Wakabayashi, N., and Nicklow, R. M. (1972). *In* "Neutron Inelastic Scattering," p. 73. IAEA, Vienna.

Soven, P. (1967). *Phys. Rev.* **156**, 809.

Taylor, D. W. (1967). *Phys. Rev.* **156**, 1017.

Timusk, T., and Klein, M. V. (1966). *Phys. Rev.* **141**, 664.

van Hove, L. (1954). *Phys. Rev.* **95**, 249.

Wallis, R., and Maradudin, A. (1960). *Progr. Theor. Phys.* **24**, 1055.

Ward, R. W., and Timusk, T. (1972). *Phys. Rev. B* **5**, 2351.

Weber, R. (1964). *Phys. Lett.* **12**, 311.

Wood, R. F., and Ganguly, B. N. (1973). *Phys. Rev. B* **7**, 1591.

Wood, R. F., and Gilbert, R. L. (1967). *Phys. Rev.* **162**, 746.

Wood, R. F., and Opik, U. (1967). *Phys. Rev.* **162**, 736.

Lattice Dynamics of Surfaces of Solids*

F. W. DE WETTE AND G. P. ALLDREDGE[†]

DEPARTMENT OF PHYSICS
THE UNIVERSITY OF TEXAS
AUSTIN, TEXAS

I. Introduction

A. General Remarks

EXPERIMENTAL AND THEORETICAL STUDIES of crystal surface phenomena have undergone a major development during the past decade. This development is due to important refinements in experimental surface techniques, to a concomitant development in theoretical surface studies, and to the strong incentive provided by the technological importance of surface phenomena in a variety of applications. This article deals with a very limited but well-defined area of surface studies—namely, numerical calculations of static, dynamic, thermodynamic, and some scattering properties of idealized simple but realistic crystal surfaces—all associated with the vibrational properties of

* Work supported in part by the U.S. Air Force Office of Scientific Research under Grant No. AFOSR-71-1973 and in part by the Robert A. Welch Foundation.

† Present address: Graduate Center for Materials Research, University of Missouri, Rolla, Missouri 65401.

such surfaces; surface electronic problems as such will not be discussed.

The theory of surface vibrations, both in continuum and lattice models, has been the subject of a number of recent reviews (Farnell, 1970; Maradudin *et al.*, 1971; Lenglart *et al.*, 1972; Wallis, 1973); these reviews are recommended for general background and for details falling outside the scope of this article. Also, for an experimentalist's perspective of surface vibrations and their experimental manifestations, we cite the forthcoming review by Lagally (1975).

In spite of the increased interest in surface dynamics of discrete lattices, the major amount of work on this subject has been performed for elastic continuum models or for highly simplified lattice models. The continuum approximation is, of course, valid only in the long wavelength limit, where the microscopic discreteness of the solid is unimportant. For wavelengths of the order of a few atomic spacings, elastic-continuum theory is no longer valid and the methods of lattice dynamics are necessary. In fact, even at arbitrarily long wavelengths, it has recently been found that there are low-frequency acoustic surface modes which cannot be obtained from elastic-continuum theory (Alldredge, 1972; Zoth *et al.*, 1974).

A complete understanding of surface vibrations therefore requires detailed lattice-dynamics calculations for realistic models. One reason for the interest in such an understanding is the influence of surface vibrations on other surface properties. These vibrations affect low-energy electron-diffraction (LEED) intensities in a number of ways and have an important effect on the scattering of atoms and molecules from surfaces at sufficiently low energies. The vibrational spectrum also determines, in part, the surface specific heat, surface free energy, and other thermodynamic functions. Moreover, the surface phonons should play an important role in determining the electron–phonon interaction in sufficiently thin films; two effects of this electron-surface vibration interaction are a significant modification of superconducting-transition temperatures T_c in very thin films and a sensitivity of T_c to adsorption of inert gases on thin films (Naugle *et al.*, 1973). As a final example, the structure of the surface is in some cases dependent upon temperature, and this fact implies that the structure is affected by the lattice vibrations. Even if the structure does not change, the thermal expansion at a surface is determined by the lattice vibrations.

The presence of vibrational states localized at the surface (surface modes) results from the symmetry breaking property of the surface. In particular, the absence of translational symmetry in the direction perpendicular to the surface prevents the existence, in that direction, of wavelike solutions for the vibrational states of the crystal. This results in a significant complication of the numerical problem of calculating the vibrational states of a crystal with surfaces. For instance, for a crystal with s atoms per unit cell, in order to find the bulk states for a given (three-dimensional) wave vector $\mathbf{q}$, one has to diagonalize a $3s \times 3s$ dynamical matrix; however, to find the vibrational states of a slab with N_3 layers of the same crystal, one has to diagonalize a $3sN_3 \times 3sN_3$ dynamical matrix for each two-dimensional wave vector. This added computational complication has emphasized the reliance on high speed, large memory computers for these kinds of calculations.

In keeping with the theme of *computational* physics, we focus our discussion in this article on methods which have been successfully applied to the study of surface vibrations (and their manifestations) in what we shall term "realistic" models of crystal surfaces. "Realism" in this usage means that such complications as bulk anisotropy, crystal structures that actually occur in nature, and arbitrary wave vector in the surface Brillouin zone (SBZ) must be dealt with. The ideal of realism cannot, however, be carried too far at present, and we limit ourselves here to perfect surfaces for which the symmetry parallel to the surface plane is the same as similar planes in the bulk (but allowing relaxation to new equilibrium positions consistent with this symmetry); thus we shall only mention in passing the problems of vibration of surface defects, reconstructed surfaces, and adsorbed molecules. Even this limited dose of realism carries us out of what may be characterized as the domain of the *mathematical* physics of surface lattice dynamics, in which answers to conceptual, qualitative questions are sought by highly simplifying physical models to allow approximate "analytic" solutions (in many cases, of course, these may in fact require nontrivial computational efforts to evaluate the solutions). Although such efforts have been of great value in developing our concepts of surface modes, are still fruitful lines of inquiry, and provide test cases for computational schemes, they lie outside the scope of this paper. Fortunately, the review by Wallis (1973) gives a comprehensive overview of recent developments in this field. In addition, the interested reader should consult relevant sections of the monographs by Maradudin *et al.* (1971) and by Ludwig (1967).

Before taking up our discussion of the computational methods, we must acknowledge one further issue that is particularly sticky—it is in fact a major reason for interest in surface vibrations: What *is* the interatomic interaction in the neighborhood of the surface? Surface physics at present suffers the lack of experimental methods capable of the precision obtainable in the bulk state by X-ray crystallography for particle positions (especially coordinates normal to the surface) and by inelastic neutron scattering for dynamics. The problem is particularly acute for the covalently bonded crystals and for metals. The desaturation of the highly directional covalent bonds at the surface usually leads to a significant reconstruction of the surface, so that the surface geometry is manifestly changed from the bulk and presumably the force constants are likewise significantly modified in ways difficult to predict from bulk considerations. In the case of metals, particularly the sp-bonded metals, the effective interionic interaction arises from the sea of delocalized conduction electrons. When a surface is created, the resulting imbalance of the Madelung force on the ions can give rise to significant inward forces on the outermost ions (Finnis and Heine, 1974), as can the extended dipole distribution set up by electrons forming a Friedel peak which can be enhanced and supplemented by crystalline effects (Alldredge and Kleinman, 1974a,b). This inward relaxing force, predicted on the basis of fundamental principles of metals, is in distinct contrast to the outward relaxation of surface ions that is almost invariably predicted from models assuming a simple pairwise description of interatomic interactions. This result suggests that simple pairwise interatomic potentials fitted to bulk properties are of questionable value in predicting near-surface force constants for metals. It is to be hoped that as theoretical work progresses on the self-consistent electron systems of covalent and metallic crystal surfaces, improved models of the interatomic interactions can be developed. In the meantime, some progress can continue to be made by Born–von Kármán force constant parameterizations in which a few force constants near the surface are modified in an ad hoc, but consistent, manner to seek to reproduce as best possible the various kinds of experimental results.

Because of their closed-shell electronic configurations, the noble gas and simple ionic crystals offer a significant contrast to metals and covalent crystals. Here we have some confidence that relatively simple, semiempirical models of the interionic potential energy fitted to bulk properties retain their validity in the neighborhood of surfaces, so that such models provide us "first principles" from which to derive surface relaxation and the surface-induced changes in near-surface elements of the dynamical matrix. Consequently, in this article we shall draw most of our examples illustrating the computational results from work dealing with noble gas and ionic crystals.

In the remainder of Section I we give a brief overview of the computational methods and of the crystal models used. In Section II we set out and discuss the formulations of the computational methods. Finally, in Section III we present a selected array of illustrative results and discuss some comparisons with experiment.

B. Overview of the Computational Methods

Consider a crystal bounded by a planar surface but extending to infinity in the directions parallel to the surface. This latter condition is taken to mean that we may apply periodic boundary conditions in the parallel directions, enclosing within one period of these boundary conditions $\bar{N}$ surface unit meshes of the two-dimensional lattice periodicity remaining. Two principal approaches to determining the dynamics of such a sample are those of molecular dynamics and of the more traditional lattice dynamics; the latter can be considered in a sense to be the linearized form of the former.

Lattice dynamics (LD) in the quasiharmonic approximation is valid for the treatment of surface vibrations at low temperatures; for instance, for Lennard-Jones (LJ) crystals it has been shown to be valid at temperatures below about one sixth of the melting temperature T_M (as compared to about one third T_M for bulk calculations). This more limited range of validity of surface lattice dynamics is a result of the larger amplitudes of vibration at the surface. *Molecular dynamics* (MD), on the other hand, is a classical method in which the Newtonian equations of motion of a system of interacting particles are numerically solved for a series of closely spaced successive time steps. In this method the system is allowed to seek its own particle equilibrium positions, and the computed particle motions are not necessarily harmonic. Thus the method takes into account thermal expansion and anharmonic effects in a completely natural and automatic fashion; however, being classical (it neglects the zero point motions), it is a high-temperature method. For the example of LJ crystals, MD is valid for temperatures above about one half of the Debye temperature. It is thus seen that LD and MD are complementary methods. Again, for the example of LJ crystals, it is found that both methods, in the intermediate temperature range, lead to the same results for the particle motions deep inside the crystal.

To implement either of these approaches we must choose specific computational methods. There have been three major classes of methods successfully applied to the study of dynamics of realistic models of crystals, which we may conveniently label as follows: (1) *direct methods* of numerical solution of the dynamical equations for slabs of finite thickness (both LD and MD); (2) *matching methods*, in which bulk LD solutions are analytically continued to complex wave vector perpendicular to the surface and solutions sought which satisfy surface boundary conditions; and (3) *Green function (GF) methods* analogous to those applied to the LD of point defects (e.g., Maradudin *et al.*, 1971; Wood, this volume). Of these three classes, the last two have been most widely used in analytical studies of very simple models of surfaces, with the GF methods more highly favored; the direct numerical methods are hardly represented.

But in the applications to realistic systems, the emphasis to date has been exactly the opposite. That reversal of emphasis is reflected in this article, where most of our attention is focused on a crystal slab containing a limited number of ionic planes parallel to its two free surfaces. In lattice dynamics, the vibrational states of this model crystal, including the surface states, can be found by straightforward diagonalization of the dynamical matrix for a representative set of (two-dimensional) wave vectors. The main limitation of this method is that, because of the finite thickness of the slab, it does not correctly reproduce the deeply penetrating surface modes of a macroscopic crystal in the extreme long wavelength limit. Also because of its numerical character, certain analytic results are not brought out explicitly, such as the temperature dependences in the limits of very low and high temperature of, for instance, the thermodynamic quantities. These disadvantages are offset by the fact that the method is computationally straightforward and gives excellent results (except for extremely long wavelengths) for slabs of very limited thicknesses (about 20 layers); in fact, the complete richness of detail that is contained in the original model is obtained, since no analytical approximations are made in the calculation. An additional advantage in the case of crystals having a reasonably approximate model of the interatomic potential (e.g., the LJ model of noble gases) is that molecular dynamics, a totally different direct method valid at high temperatures, can be applied to the same system, so that high temperature results (including anharmonicity and thermal expansion) can be obtained, without resort to perturbation theory or the self-consistent phonon approximation.

C. Crystal Models Used

1. *Noble Gas Crystals (LJ Crystals)*

Very extensive calculations have been performed on the surface dynamics of fcc noble gas crystals in a model in which the particles interact through a Lennard-Jones potential

$$\phi(r) = 4\varepsilon[(\sigma/r)^{12} - (\sigma/r)^{6}], \tag{1}$$

where ε and σ are potential parameters. A particle interacts with all of its neighbors, and the displacements of the particles near the surface from their positions in the bulk are taken into account. An advantage of this potential is that it yields results that depend only on the shape of the potential and not on the potential parameters. Some other advantages are discussed elsewhere (Allen and de Wette, 1969c). We regard this model as the best that can be used for general studies of surface vibrational properties, i.e., a more accurate interaction model would involve specializing to a particular material.

2. *Ionic Crystals*

In the Kellermann rigid-ion model (KRIM) (Kellermann, 1940) the ionic crystal is modeled as a lattice of positive and negative point ions which interact with the short-range repulsive forces (Born–Mayer potential) acting between nearest neighbors only, and long-range Coulomb forces acting between all the particles in the crystal. It is well known that the KRIM is unsatisfactory in that it neglects the polarizations of the ions. This is overcome in the shell model (SM), which takes polarization into account in an approximate way. For the bulk the SM yields results which are in much better agreement with, for instance, the experimentally determined phonon dispersion curves, and it may be expected that this will be even more true for surface calculations, since the presence of the surface is an additional cause of polarization of the surface ions. Chen *et al.* (1971a,b, 1972a,b,c, 1973) were the first to apply the shell model to surface dynamics calculations for a variety of ionic crystals. For the surface dispersion curves the differences between the results of the KRIM and the SM are indeed found to be substantial (see Section III, B).

3. *Force Constant Models*

As mentioned earlier, when no reasonably reliable approximations to the interatomic potential are available, significant progress can still be made by assuming force constant models, for which a small number of the parameters near the surface can be modified from their bulk values in attempts to attain agreement with the limited experimental results, such as the mean-square amplitudes of vibration inferred from LEED. In the use of such models, care must be taken to ensure that the set of force constants satisfies conditions of rigid-body rotational invariance. Ludwig and Lengeler (1964) pointed out that this invariance is necessary for consistency between the long wavelength limit of the acoustic surface modes in the lattice and the Rayleigh waves in continuum theory; Musser and Rieder (1970) have reported a study of the effects of violation of rotational invariance for a force constant model of the (100) surface of nickel. A particularly useful way to ensure rotational invariance of a force constant parameterization is to derive the Born–von Kármán force constants from a parameterization of the harmonic potential in terms of deformations of invariant internal coordinates, such as bond lengths and interbond angles (e.g., Gazis *et al.*, 1960; Gazis and Wallis, 1966; Brady, 1971); this approach is essentially that of the valence-force-field potential widely used in molecular vibrations and more recently developed for the (bulk) lattice dynamics of crystals containing significant covalent bonding [for extensive references see Section 3 of the review by Sinha (1973)].

II. Formulations

In this section we develop the computational formalisms of the methods introduced above and discuss techniques of solution. We first deal with the direct methods, and in so doing establish notations useful for the subsequent, more brief discussions of the matching and Green function methods.

A. Lattice Dynamics of a Crystal Slab and Its Direct Solution

1. *The Dynamical Matrix Eigenproblem*

a. Formulation and Notation. Allen *et al.* (1971a) have presented a general formulation and notation for the lattice dynamics of crystal slabs which we adopt here. We consider an orthogonal coordinate system in which the x and y directions are chosen in suitable directions parallel to the surface; the z direction is perpendicular to the surface. The number of bulk lattice planes parallel to the surface is N_3. The lattice planes will be labeled by the index m in the special case that $m = 1$ refers to a surface plane; in general, however, the lattice planes are labeled by the index l_3 (see below). The instantaneous position of a particle (atom or ion) is represented by $\mathbf{r}(\mathbf{l}, \kappa)$, such that

$$\mathbf{r}(\mathbf{l}, \kappa) = \mathbf{r}_0(\mathbf{l}, \kappa) + \mathbf{u}(\mathbf{l}, \kappa) \tag{2}$$

where $\mathbf{r}_0(\mathbf{l}, \kappa)$ gives the mean position of the particle and $\mathbf{u}(\mathbf{l}, \kappa)$ is the time-dependent displacement. The set of numbers $\mathbf{l} = (l_1, l_2, l_3)$ and the index κ specify a particular particle, l_3 labels a lattice plane parallel to the surface, and (l_1, l_2) specifies the points in a two-dimensional lattice which spans one of these planes, κ distinguishes the different particles in the unit cell associated with a particular $\mathbf{l}$. We assume the same two-dimensional lattice in all the layers but not necessarily the same two-dimensional unit cells for all layers.

As is customary we assume that the total potential energy of the system can be expanded in a Taylor series

$$\Phi - \Phi_0 = \sum_{l\kappa\alpha} \Phi_\alpha(\mathbf{l}\kappa)u_\alpha(\mathbf{l}\kappa) + \frac{1}{2}\sum_{\substack{\alpha\beta \\ l\kappa l'\kappa'}} \Phi_{\alpha\beta}(\mathbf{l}\kappa; \mathbf{l}'\kappa')u_\alpha(\mathbf{l}\kappa)u_\beta(\mathbf{l}'\kappa') + \cdots \tag{3}$$

where

$$\Phi_\alpha(\mathbf{l}\kappa) = \left(\frac{\partial\Phi}{\partial u_\alpha\,(\mathbf{l}\kappa)}\right)_0 \tag{4}$$

$$\Phi_{\alpha\beta}(\mathbf{l}\kappa; \mathbf{l}'\kappa') = \left(\frac{\partial^2\Phi}{\partial u_\alpha\,(\mathbf{l}\kappa)\,\partial u_\beta\,(\mathbf{l}'\kappa')}\right)_0, \qquad (\mathbf{l}\kappa) \neq (\mathbf{l}'\kappa'), \tag{5a}$$

and the self-interaction force constants are given by

$$\Phi_{\alpha\beta}(\mathbf{l}\kappa;\mathbf{l}\kappa) = \Phi_{\beta\alpha}(\mathbf{l}\kappa;\mathbf{l}\kappa) = -\sum_{\mathbf{l}'\kappa'}{}' \Phi_{\alpha\beta}(\mathbf{l}\kappa;\mathbf{l}'\kappa') \tag{5b}$$

The zero subscript indicates that these quantities are evaluated with all particles at their mean positions, and the prime on the summation in Eq. (5b) signifies that the $(\mathbf{l}'\kappa') = (\mathbf{l}\kappa)$ term is omitted.

In the quasiharmonic approximation (QHA) one neglects all terms but the second one on the right-hand side of (3), the other terms being considered as a small perturbation. In this approximation the equations of motion are

$$M_\kappa(l_3)(d^2/dt^2)u_\alpha(\mathbf{l}\kappa) = -\sum_{\mathbf{l}'\kappa'\beta} \Phi_{\alpha\beta}(\mathbf{l}\kappa;\mathbf{l}'\kappa')u_\beta(\mathbf{l}'\kappa') \tag{6}$$

We note that in the full QHA the force constants $\Phi_{\alpha\beta}(\mathbf{l}\kappa;\mathbf{l}'\kappa')$ are to be evaluated at the mean positions of the particles, with bulk and surface thermal expansion taken into account, rather than at the positions of static equilibrium.

It is useful to introduce two-dimensional vectors with only x and y components, and indicate these with superior bars: If $\mathbf{r} = (x, y, z)$ and $\mathbf{l} = (l_1, l_2, l_3)$, then $\bar{\mathbf{r}} = (x, y)$ and $\bar{\mathbf{l}} = (l_1, l_2)$. In this notation Eq. (2) can be written

$$\mathbf{r}(\mathbf{l}\kappa) = \bar{\mathbf{r}}_0(\bar{\mathbf{l}}) + \mathbf{r}_0(l_3\kappa) + \mathbf{u}(\mathbf{l}\kappa) \tag{7}$$

Here $\mathbf{r}_0(l_3\kappa)$ is the "basis vector" that gives the mean position of a particle (of the κth type in the l_3th lattice plane) within the large unit cell associated with the lattice point $\bar{\mathbf{l}}$; its projection on the xy plane is $\bar{\mathbf{r}}_0(l_3\kappa)$, and its z component is $z_0(l_3\kappa)$ [this notation allows for $z_0(l_3\kappa) \neq z_0(l_3\kappa')$ near the surface even if $z_0(l_3\kappa) = z_0(l_3\kappa')$ in the bulk].

Since the crystal is invariant under a translation through a two-dimensional lattice vector the force constants $\Phi_{\alpha\beta}(\mathbf{l}\kappa;\mathbf{l}'\kappa')$ depend only on the difference of $\bar{\mathbf{l}}'$ and $\bar{\mathbf{l}}$:

$$\Phi_{\alpha\beta}(\mathbf{l}\kappa;\mathbf{l}'\kappa') = \Phi_{\alpha\beta}(l_3;l_3'\kappa';\bar{\mathbf{l}}'-\bar{\mathbf{l}}) \tag{8}$$

This two-dimensional translational invariance of the force constants implies that the normal-mode solutions to Eq. (6) have the form of two-dimensional Bloch functions

$$u_\alpha(\mathbf{l}\kappa) = [\bar{N}M_\kappa(l_3)]^{-1/2}Q_0\xi_\alpha(l_3\kappa)\exp[i(\bar{\mathbf{q}}\cdot\bar{\mathbf{r}}_0(\bar{\mathbf{l}}) + \bar{\mathbf{q}}\cdot\bar{\mathbf{r}}_0(l_3\kappa) - \omega t)], \tag{9}$$

where $\bar{N}$ is the number of two-dimensional lattice points ($\bar{N} \to \infty$), $M_\kappa(l_3)$

the mass of a particle with index κ in the l_3th plane, Q_0 gives the amplitude of vibration, $\xi_\alpha(l_3\kappa)$ is normalized to unity (see below), $\bar{\mathbf{q}}$ is a two-dimensional wave vector in the xy plane, and ω is the vibrational frequency. If Eq. (9) is substituted in Eq. (6) we obtain the dynamical matrix eigenvalue equation

$$\sum_{l'_3\kappa'\beta} D_{\alpha\beta}(l_3\kappa; l'_3\kappa'; \bar{\mathbf{q}})\xi_\beta(l'_3\kappa'; \bar{\mathbf{q}}p) = \omega^2(\bar{\mathbf{q}}p)\xi_\alpha(l_3\kappa; \bar{\mathbf{q}}p) \tag{10}$$

where the elements of the dynamical matrix (DM) are defined by

$$D_{\alpha\beta}(l_3\kappa; l'_3\kappa'; \bar{\mathbf{q}}) = [M_\kappa(l_3)M_{\kappa'}(l'_3)]^{-1/2} \sum_{\bar{\mathbf{l}}'} \Phi_{\alpha\beta}(l_3\kappa; l'_3\kappa'; \bar{\mathbf{l}}' - \bar{\mathbf{l}}) \times \exp\{i\bar{\mathbf{q}} \cdot [\bar{\mathbf{r}}_0(\bar{\mathbf{l}}' - \bar{\mathbf{l}}) + \bar{\mathbf{r}}_0(l'_3\kappa') - \bar{\mathbf{r}}_0(l_3\kappa)]\} \tag{11}$$

[Note that the DM defined by Eq. (11) is not periodic under a translation by a two-dimensional reciprocal lattice vector $\bar{\mathbf{G}}$.] In Eq. (10) the index p distinguishes the different modes corresponding to a particular $\bar{\mathbf{q}}$. If there are $s(l_3)$ particles per unit cell in the l_3th layer [i.e., $\kappa = 1, 2, \ldots, s(l_3)$], then $p = 1, 2, \ldots, 3\mathcal{N}$ where $\mathcal{N} = \sum_{l_3} s(l_3)$. The eigenvectors $\xi_\alpha(l_3\kappa; \bar{\mathbf{q}}p)$ have $3\mathcal{N}$ components; since they are eigenvectors of an Hermitian matrix they can be chosen to satisfy the usual orthonormality condition

$$\sum_{l_3\kappa\alpha} \xi_\alpha(l_3\kappa; \bar{\mathbf{q}}p)\xi^*_\alpha(l_3\kappa; \bar{\mathbf{q}}p') = \delta_{pp'}, \tag{12}$$

i.e., the eigenvectors are normalized to unity over the whole thickness of the crystal slab. In the special case of a crystal whose bulk structure coincides with a Bravais lattice the indices κ and κ' can be suppressed and Eq. (10) becomes

$$\sum_{l'_3\beta} D_{\alpha\beta}(l_3l'_3; \bar{\mathbf{q}})\xi_\beta(l'_3; \bar{\mathbf{q}}p) = \omega^2(\bar{\mathbf{q}}p)\xi_\alpha(l_3; \bar{\mathbf{q}}p) \tag{13}$$

In this case $\mathcal{N} = N_3$, the number of lattice planes in the slab, so $p = 1, 2, \ldots, 3N_3$.

b. Use of Symmetry. In surface problems the amount of crystal symmetry available to simplify and organize the solutions of the dynamical equation is greatly reduced from that available in bulk problems. If no symmetry operations involving the z direction exist, the creation of a surface takes us from the possibilities of 230 triperiodic space groups on the 14 Bravais lattices to 17 *diperiodic* plane groups on 5 plane nets; in considering slabs which allow symmetry operations involving z (inversion and mirror and glide reflections in the median plane), we expand the symmetry possibilities to 80 diperiodic groups in three dimensions on the same 5 plane nets (e.g., Wood, 1964a). With the (usual) exception of boundaries of the two-dimensional surface Brillouin zone (SBZ), we can think of the high symmetry points and lines of the SBZ to be the projection on the plane of perpendicular three-dimensional symmetry lines and planes, respectively; thus the threefold degeneracies associated with three-dimensional *points* of high symmetry will be missing in surface problems. Those portions of the SBZ boundaries which are equivalent by virtue of a translation by a two-dimensional reciprocal lattice vector $\bar{\mathbf{G}}$ and which are mapped into each other by operations from the plane group of a particular surface are lines of symmetry less obviously related to the more familiar high symmetry wave vectors of the triperiodic bulk.

The matrix eigenvalue problem of Eq. (10) is in general that of a complex Hermitian matrix of dimension $3sN_3$. Symmetry should be used as much as possible to reduce the computational requirements of storage $[\frac{1}{2}3sN_3(3sN_3 + 1)$ words for real part and $\frac{1}{2}3sN_3(3sN_3 - 1)$ for imaginary$]$ and time (full diagonalization requires time proportional to the cube of dimension). A corollary benefit of symmetry reduction is that, in many cases, the reduction separates into different symmetry classes some eigenvectors whose eigenvalues are very close, and this separation enables a more precise computation of such quasidegenerate eigenvectors. We may illustrate some principal uses of symmetry by reference to Fig. 1 which displays the SBZ's and surface unit cells (SUC) for the fcc(100), (110) and (111) surfaces. The

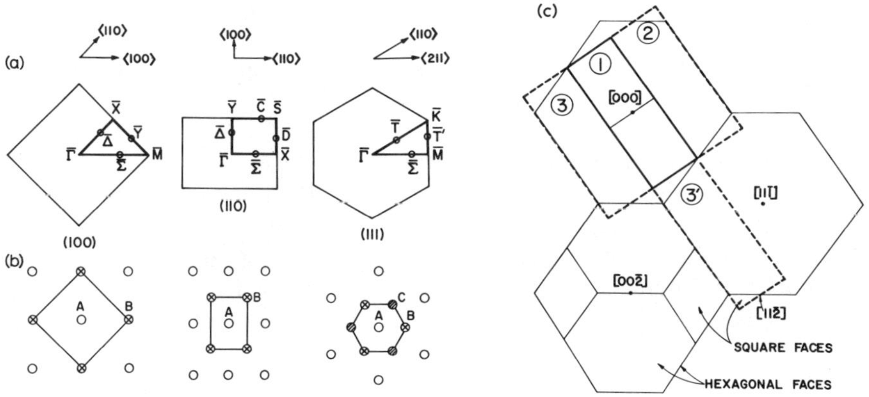

FIG. 1. (a) Surface Brillouin zones for the fcc (100), (110), and (111) surfaces; an irreducible element is outlined in each case. (b) Surface unit cells generating (a); projections of lattice points of the stacking sequences ABAB . . . and ABCABC . . . are shown. (c) $(1\bar{1}0)$ cross section of the first three-dimensional BZ and neighboring regions, illustrating the slab-adapted BZ scheme. After Allen *et al.* (1971b). Reprinted with permission.

first simplifications to be sought are those holding for general $\bar{\mathbf{q}}$. If every atom is on a twofold axis perpendicular to the surface [e.g., fcc(100), (110)], then arguments analogous to those for centers of inversion in the bulk show that all elements of the DM defined by Eq. (11) are real, except that those coupling the z direction to the parallel directions are imaginary; then the unitary transformation signified by $\xi_{x,y} \to \xi_{x,y}$, $\xi_z \to i\xi_z$ makes the DM real and symmetric (and the transformed eigenvectors can of course be assumed real), thereby reducing the storage and associated time requirements. If the median plane of a slab is a mirror plane [e.g., fcc(100), (110), N_3 = odd] or a glide-reflection plane [e.g., fcc(100), (110), N_3 = even], then the DM can be reduced to two diagonal blocks belonging to the symmetric (+) or antisymmetric (−) irreducible representations (irreps) of the mirror group by the symmetry coordinates signified by $\| l_3|, \kappa, \alpha, \pm\rangle = 2^{-1/2}[|l_3, \kappa, \alpha\rangle \pm |-l_3, \kappa, \alpha\rangle]$ ($\alpha = x,y$) and $\| l_3|, \kappa, z, \pm\rangle = 2^{-1/2}[|l_3, \kappa, z\rangle \mp |-l_3, \kappa, z\rangle]$; here this notation presupposes that the "origin" of the lattice plane index l_3 is chosen in the median plane ($l_3 = \pm 1, \pm 2, \ldots$, for N_3 = even; $l_3 = 0, \pm 1, \ldots$, for N_3 = odd), and for N_3 = odd straightforward modifications must be made in the symmetry coordinates in the case of $l_3 = 0$. If the median plane contains a center of inversion [e.g., fcc(111), N_3 = odd], the DM may be transformed to a real symmetric matrix of the same dimension [Allen and de Wette, 1969a; back transformation corrected in Appendix A (Allen *et al.*, 1971a)]. In addition to the symmetry benefits for general $\bar{\mathbf{q}}$, there are reductions for the high symmetry lines and points in the SBZ. The most straightforward of these cases occurs when the sagittal plane (defined by $\bar{\mathbf{q}}$ and the surface normal) coincides with a reflection plane, in which case the DM can be reduced into two blocks: the one belonging to the symmetric irrep contains two thirds of the modes, and the polarizations are ellipses parallel to the sagittal plane; the other block, belonging to the antisymmetric irrep, contains the remaining one third modes which are linearly polarized normal to the sagittal plane. The latter modes will be labeled SH (shear-horizontal), and the former, SP (sagittal-plane). In Fig. 1, the SH–SP partitioning occurs along $\bar{\Delta}$, $\bar{\Sigma}$ for fcc(100) and fcc(110) and along $\bar{\Sigma}$ for fcc(111). A somewhat more complicated reduction, similar to the SH–SP partitioning, occurs along SBZ boundaries related by reflection planes parallel to the boundary but passing through the origin [e.g., $\bar{Y}$ for fcc(100) and $\bar{C}$, $\bar{D}$ for fcc(110)]; here the two reduced blocks of the DM are of *equal* dimension rather than in the proportion 2:1, and the elliptical polarization and linear polarization occur in *every* mode. These two reduced blocks are distinguished only by the fact that for a particular pair of particles (l_3, κ), $(l_3 + 1, \kappa)$ any mode from one block will have (l_3, κ) undergoing elliptical motion parallel to the symmetry plane and $(l_3 + 1, \kappa)$ undergoing linear motion normal to that plane, while in any mode from the other block $(l_3\kappa)$ undergoes the linear motion and $(l_3 + 1, \kappa)$ undergoes the elliptical. For further details on this reduction, as well as those corresponding to twofold axes lying *in* the median plane [e.g., for $\bar{\mathbf{q}}$ along $\bar{T}$ and $\bar{T}'$ for fcc(111) with N_3 odd] and the high symmetry points, we must refer to work to be published elsewhere (Alldredge, 1975; see also Trullinger and Maradudin, 1974).

Before leaving the subject of symmetry we should also comment on the connection between the customary bulk dispersion curves and the results obtained in a surface problem. To make the connection, we have found it convenient to take a *slab-adapted* viewpoint for the bulk spectra (Allen *et al.*, 1971b). We take the same two-dimensional primitive translation vectors chosen for the particular surface of interest, and for the third basic translation vector $\mathbf{a}_z$ take the shortest translation vector of the bulk that is normal to the plane defined by the first two. This usually corresponds to a nonprimitive choice of bulk unit cell which contains n_{zp} times more ions than the conventional primitive unit cell; here n_{zp} is the length of the period of the stacking sequence of *lattice* (*not* ionic!) planes parallel to the surface of interest [e.g., $n_{zp} = 2$ for fcc(100) and fcc(110), $n_{zp} = 3$ for fcc(111), $n_{zp} = 1(!)$ for hcp(0001)]. The corresponding slab-adapted (first) BZ has a thickness $\mathbf{b}_z = 2\pi\mathbf{a}_z/|\mathbf{a}_z|^2$, a volume only $(1/n_{zp})$th that of the primitive bulk BZ, and in the slab-adapted reduced zone scheme must contain n_{zp} times as many branches as the primitive BZ. These extra branches may be seen to come from a slab-adapted extended zone scheme that covers much of the primitive BZ and some portions of its neighbors. This point is illustrated in Fig. 1c, which displays for fcc(111) an edge-on view of the slab-adapted first BZ (1) and its two extensions 2 and 3 which suffice to "cover" the primitive BZ. (Note that those portions of the primitive BZ left out of 1 + 2 + 3 are compensated for by portions of adjacent BZ's that are equivalent under translations by slab-adapted reciprocal lattice translations. Note also that 2 and 3 each having the same thickness as 1 is a special consequence of $n_{zp} = 3$, i.e., each of them contains one half of each of the second and third slab-adapted BZ's.) Thus for a given free-surface slab of N_3 lattice planes, we can visualize a comparison *bulk* slab of N'_3 lattice planes (N'_3 the integral multiple of n_{zp} closest to N_3) having the surface effects removed by periodic boundary conditions applied across its faces. The $3sN_3$ free-surface frequencies calculated from Eq. (10) for a given $\bar{\mathbf{q}}$ in the SBZ can be compared to the bulk frequencies calculated from the standard bulk DM equation at the array of three-dimensional wave vectors given by $\mathbf{q} = \bar{\mathbf{q}} + q_z\hat{\mathbf{i}}_z$, where $q_z = (n_{zp}b_z/N'_3)j$, $(j = 0, 1, \ldots, N'_3 - 1)$. Such constructions delineate the bulk subbands of frequencies and the gaps occuring in them at each $\bar{\mathbf{q}}$ [e.g., Allen *et al.* (1971b), Fig. 7].

c. Diagonalization Techniques. Despite occasional references still cropping up in the physics literature to "solving the secular determinant" of an algebraic eigenproblem of the form Eq. (10), the informed consensus is that direct matrix diagonalization procedures are the most reliable and efficient; a useful standard reference is Volume II, "*The Handbook of Automatic Computation*" (HAC-II) (Wilkinson and Reinsch, 1971). The matrix eigenproblem procedures in HAC-II (1971), with the addition of procedures to deal directly with complex matrices, have recently become available in a well-tested, certified package of FORTRAN subroutines called EISPACK (Garbow, 1974). We recommend that anyone initiating research involving the matrix eigenproblem consider the use of EISPACK.

Our own work has made much use of Businger's (1965) version of the QR algorithm. [We have not yet examined the improvements to the QR method described by Faulkner (1971).] With the 60-bit single precision word length of the CDC 6000 series computers, we find that eigenvalues ω^2 of Eq. (10) are produced to an adequate precision with no special efforts required. The eigenvalues at $\bar{\mathbf{q}} = 0$ provide a stringent test, since in principle they range from the maximum of the bulk, ω^2_{max}, down to $\omega^2 = 0$ for the uniform translation modes; for the SP class of modes of a 15-layer (001) slab of a rocksalt crystal, the computed value of the lowest of the 60 modes, while rarely zero, is typically some eight orders of magnitude smaller than the computed ω^2_{max}, which is the highest value. We have also experimented with the Givens–Householder subroutine widely used in quantum chemistry (Prosser, 1967), with roughly comparable results.*

d. Sampling the SBZ. In addition to computing dispersion curves along the high symmetry directions of the SBZ, we need to compute the eigensystem of Eq. (10) over an appropriate sample of $\bar{\mathbf{q}}$ points in order to compute the average physical quantities which are more directly accessible to experiments. In Section III we discuss such quantities at some length; here we introduce two specific quantities to illustrate our discussion of sampling problems. The mean-square amplitudes of vibration (MSA) of the $(l_3\kappa)$ particle are given in the QHA by

$$\langle u_\alpha^{\,2}(l_3\kappa)\rangle = (\hbar/2M_\kappa) \sum_{\bar{\mathbf{q}},p}{}' \bar{N}^{-1}|\xi_\alpha(l_3\kappa;\bar{\mathbf{q}}p)|^2 \coth[\hbar\omega(\bar{\mathbf{q}}p)/2k_{\mathrm{B}}T]/\omega(\bar{\mathbf{q}}p) \quad (14)$$

where M_κ is the mass of particle κ, and $\bar{N}$ is the number of $\bar{\mathbf{q}}$ values gen-

* The specific form used is the highly optimized routine SPEED written by Peter K. Bischof (University of Texas at Austin, 1972) in the CDC 6600 assembly language COMPASS. T. L. Hardgrave (UT-Austin, 1971) has also achieved a significant optimization in the FORTRAN version for the CDC 6600.

erated by the application of periodic boundary conditions parallel to the surface and containing $\bar{N}$ SUC's in one period. The prime on the summation signifies the omission of the three zero-frequency modes corresponding to the uniform rigid-body translations of the slab (the rigid-body rotations do not occur because of the assumption of parallel periodic boundary conditions). The surface spectral density can be defined as the surface-excess distribution of frequencies (e.g., Allen and de Wette, 1969c):

$$f^{s}(\omega) = (N_3/2)[f^{\text{slab}}(\omega) - f^{\text{bulk}}(\omega)] \tag{15}$$

where f^{slab} and f^{bulk} are computed spectral densities of the slab and bulk, and each is normalized to unity ($\int d\omega\, f(\omega) = 1$). The factor $N_3/2$ is the ratio of the number of lattice planes in the slab to the number of surface lattice planes. Of course, in computing quantities such as these one may confine the larger effort of computing eigensystems to one irreducible element (IE) of the SBZ; the eigensystems in the rest of the SBZ may be obtained from the computed ones by the less costly procedure of applying transformations from the point group of the plane group of the slab.

To a certain extent, sums over $\bar{\mathbf{q}}$ such as in Eq. (14) can be regarded as a numerical quadrature approximation to an integral over the SBZ, with $\bar{N}$ quadrature coordinates distributed on the regular, periodic boundary condition mesh and with $\bar{N}^{-1}$ for the quadrature weights. From this viewpoint, in connection with a lattice *statics* calculation for a slab, Schulze (1973) has replaced sums over the periodic boundary condition mesh in the SBZ (and its weights) by Gaussian-quadrature mesh points and weights, with very good results. But whether one proceeds by means of more efficient numerical quadratures or by the brute-force procedure of increasing the number of points ($\bar{N}$) in the periodic boundary condition mesh, for a given thickness of the slab, N_3, one reaches effective limits to the amount of improvement attainable in a calculation of surface properties by the slab method.* Such limits depend on the specific quantity of interest, i.e., how that quantity depends on the low-frequency portion of the spectrum, but the general considerations may be seen as follows. Fixing N_3 effectively fixes a frequency $\omega_4(N_3) \approx \min\{\omega(\bar{q} = 0,\, q_z = \pi/N_3 a_{3z})_{\text{bulk}}\}$ below which the only frequencies are those from the three acoustical branches that go to $\omega = 0$ as $\bar{\mathbf{q}} \to 0$ (see, e.g., Figs. 2, 4, 6, and 7 in Section III). For two of these branches, a two-dimensional Debye approximation is appropriate for small $|\bar{\mathbf{q}}|$, with the result that their contributions to $f^{\text{slab}}(\omega)$ vary with ω as ω, instead of as the

* Unless, of course, one is literally interested in the properties of ultrathin films, in which case one must deal with the film having at most only one free surface with the other surface in contact with a supporting substrate.

ω^2 variation of the bulk; this result leads to a logarithmic divergence of the high-temperature MSA as $\bar{N} \to \infty$ at fixed N_3 (Allen and de Wette, 1969a, Appendix D). The third of these branches actually makes matters even worse; it is the branch of flexural modes of a thin plate (e.g., Gazis and Wallis, 1965), and at small $|\bar{\mathbf{q}}|$, $\omega_{\text{flex}} \propto |\bar{q}|^2$ which leads to a *constant* contribution to $f^{\text{slab}}(\omega)$ at low frequencies and an even stronger divergence in the high-temperature MSA as $\bar{N} \to \infty$. Other quantities are less sensitive to this thinness effect [e.g., the low temperature MSA, since fewer frequencies satisfy $h\omega/2k_{\text{B}}T \ll 1$ (Allen and de Wette, 1969a)], but as a practical rule of thumb the sampling of the SBZ should not proceed to the point where more than a very few nonzero frequencies lie below $\omega_4(N_3)$ as defined above.*

2. *Calculation of DM Elements*

a. Dynamical Matrix for Noble-Gas Crystals. Since pairs of particles in the crystal interact through the LJ potential [Eq. (1)], the total potential energy of the crystal is given by

$$\Phi = \tfrac{1}{2}N \sum_{l,l'}{}' \, 4\varepsilon \left[\left(\frac{\sigma}{r_0(ll')} \right)^{12} - \left(\frac{\sigma}{r_0(ll')} \right)^{6} \right] \tag{16}$$

where $r_0(ll')$ is the distance between the particles $\mathbf{l} = (l_1, l_2, l_3)$ and $\mathbf{l}' = (l'_1, l'_2, l'_3)$. The elements of the DM are obtained by taking the derivatives of Φ according to Eq. (5) and substituting these into Eq. (11). The lattice sums that have to be evaluated are simple and sufficiently convergent so that they can be computed directly on the computer without recourse to special summation methods.

b. Dynamical Matrix for Ionic Crystals. In the *rigid-ion model* the DM consists of two parts, one describing the contributions of the short-range interactions, the other the contributions of the long-range Coulomb interactions. Lack of space prevents us from presenting the relevant expressions for the elements of the DM for both these contributions; we refer to Maradudin *et al.* (1971) for details. Here we just mention the following. For the short-range interaction, the Born–Mayer potential $v(\mathbf{r}) = \lambda \exp(-|\mathbf{r}|/\rho)$ is used, and only nearest neighbor interactions are taken into account. The Coulomb interaction matrix elements contain lattice sums which converge extremely slowly in direct space, and special methods have to be used to improve the convergence. Tong and Maradudin (1969), who were the first to

* *Note added in proof*: The method of "special" sample points recently described for two-dimensional Brillouin zones by S. L. Cunningham (1974), *Phys. Rev. B* **10**, 4988, shows considerable promise and will no doubt be well explored in the future.

do a complete LD calculation of an NaCl slab, used a modified Bessel function (MBF) transformation to bring the sums into a rapidly convergent form. For the special case of the Coulomb coefficients coupling ions in the *same* plane, Chen *et al.* (1970) reported that a substantial improvement in convergence over the MBF method is obtained by using the method of plane-wise summation (de Wette and Schacher, 1965) which uses an incomplete gamma function (IGF) transformation. The details of this method for evaluating the Coulomb matrix elements of a slab are at this moment only available in the thesis of Chen (1971). Lucas (1968) presents in a very compact form comparable formulas including the IGF transformation; there appear to be some errors of sign in some of his formulas as printed, but the errors do not appear to be propagated to his computed results.

The *shell model* takes into account the polarization of the ions due to the long-range Coulomb interaction, as well as to the short-range repulsive interaction. The elements of the DM become rather complex. For a review of the SM for bulk calculations we refer to Maradudin *et al.* (1971) and Cochran (1971); the formalism for slab calculations is in principle not different from the formulations for the bulk; it has been given in Chen (1971).

3. *Two Special Direct Methods for Mean-Square Amplitudes*

a. Matrix Inversion Method. Wallis and co-workers (Clark *et al.*, 1965; Wallis *et al.*, 1968a, 1969a,b) have made use of a theorem on the representation of an analytic function of a matrix in terms of the spectral representation of that matrix [here $\mathbf{D}(\bar{\mathbf{q}})$] (e.g., Born, 1942; Friedman, 1956) to cast Eq. (14) in the form

$$\langle u_\alpha{}^2(l_3\kappa)\rangle = (\hbar/2M_\kappa)\Big\{\bar{N}^{-1}\sum_p{}' |\xi_\alpha(l_3\kappa;\bar{\mathbf{0}}p)|^2 \coth[\hbar\omega(\bar{\mathbf{0}}p)/2k_B T]/\omega(\bar{\mathbf{0}}p)$$
$$+ \bar{N}^{-1}\sum_{\bar{q}}{}' [\mathbf{D}(\bar{\mathbf{q}})^{-1/2}\coth(\hbar\mathbf{D}(\bar{\mathbf{q}})^{1/2}/2k_B T)]_{l_3\kappa\alpha;l_3\kappa\alpha}\Big\} \qquad (17)$$

where $\mathbf{D}(\bar{\mathbf{q}})$ is the DM of Eq. (11), the prime on the first sum means that the three uniform translational modes are omitted, the prime on the second sum means that $\bar{q} = 0$ is omitted, and the $(l_3\kappa\alpha; l_3\kappa\alpha)$ subscript on the last summation denotes that element of the enclosed matrix-valued function. (Note that the uniform translation modes cause $\mathbf{D}(\bar{\mathbf{0}})$ to be singular, thus requiring the special treatment for the $\bar{\mathbf{q}} = 0$ term; a similar formula obtains for the mean-square velocity (MSV), but, since it involves no negative powers of $\mathbf{D}(\bar{\mathbf{q}})$, no special treatment of the $\bar{\mathbf{q}} = 0$ term is required for MSV.) The matrix function in Eq. (17) takes particularly simple forms for $T = 0°\mathrm{K}$,

$\mathbf{D}(\bar{\mathbf{q}})^{-1/2}$, and for $T \gg \theta_D$, $\mathbf{D}(\bar{\mathbf{q}})^{-1}$. These workers considered slabs of some 20 layers thickness for simple models of cubic metals, directly inverting $\mathbf{D}(\bar{\mathbf{q}})$ by Gaussian elimination for the high temperature case and in the $T = 0°\mathrm{K}$ case solving for the matrix square roots by the Newton–Raphson method. Their sample of the SBZ was compatible with their choice for N_3 (i.e., for $N_3 = 20$, $\bar{N} = N_1 N_2 = 20^2$). They report that a 10-term power series correction to the high temperature limit gives sufficient convergence for $T \geqslant \frac{1}{4}\theta_{\mathrm{D,bulk}}$.

Musser (1971) has suggested a means of removing the singularity of $\mathbf{D}(\bar{\mathbf{0}})$ by slightly breaking the rigid-body translational invariance [similar to adding a small constant to the right-hand side of Eq. (5b)], in the process generating a correction term to the MSA expression which will subtract out the error *after* the inverse power of $\mathbf{D}(\bar{\mathbf{0}})$ is taken. Musser's method is cast in the form of Green functions and makes use of matrix partitioning to reduce the dimension of matrices to be inverted; we refer to his paper for further details.

b. "Power Series" Expansion about the Einstein Approximation. Following a suggestion of Friedel, Masri, Dobrzynski, and others (e.g., Masri and Dobrzynski, 1972; Masri, 1973; Theeten *et al.*, 1973; and references therein) have considered the formal generalization of Eq. (17),

$$\langle u_{\alpha i}{}^2(\mathbf{D})\rangle = (\hbar/2M_i)[\mathbf{D}^{-1/2}\coth(\hbar\mathbf{D}^{1/2}/2k_B T)]_{i\alpha\, i\alpha}, \tag{18}$$

in a "power series" expansion in $\mathbf{R} = \mathbf{D} - \mathbf{d}$, where $\mathbf{d}$ is the diagonal of $\mathbf{D}$ [i.e., $d_{i\alpha,j\beta} = \delta_{ij}\delta_{\alpha\beta}d_{i\alpha}$, $d_{i\alpha} = D_{\alpha\alpha}(i,i)$, and $i \equiv (\mathbf{l}\kappa)$], leading to

$$\langle u_{\alpha i}{}^2(\mathbf{D})\rangle = \langle u_{\alpha i}{}^2(\mathbf{d})\rangle + \sum_{j\beta} R_{\alpha\beta}(i,j)R_{\beta\alpha}(j,i)F_2(d_{i\alpha}, d_{j\beta}, d_{i\alpha}) + \cdots. \tag{19}$$

Equation (19) is a Schafroth (1951) expansion, a generalization of the Taylor-series made necessary if $\mathbf{dR} \neq \mathbf{Rd}$, which is the case in the surface region; we refer the reader to Masri and Dobrzynski (1972) for the form of the "expansion coefficient" F_2 and for further details. The first term in Eq. (19) is the Einstein approximation, and the first two terms displayed in Eq. (19) are reported to give both bulk and surface MSA to accuracies of 10 to 20%, comparable to that of MSA determinations by LEED. Since the matrix operations involved in the right-hand side of Eq. (19) are substantially less demanding of computational effort than most other methods of computing MSA, its use in interpreting LEED experiments can yield considerable savings. Equation (19) can also readily be generalized to computing other equal-time correlation functions of interest (see, e.g., Dobrzynski and Maradudin, 1973).

B. Molecular Dynamics of a Crystal Slab

MD calculations consist in solving Newton's equations of motion for a system of interacting particles (Allen *et al.*, 1969). The calculations described here were performed for slab-shaped fcc crystals with (100), (111), and (110) orientations in which all pairs of particles interact through a LJ potential* [Eq. (1)]. The slab consisted of 11 atomic layers in each case; the total number of particles N_s differed slightly for the three cases: N_s was 550, 539, and 528 for the (100), (111), and (110) surfaces, respectively. Periodic boundary conditions were imposed with respect to translations parallel to the free surface, so that the slab is surrounded by replicas of itself, extending to infinity.

Let M, t, and $\mathbf{r}(l)$ be the atomic mass, the time, and the position of the lth particle, then adopting σ as the unit length and $(M\sigma^2/\varepsilon)^{1/2}$ as the unit time we obtain the dimensionless quantities $\mathbf{s}(l) = \mathbf{r}(l)/\sigma$ and $\tau = (M\sigma^2/\varepsilon)^{-1/2}t$. Using the LJ pair potential we can then write Newton's equation of motion for particle l as

$$\frac{d^2\mathbf{s}(l)}{d\tau^2} = 24 \sum_{l' \neq l} \left(\frac{1}{s(ll')}\right)^8 \left[2\left(\frac{1}{s(ll')}\right)^6 - 1\right] \mathbf{s}(ll') \tag{20}$$

where $\mathbf{s}(ll') = \mathbf{s}(l) - \mathbf{s}(l')$, $s(ll') = |\mathbf{s}(ll')|$.

This system of $3N_s$ coupled equations of motion are solved numerically so as to give the positions and velocities of the N_s particles in the x, y, and z directions as functions of time.

It is often useful to add to Eq. (20) a *damping term* of the form $-\gamma(d\mathbf{s}(l)/d\tau)$. This term can be used to add ($\gamma < 0$) to or remove ($\gamma > 0$) kinetic energy from the system—in other words, to raise or lower the temperature; this device is particularly useful to obtain static relaxation by allowing the temperature to go to zero. In cases where the total energy is constant, γ is simply set to zero in the calculation.

In a classical system in equilibrium, the temperature is a measure of the mean-square velocity, i.e., the mean kinetic energy of the particles. If $\mathbf{v}(l)$ denotes the velocity of particle l, we have

$$\frac{3}{2} k_B T = \frac{1}{N_s} \sum_l \frac{1}{2} M|\mathbf{v}(l)|^2 \tag{21}$$

* MD calculations are best carried out for systems of particles interacting through a simple, analytic potential such as the LJ potential. Although MD calculations for systems of particles interacting with long-range interactions (such as the dipole forces of water) have been performed, the demands on storage capacity and computer time become excessive and for these reasons such calculations are feasible only at very large scale computational facilities.

Using ε we can express T in terms of a dimensionless quantity $T^{**} = (k_B/\varepsilon)T$ so that

$$T^{**} = (3N_s)^{-1} \sum_l \left(\frac{d\mathbf{s}(l)}{d\tau}\right)^2 \tag{22}$$

The $3N_s$ coupled equations of motion given by Eq. (20) were solved numerically so as to give the positions and velocities of the N_s particles in the x, y, and z directions as functions of time. The specific integration algorithm used was a Nordsieck-type method developed by Gear (1966, 1967). The solutions were obtained for $\tau = n\,\Delta\tau$, where n is an integer, and in the calculations reported here, $\Delta\tau = 0.01$ (corresponding to $\Delta t \sim 10^{-14}$ sec for Ar). The initial positions and velocities of the particles can be chosen in a variety of ways. One can, for instance, use as initial positions those obtained from minimizing the static energy. Kinetic energy is imparted to the system by giving the particles either random initial velocities or small random displacements from their equilibrium portions. The potential energy so imparted to the system is gradually released as the particles move, till eventually thermal equilibrium is achieved. Although the time to reach equilibrium depends on the initial conditions, the final state of equilibrium does not. Starting with random displacements of the particles, about 100 steps of $\Delta\tau = 0.01$ were sufficient to establish thermal equilibrium.

It should be pointed out here that in a MD calculation the equilibrium physical quantities of interest are always obtained as averages over particles and time of the appropriate dynamical quantity. The method is extremely useful for obtaining mean positions (static and dynamic displacements), MSA, MSV, etc., as functions of temperature in the classical region up to melting. It allows us to study the effects of anharmonicity on these quantities without any assumption about magnitude of the anharmonicity. The only assumptions embodied in the calculation are that the particles interact through the LJ potential and that the system can be described by classical equations of motion.

C. Matching Method

The matching approach makes use of the full complex band structure (i.e., the evanescent and Bloch waves) belonging to the lattice-periodic dynamical equation in the bulk region; a linear combination of all elementary solutions of the dynamical equation at a given $\bar{\mathbf{q}}$ and a guessed value of ω^2 is tested to see whether it can fit the boundary conditions of a bounded crystal. The first application to lattices capable of dealing with three-dimensional anisotropic crystals was that of Gazis *et al.* (1960). Feuchtwang (1967) has presented a very general formulation of the method. The full capability of Feuchtwang's formulation has yet to be used in realistic computations. We shall merely sketch the method as it has been applied and refer the reader for fuller details to the papers of Gazis *et al.* (1960), Gazis and Wallis (1966), Brady (1971), Zoth (1973), and Zoth *et al.* (1975); the recent paper by Armand and Theeten (1974) dealing with reconstructed surfaces and adsorbed layers should also be consulted.

Let us cast the dynamical equation Eq. (6) into the following harmonic form after Fourier transforming away $\bar{\mathbf{l}}$ and suppressing explicit display of $\bar{\mathbf{q}}$

$$\omega^2\mathbf{M}(l_3)\mathbf{u}(l_3) = -\mathbf{F}(l_3,[\mathbf{u}]), \tag{23}$$

where $\mathbf{M}(l_3)$ is the diagonal mass matrix $[\mathbf{M}(l_3)]_{\alpha\kappa,\beta\kappa'} = \delta_{\alpha\beta}\delta_{\kappa\kappa'}M_\alpha(l_3)$, $[\mathbf{u}(l_3)]_{\alpha\kappa} = u_\alpha(l_3\kappa;\bar{\mathbf{q}})$, and $[\mathbf{F}(l_3,[\mathbf{u}])]_{\alpha\kappa} = F_\alpha(l_3\kappa,[\mathbf{u}];\bar{\mathbf{q}})$, which is the linear difference functional $-\sum_{l_3'\kappa'\beta}\Phi_{\alpha\beta}(l_3\kappa, l_3'\kappa';\bar{\mathbf{q}})u_\beta(l_3'\kappa';\bar{\mathbf{q}})$. We may write $\mathbf{M}$ and $\mathbf{F}$ in terms of their bulk values $\mathbf{M}^0$ and $\mathbf{F}^0$ and the deviations from these values near the surface: $\mathbf{F} = \mathbf{F}^0 - \delta\mathbf{F}$, $\mathbf{M} = \mathbf{M}^0 + \delta\mathbf{M}$; $\delta\mathbf{F}(l_3)$ signifies the removal of forces on the particles at l_3 by the creation of the surface (as well as the effects of any relaxation or other surface-induced modifications of forces). In the bulk region the dynamical equation reduces to the customary DM equation

$$[\omega^2\mathbf{M}^0 - \mathbf{\Phi}^0(\bar{\mathbf{q}}, q_z)]\mathbf{u}(\bar{\mathbf{q}}, q_z) = 0 \tag{24}$$

with the secular equation

$$\det[\omega^2\mathbf{M}^0 - \mathbf{\Phi}^0(\bar{\mathbf{q}}, q_z)] = 0 \tag{25}$$

and solutions $\mathbf{u}(\bar{\mathbf{q}}, l_3, \omega^2) \propto \mathbf{u}(\bar{\mathbf{q}}, q_z, \omega^2)\exp[iq_z x_z(l_3)]$.

In the infinite crystal the boundary condition that $\mathbf{u}(\bar{\mathbf{q}}l_3\omega^2)$ be bounded for all l_3 leads to the rejection of the evanescent waves which have complex q_z; then q_z can be regarded as a parameter just as is $\bar{\mathbf{q}}$, and Eq. (24) suffices to determine the eigenvalues ω^2. When a surface is present, l_3 is bounded (in at least one direction), the evanescent waves must be considered, and additional boundary conditions must be applied. In this case the left-hand side of Eq. (25) is, for fixed $\bar{\mathbf{q}}$ and any given real ω^2, reducible to a polynomial in the quantity $w = \exp[iq_z(\bar{\mathbf{q}}, \omega^2)a_{3z}]$. The degree of this polynomial increases with the range of interplanar forces, and its roots w_j are such that the $q_z^{(j)}(\bar{\mathbf{q}}, \omega^2)$ are either real or occur in complex conjugate pairs. Thus the general solution for fixed $\bar{\mathbf{q}}$ and given ω^2 is the linear combination

$$\mathbf{u}(l_3, [q_z], [C]) = \sum_j C_j\mathbf{u}^{(j)}(q_z^{(j)}(\omega^2))\exp[iq_z^{(j)}(\omega^2)x_z(l_3)] \tag{26}$$

Part of the additional boundary conditions to which this general solution is subject are the dynamical equations [Eq. (23)] in the selvage region (Wood, 1964b), i.e., the neighborhood of the surface where $\delta\mathbf{F} \neq 0$, $\delta\mathbf{M} \neq 0$; Eq. (23) becomes an incomplete set of linear homogeneous equations in the weighting coefficients C_j. The remaining additional boundary conditions required to complete the conditions on the set of C_j are either of two kinds. (1) The kind most used is the asymptotic condition at large distances from a single surface; in this case if $x_z(l_3) \to +\infty$ denotes the deep interior of the crystal, the remaining additional boundary conditions usually invoked are that $C_j = 0$ for any $\mathrm{Im}(q_z^{(j)}) > 0$ (for generalization, see Feuchtwang, 1967). (2) Two-surface boundary conditions are also sometimes used (e.g., Gazis and Wallis, 1965; Wallis *et al.*, 1968b; Brady, 1971), in which the remaining boundary conditions are equations like Eq. (23) for the selvage of the second surface; in those cases in which the two surfaces are related by a symmetry operation, the additional boundary conditions can be cast into the form of a partition of the general solutions into the two classes of symmetric and antisymmetric solutions plus a symmetrized form of the selvage dynamical equation. In either case, the selvage dynamical equation becomes a set of linear homogeneous equations for the weighting coefficients

$$\sum_j T_{kj}(\omega^2, [q_z], [\mathbf{u}(q_z)])C_j = 0, \tag{27}$$

for which $\det(T_{kj})$ must vanish if nontrivial solutions obtain; here j is, of course, the index for the solutions of Eqs. (24) and (25) and k is the index of the selvage dynamical equations used as boundary conditions.

In practice the simultaneous solution of Eqs. (24) and (27) for $\omega^2(\bar{q})$ and $\mathbf{u}(l_3, [q_z(\bar{\mathbf{q}})], [C(\bar{\mathbf{q}})])$ proceeds by a search on ω^2 at fixed $\bar{\mathbf{q}}$, solving (25) for the $q_z^{(j)}$ then (24) for the $\mathbf{u}^{(j)}(q_z^{(j)})$, testing $\det(T_{kj})$ for zero, and finally if $\det(T_{kj}) = 0$ solving (27) for $C_j(\bar{\mathbf{q}})$. A search might proceed by first examining a uniform mesh of values of ω or ω^2 (at fixed $\bar{\mathbf{q}}$) for minima of $|\det(T_{kj})|$; then a refined search in the neighborhood of the promising minima can be made, e.g., by the golden-section procedure or other efficient extrema search procedures (Wilde, 1964).

Several comments are in order. (1) Since the columns of the boundary condition matrix T_{kj} are linear in the eigenvectors $\mathbf{u}^{(j)}$ which in turn are each arbitrary up to a complex constant factor of arbitrary phase, the phase of $\det(T_{kj})$ does not have a straightforward interpretation, unlike the phase of the resonance determinant occurring in the Green function method. (2) In the case that $\delta\mathbf{M} = 0$ in the selvage, the selvage boundary condition [Eq. (23)] reduces to the simple form $\delta\mathbf{F}(l_3, [\mathbf{u}[C]]) = 0$; this is the form in which the selvage boundary condition has most often been invoked, and it is the lattice generalization of the stress-free surface boundary condition used in surface wave calculations in elastic continuum theory (see Farnell, 1970). (3) The simultaneous solution of Eqs. (24), (25), and (27) may possibly yield, not a surface wave, but a bulk wave which happens to satisfy the selvage boundary condition; this is unlikely except in the case of SH waves in the long wavelength *limit* (not merely the long wavelength *regime*), and in any case an examination of the imaginary parts of the $q_z^{(j)}$ serves to distinguish between surface waves and bulk waves. (4) For short range, but otherwise realistic, force constant models, the above equations can be expanded to low order in $\bar{\mathbf{q}}$ and solved algebraically along high symmetry directions, thereby permitting not only the long wave limit to be examined but also the neighborhood of the limit (the long wave regime). In this way, long wave low frequency SH surface waves have been found to exist in the long wave regime ($|\bar{\mathbf{q}}|a \ll 1$) but not in the $\bar{\mathbf{q}} = 0$ limit (Alldredge, 1972; Zoth *et al.*, 1974). (5) As outlined above, the matching method deals only with pure surface modes; it can be generalized to deal with resonances (pseudosurface waves) by means of assuming spatial damping. In the spatial damping approach, the components of $\bar{\mathbf{q}}$ are allowed to have small imaginary parts, and the search outlined above is extended to the three variables ω^2 (real), $\mathrm{Im}(q_1)$, and $\mathrm{Im}(q_2)$; this approach is used in elastic continuum studies of pseudosurface waves (e.g., Lim and Farnell, 1969; Farnell, 1970).

D. Green Function Method

The treatment of the surface perturbation by the Green function method is in most respects similar to the well-known GF procedures used to treat point defects (see, e.g., Maradudin *et al.*, 1971; Wood, this volume). It has great potential power to deal in the most coherent way with the wide variety of concepts such as localized and pseudolocalized modes, perturbed spectral density, transition matrix, and correlation functions such as the MSA so important in LEED experiments. Since the early papers by Lifshitz and Rosenzweig (Lifshitz and Rosenzweig, 1948; Rosenzweig, 1950; Lifshitz, 1956), GF methods have been applied to a large number of simple model studies of surface problems (reviewed in Maradudin *et al.*, 1971; Wallis, 1973) reaching a rather high level of refinement recently in the work of a number of French workers (see, e.g. Masri and Dobrzynski, 1973, and references therein). But in applications to computations for realistic crystal models, only the barest start has been made, notably the work of Musser and Rieder (1970; Musser, 1971) and of Benedek (1973, 1974). Consequently, we shall give only a very brief discussion of the GF method as it applies to the problem of surface phonon spectra; further details may be found in the papers cited above, particularly the latter four.

In almost all applications of the GF method to surfaces, the surface is assumed to be created in a manner equivalent to the consideration of a crystal slab of $N_3 = n_{zp}N_z$ lattice planes, where N_z denotes an integral number of planar stacking sequences and may be arbitrarily large. This assumption truncates the system of equations represented by Eq. (23), which we may write in the matrix form

$$\mathbf{L}(\omega_+{}^2)\mathbf{u} = [\mathbf{L}^0(\omega_+{}^2) + \boldsymbol{\delta}\mathbf{L}(\omega_+{}^2)]\mathbf{u} = 0, \tag{28}$$

where

$$[\mathbf{L}^0](l_3\kappa\alpha;\, l'_3\kappa'\beta) = \omega_+{}^2 M_\kappa{}^0\, \delta(l_3\kappa\alpha;\, l'_3\kappa'\beta) - \Phi_{\alpha\beta}{}^0(l_3\kappa;\, l'_3\kappa';\, \bar{\mathbf{q}}),$$

$$[\boldsymbol{\delta}\mathbf{L}](l_3\kappa\alpha;\, l'_3\kappa'\beta) = \omega_+{}^2\, \delta M_\kappa(l_3)\, \delta(l_3\kappa\alpha;\, l'_3\kappa'\beta) - \delta\Phi_{\alpha\beta}(l_3\kappa;\, l'_3\kappa';\, \bar{\mathbf{q}}),$$

$$[\mathbf{u}](l_3\kappa\alpha) = u_\alpha(l_3\kappa;\, \bar{\mathbf{q}}),$$

and $\omega_+{}^2 = \omega^2 + i\eta$, η = positive infinitesimal. These equations may be reduced to blocks corresponding to the irreps of the available slab symmetry at the particular $\bar{\mathbf{q}}$ of interest; in the subsequent discussion we shall assume we are discussing a particular irrep, although the irrep label is suppressed. If ω^2 is not in the bulk subband of this irrep, then Eq. (28) may be written

$$(\mathbf{I} + \mathbf{G}^0\, \boldsymbol{\delta}\mathbf{L})\mathbf{u} = 0, \tag{29}$$

where $\mathbf{G}^0(\omega_+{}^2; \bar{\mathbf{q}}) = \mathbf{L}^{0^{-1}}(\omega_+{}^2; \bar{\mathbf{q}})$ is the GF of the perfect slab with periodic boundary condition applied across the two faces; if ω^2 lies in the bulk subband, the right-hand side of Eq. (29) may be supplemented by an arbitrary combination of bulk subband eigenvectors having the eigenvalue ω^2.

In addition to the above assumption of the boundary condition, it is usually assumed that the surface perturbation $\boldsymbol{\delta}\mathbf{L}$ is confined to finite selvage regions with l_3 ranging over n_σ ($\ll N_3$) values near each surface, so that the total dimension of the selvage defect space is $3s \times 2n_\sigma$. The assumption of finite $\boldsymbol{\delta}\mathbf{L}$ is not fully justified for ionic crystals as $\bar{\mathbf{q}} \to 0$, since the infinite range of Coulomb interactions gives rise to the Fuchs–Kliewer dielectric surface modes discussed in Section III, B; however, even for ionic crystals, $\boldsymbol{\delta}\mathbf{L}$ has an effective short range over most of the SBZ. With the assumption of short range for $\boldsymbol{\delta}\mathbf{L}$, we can arrange the components of vectors and matrices so that they can be partitioned as follows:

$$\boldsymbol{\delta}\mathbf{L} = \begin{pmatrix} \boldsymbol{\delta}\mathbf{l} & \mathbf{0} \\ \mathbf{0} & \mathbf{0} \end{pmatrix}, \qquad \mathbf{G}^0 = \begin{pmatrix} \mathbf{g} & \mathbf{G}_{12}{}^0 \\ \mathbf{G}_{21}{}^0 & \mathbf{G}_{22}{}^0 \end{pmatrix}, \qquad \mathbf{u} = \begin{pmatrix} \mathbf{u}_1 \\ \mathbf{u}_2 \end{pmatrix} \tag{30}$$

with Eq. (29) taking the form

$$(\mathbf{I} + \mathbf{g}\,\boldsymbol{\delta}\mathbf{l})\mathbf{u}_1 = 0 \tag{31a}$$

$$\mathbf{u}_2 = \mathbf{G}_{21}{}^0\,\boldsymbol{\delta}\mathbf{l}\,\mathbf{u}_1 \tag{31b}$$

Thus, the essential problem has been reduced to Eq. (31a) which is of fairly low dimension—even for no symmetry at all, no more than 6 sn_σ (complex) dimensions. A principal focus is on the "resonance" determinant

$$\Delta(\omega_+{}^2) = \det(\mathbf{I} + \mathbf{g}\,\boldsymbol{\delta}\mathbf{l}) \tag{32}$$

of our particular irrep. Pure surface modes occur at those ω for which $\Delta(\omega_+{}^2) = 0$; resonance modes occur for $\Delta_r = \mathrm{Re}(\Delta) = 0$ and $\Delta_i = \mathrm{Im}(\Delta) \neq 0$, with the width of the resonance proportional to $-\Delta_i/(d\Delta_r/d\omega^2)$. The surface-induced change in the cumulative spectral density, $\Delta n(\omega) = \int_0^\omega d\omega'\,\Delta f(\omega')$ (given $\bar{\mathbf{q}}$ and irrep), is proportional to $\mathrm{Im}(ln\,\Delta(\omega_+{}^2))$. Unlike the point defect problem, the site-representation of the perfect crystal GF needed to solve Eq. (31a) involves only a one-dimensional sum over normal modes of the perfect crystal:

$$g_{\alpha\beta}(l_3\kappa, l'_3\kappa'; \bar{\mathbf{q}}, \omega_+{}^2) \propto N_3{}^{-1} \sum_{q_z}^{\mathrm{ESBZ}} \times \exp\{iq_z \cdot [x_z(l_3) - x_z(l'_3)]\}\, g_{\alpha\beta}(\kappa, \kappa'; \bar{q}, q_z, \omega_+{}^2) \tag{33}$$

Here $g_{\alpha\beta}(\kappa, \kappa'; q, q_z, \omega_+{}^2) = [\omega_+{}^2\mathbf{M}^0 - \boldsymbol{\Phi}^0(\bar{\mathbf{q}}, q_z)]_{\alpha\beta}^{-1}(\kappa, \kappa')$ can be obtained by the indicated inversion (dimension $3s$ complex), but since it normally will be needed at a large number of ω on a mesh, its calculation by use of eigensystems is preferred. The sum over q_z is over the extended slab-adapted BZ scheme described in Section II, A and $\bar{\mathbf{q}}$ is confined to the first SBZ. [For a cubic (001) surface the sum over q_z is symmetric about $q_z = 0$; Benedek (1973) made use of 97 uniformly spaced q_z's in the half-domain to evaluate his **g** matrix. We briefly discuss Benedek's results in comparison with direct slab calculations in Section III, B.]

III. Results

In this section we present and briefly discuss an illustrative sample of results obtained for the structure and dynamics of surfaces and their experimental manifestations.

A. Surface Structure (Relaxation)

Before reviewing the results of the surface dynamical calculations, we must first discuss the static properties of surfaces, because they have a direct bearing on the dynamics of the surface. It is well known that at absolute zero the mean positions of the particles near a surface of a crystal are displaced from their bulk position; this so-called *surface relaxation* results from the asymmetry in the forces acting upon the particles near the surface. The surface relaxation is evaluated by minimizing the static energy of the crystal with respect to the interplanar distances near the surface; the shifts in equilibrium positions near the surface as compared to the bulk positions are called *static displacements*.

In Section III, C, 2 we will discuss the total displacements (*dynamic displacements*) which obtain at finite temperatures; these, of course, are obtained by minimizing the total Helmholtz free energy (static and vibrational) with respect to the interplanar spacings. These displacements (static or dynamic) are indicated by the fractional changes δ_m in the distance d_m between the mth and the $(m + 1)$th plane: i.e., $d_m = d(1 + \delta_m)$, where d is the interplanar spacing in the bulk.

Calculations of the static displacements in semiinfinite crystals with an LJ potential have previously been carried out by Shuttleworth (1949), Alder *et al.* (1959), and Benson and Claxton (1964). Allen and de Wette (1969a) have calculated displacements in slabs ranging from 11 to 51 layers in thickness. Results for the (100), (110), and (111) surfaces of LJ crystals are shown in Section III, C, 2 (Fig. 10), together with results for the dynamic displacements at a higher temperature. Here we just mention that the displacements for all three surfaces fall off approximately as the inverse cube of the distance from the surface. In fact, these authors show that if the particles in a semiinfinite crystal interact through a potential $\phi(\mathbf{r})$ such that $\phi(\mathbf{r}) \propto |\mathbf{r}|^{-p}$ for large $|\mathbf{r}|$, where p is an integer larger than 3, then $\delta_m \propto m^{p-3}$ for large m (in the LJ potential $p = 6$).

B. Surface Phonon Spectra

1. *Surface Modes of Noble Gas Solids* (Allen *et al.*, 1971b)

We now describe results of calculations for monatomic fcc crystals with (111), (100), and (110) surfaces. The model used was a 21-layer slab-shaped crystal in which the particles interact through a LJ potential. The normal mode solutions have the form given in Eq. (9) for the case where there is one particle per unit cell. Distinct solutions $u_\alpha(\mathbf{l})$ correspond to values of the two-dimensional wave vector $\bar{\mathbf{q}}$ lying within the first SBZ, which is determined by the crystal structure and the surface orientation. The modes

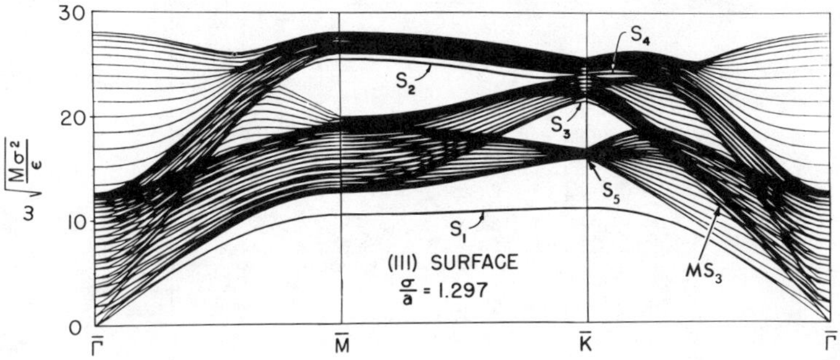

FIG. 2. Dispersion curves $\omega(\bar{\mathbf{q}}p)$ for a 21-layer fcc (111) slab, including surface relaxation, of an LJ crystal at the static density ($\sigma/a = 1.297$). Here a is one-half the usual cubic lattice constant. The pairs of surface mode branches are labeled S_i ($i = 1, 2, \ldots, 5$) in an arbitrary order. MS_3 labels a pseudosurface mode branch (locus of mixed modes). After Allen *et al.* (1971b). Reprinted with permission.

in the whole SBZ are determined by those in the irreducible element (IE) through simple symmetry transformations. In Fig. 1 the SBZ's for the (100), (110), and (111) surfaces of an fcc crystal are shown.

a. (111) Surface. In Fig. 2 the ω vs. $\bar{\mathbf{q}}$ relations are shown for the (111) slab. The dense bands of dispersion curves are the so-called bulk bands; they describe the (acoustic) bulk vibrations of the slab. Since in this case the polarization index p can take on $3N_3$ values (N_3 is number of layers), there are 63 frequencies for each $\bar{\mathbf{q}}$ value for this 21-layer slab; most of these are bulk frequencies. But there are gaps between and below the bulk bands and in these we find isolated frequencies which correspond to surface modes, e.g., S_1, S_2, S_3, S_4, and S_5. That these modes are indeed localized at the surface is shown by the fact that their calculated eigenvectors are large near the surface and show a rapid decrease with increasing distance from the surface. In Fig. 3 the squared (mass-weighted and normalized) amplitude

$$|\boldsymbol{\xi}(m)|^2 = |\xi_x(m)|^2 + |\xi_y(m)|^2 + |\xi_z(m)|^2 \tag{34}$$

for S_1, S_2, S_3, and S_4 is plotted as a function of layer index m, with $m = 1$ for the surface layer. Both the frequencies ω and the eigenvectors $\xi_\alpha(l_3)$ are obtained by direct numerical solution of Eq. (13). Recall that the displacement amplitude $|u_\alpha(l_3)|$ is proportional to $|\xi_\alpha(l_3)|$ according to Eq. (9), so $|\boldsymbol{\xi}(m)|^2$ is a measure of the amplitude of vibration in the mth layer when the crystal is vibrating in a particular mode.

Notice that the vertical scale in Fig. 3 is logarithmic and that all of the modes S_i decrease by more than an order of magnitude between the surface and the center of the crystal slab.

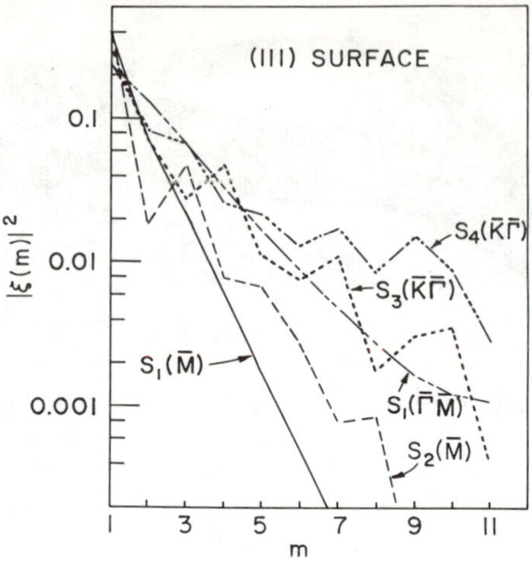

FIG. 3. Surface mode attenuation curves: $|\xi(m)|^2$ vs. layer index m for surface modes shown in Fig. 2. Here $m = 1$ for a surface plane and $m = 11$ for the center layer. $S_1(\bar{\Gamma}\bar{M})$ corresponds to the midpoint of the S_1 branch between $\bar{\Gamma}$ and $\bar{M}$. $S_3(\bar{K}\bar{\Gamma})$ and $S_4(\bar{K}\bar{\Gamma})$ correspond to these branches just off the point $\bar{K}$, and $S_1(\bar{M})$ and $S_2(\bar{M})$ correspond to the $\bar{M}$ point. After Allen *et al.* (1971b). Reprinted with permission.

The mode S_1 is typical of the "generalized Rayleigh waves" found in continuum theory, in that it lies below the bulk bands and persists into the long-wavelength limit and shows an approximate exponential decay in amplitude with increasing distance from the surface (approximate straight line in Fig. 3). None of these statements is true for the "gap modes" S_2, S_3, and S_4; their amplitudes decay in a very complicated fashion away from the surface yet they all have their largest amplitude in the surface layer. In the surface layer S_1 is primarily a "shear-vertical" (SV) mode, i.e., it is primarily associated with vibrations normal to the surface. S_3 and MS_3 are primarily "shear-horizontal" (SH) modes, i.e., associated with vibrations transverse to $\bar{\mathbf{q}}$ and parallel to the surface. S_2 and S_4 are primarily "longitudinal" or "P" modes, i.e., vibrations in the direction of $\bar{\mathbf{q}}$.

b. (100) Surface. In Fig. 4 are shown the dispersion relations $\omega(\bar{\mathbf{q}}p)$ for a 21-layer slab with (100) orientation. Two cases are shown: (a) without, and (b) with static relaxation taken into account. Notice the following features: (i) Surface modes exist below the bulk bands (S_1, S_2, S_4, S_5) and in the gaps (S_6, S_7, S_8, S_9, S_{10}), the latter generally at short wavelengths. (ii) In this case surface modes exist *within* the bulk subbands (S_3, S_4) for wave vectors $\bar{\mathbf{q}}$ along symmetry directions, as a result of symmetry decoupling of these modes from the surrounding bulk bands. (iii) Comparing Fig. 4a and b, we find that the qualitative features of the surface phonon spectrum are

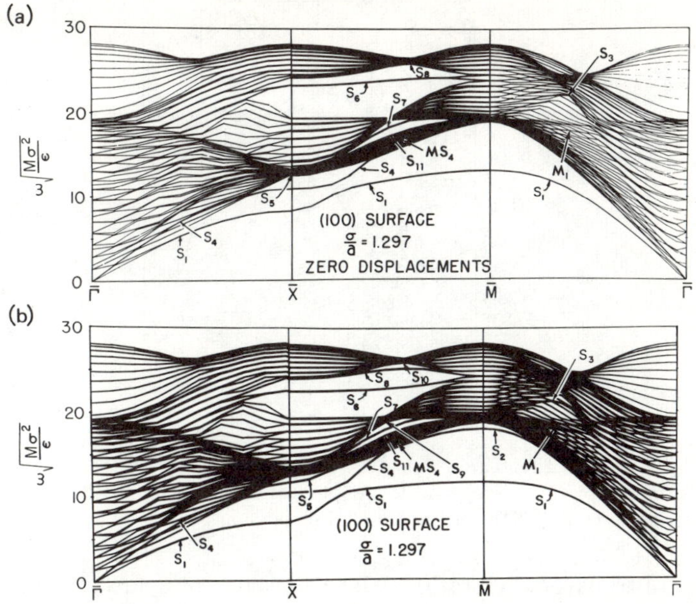

FIG. 4. Dispersion curves $\omega(\bar{\mathbf{q}}p)$ for a 21-layer fcc (100) slab of an LJ crystal: (a) without and (b) with surface relaxations taken into account. Surface (S_i) and pseudosurface (MS_i or M_i) mode branches are labeled according to an arbitrary scheme. After Allen *et al.* (1971b). Reprinted with permission.

sensitive to changes in the surface force constants (e.g., the modes S_2, S_9, and S_{10} are not present in Fig. 4a). (iv) Of all the surface modes mentioned, only S_1 and S_4 persist in the long-wavelength regime. Along $\bar{\Gamma}\bar{M}$ S_1 has the character of a so-called "generalized Rayleigh wave;" its polarization is SV and its amplitude shows an approximately exponential decay with increasing distance from the surface. (v) Along $\bar{X}\bar{M}$ the modes S_1 and S_4 show "hybridization." They approach each other closely but instead of crossing there is a sudden interchange in character; going from left to right, S_1 changes from SH to SV, and S_4 from SV to SH.

Figure 5 shows $|\xi(m)|^2$ vs. m for the modes S_1, S_4, S_6, S_7, S_8. Notice that S_8 has alternately larger and smaller amplitudes in successive layers and that it has its largest amplitude in the second layer; S_7 has its largest amplitude in the third layer.

c. Systematics of Surface Modes. Despite the apparent complexity of these surface mode spectra, their general features can be understood in the context of a simple phenomenological scheme.

For a monatomic crystal *without* surfaces there exist three intertwining bulk bands as indicated in Figs. 2 and 3 (corresponding to longitudinal and

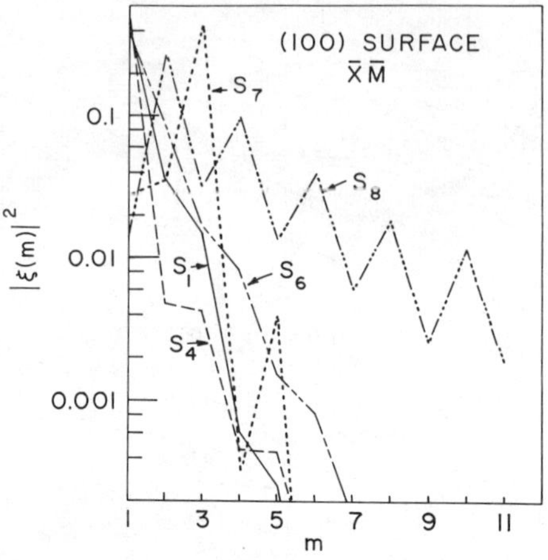

FIG. 5. Surface mode attenuation along the SBZ boundary for surface modes shown in Fig. 4b. The modes plotted correspond to the points along $\bar{X}\bar{M}$ indicated by the arrows in Fig. 4b. After Allen *et al.* (1971b). Reprinted with permission.

two types of transverse vibrations). The presence of a surface introduces a *perturbation* consisting of two parts: a large "first-order" perturbation due simply to the truncation of the crystal, and a "second-order" perturbation due to the changes in the force constants near the surface. The strength of the total perturbation depends on the location of $\bar{\mathbf{q}}$ in the BZ. If the perturbation is strong enough it will peel one or more surface modes off a given bulk band, and since the perturbation usually entails a *softening* of the force constants, the surface modes are peeled off the *bottom* of the bulk band. [If for some reason the force constants are stiffened, the surface modes are peeled off the top of the bulk band; see, e.g., Musser and Rieder (1970).]

Ordinarily, the total perturbation first peels off (from a given bulk band) a mode primarily localized in the first layer. If strong enough, the perturbation then peels off a mode primarily localized in the second layer; and so on. For instance, in Fig. 4b the modes S_6, S_8, and S_{10} all have the same polarization character, but S_6 is mainly localized in the surface layer, S_8 in the second layer, and S_{10} in the third.

When a mode is peeled off, one of four things will happen: (a) It may fall under all of the bulk bands, in which case it will necessarily be a surface mode. (b) It may fall into a gap inside the bulk bands, in which case it again will necessarily be a surface mode. (c) Along a symmetry line associated with a reflection plane, it may fall into a region occupied only by a subband of bulk modes to which it is automatically orthogonal by virtue of symmetry. In this case, once more, it will necessarily be a surface mode. (d) It may fall into a region occupied by bulk modes to which it is not automatically orthogonal. In this case it will not be able to survive as a pure surface mode and will be a mixed mode instead.

Figure 4a represents the situation for an unrelaxed slab. In this case the force constants are less weakened relative to the relaxed case, and the surface mode frequencies are nearer the bulk subbands from which they derive. For instance, S_{10} is raised into the bulk band and is no longer a surface mode.

2. *Surface Modes of Ionic Crystals*

a. NaCl (100). As was pointed out in Section II, A, surface dynamical calculations for the (100) surface of NaCl have been performed with the Kellermann rigid-ion model and the shell model. The earlier KRIM calculations (Lucas, 1968; Tong and Maradudin, 1969) had led to some apparent discrepancies which were cleared up by the work of Chen *et al.* (1970, 1971a, 1972c) (on which our discussion will be based) and of Jones and Fuchs (1971).

KRIM results for NaCl (100) are shown in Fig. 6a, calculated for a 15-layer slab. The modes labeled S_i are surface modes; for the slab, each label stands for a pair S_{i+}, S_{i-} due to the existence of two surfaces. Note the existence of a very large number of surface modes. Of these the long wavelength acoustical modes S_1 (Rayleigh modes), the long wavelength optical

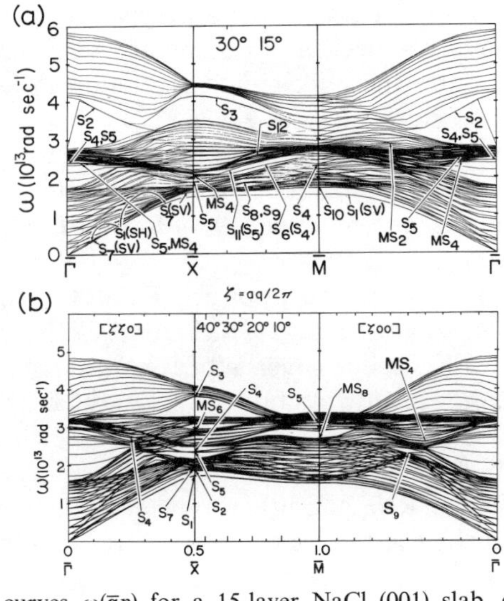

FIG. 6. Dispersion curves $\omega(\bar{\mathbf{q}}p)$ for a 15-layer NaCl (001) slab, calculated (a) with the Kellermann (1940) rigid-ion model and (b) with an eleven-parameter shell model. Modes with the same index in (a) and (b) are not necessarily related. The curves are computer-generated on a finite mesh of $\bar{\mathbf{q}}$, and crossovers between branches are not always taken into account. After Chen *et al.* (1971a). Reprinted with permission.

modes S_2 [Fuchs–Kliewer (1965) modes (FK)], and the microscopic transverse optical surface mode (TOSM) pairs S_4 and S_5 [Lucas (1968) modes] had been described earlier. All the other surface modes, existing mainly in the gaps in the bulk bands at short wavelengths, have been found by Chen *et al.* (1970, 1972c). Modes labeled MS_i are pseudosurface modes; they are continuations into bulk bands (of the same symmetry) of the corresponding surface modes S_i existing in gaps at other wavelengths.

A remark on the FK modes (S_2) is in order here. As the origin is approached, these modes penetrate more and more deeply, and their degeneracy is broken by the finite thickness of the slab. The lower member of the FK pair goes to the frequency ω_{TO} [bottom of transverse optical (TO) band] and the upper member to the frequency ω_{LO} [top of longitudinal optical (LO) band], in agreement with the results of Fuchs and Kliewer (1965). This behavior is not shown in Fig. 6a, whose graphs are computer generated for a finite mesh of $\bar{\mathbf{q}}$ values.

There is one significant difference between the continuum and lattice-dynamic results for the FK modes at large wavelengths. In the continuum results they are pure surface modes, whereas in the lattice-dynamical results they are pseudosurface modes in the region where they lie within the bulk bands. Each FK branch actually becomes a locus of interaction among hybridizing modes as it passes into the bulk bands. (See Fig. 7 where the proper connectivity of the FK modes is displayed.) Such behavior is typical of a "branch" of pseudosurface modes. At short wavelengths (large $\bar{\mathbf{q}}$), the FK modes S_2 again enter the bulk bands as pseudosurface modes, labeled MS_2 on the right-hand side of Fig. 6a.

SM results for NaCl (100) are shown in Fig. 6b; they were calculated with an eleven parameter shell model with the values of the parameters determined by Schmunk and Winder (1970) from a fit to inelastic neutron scattering results.

As can be seen from a comparison of Fig. 6a and b, there are a number of significant differences between the SM and KRIM results. The main difference is that the FK modes never appear as pure surface modes in the SM results. Instead, they are always buried within the bulk bands as pseudosurface modes, whose existence is indicated in Fig. 6b by the disturbance near $\bar{\Gamma}$ in the LO bulk band. On the basis of these results for NaCl, and results of calculations for RbF and other ionic crystals (Chen *et al.*, 1972a,b, 1973), it appears that the FK modes will ordinarily be pseudosurface modes, rather than surface modes, over most or all of the SBZ.

The Rayleigh modes S_1 also show different behavior in the SM results. They retain the same polarization (primarily SV) along $\bar{X}\bar{M}$, and do not cross S_7 along $\bar{\Gamma}\bar{X}$. (In the KRIM, S_1 changes from SV to SH along $\bar{M}\bar{X}$ and crosses S_7 along $\bar{\Gamma}\bar{X}$.)

There are considerable differences between the SM and KRIM results with respect to the gaps within the bulk bands and, as a consequence, with respect to the surface modes lying within these gaps. This indicates the need to adopt the best bulk description of a particular crystal when attempting a realistic calculation of surface vibrations.

Benedek (1973) has reported a study of the surface phonons of NaCl (001), using the Green function method and the breathing-shell model (BSM) of Nüsslein and Schröder (1967). There are significant differences in the bulk bands and gaps because of the differences in the models. It is not clear which model gives the best representation of the bulk bands, since this BSM was not fitted to phonon frequencies, and, in any case, fitted models are fitted only along high symmetry directions, whereas the gap structure is greatly influenced by off-symmetry behavior of the bulk dispersion curves. The surface perturbation matrix $\boldsymbol{\delta}\mathbf{L}$ was not calculated directly in terms of the model interactions; instead, approximate, effective first nearest-interplanar-neighbor force constants were constructed from bulk eigenvalues and eigenvectors in a manner somewhat similar to that of Foreman and Lomer (1957). The greater power of the GF method to investigate pseudosurface modes was made use of. In many respects there is at least qualitative agreement between the GF–BSM results and the slab-SM results of Fig. 6b. There are, however, grave discrepancies. Benedek's (1973) surface dispersion curves appear to be in error (1) in not having the Lucas TOSM branches, S_4 and S_5, degenerate at $\bar{\Gamma}$ and (2) in not obtaining the long-wave low-frequency SH branch, S_7, as a *pure* SH surface branch embedded in the SP bulkbands along $\bar{\Gamma}\bar{X}$ (Alldredge, 1972). The source of these discrepancies has not been determined as of this writing.*

b. Selected SM Results for Other Ionic Crystals. The surface phonon spectrum of NaCl (001) is considerably complicated by the overlap and interaction of the acoustical and optical bulkbands. To avoid this complication, K. H. Rieder (private communication) suggested the study of crystals for which the ratio of anion-to-cation mass takes extreme values. The Chen *et al.* (1972a) results for a 15-layer RbF (001) slab are shown in Fig. 7; the results do display the desired clarification. The Lucas TOSM branches exist clear of bulkbands over the entire SBZ. In addition, a microscopic *longitudinal* optical surface mode S_2 was obtained for the first time. The separation of bulkbands allows a fairly close correspondence to be drawn between surface mode branches in the SBZ and peaks in the surface spectral

* *Note added in proof*: Dr. Benedek has informed us that he has corrected his calculations and that the serious discrepancies noted above have disappeared. It must be noted that the speculations in his 1973 paper, to the effect that the thin-slab calculations of Chen *et al.* are in error, are not valid.

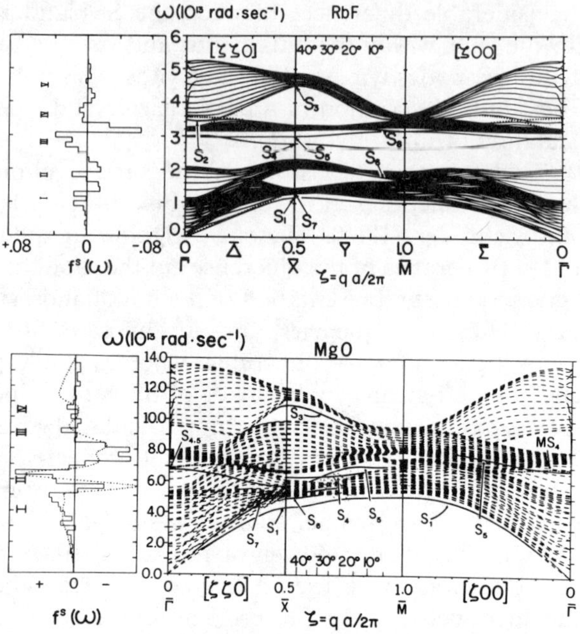

FIG. 7. Dispersion curves and surface-excess spectral density $f^s(\omega)$ for eleven-parameter shell models of 15-layer (001) slabs of RbF (Chen *et al.*, 1972a) and MgO (Chen *et al.*, 1973). For RbF dotted lines delineate bulk band edges where these are not well defined for the slab. For MgO, the bulk-like branches are dashed lines, and surface branches are emphasized by bold lines. The dotted line curve for $f^s(\omega)$ in MgO is the experimental curve of Rieder and Hörl (Rieder, 1971). Reprinted with permission from T. S. Chen *et al.*, *Solid State Commun.* **10**, 941 (1972).

density $f^s(\omega)$. Finally, note that the FK dielectric surface modes can be observed as a locus of interacting *lattice* modes.*

A fairly direct measurement of $f^s(\omega)$ has been made by inelastic neutron scattering from high surface area samples of MgO (Rieder and Hörl, 1968; Rieder, 1971). Figure 7 also displays the results of a SM calculation for a 15-layer MgO (001) slab (Chen *et al.*, 1973), including a comparison between the calculated $f^s(\omega)$ (histogram) and the best experimental $f^s(\omega)$ (dotted lines; Rieder, 1971). With one exception, there is gratifying agreement in the location of peaks and depletion valleys in the two $f^s(\omega)$. The exception

* *Note added in proof*: The RbF results displayed in Fig. 7 were obtained with the shell model parameters published by G. Raunio and S. Rolandson (1970), *Phys. Rev. B* **2**, 2098; however, the small second-neighbor short-range parameters were interchanged in that paper [see erratum (1972), *Phys. Rev. B* **6**, 2511.]. This error puts the bulk-band and surface-band frequencies in Fig. 7 too high by about 3%; the structure of bulk and surface bands is unchanged when the error is corrected [Chen *et al.*, 1975 (to be published)].

is the large peak in the experimental curve around 12×10^{13} rad/sec; this peak was originally thought to arise from the *dielectric* surface mode, but the lattice calculations seem to rule out that interpretation since the dielectric modes (FK) exist only over a very small portion of the total phase space represented by the SBZ. Comparison of the absolute amplitudes of the peaks and valleys of the experimental and theoretical $f^s(\omega)$'s is complicated by experimental difficulties in determining the specific surface area. Nevertheless, it is clear the agreement is less satisfactory in terms of relative amplitudes, and further work, both experimental and theoretical, is indicated.

Direct LD calculations for KRI models of microcrystallites containing up to 180 ions have been made for MgO (Genzel and Martin, 1972) and RbF (Martin, 1973). Not only were $f(\omega)$ and $f^s(\omega)$ obtained, but in some cases infrared absorption coefficients and corner-, edge-, and surface-projected spectral densities were computed also. These latter quantities require eigenvector processing in addition to eigenvalues, and out-of-core versions of diagonalization procedures had to be used for the larger microcrystals. Comparison of these results with those for the SM slabs is somewhat complicated by the differences between KRI and shell models and by the fact that even for the 180-ion microcrystal 73% of the ions are on the surface (including edges and corners), compared to 13% surface ions for a 15-layer slab and the estimated 8% surface ions in the experimental sample of MgO having the highest surface area.

C. Derived Physical Quantities

1. *Mean-Square Amplitudes of Vibration*

Vibrations of surface particles have an important influence on the scattering properties of surfaces such as low-energy electron diffraction and low energy atomic and molecular scattering. The vibrational quantities entering most directly into the description of surface scattering phenomena are the mean-square amplitudes of vibration. Conversely, experimentally the MSA can be directly determined from the temperature dependence of the Debye–Waller factor, notably in LEED experiments. In fact, the most direct experimental manifestations of surface vibrations and their difference from bulk vibrations are the MSA's determined in this way. In principle, the MSA can also be determined from other surface scattering phenomena such as low energy atomic scattering and the Mössbauer effect. However, these methods are less well developed than LEED.

The principal physical reason that the surface MSA differ from those in the bulk crystal is that the total force acting on a particle at or near the surface is smaller than for an equivalent bulk particle, because a certain number of neighbors are missing at the surface. This has the effect of weakening the self-interaction force constants [Eq. (5b)] at the surface, which in general leads to increased MSA's. In fact, many of the differences between static and dynamic surface effects for different surfaces of the same crystal can be qualitatively understood by a comparison of the number of bonds that are missing at these different surfaces. In addition, changes in the interparticle force constant induced by the presence of the surface (such as relaxation and, in the case of metals and covalent crystals, fundamental changes in the character of interparticle interactions) contribute to the surface-modification of MSA's.

In the QHA, the MSA of the $(l_3\kappa)$-particle, $\langle u_\alpha^2(l_3\kappa)\rangle$, are given by Eq. (14), Section II, A, 1, d. In MD the MSA are directly determined as averages [over time and particles of the same kind (κ) in a given crystal layer (l_3)] of the squares of the instantaneous displacements $u_\alpha(l_3\kappa)$ of the particles from their mean positrons.

Of the early calculations of surface MSA's we mention the work of Clark *et al.* (1965) and Wallis *et al.* (1968a, 1969a,b) which established many of the salient features of surface MSA's such as increased values toward the surface, anisotropy at the surface, and dependence on force constant changes near the surface. However, these calculations were performed with simple force constant models which did not allow for the effects of surface relaxation, many neighbor forces, etc., to be taken into account in a systematic way.* Allen and de Wette (1969a), and Allen *et al.* (1969) performed the first calculations based on an interaction potential which avoided many of these restrictions; we will illustrate the kinds of calculational results obtained by briefly reviewing this work. [See the review by Lagally (1975) for a more comprehensive survey of experiment–theory confrontations over surface MSA.]

From the results of the dynamical calculations for the noble-gas crystals one can, with Eq. (14), directly calculate the MSA for the atoms in each layer of the slab. Figure 8a shows the dependence of the MSA $\langle u_\alpha^2\rangle$ on the distance from the surface (Allen *et al.*, 1969); m labels a plane of atoms parallel with the surface, with $m = 1$ at the surface. The results shown are for a (110) oriented slab at about half the melting temperature T_M. The LD results (dashed lines) show the increase in $\langle u_\alpha^2(m)\rangle$, and increasing anisotropy, toward the surface. These effects follow from the greater vibrational freedom of the surface particles and from the lowered isotropy at the surface. The solid lines show the same quantities as obtained from a MD calculation at the same density and temperature. It is noteworthy that the

* It should be noted that the method of Wallis and co-workers, as distinct from the specific models used in their reported calculations, is sufficiently general to deal with such effects.

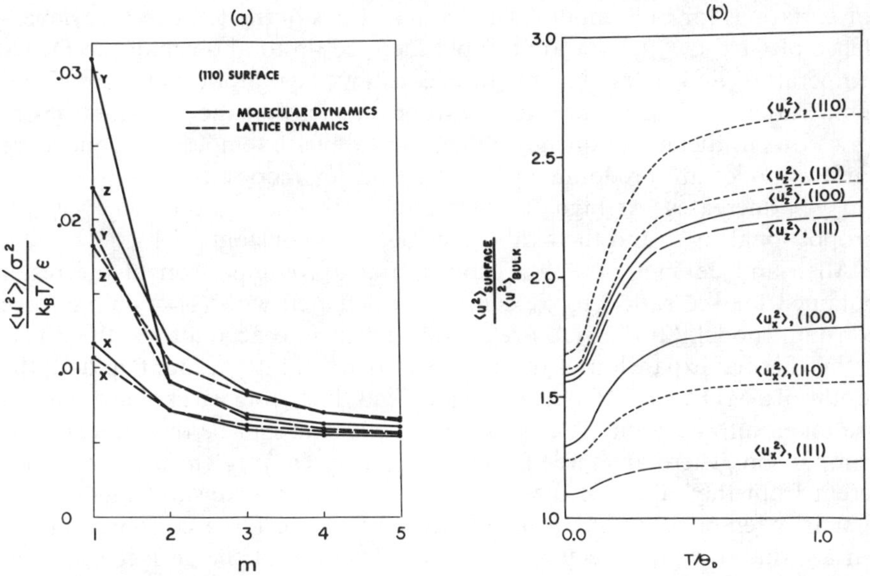

FIG. 8. Mean-square amplitudes (MSA) for fcc LJ crystals. (a) Depth dependence for the (110) surface; x is the $[\bar{1}10]$ direction, y the $[001]$, and z, $[110]$ normal to the surface. After Allen *et al.* (1969). (b) Relative temperature dependence of the surface MSA ($m = 1$) in the quasiharmonic approximation. After Allen and de Wette (1969c). Reprinted with permission.

MD and LD results are in good agreement in the center of the crystal ($m > 5$), but that there are considerable differences at the surface, particularly for $\langle u_y^2 \rangle$. These differences must be attributed to surface anharmonicity (included in MD), since they are too large to result from statistical fluctuations in the MD results. These results, therefore, indicate that near $\frac{1}{2}T_M$ anharmonic effects are small in the bulk but cause substantial increases in the MSA at the surface. On these grounds we believe that the QHA begins to breakdown for surface calculation at about $\frac{1}{3}T_M$, for the noble-gas crystals.

Next we consider the temperature dependence of the MSA. It is well known [as is easily derived from Eq. (14)] that in the QHA $\langle u_\alpha^2 \rangle$ is proportional to T for large T; at small T it has a nonzero limit due to the zero-point vibrations. The transition between the two regimes occurs around $T = \frac{1}{2}\theta_D$ (θ_D is the bulk Debye temperature).

Figure 8b shows the ratio $\langle u_\alpha^2 \rangle_{surf} / \langle u_\alpha^2 \rangle_{bulk}$ as a function of T/θ_D for surfaces of different orientation (Allen and de Wette, 1969c). It is found that this ratio is practically independent of density (within a few percent for the density range of the noble-gas solids) which makes it in practice a universal function of T/θ_D. This result follows from Eq. (14) in view of the fact that the Grüneisen parameter $\gamma(\bar{\mathbf{q}}p) = -d \ln \omega(\bar{\mathbf{q}}p)/d \ln V$ is practically the same for surface as for bulk modes. The main features of the temperature dependence of $\langle u_\alpha^2 \rangle_{\text{surf}}/\langle u_\alpha^2 \rangle_{\text{bulk}}$ are a rapid increase up to about half the Debye temperature and a leveling off to a constant asymptotic value at higher temperatures. These features follow from the fact that the contributions to $\langle u_\alpha^2 \rangle$ due to the low-frequency modes increase with temperature, and since surface modes are predominantly low-frequency modes, the ratio $\langle u_\alpha^2 \rangle_{\text{surf}}/\langle u_\alpha^2 \rangle_{\text{bulk}}$ increases. At high temperatures, both $\langle u_\alpha^2 \rangle_{\text{surf}}$ and $\langle u_\alpha^2 \rangle_{\text{bulk}}$ are proportional to T, and their ratio thus becomes constant.

Allen and de Wette (1969c, Table III) give a comparison of the result obtained for the ratio $\langle u_\alpha^2 \rangle_{\text{surf}}/\langle u_\alpha^2 \rangle_{\text{bulk}}$ calculated with (1) a simple force constant model, (2) the static relaxation taken into account, (3) the differential thermal expansion of the surface also taken into account, and (4) the results of a MD calculation in which all effects, including true anharmonicity, are taken into account. The following conclusions can be drawn from this comparison. There are three factors that cause $\langle u_\alpha^2 \rangle_{\text{surf}}/\langle u_\alpha^2 \rangle_{\text{bulk}}$ to be different from the value calculated with a simple force-constant model. The first is a temperature-independent decrease in the force constants at the surface due to static effects (i.e., due to relaxation of the surface atoms at 0°K). The second is a further temperature-dependent decrease in the surface force constants which is due to dynamical effects (differential thermal expansion at the surface). The third factor is anharmonicity, which produces increases in the MSA which are larger at the surface than in the bulk. All three factors lead to increases in the ratio $\langle u_\alpha^2 \rangle_{\text{surf}}/\langle u_\alpha^2 \rangle_{\text{bulk}}$.*

To date the only reported LEED measurements on solid noble gas surfaces with which our calculations can be compared are those on Xe (111) by Ignatiev and Rhodin (1973) and on Kr (111) by Ignatiev *et al.* (1974). The analysis of the data (Tong *et al.*, 1973) on the basis of the kinematical approximation seems to be well justified for Xe but only partly for Kr. The agreement of the results derived from the measurements with those of the calculations is self-contradictory. For instance, the ratio of surface to bulk coefficient of thermal expansion is larger than the calculated value by about a factor of 2 (cf. Section III, C, 2). On the other hand, the surface Debye temperature as a function of incident electron energy is in reasonable agreement with the calculated values as can be seen in Fig. 9. It will probably require a full nonkinematical calculation of the temperature effects in LEED from Xe (111) to clear up this apparent contradiction.

2. *Surface Thermal Expansion*

The subject of thermal expansion at the surface has received increasing attention in recent years, experimentally as well as theoretically. The first

* There is one exception: in the case of $\langle u_x^2 \rangle$ for the (111) surface, anharmonicity appears to cause a decrease.

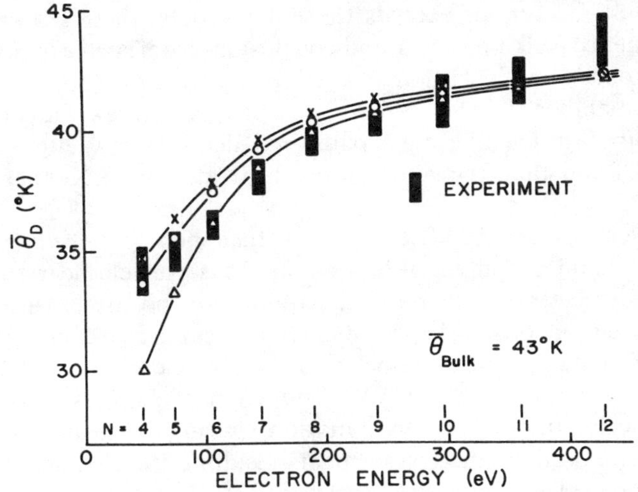

FIG. 9. Experimental (LEED) and theoretical surface Debye temperature $\bar{\theta}_D$ of Xe (111) as a function of incident electron energy for the Bragg peaks $n = 4, 5, \ldots, 12$ of the specular beam (00). Sources of theoretical results: ×, Clark *et al.* (1965); ○, Allen and de Wette (1969a); △, Allen *et al.* (1969). After Tong *et al.* (1973). Reprinted with permission.

theoretical treatments were a LD calculation involving a minimization of the Helmholtz free energy (Allen and de Wette, 1969a) and a MD calculation (Allen *et al.*, 1969), both applied to the noble-gas solids. Although these calculations only yielded thermal displacements of the surface layers at a few temperatures, rather than thermal expansion coefficients as functions of temperature, they clearly showed that the rate of thermal expansion perpendicular to the surface is significantly larger at the surface than in the bulk. These findings stimulated a detailed study by Kenner and Allen (1972, 1973) of thermal expansion for the two first surface interplanar spacings as functions of temperature for the (100) and (111) surfaces of Ar, Kr, and Xe. The first analytical treatment of a crystal slab, including cubic anharmonic effects, has been given by Dobrzynski and Maradudin (1973); numerical estimates are given for a simple pairwise potential model of α-iron (cf. the remarks on such potentials for metal surfaces in Section I, A).

Experimentally, the rate of thermal expansion near the surface can be related to the temperature-dependent shifts in the positions of the Bragg peaks in LEED experiments. Such measurements have been reported by Gelatt *et al.* (1969) for Ag (111) and Ni (111), by Woodruff and Seah (1970) for Cu (111), by Wilson and Bastow (1971) for Mo (100) and Cr (100), by Ignatiev and Rhodin (1973) for Xe (111), and by Ignatiev *et al.* (1974) for Kr (111). Although the results provide evidence that the rate of thermal expansion at the surface exceeds that for the bulk, there are other factors which influence peak positions, and conclusions from such experiments have to be approached with caution.

In the remainder of this section we will briefly review first the early evidence for *differential* surface thermal expansion (Allen and de Wette, 1969a; Allen *et al.*, 1969) and then the more complete treatment of Kenner and Allen (1973).

It has been discussed in Section III, A that the *surface relaxation* results from the asymmetry in the forces acting on a particle near the surface,* causing a displacement of the static equilibrium positions with respect to the bulk positions (static displacements). For similar reasons, the presence of the surface causes an asymmetry in the particle vibrations, as we have seen in the discussion of the MSA. A surface particle is less restricted in its outward motion than in its inward motion, causing its mean position to shift outward. The total mean displacements resulting from static and dynamic causes are called *dynamic displacements.* With LD, the dynamic displacements (which are temperature dependent) are evaluated by minimizing the free energy (at each temperature) as a function of the interplanar distances. With MD, the displacements are found from the mean positions of the crystalline planes as functions of temperature, which follow directly from the calculations. As in Section III, A, we express displacements in terms of the quantities δ_m, the fractional change in the distance between the mth and $(m + 1)$th planes near the surface.

In Fig. 10 are shown the δ_m for the (100), (110), and (111) surfaces of the noble-gas crystals at 0°K (static displacements, Allen and de Wette, 1969a) and at about $\frac{1}{2}T_{\mathrm{M}}$ (calculated with MD, Allen *et al.*, 1969). Since the δ_m indicate *fractional* changes, the curves for 0°K and $\frac{1}{2}T_{\mathrm{M}}$ would have to coincide if the thermal expansion at the surface were the same as in the bulk. The fact that the δ_m are larger at finite temperatures than at 0°K indicates that the rate of thermal expansion is larger at the surface than in the bulk.

Figure 11 shows a comparison between LD and MD results for the (100) surface of the noble-gas solids at about $\frac{1}{2}T_{\mathrm{M}}$ ($T^{**} = 0.356$). Notice that at this temperature the LD and MD results are in complete agreement; this indicates that at this temperature the effects of anharmonicity on the displacements (contained in the MD results but not in the LD results) are negligible. However, at about $\frac{3}{4}T_{\mathrm{M}}$ ($T^{**} = 0.547$), at which LD in the QHA is no longer valid, the MD result (taken at a slightly lower density) shows that the differential thermal expansion at the surface continues up to temperatures close to melting.

* This asymmetry must, of course, include the effects of any fundamental surface-induced modifications of the electron distribution; see remarks in Section I, A.

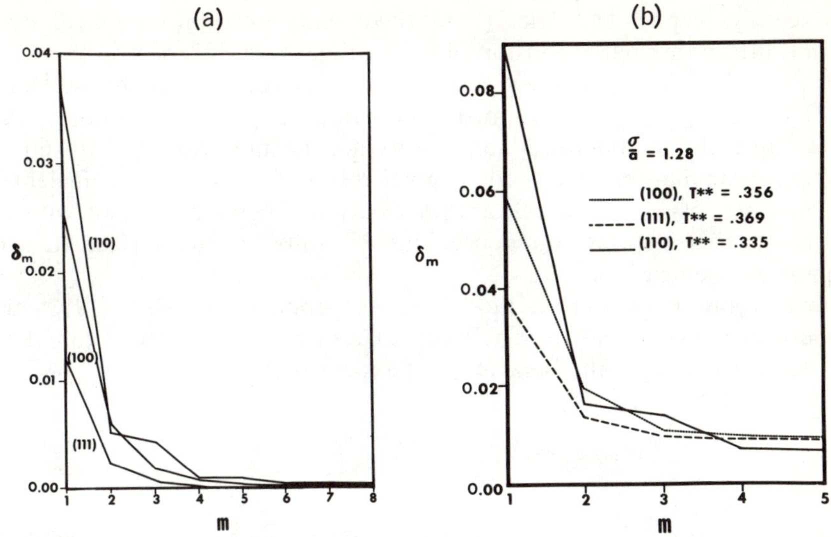

FIG. 10. Displacements δ_m for low-index surfaces of fcc LJ crystals: (a) Static displacements (after Allen and de Wette, 1969a); (b) dynamic displacements at about half the melting temperature. After Allen *et al.* (1969). Reprinted with permission.

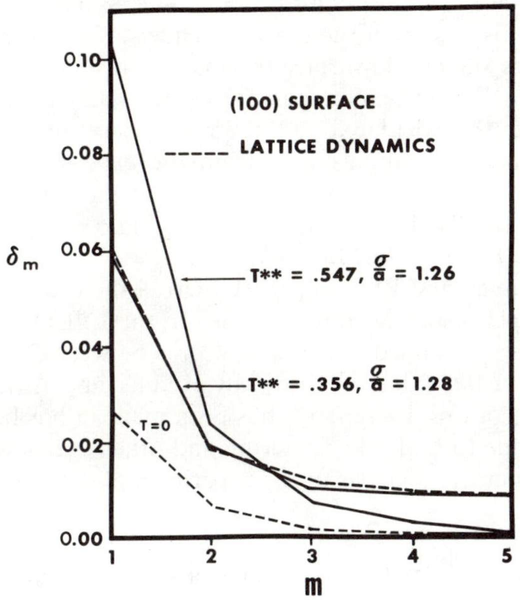

FIG. 11. Comparison between dynamic displacements for a fcc (100) LJ surface, as obtained by QHA lattice dynamics and anharmonic molecular dynamics (MD) at about $\frac{1}{2}T_M$ (T^{**} = .356) and by anharmonic MD at a higher temperature. The static harmonic displacement is given for comparison ($T = 0$). After Allen *et al.* (1969). Reprinted with permission.

Recently, Kenner and Allen (1973) have made a careful numerical study of the surface thermal expansion of the noble-gas solids based on the QHA calculations of Allen and de Wette (1969a). The minimization of the Helmholtz free energy F was performed on a truncated Taylor expansion, rather than on the full temperature- and density-dependent free energy. In spite of these approximations, the method gives results for the bulk coefficient of thermal expansion α_{bulk} which are in remarkably good agreement with experiment; this inspires confidence in the results for the surface thermal expansion coefficient, α_{surf}.

In the context of a crude model, earlier proposed by Allen (1972; also unpublished work) the results imply that in the high temperature limit ($T \geqslant \theta_D$, where θ_D is the bulk Debye temperature)

$$\frac{\alpha_{\text{surf}}}{\alpha_{\text{bulk}}} \approx \frac{3}{4} \frac{\langle u_z^2 \rangle_{\text{surf}}}{\langle u_z^2 \rangle_{\text{bulk}}} \tag{35}$$

where $\langle u_z^2 \rangle$ is the MSA normal to the surface. (This relation underscores the close connection between MSA's and thermal expansion, alluded to above.) In going from high to low temperatures, $\alpha_{\text{surf}}/\alpha_{\text{bulk}}$ undergoes a large increase, passes through a peak at roughly 6% of θ_D and then decreases again to a limiting value $\alpha_{\text{surf}}/\alpha_{\text{bulk}}$ = constant for $T \to 0°\text{K}$. This behavior is closely correlated with dispersion of the low frequency surface modes.

In Fig. 12a are shown the surface thermal expansion for the first ($m = 1$) and second ($m = 2$) interplanar spacings for the (100) and (111) surfaces of Xe, together with the calculated bulk thermal expansion (on the scale of the figure the latter is indistinguishable from the experimental results). It is evident that the increased thermal expansion at the surface is a very pronounced effect. In Fig. 12b, the ratios $\alpha_{\text{surf}}/\alpha_{\text{surf}}$ and $\alpha_2/\alpha_{\text{bulk}}$ are plotted for the (100) and (111) surfaces of Xe.

Recently, Ignatiev and Rhodin (1973) inferred the rate of thermal expansion for the Xe (111) surface from the shifts in the LEED Bragg peaks with temperature. They obtained $\alpha_{\text{surf}}/\alpha_{\text{bulk}} \approx 4$ or 5 between 55° and 75°K; this is compared with the value 1.9 found by Kenner and Allen. Although no detailed analysis of this discrepancy has been made, it has been conjectured that it may be due to multiple scattering and other effects which invalidate the simple kinematical analysis of the experimental data.

3. *Surface Thermodynamic Quantities*

Thermodynamic properties of surfaces are of importance because they control the formation and stability of surfaces, play an important role in physical adsorption and chemisorption processes, nucleation, and a number of other surface processes. In the context of lattice theories, of the various

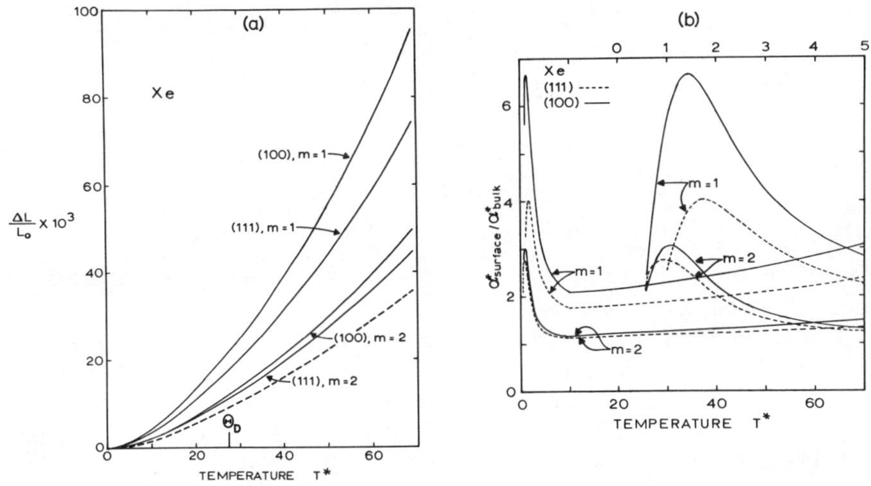

FIG. 12. (a) Surface thermal expansion for the first ($m = 1$) and second ($m = 2$) interplanar spacings for two surfaces of xenon. The dashed line gives the bulk result. $T^* = k_B T/\hbar(\varepsilon/M\sigma^2)^{1/2}$. (b) Relative surface thermal expansion coefficient for the two outermost interplanar spacings. An expanded temperature scale (*top*) shows the low temperature behavior. After Kenner and Allen (1973). Reprinted with permission.

thermodynamic quantities, the surface specific heat C_v^s has received much of the theoretical interest in the past. In these theories, two approaches for the treatment of thermodynamic quantities have been followed: analytic treatments of simple models and direct numerical calculations. While the former have in principle the advantage of bringing out an analytic temperature dependence of the various quantities, in practice this can only be achieved at the cost of rather limiting assumptions. For instance, for C_v^s the discussion is limited to very low temperatures (T^2-region) and isotropic crystals (see the reviews by Maradudin *et al.*, 1971; Wallis, 1973). However, temperatures immediately above the Debye (T^2) region are of particular interest for C_v^s, and, of course, actual crystals are not isotropic. We will therefore not review these theories here, but limit ourselves to the direct numerical approach in which these restrictive assumptions do not have to be made.

Within the harmonic approximation, one can express the vibrational contributions to the energy E, the entropy S, the Helmholtz free energy F, and the specific heat at constant volume C_v, in terms of the vibrational frequencies of the crystal. If we define $x = \hbar\omega(\bar{\mathbf{q}}p)/k_B T$, then

$$E = k_B T \sum_{\bar{q},p} \{\tfrac{1}{2}x + [x/(e^x - 1)]\}, \tag{36}$$

$$S = k_B T \sum_{\bar{q},p} \{-\ln(1 - e^{-x}) + [x/(e^x - 1)]\}, \tag{37}$$

$$F = k_B T \sum_{\bar{q},p} [\tfrac{1}{2}x + \ln(1 - e^{-x})], \tag{38}$$

$$C_v = k_B \sum_{\bar{q},p} [x^2 e^x/(e^x - 1)^2], \tag{39}$$

As before, $\bar{\mathbf{q}}$ and p are the two-dimensional wave vector and polarization index, respectively.

The thermodynamic functions for an infinite, three-dimensional crystal can be calculated by means of equations similar to Eqs. (36)–(39), with the only difference that $\bar{\mathbf{q}}$ is replaced by the three-dimensional wave vector $\mathbf{q}$, and that p takes on $3s$ values, where s is the number of particles in the unit cell. Once the thermodynamic functions for the crystal with and without surface have been calculated, the surface contributions to the thermodynamic function can be obtained in the following way. Let u be the value per particle of some thermodynamic function U for a crystal with surfaces, i.e., $u = U/N$, where N is the number of particles in the crystal (U stands for E, S, F, or C_v). Let u^b and U^b be the same quantities for an infinite crystal without surfaces. Then if U^s, the surface contribution to U, is written in the form $U^s = Au^s$, where A is the surface area of the crystal (twice the area of one surface for a slab-shaped crystal), one has

$$u^s = (N/A)[u - u^b]. \tag{40}$$

An alternative but equivalent way of expressing the surface thermodynamic functions obtains by integrating the summands in Eqs. (36)–(39), multiplied by the surface frequency distribution $f^s(\omega)$ [see Eq. (15)], e.g.;

$$C_v{}^s = \frac{3sk_B}{A_0} \int_0^{\omega_{max}} d\omega f^s(\omega) \frac{x^2 e^x}{(e^x - 1)^2}, \tag{41}$$

where $A_0 = A/n$ is the area per surface particle. In actual calculations $f^s(\omega)$ will be available in terms of a table (histogram) obtained from the dynamical calculations. Since the formation of a table for $f^s(\omega)$ involves a loss of information about the exact values of the frequencies,* it is more accurate to carry out the summations of Eqs. (36)–(39) for the crystals with and

* To date, the calculation of surface spectral densities does not yet begin to approach the resolution attained for the bulk (Gilat, this volume). The highest resolution reported so far for $f^s(\omega)$ is that for LiF (001) (Alldredge *et al.*, 1974).

without surfaces, and then use Eq. (40). In the following we briefly review the kind of results obtained for noble-gas and ionic crystals.

a. Noble-Gas Crystals (Allen and de Wette, 1969c). The results for the noble-gas crystals have been derived from dynamical calculations on 11-layer slabs with fcc (100), (110), and (111) orientations; surface relaxation has been taken into account. The dynamical calculations were carried out for a density corresponding to $\sigma/a = 1.27$ (σ is the Lennard–Jones parameter, a is one-half the usual cubic lattice constant) for the (100) and (110) surfaces, and $\sigma/a = 1.29724$ for the (111) surface. Thermal expansion was taken into account assuming (1) that the coefficient of thermal expansion is the same at the surface as in the bulk, and (2) that the density dependence of the phonon frequencies could be obtained from a Grüneisen parameter $\gamma(\bar{\mathbf{q}}p) = -d[\log \omega(\bar{\mathbf{q}}p)]/d(\log V)$ which is the same for all frequencies, namely $\gamma(\bar{\mathbf{q}}p) = 3.1$. The variation of the density with temperature was obtained from experimental data. [In light of the results on surface thermal expansion discussed above, assumption (1) involves some error which has not yet been assessed. On the other hand, assumption (2) appears to be fairly reasonable for bulk and surface modes alike on the basis of test calculations at different densities.]

In Figs. 13 and 14 are shown the static surface energy Φ^s, and E^s, S^s, F^s, and $C_v{}^s$ for Ar between 0°K and the melting temperature (84°K). The calculated surface thermodynamic functions for different surfaces are qualitatively alike but quantitatively different. The static surface energy is smallest for the (111) surface and largest for the (110) surface. The dynamical quantities are approximately equal for the (100) and (110) surfaces but somewhat smaller

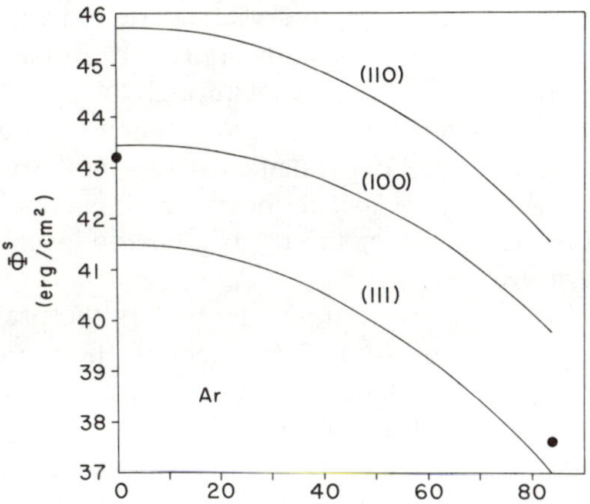

FIG. 13. Temperature dependence of the (QHA) static surface energy Φ^s for argon. After Allen and de Wette (1969b). Reprinted with permission.

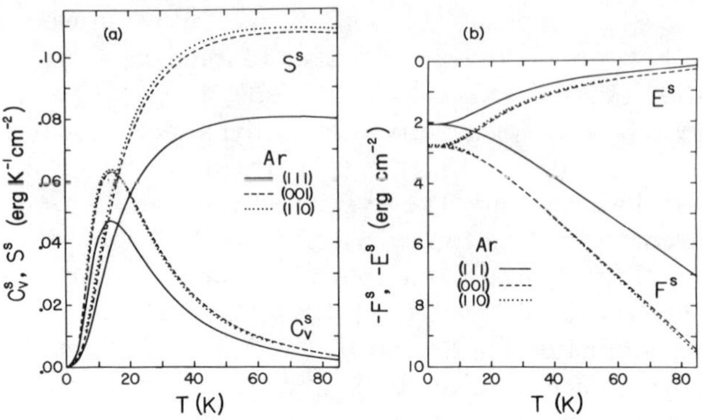

FIG. 14. Surface-excess *vibrational* contributions to thermodynamic functions of argon for three surfaces: (a) $C_v{}^s$, specific heat at constant volume, and S^s, entropy; (b) E^s, internal energy, and F^s, Helmholtz free energy. After Allen and de Wette (1969b). Reprinted with permission.

for the (111) surface. The differences among the surfaces are qualitatively explained by the number of bonds broken by the surface.

The qualitative features of the temperature dependence of the surface thermodynamic functions can be explained with simple arguments: Φ^s decreases slightly as the temperature increases because of thermal expansion. The fact that the vibrational frequencies are lowered by the presence of a surface causes E^s and F^s to be negative and S^s and $C_v{}^s$ to be positive. E^s decreases rapidly in absolute value as the temperature T increases because the energy per particle at high temperature is $3k_BT$ in crystals both with and without surfaces. Similarly, $C_v{}^s$ decreases to zero as T becomes large because the specific heat per particle is $3k_B$ in crystals with and without surfaces. At high temperatures S^s is nearly constant, and F^s decreases linearly with T because of the relation $F^s = E^s - TS^s$. Although Φ^s is much larger than F^s, F^s changes rapidly with temperature and is therefore more important in determining the temperature dependence of the total surface free energy ($\Phi^s + F^s$). Finally, the surface specific heat, which is equal to zero both at 0°K and in the high-temperature limit, has a narrow peak at about 15% of the Debye temperature.

The only important approximations involved in the present calculations are the following: the use of a LJ 12-6 potential, the use of the QHA, the assumption that the surface does not undergo a change of structure, and the assumption that thermal expansion is uniform throughout the crystal, including the surface region. All of these approximations are probably rather good at low temperatures for the heavier noble gases, but the neglect of anharmonic effects and differential thermal expansion at the surface probably leads to substantial errors at high temperatures ($T \geq \frac{1}{2}T_M$).

b. Ionic Crystals. Although the noble-gas solids are excellent prototypes for theoretical studies of surface vibrations, they are difficult materials to use in experimental surface studies, particularly in calorimetric experiments where one would like samples having both well-characterized surfaces *and* high specific surface areas. Ionic crystals, especially those of the NaCl structure, are more favorable materials in both regards, and measurements of specific heat of small-particle samples of MgO and NaCl were made long ago. Unfortunately, the specific surface area was not measured for the MgO sample, which makes the comparison of theoretical results to the experiment uncertain; we refer the reader to the analysis by Rieder (1971) for further discussion and references for the case of MgO.

Two sets of data have been published for small-particle samples of NaCl for which the specific surface areas were measured. That of Morrison and Patterson (1956) displayed a peak in the "surface"-excess heat capacity at a temperature about one-sixth of the (bulk) Debye temperature (i.e., at about 40°K); however, this excess heat capacity did not appear to be proportional to the reported specific surface area, and the only theory available at the time (the "Montroll correction") predicted that the peak should not be at 40°K but at 75°K. Later data of Barkman *et al.* (1965) did display a proportionality to specific surface area, but did not extend beyond 20°K and hence did not display the maximum in the excess heat capacity. Subsequent to the "Montroll correction" and prior to the lattice calculation of Chen *et al.* (1971b), all theories developed had been specialized to the low-temperature limit and were also deficient in one or more respects (such as ignoring the anisotropy of the crystal or ignoring the dispersion of the surface modes).

In Fig. 15a we show the comparison between C_v^s results drawn from the better of the Morrison and Patterson samples and those drawn from the KRIM calculation of Chen *et al.* (1971b). There is fairly satisfactory agreement in the location of the peak in C_v^s near 40°K, but there is a large discrepancy in the peak height. Figure 15b displays the comparison between theory and experiment in a lower temperature regime. The KRIM results of Chen *et al.* agree with the data of Barkman *et al.* to within the envelope of their points. Note that the data for the two samples of Barkman *et al.* do display a proportionality to surface-excess heat capacity ($A_0 C_v^s$), whereas the data for the three earlier samples of Patterson *et al.* (1955) and Morrison and Patterson do not. The fact that there was some controversy in the mid-1950's over the standard BET measurement of surface area for very pure ionic surfaces (see MacIver and Emmett, 1956) suggests that the reported lack of proportionality to surface area in the early data may derive from errors in their surface area determination. When the surface areas reported for the early data are adjusted to make the C_v^s/T^2 agree with that of the

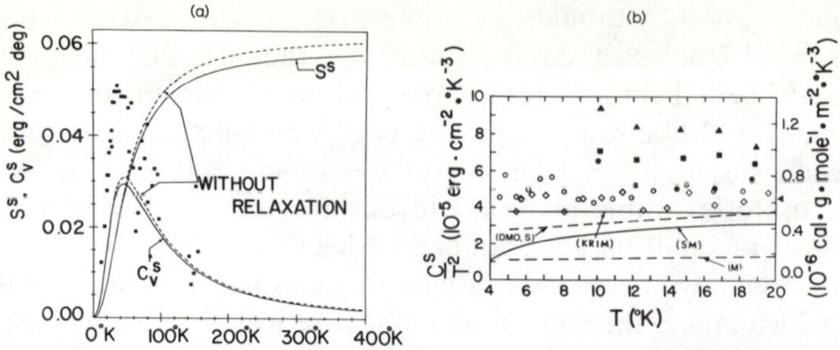

FIG. 15. (a) Vibrational contributions to surface entropy S^s and surface specific heat $C_v{}^s$ for the KRI model of the (001) surface of NaCl. Experimental points for sample B of Morrison and Patterson (1956). (b) $C_v{}^s/T^2$ at low temperatures for NaCl (001). Open data points from Barkman *et al.* (1965); closed data points from Patterson *et al.* (1955) and Morrison and Patterson (1956). Solid lines are the results of lattice calculations; dashed lines represent the continuum theories of Dupuis *et al.* (1960) and Stratton (1953) (DMO, S) and of Montroll (1950) (M). After Chen *et al.* (1971b). Reprinted with permission.

later data, the experimental peak in Fig. 15a is brought down to within 15% of the KRIM peak.

There is obviously a need for further experimental work, covering both the low-temperature regime and the temperature regime in the neighborhood of the peak in $C_v{}^s$ and with surface area determinations made with great care—preferably by several different methods. There is also a need for further theoretical work, since preliminary estimates of $C_v{}^s$ from (unrelaxed) shell model calculations (Alldredge *et al.*, 1974, shown as SM in Fig. 15b) agree with experiment less well than the KRIM results.

Acknowledgments

We gratefully acknowledge useful discussions with our colleagues, especially R. E. Allen, T. S. Chen, and V. L. Zoth. This work was supported in part by the U.S. Air Force Office of Scientific Research under Grant No. AFOSR-71-1973 and in part by the Robert A. Welch Foundation.

References

Alder, B. J., Vaisnys, J. R., and Jura, G. (1959). *J. Phys. Chem. Solids* **11**, 182.
Alldredge, G. P. (1972). *Phys. Lett. A* **41**, 281.
Alldredge, G. P. (1975). To be published.
Alldredge, G. P., and Kleinman, L. (1974a). *Phys. Lett. A* **48**, 337.
Alldredge, G. P., and Kleinman, L. (1974b). *J. Phys. F* **4**, L207.
Alldredge, G. P., Chen, T. S., and de Wette, F. W. (1974). *Proc. Int. Conf. Low Temp. Phys., 13th, 1972* Vol. IV, p. 441.

Allen, R. E. (1972). *J. Vac. Sci. Technol.* **9**, 934.
Allen, R. E., and de Wette, F. W. (1969a). *Phys. Rev.* **179**, 873.
Allen, R. E., and de Wette, F. W. (1969b). *J. Chem. Phys.* **51**, 4820.
Allen, R. E., and de Wette, F. W. (1969c). *Phys. Rev.* **188**, 1320.
Allen, R. E., de Wette, F. W., and Rahman, A. (1969). *Phys. Rev.* **179**, 887.
Allen, R. E., Alldredge, G. P., and de Wette, F. W. (1971a). *Phys. Rev. B* **4**, 1648.
Allen, R. E., Alldredge, G. P., and de Wette, F. W. (1971b). *Phys. Rev. B* **4**, 1661.
Armand, G., and Theeten, J. B. (1974). *Phys. Rev. B* **9**, 3969.
Barkman, J. H., Anderson, R. L., and Brackett, T. E. (1965). *J. Chem. Phys.* **42**, 1112.
Benedek, G. (1973). *Phys. Status Solidi B* **58**, 661.
Benedek, G. (1974). *In* "Lectures of International School of Physics 'Enrico Fermi,' LVIII Course, Varenna, 1973." Editrice Compositori, Bologna.
Benson, G. C., and Claxton, T. A. (1964). *J. Phys. Chem. Solids* **25**, 367.
Born, M. (1942). *Rep. Progr. Phys.* **9**, 294.
Brady, K. J. (1971). *Int. J. Solids Struct.* **7**, 941.
Businger, P. A. (1965). *Commun. ACM* **8**, 218.
Chen, T. S. (1971). Ph.D. Thesis, University of Texas, Austin (unpublished). Available from Univ. Microfilms, Inc., Ann Arbor, Michigan 48106 (order No. 72-15726, T. S. Chen).
Chen, T. S., Allen, R. E., Alldredge, G. P., and de Wette, F. W. (1970). *Solid State Commun.* **8**, 2105; corrigendum: *ibid.* **9**, No. 16, xi (1971).
Chen, T. S., Alldredge, G. P., de Wette, F. W., and Allen, R. E. (1971a). *Phys. Rev. Lett.* **26**, 1543.
Chen, T. S., Alldredge, G. P., de Wette, F. W., and Allen, R. E. (1971b). *J. Chem. Phys.* **55**, 3121.
Chen, T. S., Alldredge, G. P., and de Wette, F. W. (1972a). *Solid State Commun.* **10**, 941.
Chen, T. S., Alldredge, G. P., and de Wette, F. W. (1972b). *Phys. Lett. A* **40**, 401.
Chen, T. S., Alldredge, G. P., de Wette, F. W., and Allen, R. E. (1972c). *Phys. Rev. B* **6**, 627.
Chen, T. S., Alldredge, G. P., and de Wette, F. W. (1973). *Phys. Lett. A* **46**, 91.
Clark, B. C., Herman, R., and Wallis, R. F. (1965). *Phys. Rev.* **139**, A860.
Cochran, W. (1971). *Crit. Rev. Solid State Sci.* **2**, 1.
de Wette, F. W., and Schacher, G. E. (1965). *Phys. Rev.* **137**, A78; erratum: *ibid.* **138**, AB4.
Dobrzynski, L., and Maradudin, A. A. (1973). *Phys. Rev. B* **7**, 1207.
Dupuis, M., Mazo, R., and Onsager, L. (1960). *J. Chem. Phys.* **33**, 1452.
Farnell, G. W. (1970). *In* "Physical Acoustics" (W. P. Mason and R. N. Thurston, eds.), Vol. 6, p. 109. Academic Press, New York.
Faulkner, R. A. (1971). *In* "Computational Methods in Band Theory" (P. M. Marcus, J. F. Janak, and A. R. Williams, eds.), p. 16. Plenum, New York.
Feuchtwang, T. E. (1967). *Phys. Rev.* **155**, 715 and 731.
Finnis, M. W., and Heine, V. (1974). *J. Phys. F* **4**, L37.
Foreman, A. J. E., and Lomer, W. H. (1957). *Proc. Phys. Soc., London, Sect. B* **70**, 1143.
Friedman, B. (1956). "Principles and Techniques of Applied Mathematics." Wiley, New York.
Fuchs, R., and Kliewer, K. L. (1965). *Phys. Rev.* **140**, A2076.
Garbow, B. S. (1974). *Comput. Phys. Commun.* **7**, 179.
Gazis, D. C., and Wallis, R. F. (1965). *Acta Mech.* **1**, 253.
Gazis, D. C., and Wallis, R. F. (1966). *Surface Sci.* **5**, 482.
Gazis, D. C., Herman, R., and Wallis, R. F. (1960). *Phys. Rev.* **119**, 533.
Gear, C. W. (1966). AEC Res. Develop. Rep. No. ANL-7126. Argonne Nat. Lab., Lemont, Illinois (unpublished). Available from National Technical Information Service, Springfield, Virginia 22151.
Gear, C. W. (1967). *Math. Comput.* **21**, 146.

Gelatt, C. D., Lagally, M. G., and Webb, M. B. (1969). *Bull. Amer. Phys. Soc.* [2] **14**, 793.
Genzel, L., and Martin, T. P. (1972). *Phys. Status Solidi B* **51**, 101.
Ignatiev, A., and Rhodin, T. N. (1973). *Phys. Rev. B* **8**, 893.
Ignatiev, A., Rhodin, T. N., and Tong, S. Y. (1974). *Surface Sci.* **42**, 37.
Jones, W. E., and Fuchs, R. (1971). *Phys. Rev. B* **4**, 3581.
Kellermann, E. W. (1940). *Phil. Trans. Roy. Soc. London, Ser. A* **238**, 513.
Kenner, V. E., and Allen, R. E. (1972). *Phys. Lett. A* **39**, 245.
Kenner, V. E., and Allen, R. E. (1973). *Phys. Rev. B* **8**, 2916.
Lagally, M. G. (1975). *In* "Surface Physics of Crystalline Solids" (J. M. Blakely, ed.). Academic Press, New York (in press).
Lenglart, P., Dobrzynski, L., and Leman, G. (1972). *Ann. Phys. (Paris)* [14] **7**, 407.
Lifshitz, I. M. (1956). *Nuovo Cimento* [10] **3**, Suppl., 716.
Lifshitz, I. M., and Rosenzweig, L. N. (1948). *Zh. Eksp. Teor. Fiz.* **18**, 1012. English translation available from National Translations Center, John Crerar Library, 35 West 33rd St., Chicago, Illinois 60616.
Lim, T. C., and Farnell, G. N. (1969). *J. Acoust. Soc. Amer.* **45**, 845.
Lucas, A. A. (1968). *J. Chem. Phys.* **48**, 3156.
Ludwig, W. (1967). *Springer Tracts Mod. Phys.* **43**, 1.
Ludwig, W., and Lengeler, B. (1964). *Solid State Commun.* **2**, 83.
MacIver, D. S., and Emmett, P. H. (1956). *J. Phys. Chem.* **60**, 824.
Maradudin, A. A., Montroll, E. W., Weiss, G. H., and Ipatova, I. P. (1971). "Theory of Lattice Dynamics in the Harmonic Approximation," 2nd ed. Academic Press, New York.
Martin, T. P. (1973). *Phys. Rev. B* **7**, 3906.
Masri, P. (1973). *Surface Sci.* **39**, 51.
Masri, P., and Dobrzynski, L. (1972). *Surface Sci.* **32**, 623.
Masri, P., and Dobrzynski, L. (1973). *J. Phys. Chem. Solids* **34**, 847.
Montroll, E. W. (1950). *J. Chem. Phys.* **18**, 183.
Morrison, J. A., and Patterson, D. (1956). *Trans. Faraday Soc.* **52**, 764.
Musser, S. W. (1971). *J. Phys. Chem. Solids* **32**, 115.
Musser, S. W., and Rieder, K. H. (1970). *Phys. Rev. B* **2**, 3034.
Naugle, D. G., Baker, J. W., and Allen, R. E. (1973). *Phys. Rev. B* **7**, 3028.
Nüsslein, V., and Schröder, U. (1967). *Phys. Status Solidi* **21**, 309.
Patterson, D., Morrison, J. A., and Thompson, F. W. (1955). *Can. J. Chem.* **33**, 240.
Prosser, F. (1967). "GIVENS—Eigenvalues and Eigenvectors by the Givens Method," Program 62.3. Quantum Chemistry Program Exchange, Indiana University, Bloomington (unpublished).
Rieder, K. H. (1971). *Surface Sci.* **26**, 637.
Rieder, K. H., and Hörl, E. M. (1968). *Phys. Rev. Lett.* **20**, 209.
Rosenzweig, L. N. (1950). *Tr. Fiz. Otd. Fiz.-Mat. Fakul'teta Khark. Gos. Univ.* **2**, 19.
Schafroth, M. R. (1951). *Helv. Phys. Acta* **24**, 645.
Schmunk, R. E., and Winder, D. R. (1970). *J. Phys. Chem. Solids* **31**, 131.
Schulze, P. D. (1973). *Surface Sci.* **34**, 136.
Shuttleworth, R. (1949). *Proc. Phys. Soc., London, Sect. A* **62**, 167.
Sinha, S. K. (1973). *Crit. Rev. Solid State Sci.* **3**, 273.
Stratton, R. (1953). *Phil. Maq.* [7]. **44**, 519. See also the later note containing corrections: *J. Chem. Phys.* **34**, 2972 (1962).
Theeten, J. B., Domange, J. L., and Hurault, J. P. (1973). *Surface Sci.* **35**, 145.
Tong, S. Y., and Maradudin, A. A. (1969). *Phys. Rev.* **181**, 1318.
Tong, S. Y., Rhodin, T. N., and Ignatiev, A. (1973). *Phys. Rev. B* **8**, 906.
Trullinger, S. E., and Maradudin, A. A. (1974). *Phys. Rev. B* **10**, 1350.

Wallis, R. F. (1973). *Progr. Surface Sci.* **4**, 233.
Wallis, R. F., Clark, B. C., and Herman, R. (1968a). *Phys. Rev.* **167**, 652.
Wallis, R. F., Mills, D. L., and Maradudin, A. A. (1968b). *In* "Localized Excitations in Solids" (R. F. Wallis, ed.), p. 403. Plenum, New York.
Wallis, R. F., Clark, B. C., Herman, R., and Gazis, D. C. (1969a). *Phys. Rev.* **180**, 716.
Wallis, R. F., Clark, B. C., and Herman, R. (1969b). *In* "Structure and Chemistry of Solid Surfaces" (G. A. Somorjai, ed.), Paper No. 17. Wiley, New York.
Wilde, D. J. (1964). "Optimum Seeking Methods." Prentice-Hall, Englewood Cliffs, New Jersey.
Wilkinson, J. H., and Reinsch, C. eds. (1971). "Handbook for Automatic Computation," Vol. II. Springer–Verlag, Berlin and New York.
Wilson, J. M., and Bastow, T. J. (1971). *Surface Sci.* **26**, 461.
Wood, E. A. (1964a). *Bell Syst. Tech. J.* **43**, 541.
Wood, E. A. (1964b). *J. Appl. Phys.* **35**, 1306.
Woodruff, D. P., and Seah, M. P. (1970). *Phys. Status Solidi A* **1**, 429.
Zoth, V. L. (1973). M.A. Thesis, University of Texas, Austin (unpublished). Available as technical report AFOSR-TR-74-1256 from National Technical Information Service, Springfield, Virginia 22151.
Zoth, V. L., Alldredge, G. P., and de Wette, F. W. (1974). *Phys. Lett. A* **47**, 247.
Zoth, V. L., Alldredge, G. P., and de Wette, F. W. (1975). *Phys. Rev. B* (to be published).

Vibrational Properties of Amorphous Solids

R. J. BELL

DIVISION OF QUANTUM METROLOGY
NATIONAL PHYSICAL LABORATORY
TEDDINGTON, MIDDLESEX, ENGLAND

I. Introduction

ALTHOUGH THE ATOMISTIC THEORY of lattice dynamics was originally proposed as long ago as 1912 by Born and von Kármán, and by the mid-1930's had already been fully accepted as providing a realistic basis for the calculation of vibrational properties of solids, applications of the theory to amorphous or disordered materials were at first slow to develop. The enormous simplification which can be effected in the equations of vibrational motion for perfect lattices made it more profitable and more practicable for workers to apply the theory to crystalline structures, and so for many years the area of atomic vibrations in disordered solids remained something of a lacuna in our general body of solid state knowledge. Nonetheless, the awkward fact remained that the majority of materials of technical importance depart significantly from crystallinity—in fact, many, such as glasses, are very disordered indeed—and it was bound to be only a matter of time before suitable theoretical techniques were developed for the treatment of these systems.

The first substantial advance in the area came with the work of Lifschitz and colleagues in the Soviet Union in the mid-1940's. However, this work remained unknown in the West for a number of years (Lifschitz, 1956) and essentially similar results were produced independently by Montroll and Potts (1955, 1956) and by Maradudin and co-workers (Maradudin, 1963, 1965; Maradudin *et al.*, 1963). The treatment, based on the use of Green's functions, was essentially a perturbation analysis, vibrational properties of the perfect lattice being taken as the starting point and disorder being introduced as a perturbation. A different but related perturbative scheme was formulated (originally for electron states in solids) by Anderson (1958), in which the starting point was taken to be a set of isolated oscillators localized on the atomic sites of the solid, while the perturbation consisted of the coupling between atoms. In this approach, an element of disorder could be introduced at the outset by allowing the individual localized oscillators to vary randomly from one atomic site to another.

The Green's function method possesses an attractive overall elegance and generality and has the outstanding advantage that it can accommodate within one compact unifying formalism a great many of the important processes of solid state physics. One can, for example, derive such physical properties as frequency spectra, neutron scattering cross sections, and infrared absorption coefficients as the trace of a Green's function or some related operator. An appreciable quantity of interesting and significant qualitative results has been deduced for simple model systems through the medium of the perturbative methods. It is fair to point out, though, that the accuracy (and even the precise interpretation) of many of the analytical results has often been open to question. Indeed there remain fundamental difficulties still largely unresolved concerning the analytical behavior of many of the operators involved. A more down-to-earth problem in applying the methods to more grossly disordered materials has been that there is real difficulty in deciding just what constitutes a meaningful choice of unperturbed system.

The other major source of progress in the last two decades has been the success of the direct numerical method. The idea underlying this approach is simply to analyze the lattice dynamical equations of motion directly, in their dynamical matrix form, without reformulation or approximation. As long as this approach consisted simply of straightforward diagonalization of the dynamical matrix concerned, as it largely did prior to the 1960's, the restricted computer storage and speeds of that era placed severe limitations on the size of sample system that could be considered. However in the early 1960's, Dean *et al.* (Dean and Martin, 1960; Dean, 1960, 1961) developed a method based on the negative eigenvalue theorem, which enabled an accurate estimate of the eigenvalue distribution to be obtained for certain types of dynamical matrices, using considerably fewer arithmetical operations than were involved in a full diagonalization procedure. Later elaborations of the numerical method enabled selected eigenfrequencies and normal mode vectors of the vibrational system to be calculated with great precision.

The negative eigenvalue theorem (NET), no less than the Green's function method, revolutionized the calculation of vibrational properties of disordered materials. It found application to realistic three-dimensional materials such as disordered alloys and glasses; and the results of its application to simpler model systems provided a yardstick against which some of the more approximate analytical methods could be compared and tested.

There have been relatively few difficulties associated with the numerical accuracy of the NET-based method, and the most frequently occurring problem has generally concerned the availability of suitable efficient software for successive ranges of computer. However, the method is not without its inherent disadvantages. Unlike many of the analytical methods, the numerical method leads in the first instance to information on normal mode displacements in a rather bulky and unsophisticated form, resulting in very real practical problems of interpretation. Not until quite recently has attention been given to translating this detailed information into more manageable terms, with the help of suitable descriptive parameters such as the participation ratio and the phase quotient.

Earlier reviewers have dealt with a number of aspects of the atomic vibrations of disordered solids. Thus Lifschitz (1956, 1964), Maradudin (1963, 1965), and Elliott *et al.* (1974) have considered application of the Green's function method to the problem of crystal lattices with isolated defects or finite impurity concentrations. Dean (1972) has considered numerical studies and Bell (1972a), numerical and analytical studies of randomly disordered lattices and glasses, mainly from the point of view of the calculation of vibrational densities of states, while recent articles by Bottger (1974), Dean (1974), and Wong and Angell (1971, 1974) contain useful references to experimental studies of atomic vibrations in glasses.

Our objective in this article will be to review, selectively, contributions to progress in the lattice dynamics of disordered materials over the last ten years or so. Our main preoccupation will be with the numerical method and its predictions, although we shall have occasion to touch on the results of analytical studies where these bear strongly on our main line of reasoning. We shall be concerned not only with squared frequency spectra, but with the description of normal mode atomic motions and their interpretation in terms of physical properties of disordered materials.

We begin, in Section II, by enunciating the basic lattice dynamical equations of motion, introducing the concept of phonons, and showing the essential equivalence, in the harmonic approximation, of quantum and classical formulations. Next, in Section III, we list briefly those results for regular or slightly imperfect lattices that are most relevant to an understanding of the spectral properties of more grossly disordered systems. In Section IV we introduce the negative eigenvalue theorem and describe its role in deriving spectral properties of vibrational systems. Vibrational spectra for individual systems, obtained by the negative eigenvalue theorem method, are considered in detail in Section V. Finally we deal with the nature of atomic motions in disordered solids. The way in which vibrational modes interact with radiation to give rise to observed physical effects is considered in Section VI, and in Section VII we discuss some of the descriptive schemes which have been used in connection with the normal modes of disordered solids, in particular, glasses.

II. Equations of Vibrational Motion

In this section the basic lattice dynamical equations of motion for harmonic systems are presented in their most general form, valid for crystalline or disordered solids. We demonstrate, too, the essential equivalence of quantum mechanical and classical formulations, and outline a convenient matrix notation for the equations.

A. Quantum-Mechanical Equations of Motion

Consider a system of point atomic masses m_i $(i = 1, 2, \ldots, N)$ undergoing small amplitude displacements $\mathbf{x}_i$ about equilibrium positions $\mathbf{r}_i$. (We note here that the assignment of a specific equilibrium position to each atom implies that we are treating the atoms as distinguishable.) Suppose the potential energy of the system can be expanded as a Taylor's series about the equilibrium configuration

$$V(\mathbf{x}_1, \mathbf{x}_2, \ldots) = V_0 + V_1 + V_2 + \cdots \tag{1}$$

where V_0, V_1, V_2, etc., represent terms which are constant, linear, quadratic, etc., in the displacements, and each term depends parametrically upon the equilibrium configuration $(\mathbf{r}_1, \mathbf{r}_2, \ldots)$. We neglect the constant term V_0 which has no bearing on the dynamics of the situation, and V_1 which is formally zero for motion about the equilibrium configuration, and retain only the first of the remaining terms

$$V_2 = \tfrac{1}{2} \sum_{i\alpha j\beta} V_{i\alpha, j\beta} x_{i\alpha} x_{j\beta} \tag{2}$$

where $x_{i\alpha}$ ($\alpha = 1, 2, \ldots, d$) are d-dimensional cartesian components of the displacement $\mathbf{x}_i$ of atom i. The coefficients, or coupling constants, $V_{i\alpha,j\beta}$ form the elements of a non-negative symmetric matrix. Truncation of the potential as described above leads to the harmonic approximation for motion of the system. The effects of higher order terms may be partially taken into account within the framework of the harmonic approximation by allowing the quadratic potential term V_2 to depend parametrically upon temperature (the temperature-dependent contribution representing some averaged-out effect of the higher terms).

In atomic units, with $e = \hbar = m_e = 1$, the Schrödinger equation for the system is given by

$$\left(-\frac{1}{2} \sum_{i\alpha} \frac{1}{m_i} \frac{\partial^2}{\partial x_{i\alpha}{}^2} + \frac{1}{2} \sum_{i\alpha j\beta} V_{i\alpha, j\beta} x_{i\alpha} x_{j\beta} - E \right) \Psi_E = 0 \tag{3}$$

It is convenient to incorporate the masses m_i into the coordinates $x_{i\alpha}$ and the coupling constants $V_{i\alpha,j\beta}$ by the transformation

$$s_{i\alpha} = m_i{}^{1/2} x_{i\alpha} \tag{4}$$

$$W_{i\alpha, j\beta} = (m_i m_j)^{-1/2} V_{i\alpha, j\beta} \tag{5}$$

With this substitution, the Schrödinger equation becomes

$$\left(-\frac{1}{2} \sum_{i\alpha} \frac{\partial^2}{\partial s_{i\alpha}{}^2} + \frac{1}{2} \sum_{i\alpha j\beta} W_{i\alpha, j\beta} s_{i\alpha} s_{j\beta} - E \right) \Psi_E = 0 \tag{6}$$

We now consider the solutions of the eigenvalue problem

$$\sum_{j\beta} W_{i\alpha,j\beta} u_{j\beta,l} = \omega_l^2 u_{i\alpha,l} \qquad l = 1, 2, \ldots, L \tag{7}$$

$L = Nd$ being the total number of degrees of freedom in the dynamical system. We adopt the normalization

$$\sum_{i\alpha} u_{i\alpha,l} u_{i\alpha,l'} = \delta_{ll'} \tag{8}$$

so that

$$\sum_{i\alpha j\beta} u_{i\alpha,l} W_{i\alpha,j\beta} u_{j\beta,l'} = \omega_l^2 \delta_{ll'} \tag{9}$$

The eigenvectors $u_{i\alpha,l}$ may be used as the basis of an orthogonal transformation from the mass-dependent coordinates $s_{i\alpha}$ to a system of *normal coordinates* q_l

$$s_{i\alpha} = \sum_l u_{i\alpha,l} q_l$$

$$q_l = \sum_{i\alpha} u_{i\alpha,l} s_{i\alpha} \tag{10}$$

which leaves the Schrödinger equation in the form

$$\left[\frac{1}{2} \sum_l \left(-\frac{\partial^2}{\partial q_l^2} + \omega_l^2 q_l^2 \right) - E \right] \Psi_E = 0 \tag{11}$$

A further substitution

$$\Psi_E = \prod_l \psi_l(q_l)$$

$$E = \sum_l E_l \tag{12}$$

then enables the Schrödinger equation to be decoupled into L separate equations

$$\left[\frac{1}{2}\left(-\frac{\partial^2}{\partial q_l^2} + \omega_l^2 q_l^2\right) - E_l\right]\psi_l(q_l) = 0 \qquad l = 1, 2, \ldots, L \tag{13}$$

Equation (13) is the Schrödinger equation for a linear harmonic oscillator, with wave functions and quantized energy levels

$$\psi_l(q_l) = \phi_{n_l}(\omega_l, q_l) = \left(\frac{(\omega_l/\pi)^{1/2}}{2^{n_l} n_l!}\right)^{1/2} H_{n_l}(\omega_l^{1/2} q_l) \exp(-\tfrac{1}{2}\omega_l q_l^2)$$

$$E_l = (n_l + \tfrac{1}{2})\omega_l, \qquad n_l = 0, 1, 2, \ldots \tag{14}$$

where H_{n_l} is the hermite polynomial of order n_l (cf. Abramowitz and Stegun, 1964).

In the harmonic approximation, then, the quantum states of the system are uncoupled oscillator states, the independent coordinates being, not the original atomic coordinates $x_{i\alpha}$, but certain combinations of these, defined by (10). The individual oscillators of this system are, like the atoms, distinguishable and their occupation numbers are mutually independent. There is no energy exchange between the normal mode oscillators unless some interaction (external or internal) beyond the harmonic potential intervenes. Transfer of energy to or from the system, accompanying such interactions, is most conveniently expressed in terms of transfer to or from the normal mode oscillators, as distinct from the individual vibrating atoms.

An oscillator in its nth quantum state is said to contain n *phonons* or quanta of vibrational energy, and the quantum number n is called the phonon occupation number of the state. When the oscillator changes its quantum state, phonons are said to be created (n increased) or destroyed (n decreased), and vibrational excitations of the solid are generally classified according to the number of phonon creations or destructions involved.

B. Classical Equations of Motion

The classical mechanical expression for the total energy, in mass dependent coordinates, is given by

$$E = \tfrac{1}{2}\sum_{i\alpha} \dot{s}_{i\alpha}^2 + \tfrac{1}{2}\sum_{i\alpha j\beta} W_{i\alpha,j\beta} s_{i\alpha} s_{j\beta} \tag{15}$$

leading to the classical equations of motion

$$\ddot{s}_{i\alpha} + \sum_{j\beta} W_{i\alpha,j\beta} s_{j\beta} = 0 \tag{16}$$

Like the quantum mechanical equations, these are brought to diagonal form by the transformation (10)

$$E = \tfrac{1}{2}\sum_{l} (\dot{q}_l^{\,2} + \omega_l^{\,2} q_l^{\,2}) \tag{17}$$

$$\ddot{q}_l + \omega_l^{\,2} q_l = 0 \qquad l = 1, 2, \ldots, L \tag{18}$$

Each l value in (18) corresponds to a classical simple harmonic oscillator with time dependence

$$q_l = a_l \cos(\omega_l t + \varepsilon_l) \tag{19}$$

where ω_l is the frequency and a_l, ε_l are integration constants defining the amplitude and phase. If just one classical oscillator, the lth one, is active in the solid, we may take all a's to be zero except the lth one $a_l = 1$. The cartesian displacements of atoms in this mode are given by

$$s_{i\alpha,l} = u_{i\alpha,l} \cos(\omega_l t + \varepsilon_l) \tag{20}$$

A more general motion of the system may be written as a superposition of terms of the form (20).

The close relation between classical frequency ω_l and quantum energy levels (14), and the fact that the same transformation (10) decouples both quantum and classical mechanical equations—giving the normal mode displacements in the latter case—are points of central importance in the harmonic approximation. They mean that many of the purely dynamical properties of vibrational systems may be computed and discussed within the conceptually simpler, classical framework, without sacrificing the essential rigor of the quantum mechanical approach. It is important to stress, however, that as far as the radiative properties of the system are concerned (infrared absorption, Raman scattering, etc.), a full quantum mechanical calculation is generally necessary to achieve reliable results.

C. Matrix Notation for Equations of Motion

It is useful for the purpose of our subsequent discussions to re-express Eq. (4) to (18) in matrix notation. We define a column vector $\mathbf{x}$ with elements $x_{i\alpha}$, and mass and force constant matrices $\mathbf{M}$, $\mathbf{V}$ with elements $m_i\delta_{ij}\delta_{\alpha\beta}$ and $V_{i\alpha,j\beta}$, respectively. Then the vector $\mathbf{s}$ containing mass dependent amplitudes $s_{i\alpha}$ and the *dynamical matrix* $\mathbf{W}$ with elements $\mathbf{W}_{i\alpha,j\beta}$ are given by

$$\mathbf{s} = \mathbf{M}^{1/2}\mathbf{x} \tag{4$'$}$$

$$\mathbf{W} = \mathbf{M}^{-1/2}\mathbf{V}\mathbf{M}^{-1/2} \tag{5$'$}$$

In this notation, the coupled Schrödinger equation (6) takes the form

$$(-\tfrac{1}{2}\mathbf{\nabla}_s{}^T\mathbf{\nabla}_s + \tfrac{1}{2}\mathbf{s}^T\mathbf{W}\mathbf{s} - E)\Psi_E = 0 \tag{6$'$}$$

where $\mathbf{\nabla}_s$ is a vector operator containing elements $\partial/\partial s_{i\alpha}$.

The eigenvalue equation (7), normalization condition (8), and diagonalization equation (9) become

$$\mathbf{W}\mathbf{u}_l = \omega_l{}^2\mathbf{u}_l \tag{7$'$}$$

$$\mathbf{u}_l{}^T\mathbf{u}_{l'} = \delta_{ll'} \tag{8$'$}$$

$$\mathbf{u}_l{}^T\mathbf{W}\mathbf{u}_{l'} = \omega_l{}^2\delta_{ll'} \tag{9$'$}$$

In terms of the matrix $\mathbf{U}$ whose columns are $\mathbf{u}_l$, these equations take the compact form

$$\mathbf{W}\mathbf{U} = \mathbf{U}\mathbf{\Omega}^2 \tag{7$''$}$$

$$\mathbf{U}^T\mathbf{U} = \mathbf{I} \tag{8$''$}$$

$$\mathbf{U}^T\mathbf{W}\mathbf{U} = \mathbf{\Omega}^2 \tag{9$''$}$$

where $\mathbf{\Omega}$ is a diagonal matrix with elements $\omega_l\delta_{ll'}$ and $\mathbf{I}$ is a unit matrix of order L. The normal coordinate displacement vector $\mathbf{q}$ with components q_l is related to $\mathbf{s}$ by

$$\mathbf{s} = \mathbf{U}\mathbf{q}$$

$$\mathbf{q} = \mathbf{U}^T\mathbf{s} \tag{10$'$}$$

and in this matrix representation the diagonalized Schrödinger equation reduces to

$$(-\tfrac{1}{2}\mathbf{\nabla}_q{}^T\mathbf{\nabla}_q + \tfrac{1}{2}\mathbf{q}^T\mathbf{\Omega}^2\mathbf{q} - E)\Psi_E = 0 \tag{11$'$}$$

$\mathbf{\nabla}_q$ being a vector with elements $\partial/\partial q_l$.

Transformation of the classical equations follows a similar pattern, with (15) and (16) being replaced by the matrix versions

$$E = \tfrac{1}{2}\dot{\mathbf{s}}^T\dot{\mathbf{s}} + \tfrac{1}{2}\mathbf{s}^T\mathbf{W}\mathbf{s} \tag{15$'$}$$

$$\ddot{\mathbf{s}} + \mathbf{W}\mathbf{s} = 0 \tag{16$'$}$$

and the diagonalized forms (17) and (18) by

$$E = \tfrac{1}{2}\dot{\mathbf{q}}^T\dot{\mathbf{q}} + \mathbf{q}^T\mathbf{\Omega}^2\mathbf{q} \tag{17$'$}$$

$$\ddot{\mathbf{q}} + \mathbf{\Omega}^2\mathbf{q} = 0 \tag{18$'$}$$

For most of the disordered systems we shall be concerned with, it is a valid approximation to regard the interatomic forces as being short range in character. In this case the dynamical matrix **W** assumes a simple banded form, with elements beyond a certain distance from the main diagonal being zero. This leads to considerable computational advantages, as explained in Section IV, D.

D. The Green's Function

Another matrix quantity of considerable primary significance is the *Green's function*, or *resolvent* of a vibrational system. The Green's function $\mathbf{G}(\omega^2)$ for a system with dynamical matrix **W** obeys the equation

$$(\mathbf{W} - \omega^2 \mathbf{I})\mathbf{G}(\omega^2) = \mathbf{I} \tag{21}$$

and may be written symbolically as

$$\mathbf{G} = (\mathbf{W} - \omega^2 \mathbf{I})^{-1} \tag{22}$$

This definition (22) causes no formal difficulty as long as ω does not coincide with one of the natural frequencies ω_l of **W**. However if ω approaches an individual ω_l value, or lies in a spectral region densely occupied by ω_l values, some care must be exercised in the interpretation of (22). In these circumstances it is conventional to replace ω^2 by an infinitesimally different quantity $\omega^2 + i0$ (i.e., one replaces ω^2 by $\omega^2 + i\delta$ and takes the limit as $\delta \to 0+$).

Like other operators which are expressible as simple algebraic functions of **W**, the Green's function is diagonalized by the transformation analogous to (9″) and may be written in terms of the eigenvectors and eigenfrequencies of **W** as

$$\mathbf{G} = \sum_l \frac{\mathbf{u}_l \mathbf{u}_l^T}{\omega_l^2 - \omega^2} \tag{23}$$

The function $\mathbf{G}(\omega^2)$ defined above stems essentially from the classical equations of motion. A related Green's function arising from the quantum mechanical equations may also be defined. In either case one may work instead in terms of the time-dependent Green's function, or propagator, which is essentially a Fourier frequency transform of the resolvent. And when physical processes at finite temperature are being considered an element of thermodynamic averaging may also be introduced.

In principle a Green's function such as (22) contains precisely the same information as the dynamical matrix **W**. It can therefore be manipulated to determine various dynamical properties of the system such as the frequency distribution (cf. Section II, E). Equally, it can be used as a tool to investigate the effect on a system of various types of perturbation. Two distinct categories of perturbation are most frequently encountered in lattice dynamical theory. The first is related to the introduction of internal changes, such as anharmonicity or some element of disorder, into the equations of motion. The second arises from external disturbances such as incident radiation, and in this case the Green's function treatment leads to an expression for the response of the system, in the form, for example, of an absorption coefficient or cross section.

E. Density of States

A quantity of particular interest in connection with vibrational systems is the fundamental frequency spectum $g(\omega)$; here $g(\omega)\,d\omega$ represents the number of fundamental vibrational frequencies of the system lying between ω and $\omega + d\omega$. It is preferable in some situations to deal with the squared frequency spectrum $\mathscr{G}(\omega^2)$, defined so that $\mathscr{G}(\omega^2)\,d(\omega^2)$ gives the number of squared frequencies between ω^2 and $\omega^2 + d(\omega^2)$. The two distributions are related by

$$g(\omega) = 2\omega\mathscr{G}(\omega^2) \tag{24}$$

and it is convenient in the present context to have the normalization

$$\int_0^\infty d\omega\, g(\omega) = L = \int_0^\infty d(\omega^2)\mathscr{G}(\omega^2) \tag{25}$$

For the finite model systems often treated in disordered lattice dynamics, it is usually necessary to represent $g(\omega)$ and $\mathscr{G}(\omega^2)$ as histograms. Only in the limit of a very large number of degrees of freedom can one think of these distributions as having the form of continuous curves.

Although $g(\omega)$ and $\mathscr{G}(\omega^2)$ cannot generally be observed directly, they are closely related to various thermodynamic and radiative properties of the system and form a convenient starting point for intercomparison and discussion of the spectral properties of solids.

The density of states $\mathscr{G}(\omega^2)$ can be directly related to matrix operators introduced earlier in the section. Thus, for example,

$$\eta(\mathbf{W} - \omega^2\,\mathbf{I}) = \int_0^{\omega^2} d(\omega^2)\mathscr{G}(\omega^2) \tag{26}$$

where $\eta(\mathbf{A})$ denotes the number of negative eigenvalues of matrix $\mathbf{A}$. Alternatively, formula (23) for $\mathbf{G}$ may be combined with the identity

$$(x - \mathrm{i}0)^{-1} = \mathscr{P}\frac{1}{x} + \mathrm{i}\pi\,\delta(x)$$

to establish the relationship

$$\operatorname{Tr}\mathbf{G} = \tilde{\mathscr{G}}(\omega^2) + \mathrm{i}\pi\mathscr{G}(\omega^2) \tag{27}$$

where $\tilde{\mathscr{G}}(\omega^2)$ is the Hilbert transform of the squared frequency spectrum, given by the principal value integral

$$\tilde{\mathscr{G}}(\omega^2) = \mathscr{P}\int \frac{d(\omega'^2)}{\omega'^2 - \omega^2}\mathscr{G}(\omega'^2) \tag{28}$$

III. Properties of Regular Lattices and Lattices with Point Defects

A number of phenomena in amorphous materials can be elucidated by referring to the phonon properties of less seriously disordered materials, and these, in turn, are most readily understood as departures from the properties of the purely crystalline solid. To this end we recount briefly some of the essential features of atomic vibrations in perfect lattices and lattices with point imperfections.

A. Matrix Reduction Theorem for Perfect Lattices

If a system is periodic in one of its directions and is taken to have cyclic boundary conditions in that direction, the dynamical matrix takes the block circulant form

$$\mathbf{W} = \begin{bmatrix} \mathbf{W}_{(0)} & \mathbf{W}_{(1)} & \cdot\ \cdot\ \cdot\ \cdot & \mathbf{W}_{(N_\alpha - 1)} \\ \mathbf{W}_{(N_\alpha - 1)} & \mathbf{W}_{(0)} & \cdot\ \cdot\ \cdot\ \cdot & \mathbf{W}_{(N_\alpha - 2)} \\ \cdot & \cdot & \cdot\ \cdot\ \cdot\ \cdot & \cdot \\ \mathbf{W}_{(1)} & \mathbf{W}_{(2)} & \cdot\ \cdot\ \cdot\ \cdot & \mathbf{W}_{(0)} \end{bmatrix} \tag{29}$$

where $\mathbf{W}_{(t)}$ are square matrices of equal order and N_α is the number of repeat cells in the direction of periodicity, α. In this case we may make use of a theorem due to Friedman (1961) which states that the eigenvalues of $\mathbf{W}$ are the eigenvalues of the *reduced matrices*

$$\mathbf{W}^{(R)}(k) = \sum_{t=0}^{N_\alpha - 1} \mathbf{W}_{(t)} \exp(2ikt) \tag{30}$$

where k may take the values

$$k = \pi s/N_\alpha, \qquad s = 0, 1, 2, \ldots, N_\alpha - 1$$

Moreover, any eigenvector $\mathbf{u}^{(R)}(k)$ of $\mathbf{W}^{(R)}(k)$ with eigenvalue $\lambda(k)$ corresponds to an eigenvector

$$\mathbf{u}(k) = \begin{bmatrix} \mathbf{u}^{(R)}(k) \\ \mathbf{u}^{(R)}(k) \exp(2ik) \\ \cdot \\ \cdot \\ \mathbf{u}^{(R)}(k) \exp(2[N_\alpha - 1]ik) \end{bmatrix} \tag{31}$$

of $\mathbf{W}$, also with eigenvalue $\lambda(k)$. In the case of very large systems ($N_\alpha \to \infty$), we may take k to be a continuous variable, ranging between 0 and π or, equivalently, between $-\pi/2$ and $+\pi/2$.

If the system possesses periodicity and a cyclic boundary condition in a second direction, $\mathbf{W}^{(R)}(k)$ itself has block circulant form and the process may be repeated. For a d-dimensional periodic crystal, Friedman's theorem may be applied d times, leading to a d-fold reduced matrix

$$\mathbf{W}^{(R)}(\mathbf{k}) = \sum_{\mathbf{t}} \mathbf{W}_{(\mathbf{t})} \exp(2i\mathbf{k} \cdot \mathbf{t}) \tag{32}$$

here $\mathbf{k}$ is a d-component vector whose components k_α can take values $k_\alpha = \pi s_\alpha/N_\alpha$ ($s_\alpha = 0, 1, 2, \ldots, N_\alpha - 1$), and the summation is over all d-component vectors $\mathbf{t}$ with elements of the form $t_\alpha = 0, 1, 2, \ldots, N_\alpha - 1$. Again, for large N_α, the elements may be taken to vary smoothly between 0 and π, or $-\pi/2$ and $+\pi/2$. The eigenvalues of $\mathbf{W}$ are thus those of the $\prod_\alpha N_\alpha$ reduced matrices $\mathbf{W}^{(R)}(\mathbf{k})$, (whose dimension is, in each case, equal to the number of degrees of freedom in a unit cell of the crystal) and each eigenvector $\mathbf{u}^{(R)}(\mathbf{k})$ of the reduced matrix gives rise to an eigenvector $\mathbf{u}(\mathbf{k})$ of the full matrix, whose components differ only by a multiplicative phase factor from those of $\mathbf{u}^{(R)}(\mathbf{k})$. Specifically, that portion of the full eigenvector corresponding to a unit cell whose position is labeled by $\mathbf{t}$ is just $\mathbf{u}^{(R)}(\mathbf{k}) \exp(2i\mathbf{k} \cdot \mathbf{t})$.

We should note that, for systems with cyclic boundary conditions, the solutions occur in degenerate pairs, with equal and opposite wave vectors

$\pm\mathbf{k}$ corresponding to the same eigenvalue $\omega^2(\pm\mathbf{k})$. This enables the complex vectors (31) to be combined to give real normal mode displacements.

B. Linear Chains and Simple Lattices

The simplest example of a periodic vibrational system is the 1-dimensional linear chain with nearest-neighbor restoring forces. We consider a chain where each of the N atoms is coupled to its immediate neighbors by Hooke's law springs, the force constant being γ. Suppressing the α dependence on account of the 1-dimensionality of the system, the equation of motion for each atom is

$$m_i\ddot{x}_i = \gamma(x_{i+1} - x_i) + \gamma(x_{i-1} - x_i) \qquad i = 1, 2, \ldots, N \tag{33}$$

where the identification $i \pm N \equiv i$ corresponds to the assumption of cyclic boundary conditions. Then the dynamical matrix has elements

$$W_{ij} = \begin{cases} 2\gamma/m_i & i = j \\ -\gamma/(m_i m_j)^{1/2} & |i - j| = 1 \\ 0 & \text{otherwise} \end{cases} \tag{34}$$

For a monatomic chain with all masses equal to m, $\mathbf{W}$ has block circulant form, possessing simple 1×1 blocks

$$\mathbf{W}_{(t)} \begin{cases} 2\gamma/m & t = 0 \\ -\gamma/m & t = 1, \quad N - 1 \\ 0 & \text{otherwise} \end{cases} \tag{35}$$

Then (34) gives the vibrational squared frequencies as

$$\left.\begin{aligned} \omega^2(k) &= (\gamma/m)[2 - \exp(2ik) - \exp(2(N - 1)ik)] \\ &= (4\gamma/m)\sin^2 k \end{aligned}\right\} \tag{36}$$

and the displacement eigenvectors take the form

$$\mathbf{u}(k) = \begin{bmatrix} u^{(R)}(k) \\ u^{(R)}(k)\exp(2ik) \\ \cdot \\ \cdot \\ u^{(R)}(k)\exp(2[N - 1]ik) \end{bmatrix} \tag{37}$$

where $u^{(R)}(k)$ is a constant, which we take for normalization purposes to be $N^{-1/2}$.

In the limit of large N, the density of states for the monatomic linear chain may be calculated to be

$$\mathscr{G}(\omega^2) = N[\pi\omega({\omega_M}^2 - \omega^2)^{1/2}]^{-1} \qquad 0 < \omega < \omega_M \tag{38}$$

where ${\omega_M}^2 = 4\gamma/m$ defines a maximum normal mode frequency consistent with (34).

If the chain is diatomic, with alternate heavy and light masses m_H and m_L, the relevant blocks are of order 2×2, and the reduced secular equation corresponding to (34) has two roots

$$\omega^2(k) = \gamma\left[\left(\frac{1}{m_L} + \frac{1}{m_H}\right) \pm \left\{\left(\frac{1}{m_L} + \frac{1}{m_H}\right)^2 - \frac{4}{m_L m_H}\sin^2 k\right\}^{1/2}\right] \tag{39}$$

This gives a two-band spectrum, covering the ranges $0 < \omega^2 < 2\gamma/m_H$ and $2\gamma/m_L < \omega^2 < {\omega_M}^2 = 2\gamma/m_L + 2\gamma/m_H$. The explicit form of the squared frequency spectrum is

$$\mathscr{G}(\omega^2) = \frac{N}{\pi}\frac{|\omega^2 - {\omega_M}^2/2|}{[\omega^2(2\gamma/m_H - \omega^2)(2\gamma/m_L - \omega^2)({\omega_M}^2 - \omega^2)]^{1/2}} \tag{40}$$

Modes in the upper band, corresponding to the positive sign in (39), are optical modes, with neighboring atoms in the unit cell moving out of phase during vibration; those in the lower band, corresponding to the negative sign, are acoustic in character, with neighboring atoms moving in phase.

Extension to chains with more complicated unit cells, and to simple systems in higher dimensions is straightforward and has been described in the literature. Taking the 2-dimensional monatomic quadratic lattice with nearest-neighbor central and noncentral forces as an example, the equations of motion for vibration in the two main symmetry directions decouple, giving for the $\alpha = 1$ vibrations squared frequency values of

$${\omega_1}^2(\mathbf{k}) = (4\gamma_c/m)\sin^2 k_1 + (4\gamma_n/m)\sin^2 k_2 \tag{41}$$

with a corresponding expression for $\alpha = 2$; here m is the atomic mass and γ_c, γ_n are central and noncentral force constants, respectively. Montroll (1956) has given an expression for the squared frequency distribution for this system. The generalization of (41) to higher dimensional analogs, such as the simple cubic lattice is obvious.

C. Lattice Green's Functions

Given the squared frequencies of a vibrational system and details of the normal mode eigenvectors, the Green's function may, in principle at least, be evaluated from (23). Green's function matrix elements for a wide range of simple lattices vibrating under various elementary harmonic force fields have been computed by a number of authors, an extensive bibliography being given by Katsura *et al.* (1971).

The Green's function of a system exhibits the same symmetry and periodicity properties as its parent dynamical matrix **W**. In the case of the infinitely long 1-dimensional monatomic chain, for example, summation (i.e., integration) of (23) gives

$$G_{ij}(\omega^2) = -\frac{1}{\omega^2 \xi}\left(-\frac{1-\xi}{1+\xi}\right)^{|i-j|} \tag{42}$$

where we have set

$$\xi = +\left[1 - (\omega_M/\omega)^2\right]^{1/2} \tag{43}$$

The elements of **G**, like those of **W**, depend only upon the separation of atomic sites i, j involved. Note that the use of (42) together with (27) leads to the linear chain squared frequency spectrum (38).

Another property of the linear chain Green's function (42) is worth commenting on. If ω lies within the normal frequency range $0 < \omega < \omega_M$ of the lattice, ξ is imaginary and the elements G_{ij} oscillate as a function of the distance $|i-j|$ from the main diagonal; if, however, ω is outside the allowed frequency range, ξ is real and G_{ij} decays exponentially with increasing distance $|i-j|$. This behavior is an important general feature of Green's functions in crystal lattices. The extension to disordered lattices has been considered by a number of authors, including Fujita and Hori (1972, 1973), who show that an exponentially decaying Green's function is characteristic not just of frequencies outside the normal frequency range of the lattice, but also of in-band frequency regions corresponding to localized vibrations.

D. The Perturbed Green's Function

If a system is modified so that the dynamical matrix undergoes the change

$$\mathbf{W} \rightarrow \mathbf{W} + \delta\mathbf{W} \tag{44}$$

then the eigenvectors and Green's function, **u** and **G**, of the unperturbed system corresponding to frequency ω are related to their perturbed counterparts, $\mathbf{v}$ and $\mathbf{G} + \delta\mathbf{G}$, by

$$(\mathbf{I} + \mathbf{G}\,\delta\mathbf{W})\mathbf{v} = c\mathbf{u} \tag{45}$$

$$(\mathbf{I} + \mathbf{G}\,\delta\mathbf{W})(\mathbf{G} + \delta\mathbf{G}) = \mathbf{G} \tag{46}$$

If ω approaches a natural frequency ω_l of the unperturbed system, or lies in a spectral region densely occupied by ω_l values, the operator $(\mathbf{I} + \mathbf{G}\,\delta\mathbf{W})$, like $\mathbf{G}$, must be evaluated by making the replacement $\omega^2 \to \omega^2 + i0$. The operator thus obtained is a complex quantity and, in particular, has nonzero imaginary part if ω coincides with an unperturbed ω_l. In this case, $(\mathbf{I} + \mathbf{G}\,\delta\mathbf{W})$ is nonsingular and the perturbed quantities can be computed directly from (45) and (46), in the form

$$\mathbf{v} = c(\mathbf{I} + \mathbf{G}\,\delta\mathbf{W})^{-1}\mathbf{u} \tag{47}$$

$$\mathbf{G} + \delta\mathbf{G} = (\mathbf{I} + \mathbf{G}\,\delta\mathbf{W})^{-1}\mathbf{G} \tag{48}$$

Here $(\mathbf{I} + \mathbf{G}\,\delta\mathbf{W})^{-1}$ can be regarded as a *shift operator* which replaces certain of the unperturbed variables of the system by their perturbed counterparts.

There may also exist ω values, not identical to the unperturbed natural frequencies, which correspond to singularities of the operator $(\mathbf{I} + \mathbf{G}\,\delta\mathbf{W})$, giving

$$\det|\mathbf{I} + \mathbf{G}\,\delta\mathbf{W}| = 0 \tag{49}$$

These give rise to perturbed solutions obtained from the homogeneous equation

$$(\mathbf{I} + \mathbf{G}\,\delta\mathbf{W})\mathbf{v} = 0 \tag{50}$$

Although (50) may not be satisfied for ω values lying in the continuous frequency range of the unperturbed system, it may be possible to satisfy the less stringent condition

$$\mathrm{Re}(\det|\mathbf{I} + \mathbf{G}\,\delta\mathbf{W}|) = 0 \tag{51}$$

Such solutions, if they exist, give rise to resonances (i.e., enhancements) of the spectrum in the region of the frequency concerned. They are most usually associated with the presence of heavy impurities in the system, and the motion of such impurities tends to be accentuated in vibrational modes near the resonance frequency.

E. Localized Defects

When dealing with an isolated defect, it is convenient to partition the operators so as to display explicitly those parts of vectors and matrices referring to the defect sites. We write

$$\mathbf{u} = \begin{bmatrix} \mathbf{u}_{(d)} \\ \mathbf{u}_{(n)} \end{bmatrix} \qquad \mathbf{v} = \begin{bmatrix} \mathbf{v}_{(d)} \\ \mathbf{v}_{(n)} \end{bmatrix}$$
$$\delta\mathbf{W} = \begin{bmatrix} \delta\mathbf{W}_{(dd)} & \mathbf{0} \\ \mathbf{0} & \mathbf{0} \end{bmatrix} \qquad \mathbf{G} = \begin{bmatrix} \mathbf{G}_{(dd)} & \mathbf{G}_{(dn)} \\ \mathbf{G}_{(nd)} & \mathbf{G}_{(nn)} \end{bmatrix} \tag{52}$$

where (n) labels the normal crystal sites collectively, and (d) the defect sites. In partitioned notation (45) takes the form

$$c\mathbf{u}_{(d)} = (\mathbf{I} + \mathbf{G}_{(dd)}\,\delta\mathbf{W}_{(dd)})\mathbf{v}_{(d)} \tag{53a}$$

$$c\mathbf{u}_{(n)} = \mathbf{v}_{(n)} + \mathbf{G}_{(nd)}\,\delta\mathbf{W}_{(dd)}\,\mathbf{v}_{(d)} \tag{53b}$$

Once again, provided ω coincides with an unperturbed frequency ω_l, Eq. (53a) may be inverted to give

$$\mathbf{v}_{(d)} = c(\mathbf{I} + \mathbf{G}_{(dd)}\,\delta\mathbf{W}_{(dd)})^{-1}\mathbf{u}_{(d)} \tag{54}$$

with $\mathbf{v}_{(n)}$ being obtained from (53b). And if there exist ω values distinct from the unperturbed frequencies ω_l which give

$$\det|\mathbf{I} + \mathbf{G}_{(dd)}\,\delta\mathbf{W}_{(dd)}| = 0 \tag{55}$$

corresponding perturbed solutions may be obtained from the equation

$$(\mathbf{I} + \mathbf{G}_{(dd)}\,\delta\mathbf{W}_{(dd)})\mathbf{v}_{(d)} = 0 \tag{56}$$

F. Isolated Mass Defects in Simple Lattices

Consider a perturbation in which the mass matrix $\mathbf{M}$ is replaced by $\mathbf{M} + \delta\mathbf{M}$. The matrix equation of motion may be rewritten in the form

$$\mathbf{W}\mathbf{v} = \omega^2(\mathbf{I} + \mathbf{M}^{-1/2}\,\delta\mathbf{M}\,\mathbf{M}^{-1/2})\mathbf{v} \tag{57}$$

where the perturbed eigenvectors $\mathbf{v}_l$ are assumed to satisfy a perturbed orthogonality relation of the form

$$\mathbf{v}_l^T(\mathbf{I} + \mathbf{M}^{-1/2}\,\delta\mathbf{M}\,\mathbf{M}^{-1/2})\mathbf{v}_{l'} = \delta_{ll'} \tag{58}$$

This formulation leads to an ω-dependent form of the perturbation

$$\delta \mathbf{W} = -\omega^2 \mathbf{M}^{-1/2}\, \delta \mathbf{M}\, \mathbf{M}^{-1/2}. \tag{59}$$

The simplest example of a defect system is the monatomic linear chain with one of the masses m replaced by a defect mass $m_{(\mathrm{d})}$. In this case all elements of $\delta \mathbf{W}$ are zero except that in the single defect position (d, d) which takes the value

$$\delta \mathbf{W}_{(\mathrm{dd})} = \varepsilon_{\mathrm{d}} \omega^2 \tag{60}$$

where

$$\varepsilon_{\mathrm{d}} = (m - m_{\mathrm{d}})/m \tag{61}$$

may be regarded as a smallness parameter.

Using Eq. (42) for the elements of the infinitely long linear chain Green's function, the secular equation for out-of-band modes (or resonances) becomes simply

$$\mathrm{Re}[1 - (\varepsilon_{\mathrm{d}}/\xi)] = 0 \qquad \text{or} \qquad \mathrm{Re}\, \xi = \varepsilon_{\mathrm{d}} \tag{62}$$

According to (43), ξ is real and positive for $\omega > \omega_{\mathrm{M}}$, while from (61) ε_{d} is positive for a light mass defect. For light defects, therefore, there exist out-of-band solutions corresponding to $\xi = \varepsilon_{\mathrm{d}}$, at frequencies $\omega_{\mathrm{d}} > \omega_{\mathrm{M}}$ given by

$$\varepsilon_{\mathrm{d}} = +[1 - (\omega_{\mathrm{M}}/\omega_{\mathrm{d}})^2]^{1/2} \tag{63}$$

However (43) gives ξ to be imaginary for $\omega < \omega_{\mathrm{M}}$, so that there are no corresponding solutions of (62) and resonances do not occur for the case of an isolated mass defect in the linear chain.

The form of atomic displacements is of interest for out-of-band modes. This may be deduced by using Eq. (42) for the linear chain Green's function, together with Eq. (53b) with $c = 0$. Clearly the displacement amplitudes simply decay exponentially as $[-(1 - \varepsilon_{\mathrm{d}})/(1 + \varepsilon_{\mathrm{d}})]^{|d-j|}$ with distance from the defect site (d), oscillating in sign from site to site as one expects for modes in the optical region of the spectrum. As the mass of the defect atom approaches zero ($\varepsilon_{\mathrm{d}} \to 1$), the defect frequency given by (63) increases and the decay rate becomes increasingly severe. This property of localization is an important characteristic of out-of-band modes and contrasts strongly with the spatial behavior of in-band modes in the single defect system which are much more like the plane wave modes of the unperturbed lattice.

The treatment for pairs of mass defects is similar, with the matrix appearing in (55) now being of order 2 × 2, giving the secular equation

$$\left[1 - \frac{\varepsilon_d}{\xi}\right]^2 = \left[\frac{\varepsilon_d}{\xi}\left(-\frac{1-\xi}{1+\xi}\right)^s\right]^2 \tag{64}$$

where s is the separation of the defect atoms in the system. The behavior of ξ versus ε_d is shown in Fig. 1 for a single defect (the dashed line $\xi = \varepsilon_d$) and for a pair of adjacent defects (dotted lines). The solid line $\xi = \varepsilon_d{}^{1/2}$ gives the ξ value corresponding to the maximum frequency of a regular lattice composed entirely of the light (defect) atoms. In the case of a defect pair, two out-of-band modes clearly occur for sufficiently light impurity; however, as the mass increases beyond a threshold value $m_d = m/2$ ($\varepsilon_d = \frac{1}{2}$) one of the modes merges with the continuous spectrum at $\omega = \omega_M$ ($\xi = 0$). In both single and paired defect cases, the defect frequencies do not exceed the maximum light-lattice frequency.

Dean (1967), who has treated the general problem of r adjacent identical defect masses, finds that, for sufficiently small value of the light defect mass, r out-of-band modes appear. As in the single and paired defect cases, their frequencies do not exceed the maximum frequency of a lattice composed

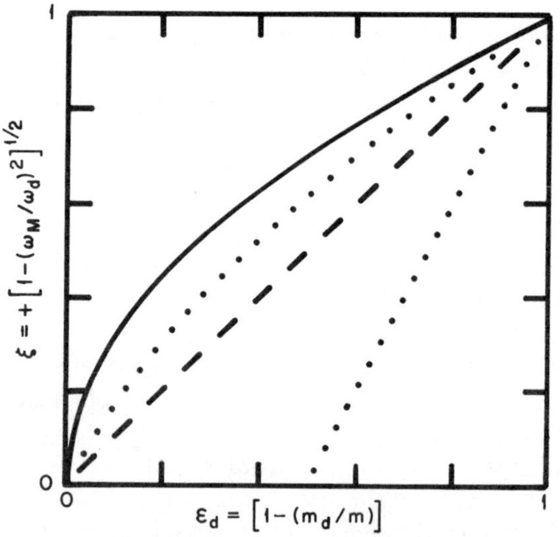

FIG. 1. Defect frequency parameter $\xi = +[1 - (\omega_M/\omega_d)^2]^{1/2}$ plotted against mass defect parameter $\varepsilon_d = [1 - (m_d/m)]$ for light defects. Dashed line, single defect case. Dotted line, pair of adjacent defects. The solid line gives $\xi = \varepsilon_d{}^{1/2}$, the ξ value corresponding to the maximum frequency of a regular lattice composed entirely of the defect atoms; this curve represents an upper bound to the ξ values which may occur for defect clusters of arbitrary size.

entirely of light atoms. As ε_d decreases toward zero, the modes coalesce one by one into the continuous spectrum of the host lattice, the last one disappearing in the limit $\varepsilon_d = 0$. Work by Economou and Papatriantafillou (1972) suggests that some of the modes that are lost by coalescing with the continuous spectrum appear in the form of in-band resonances.

The case of a mass defect in a regular diatomic chain has been studied in detail by Mazur *et al.* (1956). Here the situation is slightly more complicated, in that out-of-band modes may occur not only above the maximum frequency ω_M, but also in the gap between the acoustic and optical frequency ranges; the exact number and position of the modes depends both on the value of the defect mass and on whether a light or heavy atom of the host lattice has been replaced.

The vibrational behavior of mass defects in higher dimensional lattices is qualitatively similar to that outlined above, with impurity masses lighter or heavier than some appropriate threshold value being associated with the occurrence of out-of-band modes or resonances. The presence of force constant defects, too, leads to characteristic modifications of the spectrum. Generally speaking, the effect of a weakened force constant is similar to that of a heavy mass defect, while a strengthened constant produces an effect not unlike that of a light mass defect. A number of simple cases of force constant defects and single and multiple mass defects in higher dimensional lattices have been treated in the literature and are reviewed by Ludwig (1967).

In later stages of the article we shall go on to examine the vibrational properties of disordered lattices and glasses, as derived by the numerically based methods. As we shall see, for lattices with small and moderate defect concentrations, many of the properties may be understood fairly well on the basis of the analytical results indicated above for single defects or isolated defect clusters. However, if the analytical approach is to be applied usefully to systems with larger impurity concentrations, it is necessary to refine the Green's function approach further, allowing for interactions between neighboring defects or clusters and performing suitable ensemble averages. For these developments, which lie outside the scope of the present chapter, we refer the reader to articles by Elliott and Taylor (1967), Taylor (1967), and Elliott *et al.* (1974), in which normal modes are expressed as Bloch wave expansions, or by Anderson (1958, 1970) where a localized state expansion is used.

IV. The Numerical Determination of Frequency Spectra

As indicated earlier, there have been two main numerically based approaches to the computation of frequency spectra. What appears, at first sight, to be the most straightforward method is simply to diagonalize the dynamical matrix of the sample system concerned, obtaining all of its vibrational eigenfrequencies and eigenvectors at one fell swoop. However, limitations of computer speed and storage have, as remarked, generally placed severe restrictions on the size of system which can be treated in this way—not simply on account of the large volume of computation actually involved (the number of arithmetical operations involved in diagonalizing an $L \times L$ matrix is of order L^3), but because the analysis of very large matrices can raise serious problems of efficient program organization and computer usage. In the early 1960's Dean and his co-workers (Dean and Martin, 1960; Dean, 1960, 1961; see also Dean, 1972) introduced a technique whereby the eigenvalue distribution of a dynamical matrix for a system with short range interactions could be estimated accurately with much less computational effort than that involved in a complete diagonalization. The method, based on the negative eigenvalue theorem (NET), enabled spectra to be computed for reasonably large model samples (~ 1000 atoms) of 3-dimensional vibrational systems. While it is no longer totally impractical to diagonalize dynamical matrices of the order of several thousand, the NET method is still by far the most efficient general scheme for a fast and accurate estimation of a frequency distribution for a system of this size; for significantly larger vibrational systems, it remains the only practical method.

In this section we outline the basis and application of the NET scheme, touching on the computational aspects of the method and its use in the calculation of individual eigenvalues and eigenvectors. As far as the full diagonalization approach is concerned, the problem of symmetric matrix diagonalization has already been treated in some detail in the literature, and for this particular computational problem we refer the reader to one of the standard texts on matrix analysis (cf. Wilkinson, 1965).

A. The Negative Eigenvalue Theorem

Consider a vibrational system with dynamical matrix **W**. Then, as noted in Section II, E, the squared frequency distribution $\mathscr{G}(\omega^2)$ is related to **W** by the equation

$$\int_0^{\omega 2} d(\omega^2)\ \mathscr{G}(\omega^2) = \eta(\mathbf{A}) \tag{65}$$

where $\eta(\mathbf{A})$ denotes the number of negative eigenvalues of matrix **A**, and **A** is given by

$$\mathbf{A}(\omega^2) = \mathbf{W} - \omega^2\mathbf{I} \tag{66}$$

Evidently the specification of $\eta(\mathbf{A})$ on a suitable mesh of ω values (ω^2 values) enables one to derive the frequency (squared frequency) distribution of the system in the form of a histogram. The essential relevance of the negative eigenvalue theorem is that it provides a neat and efficient method for the evaluation of the quantities $\eta(\mathbf{A})$ for the large-order band matrices $\mathbf{A}$ occurring in the lattice vibrational problem.

In its simplest form, the theorem implies that given a partition of matrix $\mathbf{A}$ according to

$$\mathbf{A} = \begin{bmatrix} \mathbf{A}_1 & \mathbf{C}_1 \\ \mathbf{C}_1{}^T & \mathbf{B}_1 \end{bmatrix} \tag{67}$$

where $\mathbf{A}_1$, $\mathbf{B}_1$ are square matrices, we may rewrite $\eta(\mathbf{A})$ as

$$\eta(\mathbf{A}) = \eta(\mathbf{A}_1) + \eta(\mathbf{A}^{(2)}) \tag{68}$$

where $\mathbf{A}^{(2)}$ is defined in terms of the blocks of $\mathbf{A}$ by

$$\mathbf{A}^{(2)} = \mathbf{B}_1 - \mathbf{C}_1{}^T\mathbf{A}_1{}^{-1}\mathbf{C}_1 \tag{69}$$

If $\mathbf{A}^{(2)}$ is, in turn, partitioned

$$\mathbf{A}^{(2)} = \begin{bmatrix} \mathbf{A}_2 & \mathbf{C}_2 \\ \mathbf{C}_2{}^T & \mathbf{B}_2 \end{bmatrix} \tag{70}$$

the theorem may be applied again, to give

$$\eta(\mathbf{A}) = \eta(\mathbf{A}_1) + \eta(\mathbf{A}_2) + \eta(\mathbf{A}^{(3)}) \tag{71}$$

with

$$\mathbf{A}^{(3)} = \mathbf{B}_2 - \mathbf{C}_2{}^T\mathbf{A}_2{}^{-1}\mathbf{C}_2 \tag{72}$$

and so on.

In general, we may write

$$\eta(\mathbf{A}) = \sum_{i=1}^{n} \eta(\mathbf{A}_i)$$

with

$$\mathbf{A}^{(i)} = \begin{bmatrix} \mathbf{A}_i & \mathbf{C}_i \\ \mathbf{C}_i{}^T & \mathbf{B}_i \end{bmatrix} \tag{73}$$

$$\mathbf{A}^{(i+1)} = \mathbf{B}_i - \mathbf{C}_i{}^T\mathbf{A}_i{}^{-1}\mathbf{C}_i$$

$$\mathbf{A}^{(1)} \equiv \mathbf{A}, \qquad \mathbf{A}_n \equiv \mathbf{A}^{(n)}$$

In this way, the NET enables one to relate the quantity $\eta(\mathbf{A})$ for the full order matrix $\mathbf{A}$ to the set of quantities $\eta(\mathbf{A}_i)$ for smaller order matrices $\mathbf{A}_i$.

It has generally proved convenient, in applications of the theorem, to choose the partitioning so that just one row and column of $\mathbf{A}$ are separated off at each stage of the iteration process. The algorithm of arithmetical operations carried out in this case is just the Gaussian elimination algorithm used in the solution of linear simultaneous equations. The $\mathbf{A}_i \equiv A_i$ are scalars and we can write

$$\eta(\mathbf{A}) = \text{number of negative } A_i \tag{74}$$

The pivots A_i occurring here have a further significance—they are the ratios of successive principal minors of matrix $\mathbf{A}$, that is

$$A_i = p_i(\mathbf{A})/p_{i-1}(\mathbf{A}) \qquad (p_0(\mathbf{A}) \equiv 1) \tag{75}$$

where $p_i(\mathbf{A})$ is the determinant arising from the first i rows and columns of $\mathbf{A}$.

B. Calculation of Selected Eigenvalues

Certain detailed studies, such as the investigation of special lattice vibrational frequencies or normal mode displacements, call for the accurate evaluation of selected eigenvalues or eigenvectors.

A fairly crude preliminary estimate of a particular eigenvalue may be obtained by using the property that $\eta(\mathbf{A})$ changes by 1 as ω passes through an exact eigenfrequency ω_l. Starting with a suitable frequency interval which contains one or more ω_l, one may bracket the frequency of interest simply by performing successive bisections of the interval and studying the changes in $\eta(\mathbf{A})$. This is clearly a fairly slowly convergent procedure; however, once the required frequency has been located approximately, convergence may be assisted by using the property (75) of the pivots. The final pivot obtained in an elimination

$$A_L = p_L(\mathbf{A})/p_{L-1}(\mathbf{A}) \tag{76}$$

has as numerator

$$p_L(\mathbf{A}) = \det|\mathbf{W} - \omega^2\mathbf{I}| \tag{77}$$

which vanishes when ω becomes equal to an exact eigenfrequency. Then provided this value of ω does not also cause $p_{L-1}(\mathbf{A})$ to become zero, one may interpolate on A_L to hasten convergence toward the eigenfrequency. In many cases simple polynomial interpolation will suffice, but if the zeros of numerator and denominator in (76) lie particularly close together a more sophisticated scheme, such as that based on the rational approximation, may be required.

C. Eigenvectors

Given an approximation ω' to an exact eigenvalue ω_l, a corresponding eigenvector can be obtained by iteration of the equation

$$(\mathbf{W} - \omega'^2\mathbf{I})\mathbf{w}_k = \hat{\mathbf{w}}_{k-1} \qquad k = 1, 2, \ldots \tag{78}$$

where $\hat{\mathbf{w}}_k$ is some appropriately normalized version of the kth iterate

$$\mathbf{w}_k = c_k\hat{\mathbf{w}}_k \tag{79}$$

The iteration process is continued until some suitable convergence criterion is satisfied: one possible criterion is that the modulus of the vector difference between successive iterates should become sufficiently small, another that the accumulated product of normalization factors $\prod_k c_k$ exceed some suitably chosen threshold value $\prod_{\text{CRIT}}$. In practice the zero-order iterate $\hat{\mathbf{w}}_0$ may be chosen either arbitrarily or on the grounds of physical intuition. A suitable normalization scheme is to scale the vector so that the element of $\hat{\mathbf{w}}_k$ having maximum magnitude has modulus unity.

If the zero-order vector $\hat{\mathbf{w}}_0$ has projections b_l along the exact eigenvectors $\mathbf{u}_l$,

$$\hat{\mathbf{w}}_0 = \sum_{l=1}^{L} b_l\mathbf{u}_l \tag{80}$$

then K iterations of Eq. (78) produce a normalized vector

$$\hat{\mathbf{w}}_K = \left[\prod_{k=1}^{K} c_k^{-1}\right]\sum_{l=1}^{L} b_l(\omega_l^2 - \omega'^2)^{-K}\mathbf{u}_l \tag{81}$$

Clearly, if ω' is much closer to one eigenvalue ω_l than to any other, the convergence to $\mathbf{u}_l$ is very rapid.

The implementation of (78) is greatly facilitated by the use of data already obtained during the Gaussian elimination stage of the calculations, as described below.

It is easily confirmed that the sequence of operations (73) may be rewritten in a form

$$\mathbf{A} = \mathbf{T}^T\mathbf{A}_{\text{DIAG}}\mathbf{T} \tag{82}$$

where $\mathbf{A}_{\mathrm{DIAG}}$ is a block diagonal matrix with the structure

$$\mathbf{A}_{\mathrm{DIAG}} = \begin{bmatrix} \mathbf{A}_1 & & & \\ & \mathbf{A}_2 & & \mathbf{0} \\ & & \ddots & \\ & \mathbf{0} & & \mathbf{A}_n \end{bmatrix} \tag{83}$$

and $\mathbf{T}$ is a block upper triangular matrix

$$\mathbf{T} = \begin{bmatrix} \mathbf{I}_1 & \multicolumn{4}{c}{\mathbf{A}_1^{-1}\mathbf{C}_1} \\ & \mathbf{I}_2 & \multicolumn{3}{c}{\mathbf{A}_2^{-1}\mathbf{C}_2} \\ & & \ddots & & \\ & \mathbf{0} & & \mathbf{I}_{n-1} & \mathbf{A}_{n-1}^{-1}\mathbf{C}_{n-1} \\ & & & & \mathbf{I}_n \end{bmatrix} \tag{84}$$

$$= \begin{bmatrix} \mathbf{I}_1 & \multicolumn{4}{c}{\mathbf{0}} \\ & \mathbf{I}_2 & \multicolumn{3}{c}{\mathbf{0}} \\ & & \ddots & & \\ & \mathbf{0} & & \mathbf{I}_{n-1} & \mathbf{A}_{n-1}^{-1}\mathbf{C}_{n-1} \\ & & & & \mathbf{I}_n \end{bmatrix} \cdots$$

$$\begin{bmatrix} \mathbf{I}_1 & \multicolumn{4}{c}{\mathbf{0}} \\ & \mathbf{I}_2 & \multicolumn{3}{c}{\mathbf{A}_2^{-1}\mathbf{C}_2} \\ & & \ddots & & \\ & \mathbf{0} & & \mathbf{I}_{n-1} & \mathbf{0} \\ & & & & \mathbf{I}_n \end{bmatrix} \begin{bmatrix} \mathbf{I}_1 & \multicolumn{4}{c}{\mathbf{A}_1^{-1}\mathbf{C}_1} \\ & \mathbf{I}_2 & \multicolumn{3}{c}{\mathbf{0}} \\ & & \ddots & & \\ & & & \mathbf{I}_{n-1} & \mathbf{0} \\ & & & & \mathbf{I}_n \end{bmatrix} \tag{85}$$

the $\mathbf{I}_n$ being unit matrices of appropriate orders. In the usual case, where the elimination proceeds one row and column at a time, the $\mathbf{A}_i$ are just scalars and the $\mathbf{A}_i^{-1}\mathbf{C}_i$ are row vectors, so that $\mathbf{A}_{\mathrm{DIAG}}$ and $\mathbf{T}$ take the form of simple diagonal and upper triangular matrices.

If the elements A_i and the rows $\mathbf{A}_i^{-1}\mathbf{C}_i$ are stored during the final elimination performed in the course of calculating an accurate eigenvalue, the resulting triangular decomposition (82) of $\mathbf{A}$ may be used to implement (78) by the sequence of operations

$$\begin{aligned} \mathbf{T}^T\mathbf{w}_k' &= \hat{\mathbf{w}}_{k-1} \\ \mathbf{T}\,\mathbf{w}_k &= \mathbf{A}_{\mathrm{DIAG}}^{-1}\mathbf{w}_k' \end{aligned} \tag{86}$$

which, since triangular and diagonal matrices are involved, requires simple substitutions rather than any complicated matrix algebra.

D. Computational Aspects

If the vibrational behavior of the systems under consideration can be represented adequately by the use of short-range interactions (and this has proved a reasonably valid assumption for the majority of disordered systems treated to date), certain computational advantages ensue. In such a situation the dynamical matrix $\mathbf{W}$, and of course $\mathbf{A}$, assume a simple banded form, with all elements beyond a certain distance $k - 1$ from the main diagonal being zero (k is called the half bandwidth and $2k - 1$ the bandwidth of $\mathbf{W}$ or $\mathbf{A}$). It can readily be shown that the same half bandwidth k persists for each of the submatrices $\mathbf{A}^{(i)}$ and $\mathbf{B}_i$ generated during the Gaussian elimination, with (in the case of one row at a time elimination) only the leading $k - 1$ elements of vectors $\mathbf{C}_i$ being nonzero. As a result, it is necessary to concern oneself, during the ith elimination step, with only the leading $k \times k$ submatrix of $\mathbf{A}^{(i+1)}$ or $\mathbf{B}_i$, the remaining rows and columns of $\mathbf{C}_i{}^T\mathbf{A}_i{}^{-1}\mathbf{C}_i$ being identically zero.

The total number of arithmetical operations needed for a single Gaussian elimination of an $L \times L$ dynamical matrix with half bandwidth k is just $\mathcal{O}(Lk^2)$ (consisting of approximately $\frac{1}{2}Lk^2$ multiplications, a similar number of subtractions, and Lk divisions), as compared with the $\mathcal{O}(L^3)$ operations required in the conventional diagonalization process. On this basis, the time involved in computing a 40-point histogram representation of the frequency spectrum of a system having, say, $L = 2000$ and $k = 100$ is significantly less than that required for full diagonalization.

Given an approximation ω' to an eigenvalue ω_l, accurate to within an error of (say) 1×10^{-3} of the mean eigenvalue separation, one finds that some two or three iterations of Eqs. (78) or (86) tend to be sufficient in most circumstances to produce an eigenvector whose elements have converged to their final values, to within an error of about $1 \times 10^{-6} \times$ (maximum element). Apart from this, the algorithm (86) for calculation of vectors is itself very economical, with each iteration involving less than $2Lk$ arithmetical operations, so that the computation of an eigenvector *per se* represents no substantial problem in terms of computer time. The time required to obtain accurate estimates of the ω_l, on the other hand, is a very critical factor. Once a frequency spectrum has been obtained, then to determine a given eigenfrequency to within the figure mentioned above of 1×10^{-3} of the mean eigenvalue separation may entail upward of half a dozen evaluations of $\eta(\mathbf{A})$ and A_L, that is, half a dozen full Gaussian eliminations. Because of the relative amounts of time consumed by eigenvalue and eigenvector parts of the cycle, it may often be efficient to sacrifice accuracy in the initial eigenvalue evaluation, at the expense of having to increase substantially the number of eigenvector iterations (86). It is difficult to be categorical, however, since various factors combine to give accelerated convergence in the later stages of eigenvalue determination.

The Gaussian elimination approach is, clearly, not an ideal method for the calculation of large numbers of eigenvalues and eigenvectors. Thus, in the example quoted above of a system having $L = 2000$ and $k = 100$, to generate a 40-point frequency distribution histogram and to compute, with reasonable accuracy, one eigenvalue and eigenvector per histogram interval may in principle necessitate computer time not appreciably less than that required for the complete diagonalization of the dynamical matrix. It is only the fact that the full diagonalization of large matrices (banded or otherwise) has until recently raised serious practical problems of program organization and efficient computer usage, that has made the eigenvalue and eigenvector computation by Gaussian elimination worthwhile.

We should comment, finally, on the numerical stability aspect of the Gaussian elimination algorithm. The main occurrence leading to numerical difficulty is the appearance, rarely, of very small or zero values of the pivots A_i during the elimination process. This type of situation may arise, for example, in certain simple cases where dynamical matrix elements and selected mesh values of ω^2 are integers or simple ratios of integers, so that the elementary combinations of these occurring in the first few A_i may, through cancellation, lead to pivotal values of precisely zero (or zero within round-off error) in the first few steps of elimination. The occurrence of zero pivots of course causes the numerical process to break down completely. However, the appearance of small but nonzero A_i values either in the above or in more general circumstances may also result in serious numerical difficulty: in extreme cases, if the elements of $\mathbf{C}_i{}^T\mathbf{A}_i{}^{-1}\mathbf{C}_i$ exceed those of $\mathbf{B}_i$ by a factor equivalent to the number of figures retained in the computation, information from the leading $k \times k$ block of $\mathbf{B}_i$ will be lost. Clearly checks aimed at detecting such a situation should be written into the computer programs. Given an indication of the occurrence of an unacceptably small A_i, a simple but effective remedy is just to repeat the elimination with the relevant ω^2 value replaced by a slightly different one, say $\omega^2 + 10^{-6}$.

Wilkinson (1965) has suggested the use of pivoting to avoid numerical instability during the Gaussian elimination process. This involves the use of row and column interchanges to ensure that the largest (or, at least, a reasonably large) diagonal element of submatrix $\mathbf{A}^{(i)}$ is chosen as pivot A_i for the next step. Unfortunately the implementation of such a scheme leads, at best, to the need to store and manipulate additional data relating to the row and column interchanges and, at worst, to substantial inflation of the effective matrix bandwidth.

For the reasons described above, the method of pivoting has not generally been used in connection with the band matrices arising in disordered lattice dynamics. However, practical experience over a wide range of vibrational systems has indicated that the straightforward Gaussian elimination algorithm, with the incorporation of rather less sophisticated stability checks such as that outlined earlier, gives satisfactory results in relation to frequency spectra, eigenvalues, and eigenvectors. The reliability of the method has been confirmed, for example, by calculations on ordered systems, for which results may be obtained independently by the standard methods of crystal lattice dynamics (cf. Section III). Moreover, disordered lattices too can exhibit certain regularities in their vibrational properties, in the form of special frequencies of various kinds (cf. Section V) and special modes in which particular groups of atoms remain at rest. The successful prediction or confirmation of many such regularities has provided further evidence for the adequacy of the straightforward elimination algorithm in vibrational problems.

V. Vibrational Spectra of Noncrystalline Solids

We shall be concerned, in the present section, with the vibrational frequency spectra (and squared frequency spectra) of disordered solids as derived by the numerical techniques, particularly the NET method reviewed in Section IV. The earliest application of the NET method was to the 1-dimensional diatomic mass-disordered chain and, although this chain does not correspond to any disordered solid actually found in nature, there are several reasons why an examination of its properties should be worthwhile. To begin with, it is one of a small number of disordered systems for which a precise frequency spectrum can be obtained by techniques not based on the NET method, giving results against which the accuracy of the latter method may be checked. In addition, the linear chain provides an example of a reasonably simple system which yet manages to exhibit several spectral features typical of disordered solids in general. Following a discussion of the linear chain and its generalizations to higher dimensions, we proceed to an examination of more seriously disordered systems, ranging from glasslike chains to realistic models representing the structure of amorphous Si- and SiO_2-like systems.

A. Two-Component Mass Disordered Chains and Lattices

For the mass disordered chain with nearest-neighbor interactions, the dynamical matrix has a simple tridiagonal form and the algorithm (73) reduces to an essentially scalar recurrence relation between successive pivots A_i. Since the time involved in Gaussian elimination is proportional to k^2, the square of the half-bandwidth, the spectra of quite long chains may be produced very speedily, even on a fairly modest computer. The calculated vibrational squared frequency spectrum for a two-component mass disordered chain containing 32,000 atoms, obtained by Dean (1960), is shown in Fig. 2a. The chain in question contains equal numbers of heavy and light atoms, and the mass ratio between atomic types is 2:1. Results given in Fig. 2b for a later calculation on a 250,000 atom chain (Dean, 1972) show that, to graphical accuracy at any rate, the 32,000 atom chain results may be

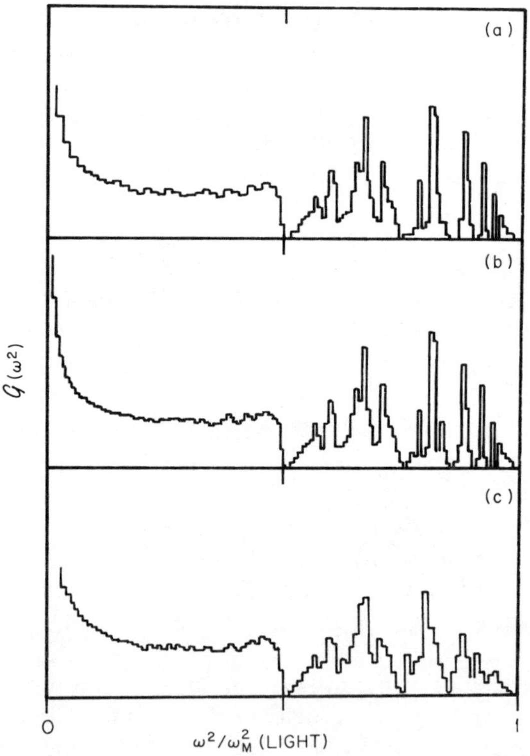

Fig. 2. Squared frequency spectra $\mathcal{G}(\omega^2)$ for mass disordered diatomic linear chains with equal numbers of light and heavy atoms, the mass ratio between the atomic species being 2:1. (a) NET calculation for a 32,000 atom chain (Dean, 1960). (b) NET calculation for a 250,000 atom chain (Dean, 1972). (c) Spectrum obtained by solving a functional equation (Agacy, 1964).

regarded as having converged to their limiting ($N \to \infty$) form. In fact unpublished calculations by Dean indicate that results for even shorter chains ($N \sim 8000$) reproduce all the essential spectral features quite well.

There are a number of additional pieces of evidence concerning the accuracy of the NET calculations, in the case of the two-component mass disordered chain. As mentioned, an extremely accurate spectrum for this system may be obtained using a completely independent technique, based on the functional equation approach of Dyson (1953; see also Schmidt, 1957). A spectrum computed by Agacy (1964), using this technique, is shown in Fig. 2c. Agreement with the NET results is quite striking. Indeed such minor discrepancies as do occur between the two sets of results can probably be attributed to inaccuracy in the solution of the functional equation.

The other decisive piece of evidence relating to the accuracy of the NET results stems from the so-called Saxon–Hutner theorem on forbidden energy ranges in mixed systems (cf. Hori, 1968). It is a consequence of this theorem that, for the mass disordered chain, the spectrum should go to zero, regardless of detailed chain arrangement and atomic concentration, at frequencies given by

$$\omega(s, n) = \omega_M(\text{LIGHT}) \sin\left(\frac{\pi}{2}\frac{s}{n}\right) \tag{87}$$

where $s < n$ are relatively prime integers, provided only the mass ratio is greater than a critical value $(m_H/m_L)_{\text{CRIT}}$, given by

$$(m_H/m_L)_{\text{CRIT}} = 1 + \cot\left(\frac{\pi}{2}\frac{s}{n}\right)\cot\left(\frac{\pi}{2}\frac{1}{n}\right) \tag{88}$$

m_H, m_L being the masses of heavy and light atoms, respectively. For the case in question, $m_H/m_L = 2$ and special frequencies occur at

$$\omega = \omega_M(\text{LIGHT}) \sin\left(\frac{\pi}{2}\frac{(n-1)}{n}\right) \tag{89}$$

The special frequencies given by (89) are indicated on the horizontal axis of Fig. 3, which reproduces the numerically evaluated spectrum shown earlier in Fig. 2b. It is quite clear that the zeros predicted by (89) coincide closely with gaps in the computed spectrum, offering further strong proof of the accuracy of the latter. In point of fact, information about the integrated density of states $\int_0^{\omega^2} d(\omega^2)\, \mathscr{G}(\omega^2)$ at the special squared frequencies may also be derived by analytical methods (cf. Hori, 1968), and here, too, the NET-based results show precise agreement with analytical predictions.

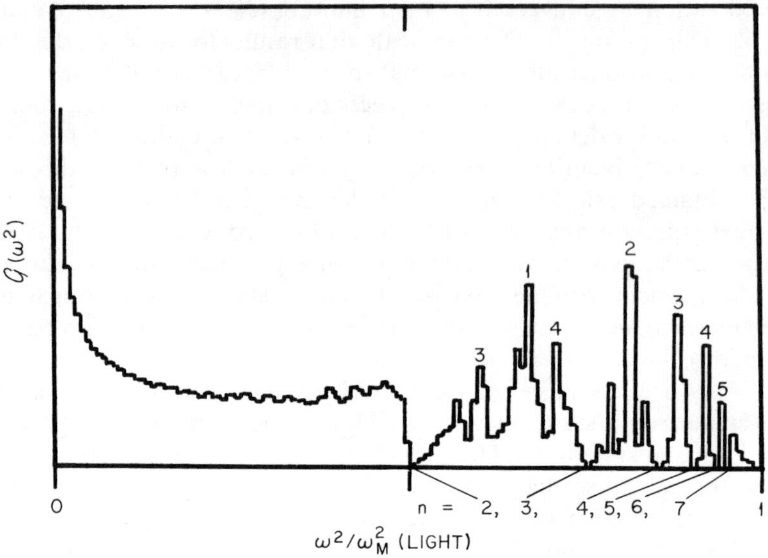

FIG. 3. Squared frequency spectrum for a mass disordered diatomic linear chain with equal numbers of light and heavy atoms, the mass ratio between atomic species being 2:1 (Dean, 1972). Designations above the individual peaks indicate the sizes of light atom clusters giving rise to the peaks. Numbers along the horizontal axis correspond to spectral gaps at the special frequencies predicted by Eq. (89) of the text.

The squared frequency spectrum of the disordered chain with equal numbers of heavy and light atoms differs considerably from the simple U-shaped curves associated with monatomic chains (or the double U-shaped spectrum of the regular diatomic chain) and some comment on the individual features in the spectrum of Fig. 3 is appropriate. By inspecting numerically calculated normal mode eigenvectors, Dean and Bacon (1963) were able to associate individual peaks in the high frequency region with localized vibrations of variously sized clusters of light atoms bounded on either side by one or more heavy atoms. The peaks in Fig. 3 are appropriately identified, with that labeled "1" arising from single light atoms bounded by heavy atoms, and so on. Such modes are essentially similar in character to those associated with isolated light atom clusters in heavy lattices, and the disposition of the labeled peaks in Fig. 3 is in substantial agreement with the predictions of Dean (1967), mentioned in Section III, for clusters of light impurities in a heavy host lattice.

The lower portion of the vibrational spectrum in Fig. 3 presents a much more continuous appearance than the high frequency region consisting of many jagged peaks. However, it is believed that the use of a coarse histogram interval in this case may well conceal a fairly intricate (if not pathological) substructure. Such a complex substructure has, in fact, been verified for the spectrum in the case of the equivalent electronic alloy problem (Gubernatis and Taylor, 1971, 1973).

While the higher frequency modes, corresponding to the various isolated peaks in the spectrum, are, as mentioned, known to be highly localized, the nature of modes in the low and medium frequency region has not been completely settled by the numerical studies. Inspection of eigenvectors indicates these vibrations to be very much more extended than their high frequency counterparts, but the evidence from numerical studies is inconclusive as regards the ultimate form of the modes in the limit of very long chains. Analytical results by Borland (1963), Dean (1964), and Economou and Cohen (1971) seem to indicate, though, that modes in any 1-dimensional system having nearest-neighbor interactions and a finite degree of disorder should be localized in the infinite chain limit.

The behavior of the linear chain spectrum with varying mass ratio and relative atomic concentration has been reported and reviewed in detail by Dean (1961, 1972), Payton (1966), and Payton and Visscher (1966, 1967a,b, 1968). Disordered two-component chains with first and second neighbor forces have been studied by Martin (1961), Brouers and Deltour (1967), Sah and Srivastava (1970), while Payton (1966) and P. Dean (unpublished) have considered the effect of allowing the nearest-neighbor force constant to vary with the type of atomic pair involved. Calculations on three-component chains, too, have been reported by Payton (1966) and Payton and Visscher (1968). The special case of $A_xB_{1-x}C$ chains (with one atomic species C occupying all even sites, and two kinds of atom A and B distributed randomly on the odd sites) is of interest in connection with mixed crystal systems and has been studied numerically by a number of authors (Payton, 1966; Hass *et al.*, 1969; C. Hall, unpublished; P. Dean, unpublished).

A number of calculations of vibrational spectra and normal modes have also been reported for 2- and 3-dimensional disordered lattices. Here again the main contributions have been made by Dean and Bacon (1962, 1965), Payton (1966), and Payton and Visscher (1967b, 1968). The spectral trends with changing mass ratio and concentration are qualitatively similar to those reported by the same authors for the linear chain. However, the practical task of distinguishing and assigning different spectral peaks becomes more difficult with increasing dimension, simply because the number of types of atomic cluster of a particular size that may occur grows rapidly as a function of the lattice dimensionality d. Apart from hindering identification, the resulting high density of peaks can cause the histogram representation of the spectrum to assume a more or less continuous appearance even, misleadingly, in the high frequency region.

Numerical evidence on mode localization is even less conclusive for higher dimensions than in the 1-dimensional case, due partly to the fact that calculations have had to be restricted to 2- and 3-dimensional lattices of fairly modest linear extent. There are, however, indications the certain changes take place in the spatial characteristics of modes and in the character of the spectrum as the relative concentration of the atomic species passes through certain critical concentrations. Mode localization, spectral continuity, and energy transfer through the lattice are all connected in a fairly complex way (cf. Bell, 1972a). The relationship between these and the percolation problem (which has its own critical concentrations) has been reviewed recently by Kirkpatrick (1973).

B. One- and Two-Dimensional Glasses

A number of authors have performed vibrational calculations on 1-dimensional chain models exhibiting some of the disordering features present in real glasses. The archetypal example of a real system which displays glasslike disorder and which yet retains the simplicity of 1 dimensionality, at least with respect to its topology, is the convoluted polymer chain. The most sophisticated calculations of vibrational spectra for such systems are those of Zerbi and co-workers, based on the NET. Exploratory calculations by Piseri and Zerbi (1968) revealed the strong sensitivity of computed spectra to chain configuration for conformationally disordered chains. In later calculations Zerbi *et al.* (1971) have managed to give a remarkably good account of observed peaks in the infrared spectrum, at room temperature, and at 160 °C.

Chains considered by Zerbi and co-workers exhibit a discrete type of geometrical disorder, in the form of random sequences of trans and gauche conformations. A polymerlike system showing continuous angular disorder of a type closer to that found in real glasses has been considered by Bermudez (1971, 1972a). Bermudez has computed spectra for a system consisting of an -ABABAB- chain laid out to form a random walklike arrangement on a plane. Individual BAB units were taken to be regular, with fixed BAB angle and bond length AB; however, successive BAB units, linked by a common B atom, were randomly oriented with respect to each other, with the ABA angle showing a Gaussian probability distribution. The spectra show a band structure similar to that found in real polymers and glasses.

Other workers have attempted to incorporate some element of continuous disorder into chain models, while retaining a strict geometrical 1 dimensionality. Thus Dean (1964) and later Payton (1966) and Payton and Visscher (1967b) have considered linear chains of identical atomic masses vibrating under nearest-neighbor forces with a force constant which varies randomly from bond to bond according to a continuous rectangular distribution. The computed spectra tend to be smooth and featureless in appearance and, except near the limit of infinitesimally narrow force constant distribution, they do not resemble the spectrum of the corresponding regular chain. Calculations by Dean on a diatomic chain, regular with respect to mass distribution but disordered with respect to force constants, show a similar trend. The two-band spectral structure characteristic of the completely ordered system tends to persist, but with the bands appreciably smeared out and, for wide force constant distributions, overlapping.

A criticism often leveled at chain models is that they do not incorporate any of the topological disorder that is present in real glasses. Bell (1972b) has, in fact, devised a dynamical model which, while 1-dimensional from a geometrical point of view, does exhibit topological disorder of a kind. The model in question consists of a monatomic chain, each of whose atoms is connected to four other atoms of the system by equal restoring harmonic forces. Disorder is introduced by choosing at random some of the four atoms with which a given atom is permitted to interact. Bell considered chains in which all four linkages were chosen at random, as well as chains in which only two were chosen at random (the other two being assigned as nearest-neighbor interactions). He found the vibrational spectra of such systems to differ considerably from those characteristic of corresponding ordered chains (e.g., the monatomic chain with equal nearest- and second-neighbor forces).

Neither the continuously disordered models of the type studied by Dean and others, nor the topologically disordered system considered by Bell, appear to exhibit the isolated high frequency peak structure arising in the case of discrete isotopic disorder. Out-of-band modes, when they do occur in glassy solids, tend to be associated not with the severity of the overall angular or topological irregularity, for example, but with local departures from short range order in the form, say, of broken bonds and incomplete atomic coordination.

Studies of the spatial characteristic of normal modes in glasslike chains have also been performed, for the systems considered by Bell (1972b) and Dean (1964). In both cases it is found that the nature of the vibrations changes fairly smoothly as one progresses through the spectrum, with modes at low frequencies having a relatively extended form and those at the upper end of the spectrum being strongly localized. As in the case of isotopically disordered systems, however, the numerical studies have not been able, of themselves, to resolve the question of the ultimate form of modes in the limit of very large systems; for a final resolution of this problem, one must again rely on analytical studies of the type cited in Section III, E.

Once the restriction of 1 dimensionality has been relaxed, geometrical disorder (continuous or discrete) and topological irregularity can be incorporated simultaneously into models of the amorphous state in a much more natural way. Thus Bermudez (1972b) has considered a planar arrangement of corner-linked AB_3 molecular units displaying both types of disorder. This system, intermediate in structure between a highly branched polymer chain and a random A_2B_3 network, is not claimed to be a rigorously correct structural representation of any given real system; however Bermudez has shown that frequency spectra computed for such a network reproduce the observed vibrational bands of layered chalcogenides such as As_2Se_3 and As_2S_3 fairly well.

Bell *et al.* (1974), too, have performed extensive calculations on 2-dimensional model systems having both angular and topological disorder. The systems concerned consist of planar arrangements of corner-linked AB_3 triangular molecules. Adjacent molecules in the model are allowed to have any one of several pre-specified relative orientations, and the whole arrangement forms a continuously connected A_2B_3 network. The aim of the authors in this instance was not to reproduce the vibrational spectrum of any given real amorphous solid, but rather, by suitable adjustment of the parameters of the system, to separate out and analyze the spectral effects of different types of disorder. Their calculations indicate that, for the kind of system concerned, possessing short range order, the introduction of topological and angular disorder can lead to a spectrum appreciably different from that of the corresponding regular lattice. Topological disorder, in particular, can affect the detailed spectral profile, influencing the number and disposition of peaks within bands, although not necessarily the frequency range covered by the bands. The introduction of an element of angular randomness increases the disparity between glass and spectra further, influencing in particular the frequency range occupied by spectral bands. Other types of imperfection, such as the breaking of interatomic bonds and the distortion of individual molecular units, are also shown to produce characteristic spectral effects. The 2-dimensional model system studied by Bell *et al.* has strong structural similarities with a number of real 3-dimensional glasses, such as SiO_2 and B_2O_3, and the results of those authors represent the strongest evidence to date against the validity of using crystal-lattice models to interpret the vibrational properties of such materials.

C. Real Three-Dimensional Glasses

In contrast with the situation for lattice-based alloys or idealized 1- and 2-dimensional models of glasses, there has generally been a real difficulty, in the case of real glasses, in providing a realistic set of equilibrium positions on which to base an atomic vibrational calculation. For this reason, all of the early work in this field tended to rely either on the use of small representative molecular units or on the assumption of a perfectly crystalline network. The first vibrational calculations for glasses to be based on realistic structural models of the amorphous state were those of Bell *et al.* (1968, 1970, 1971) for SiO_2-like materials. They solved the atomic coordinate problem by constructing physical models of sample atomic arrangements in glasses, built so as to conform to the known structural data from X-ray and neutron diffraction experiments. Measured coordinates from the models, each of which contained some hundreds of atoms (cf. Bell and Dean, 1972), were used, together with a set of suitable interatomic force constants, to generate dynamical matrices which were, in turn, analyzed by the NET method described in Section IV.

Results of such a model-based lattice dynamical calculation for vitreous SiO_2 are shown in Fig. 4 (after Bell *et al.*, 1968). Upper and lower pairs of

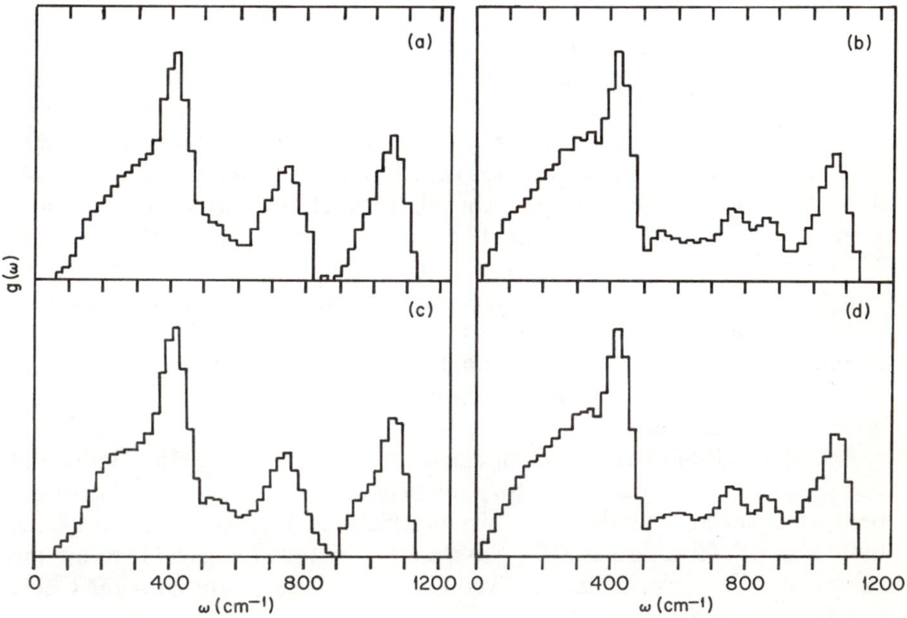

FIG. 4. Computed frequency spectra for vitreous SiO_2, based on molecular models of sample atomic arrangements in the glass (Bell *et al.*, 1968). Upper and lower pairs of diagrams correspond to two separate molecular models, differing in detail but with similar overall statistical properties. Left-hand diagrams (a) and (c) refer to a calculation in which singly connected O atoms on the model boundary are constrained to be fixed, while the right-hand diagrams (b) and (d) correspond to a calculation in which these nonbridging atoms are assumed free to vibrate.

diagrams stem from calculation on two molecular models differing in detail, but having closely similar overall statistical properties. The left-hand diagrams (a and c) in the figure refer to "fixed-end" calculations in which atoms on the model boundary (singly bonded O atoms, in this instance) were constrained to be stationary, while the right-hand diagrams (band d) correspond to "free-end" calculations, in which these boundary atoms were assumed free to vibrate.

Spectra in the upper and lower sets of diagrams show the same general features, the small differences between the spectra being attributable to statistical fluctuation. Thus, in the fixed-end case, there are bands at 1050, 750, 550 (shoulder), 400, and below 350 cm^{-1} (shoulder). These show substantial agreement with the positions of bands deduced from infrared absorption (Florinskaya and Pechenkina, 1952; Lippincott *et al.*, 1958; Miler, 1968; Hanna, 1965), Raman scattering (Bobovich and Tulub, 1958; Harrand, 1954; Hass, 1970; Krishnan, 1953) and neutron scattering experiments (Leadbetter and Stringfellow, 1974).

The free-end spectra show two features not present in their fixed-end counterparts, namely, an additional peak in the region of 850 cm^{-1} and a small enhancement at around 350 cm^{-1} at the upper end of the low frequency shoulder. Bell *et al.* (1968) have shown, by inspection of the appropriate displacement eigenvectors, that these two features are associated with vibrations of the singly bonded O atoms of the structure: this diagnosis was confirmed by the authors in their later work on normal mode assignments and localization (Bell *et al.*, 1970, 1971; cf. Section VII of this chapter). These bands are not found in the spectrum of normal vitreous SiO_2, but a counterpart of that at 850 cm^{-1} has been detected in the infrared spectrum of the glass after exposure to intense neutron irradiation. It has been suggested (Simon, 1957) that the irradiation produces extensive network disruption, breaking bonds and creating a proportion of singly bonded O atoms in the structure.

Bell *et al.* (1968) performed similar calculations for vitreous GeO_2 and BeF_2, again obtaining satisfactory agreement between computed and experimental spectral bands. These two materials are believed to have the same basic network structure as SiO_2 glass, so that Bell *et al.* were able to use the model coordinates originally derived for the SiO_2 calculation for the GeO_2 and BeF_2 cases also.

Apart from the calculation on SiO_2 type glasses mentioned above, those of Zerbi *et al.* (1971) on realistic disordered polymer models, and the results for disordered alloys described in Section V, A, the NET has also been applied to 3-dimensional systems with purely orientational disorder. Here Shawyer and Dean (1972) have performed calculations for hexagonal H_2O and D_2O ice, finding reasonable agreement between the positions of computed bands and those observed in X-ray, Raman, and neutron experiments.

Finally, one further calculation on a real 3-dimensional disordered material that we should mention is that of Alben *et al.* (1973) for amorphous Si. Like the work of Bell *et al.*, this calculation was based on the use of molecular models of sample atomic arrangements in the material. However, the molecular models used by Alben *et al.* involve a relatively smaller number of degrees of freedom, leading to dynamical matrices which may be analyzed by conventional diagonalization procedures, without the need for using the NET.

VI. Interaction with Radiation

In the previous sections we have been concerned with frequency and squared frequency distributions of disordered solids, irrespective, for the most part, of the detailed properties of the normal mode eigenvectors. As remarked earlier, however, while the frequency spectrum provides a logical and convenient starting point for the discussion and intercomparison of spectral properties of solids, this quantity is not, itself, generally observed directly in experimental work. An infrared absorption experiment or neutron scattering measurements, for example, will tend to produce not the simple density of states, but that distribution modified by some response function which reflects the strength of coupling between the radiation in question and the phonons at the frequency concerned. In this present section, we discuss the relation between the normal modes and a number of important physical processes.

An induced transition between vibrational states Ψ_E and $\Psi_{E'}$ of the solid, with absorption of a quantum of energy $E' - E$ (or the emission of a quantum of energy $E - E'$) is generally characterized by a matrix element of the form

$$\langle E'|\Upsilon|E\rangle = \int d\tau\, \Psi_{E'}^{*}\Upsilon\Psi_E \tag{90}$$

where Υ is a potential describing some effective interaction between vibrating atoms and the external radiation field. The effective potential will usually be expressed in terms of the Cartesian displacements $x_{i\alpha}$, the electronic coordinates having been averaged out. A transformation from the $x_{i\alpha}$ to the normal coordinate system will be required before the integration (90), involving oscillator wave functions, can be carried out.

The transition probability for the process under consideration is of the form

$$I_{EE'} \propto |\langle E'|\Upsilon|E\rangle|^2 \tag{91}$$

and the overall cross section or absorption coefficient for transfer of energy quanta ω to the solid has the proportionality

$$I(\omega) \propto \sum_{EE'} P_E I_{EE'}\, \delta(E' - E - \omega) \tag{92}$$

where P_E denotes the thermal occupation probability for the initial state Ψ_E.

In many processes, single phonon transitions are of dominating importance, so that only energy differences $\omega = E' - E$ corresponding to fundamental vibrational frequencies of the system need be considered. In this case, the part of Υ contributing to the transition is the term linear in atomic coordinates, so that calculation of the matrix element (90) reduces to evaluation of the integrals

$$\begin{aligned} \int dq_l\, \phi^*_{n_l+1}(\omega_l, q_l) q_l \phi_{n_l}(\omega_l, q_l) &= \left(\frac{n_l + 1}{2\omega_l}\right)^{1/2} \\ \int dq_l\, \phi^*_{n_l-1}(\omega_l, q_l) q_l \phi_{n_l}(\omega_l, q_l) &= \left(\frac{n_l}{2\omega_l}\right)^{1/2} \end{aligned} \tag{93}$$

A. Neutron Scattering

The effective potential appropriate to neutron scattering from the solid is

$$\Upsilon = \sum_i b_i \exp i\mathbf{h} \cdot (\mathbf{r}_i + \mathbf{x}_i) \tag{94}$$

Here b_i is the scattering length, $\mathbf{r}_i$ the equilibrium position and $\mathbf{x}_i$ the vibrational displacement of atom i, while the momentum transferred to the solid is

$$\mathbf{h} = \boldsymbol{\kappa} - \boldsymbol{\kappa}' \tag{95}$$

where $\boldsymbol{\kappa}$ and $\boldsymbol{\kappa}'$ are neutron momenta before and after scattering. Energy conservation implies the relation

$$\omega = (\kappa^2 - \kappa'^2)/2m_{\text{NEUT}} \tag{96}$$

where ω is the transfer of energy to the solid. If only 1-phonon contributions to the scattering cross section are retained, the cross section for transfer of energy $\omega = \omega_l$ and momentum $\mathbf{h}$ to the solid may be reduced to the form (cf. Lomer and Low, 1965)

$$I(\omega_l, \mathbf{h}) = \frac{\kappa'}{\kappa}\left(\frac{\bar{n}_l + 1}{2\omega_l}\right) g(\omega_l)\left|\sum_i b_i m_i^{-1/2} e^{-W_i} e^{i\mathbf{h}\cdot\mathbf{r}_i}(\mathbf{h}\cdot\mathbf{u}_{i,l})\right|^2 \qquad (97)$$

where

$$\bar{n}_l = (e^{\omega_l/kT} - 1)^{-1} \qquad (98)$$

is a mean occupation number for oscillator l at temperature T, and $\mathbf{u}_{i,l}$ is the 3-component vector with components $u_{i\alpha,l}$. The Debye–Waller factor e^{-W_i}, which arises from the thermodynamic averaging implicit in (92), contains the thermally averaged squared atomic displacement amplitudes

$$W_i = \frac{1}{2}\langle|\mathbf{h}\cdot\mathbf{x}_i|^2\rangle_T = \frac{1}{2}\sum_l \left(\frac{\bar{n}_l + \frac{1}{2}}{\omega_l}\right) m_i^{-1}|\mathbf{h}\cdot\mathbf{u}_{i,l}|^2 \qquad (99)$$

For large momentum transfer **h**, and for an isotropic system, the incoherent approximation to (97) gives

$$I(\omega_l, \mathbf{h}) = \frac{h^2}{3}\frac{\kappa'}{\kappa}\left(\frac{\bar{n}_l + 1}{2\omega_l}\right) g(\omega_l)\sum_i b_i^2 m_i^{-1} e^{-2W_i}|\mathbf{u}_{i,l}|^2 \qquad (100)$$

If atoms $i(t)$ of the same species t can be taken to have the same scattering length $b_{(t)}$ and the same Debye–Waller factor $e^{-W_{(t)}}$, further reduction occurs, viz

$$I(\omega_l, \mathbf{h}) = \frac{h^2}{3}\frac{\kappa'}{\kappa}\left(\frac{\bar{n}_l + 1}{2\omega_l}\right) g(\omega_l)\sum_t b_{(t)}^2 m_{(t)}^{-1} e^{-2W_{(t)}}\sum_{i(t)}|\mathbf{u}_{i(t),l}|^2 \qquad (101)$$

Using observed values of $W_{(t)}$ from the crystalline silica forms and atomic displacements computed by Bell *et al.* (1968, 1971), Leadbetter and Stringfellow (1974) have evaluated $I(\omega_l, \mathbf{h})$ for inelastic scattering by vitreous SiO_2. The results show excellent agreement with the measured cross section, as shown in Fig. 5. Here, the upper diagram (a) depicts the frequency distribution calculated by Bell *et al.* (1968), diagram (b) gives the function $I(\omega_l, \mathbf{h})$ as computed by Leadbetter and Stringfellow, and (c) shows the observed inelastic neutron scattering cross section (also due to these latter authors), with a correction for 2-phonon scattering effects. Clearly the form of the scattering cross section follows that of the frequency spectrum quite closely and, unlike the situation for infrared and Raman scattering (cf. Section VI,

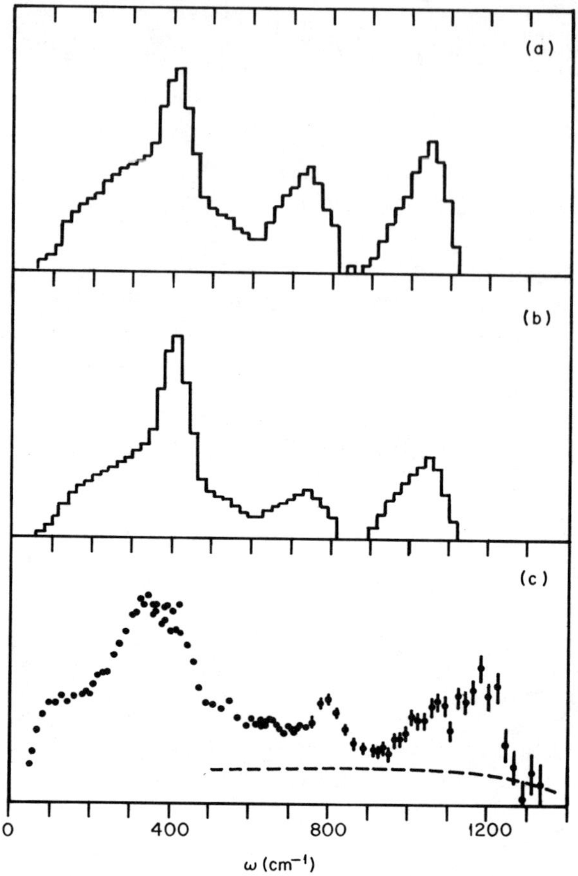

FIG. 5. Inelastic neutron scattering from vitreous SiO_2. (a) Frequency spectrum calculated by Bell *et al.* (1968). (b) Inelastic neutron scattering cross section [Eq. (101) of the text] computed by Leadbetter and Stringfellow (1974) on the basis of the calculated frequency spectrum and normal mode assignment data provided by Bell *et al.* (1968, 1971). (c) Observed inelastic neutron scattering cross section (Leadbetter and Stringfellow, 1974); here the vertical bars represent error estimates, and the broken curve gives the estimated contribution from two-phonon processes. Reprinted with permission.

B, C), the response of modes to the exciting radiation is fairly uniform throughout the spectrum.

Alben *et al.* (1974) and Weaire and Alben (1974) have computed $I(\omega_l, \mathbf{h})$ for amorphous monatomic tetrahedrally bonded semiconductors of the type Si and Ge. In this case, however, no published inelastic neutron scattering results are yet available for comparison.

B. Infrared Absorption

For the infrared light scattering process, the approximate form of the interaction is

$$\Upsilon \sim \boldsymbol{\mu} \tag{102}$$

where $\boldsymbol{\mu}$, with cartesian components μ_α is the intrinsic electric dipole moment of the vibrating system. To display the dependence of $\boldsymbol{\mu}$ upon atomic displacements specifically, we expand

$$\mu_\alpha = \mu_\alpha{}^0 + \sum_{j\beta} \mu_\alpha{}^{j\beta} x_{j\beta} + \sum_{j\beta k\gamma} \mu_\alpha^{j\beta,k\gamma} x_{j\beta} x_{k\gamma} + \cdots \tag{103}$$

The term linear in displacement, relating to 1-phonon processes, is of dominant interest, so that the relevant matrix element is

$$\langle E'|\Upsilon|E\rangle \sim \int d\tau\, \Psi_{E'}^* \left(\sum_{j\beta} \mu_\alpha{}^{j\beta} x_{j\beta} \right) \Psi_E \tag{104}$$

We transform the operator in (104) to a normal coordinate representation

$$\sum_{j\beta} \mu_\alpha{}^{j\beta} x_{j\beta} = \sum_l \bar{\mu}_\alpha{}^l q_l \tag{105}$$

where

$$\bar{\mu}_\alpha{}^l = \sum_{j\beta} \mu_\alpha{}^{j\beta} m_j{}^{-1/2} u_{j\beta,l} \tag{106}$$

is the amplitude of the dipole moment fluctuation resulting from motion in the normal mode labeled l. Consider, now, a 1-phonon transition in which the occupation number of a single mode l increases from n_l to $n_l + 1$ with absorption of a photon of energy ω_l. Making use of the result (93), the matrix element simplifies to

$$\langle n_l + 1|\Upsilon|n_l\rangle = \left(\frac{n_l + 1}{2\omega_l} \right)^{1/2} \bar{\mu}_\alpha{}^l \tag{107}$$

The single phonon absorption rate, at frequency $\omega \sim \omega_l$, for an electromagnetic field with radiation density $\rho(\omega)$ is

$$I^{\mathrm{ABSN}} = \frac{(\bar{n}_l + 1)\pi}{3\omega_l} g(\omega_l)\rho(\omega_l) \sum_\alpha |\bar{\mu}_\alpha{}^l|^2 \tag{108}$$

A similar calculation for single phonon emission at frequency $\omega \sim \omega_l$ gives

$$I^{\text{EMIS}} = \frac{\bar{n}_l \pi}{3\omega_l} g(\omega_l)\rho(\omega_l) \sum_\alpha |\bar{\mu}_\alpha{}^l|^2 \tag{109}$$

The nett absorption rate (per unit time) for photons of frequency ω_l is

$$I^{\text{NETT}} = \frac{\pi}{3\omega_l} g(\omega_l)\rho(\omega_l) \sum_\alpha |\bar{\mu}_\alpha{}^l|^2 \tag{110}$$

and the fraction of electromagnetic energy absorbed per unit thickness of the solid is proportional to

$$\frac{\omega_l I^{\text{NETT}}}{cV\rho(\omega_l)} = \frac{\pi}{3cV} g(\omega_l) \sum_\alpha |\bar{\mu}_{\alpha l}|^2 \tag{111}$$

The evaluation of (111) hinges on a knowledge of the vibrational frequency spectrum $g(\omega)$, the displacement amplitudes $u_{j\beta,l}$ and the dipole moment coefficients $\mu_\alpha{}^{j\beta}$. The problem of calculating the first two quantities has already been dealt with in previous sections. To obtain an estimate of $\mu_\alpha{}^{j\beta}$ one must now consider suitable physical models of the charge distribution in the solid, or rather its response to motion of the atoms. The simplest such model is the point charge model of the type

$$\mu_\alpha{}^{j\beta} = \mathscr{Z}_j \delta_{\alpha\beta} \tag{112}$$

where $\mathscr{Z}_j$ is an effective charge on atom j. The point charge model has been applied by a number of authors with varying degrees of success. Hass *et al.* (1969) and C. Hall (unpublished) have computed line strengths for modes in mixed crystal systems, while R. J. Bell and D. C. Hibbins–Butler (unpublished) have used the model to derive values of the infrared absorption coefficient in vitreous SiO_2, GeO_2, and BeF_2.

Some results of this last-mentioned calculation are shown in Fig. 6 for the case of SiO_2. Figure 6a shows the calculated frequency spectrum, while the absorption coefficient computed from (111) is depicted in Fig. 6b. The bottom diagram (c) shows the measured infrared absorption coefficient of vitreous SiO_2 (Gaskell and Johnson, 1975). The theoretical absorption coefficient in Fig. 6b was actually computed as a line spectrum, based on selected displacement eigenvectors, and the smooth curve shown in the diagram results from replacing each line by a Lorentz profile of halfwidth

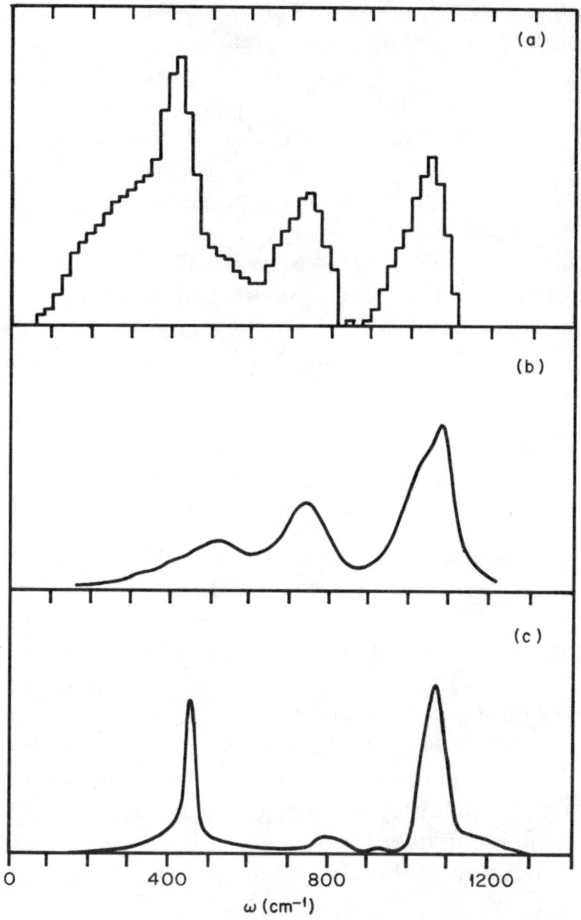

FIG. 6. Infrared absorption in vitreous SiO_2. (a) Frequency spectrum calculated by Bell *et al.* (1968). (b) Infrared absorption coefficient [Eq. (111) of the text] computed in the point charge approximation (R. J. Bell and D. C. Hibbins–Butler, unpublished). (c) Observed absorption coefficient. After Gaskell and Johnson (1975). Reprinted with permission.

30 cm^{-1}. Frequency spectra and selected eigenvectors were calculated by the methods described in Section IV.

The calculated infrared absorption differs significantly from the frequency spectrum in Fig. 6a, due to the response of modes to radiation varying considerably as a function of frequency. The two upper bands in the frequency spectrum clearly retain their identity in the absorption spectrum. At lower frequencies, however, the maximum calculated absorption coincides not with the prominent peak at 400 cm^{-1} in the frequency spectrum, but with the shoulder at about 550 cm^{-1}. It is not clear, on the basis of this comparison alone, whether the observed absorption maximum at around 500 cm^{-1} corresponds to the shoulder at 550 cm^{-1} in the calculated frequency spectrum, or whether it stems from the computed peak at 400 cm^{-1} with the point charge model giving an artificially low value of the absorption in this region. Comparison between the positions of low frequency maxima in measured infrared and neutron scattering results suggests, however, that the former is most likely to be the case.

The point charge model (112) is most likely to be valid for solids of a strongly ionic character. For materials with a covalent type of bonding, such as Si and Ge, a rather different model of the charge distribution derivative is needed. Alben *et al.* (1973) have suggested an expression essentially of the form

$$\sum_{j\beta} \mu_\alpha{}^{j\beta} x_{j\beta} = \sum_{ijk}^{\text{bond pairs}} \{\mathscr{Z}_{ijk}[(\mathbf{x}_k - \mathbf{x}_j)\cdot\hat{\mathbf{r}}_{jk} - (\mathbf{x}_j - \mathbf{x}_i)\cdot\hat{\mathbf{r}}_{ij}][\hat{\mathbf{r}}_{ij} + \hat{\mathbf{r}}_{jk}]\}_\alpha \tag{113}$$

for the linear term in the dipole moment expansion for covalent solids. Here $\mathscr{Z}_{ijk}$ is some effective charge movement parameter for the triad of atoms ijk, $\hat{\mathbf{r}}_{ij}$ is a unit vector in the direction of bond ij, subscript α denotes the cartesian component in direction α, and the summation is over pairs of adjacent bonds ij, jk. Frequency spectra and absorption coefficients for amorphous Si computed by Alben *et al.* (1973) are compared with experimental infrared spectra in Fig. 7. The theoretical curves have been somewhat smoothed by Alben *et al.*, using a Lorentzian broadening procedure. While the main features of the observed infrared profile are already easily identifiable in the calculated frequency spectrum, it is clear that proper allowance for the response of modes enhances the agreement considerably.

C. Raman Scattering

An approximate form of potential appropriate to Raman scattering is

$$\Upsilon \sim \mathbf{p} \tag{114}$$

where $\mathbf{p}$, the induced dipole moment of the solid (with components p_α) is related to the polarizability tensor $\alpha_{\alpha\alpha'}$ and electric field (with components $\mathscr{E}_\alpha$) by

$$p_\alpha = \sum_{\alpha'} \alpha_{\alpha\alpha'} \mathscr{E}_{\alpha'} \tag{115}$$

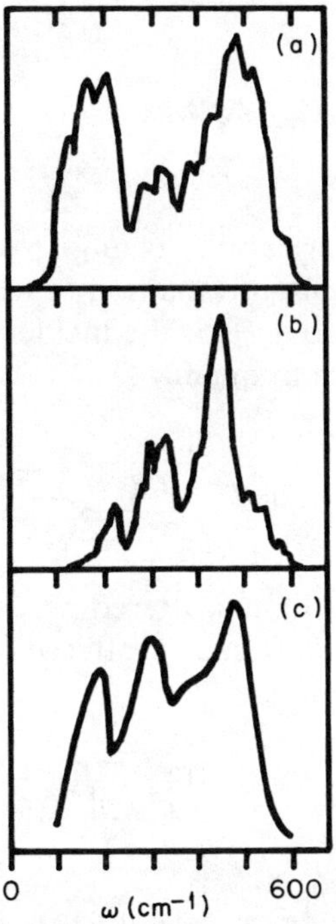

FIG. 7. Infrared absorption in amorphous Si. (a) Calculated frequency spectrum. (b) Infrared absorption coefficient, calculated in the covalent bond approximation described in the text. (c) Observed infrared absorption coefficient for amorphous Si. After Alben *et al.* (1973). Reprinted with permission.

If the polarizability tensor $\alpha_{\alpha\alpha'}$ is expanded in powers of the vibrational displacements

$$\alpha_{\alpha\alpha'} = \alpha^0_{\alpha\alpha'} + \sum_{j\beta} \alpha^{j\beta}_{\alpha\alpha'} x_{j\beta} + \sum_{j\beta k\gamma} \alpha^{j\beta,k\gamma}_{\alpha\alpha'} x_{j\beta} x_{k\gamma} + \cdots \tag{116}$$

the matrix element relevant to 1-phonon scattering is

$$\langle E'|\Upsilon|E\rangle \simeq \int d\tau\, \Psi^*_{E'} \left(\sum_{j\beta} \alpha^{j\beta}_{\alpha\alpha'} x_{j\beta} \right) \Psi_E \tag{117}$$

As in the infrared case, we transform the operator to a normal coordinate representation

$$\sum_{j\beta} \alpha^{j\beta}_{\alpha\alpha'} x_{j\beta} = \sum_{l} \bar{\alpha}^{l}_{\alpha\alpha'} q_l \tag{118}$$

where

$$\overline{\alpha}^{l}_{\alpha\alpha'} = \sum_{j\beta} \alpha^{j\beta}_{\alpha\alpha'} m_j^{-1/2} u_{j\beta,l} \tag{119}$$

is the amplitude of the polarizability change resulting from vibration of the solid in mode l. For a 1-phonon transition in which the occupation number of mode l changes from n_l to $n_l \pm 1$, the incident light beam at frequency Ω acquiring a component with frequency $\Omega \mp \omega_l$, we have from (93)

$$\langle n_l \pm 1 | \Upsilon | n_l \rangle = \left(\frac{n_l + \frac{1}{2} \pm \frac{1}{2}}{2\omega_l} \right)^{1/2} \overline{\alpha}^{l}_{\alpha\alpha'} \tag{120}$$

For unpolarized incident light scattered perpendicular to the direction of propagation, the intensities of Raman scattered components with frequencies $\Omega \pm \omega_l$ have the proportionality

$$I_{\parallel} \propto \frac{(\overline{n}_l + \frac{1}{2} \pm \frac{1}{2})}{2\omega_l} g(\omega_l)[7G_l^2 + 45A_l^2] \tag{121}$$

$$I_{\perp} \propto \frac{(\overline{n}_l + \frac{1}{2} \pm \frac{1}{2})}{2\omega_l} g(\omega_l)[6G_l^2] \tag{122}$$

where the subscripts $\parallel$ and $\perp$ refer to observed light polarized parallel and perpendicular to the plane of scattering. Similar expressions hold for the case of polarized incident light. The quantities A, G in (121) and (122) are invariants of the polarizability tensor, defined by

$$\begin{aligned} A_l &= \tfrac{1}{3}(\overline{\alpha}_{11}{}^l + \overline{\alpha}_{22}{}^l + \overline{\alpha}_{33}{}^l) \\ G_l^2 &= \tfrac{1}{2}[(\overline{\alpha}_{11}{}^l - \overline{\alpha}_{22}{}^l)^2 + (\overline{\alpha}_{22}{}^l - \overline{\alpha}_{33}{}^l)^2 + (\overline{\alpha}_{33}{}^l - \overline{\alpha}_{11}{}^l)^2] \\ &\quad + 3[(\overline{\alpha}_{12}{}^l)^2 + (\overline{\alpha}_{23}{}^l)^2 + (\overline{\alpha}_{31}{}^l)^2] \end{aligned} \tag{123}$$

As in the case of the dipole moment in infrared absorption, some physically plausible expression is required for the dependence of polarizability upon atomic displacement. Here, the bond polarizability approximation provides the basis of a suitable model. This approximation associates an axially symmetric polarizability tensor $\alpha_{\alpha\alpha'}(\boldsymbol{\rho}_{ij})$ with each bond ij. In a vector notation we may write

$$\alpha_{\alpha\alpha'}(\boldsymbol{\rho}_{ij}) = \{\alpha(\rho_{ij})\mathbf{1} + \gamma(\rho_{ij})[\hat{\boldsymbol{\rho}}_{ij}\hat{\boldsymbol{\rho}}_{ij} - \tfrac{1}{3}\mathbf{1}]\}_{\alpha\alpha'} \tag{124}$$

where **1** is a unit dyadic, the mean polarizability $\alpha(\rho_{ij})$ and the anisotropy $\gamma(\rho_{ij})$ are functions of bond length ρ_{ij}, and the subscript $\alpha\alpha'$ denotes the $\alpha\alpha'$ component of the dyadic expression within the braces. In (124), $\hat{\boldsymbol{\rho}}_{ij}$ is the unit vector derived from

$$\boldsymbol{\rho}_{ij} = \mathbf{r}_{ij} + (\mathbf{x}_j - \mathbf{x}_i) \tag{125}$$

which defines the relative position of neighboring atoms i and j. Summing (125) over all bonds and extracting the first-order term in a Taylor's expansion with respect to the displacements, we obtain

$$\begin{aligned}\sum_{j\beta} \alpha_{\alpha\alpha'}^{j\beta}\, x_{j\beta} = \sum_{ij}^{\text{bonds}} \{&\alpha'(r_{ij})[(\mathbf{x}_j - \mathbf{x}_i)\cdot\hat{\mathbf{r}}_{ij}]\mathbf{1} \\ &+ \gamma'(r_{ij})[(\mathbf{x}_j - \mathbf{x}_i)\cdot\hat{\mathbf{r}}_{ij}][(\hat{\mathbf{r}}_{ij}\hat{\mathbf{r}}_{ij}) - \tfrac{1}{3}\mathbf{1}] \\ + {r_{ij}}^{-1}\gamma(r_{ij})[(\mathbf{x}_j - \mathbf{x}_i)\hat{\mathbf{r}}_{ij} + \hat{\mathbf{r}}_{ij}(\mathbf{x}_j - &\mathbf{x}_i) - 2(\mathbf{x}_j - \mathbf{x}_i)\cdot\hat{\mathbf{r}}_{ij}(\hat{\mathbf{r}}_{ij}\hat{\mathbf{r}}_{ij})]\}_{\alpha\alpha'}\end{aligned} \tag{126}$$

where $\alpha'(r_{ij})$ and $\gamma'(r_{ij})$ are the derivatives of $\alpha(\rho_{ij})$ and $\gamma(\rho_{ij})$, evaluated at the equilibrium bond lengths. In (126), the product involving $\alpha'(r_{ij})$ contributes only to A^2, while the remaining portion of the expression contributes only to G^2. The equation is valid for materials like Si and SiO_2, whose nearest-neighbor bonds ij are all of the same kind. For solids containing more than one kind of bond, (126) needs to be modified to allow for different polarizability parameters α', γ', ${r_{ij}}^{-1}\gamma$ for each bond type.

In general, the values of the polarizability parameters α', γ', ${r_{ij}}^{-1}\gamma$ are not accurately known *a priori*. For perfect crystals, however, symmetry considerations related both to equilibrium geometry and the normal modes will often lead to considerable cancellation in (126): in the most favorable situations, the expression for the scattering intensity may contain just one of the polarizability parameters, occurring as a simple outside scaling factor. Just occasionally, Raman measurements on crystals will provide information on the relative magnitudes of α', γ', and ${r_{ij}}^{-1}\gamma$, which may be utilized in calculating Raman intensities for the corresponding amorphous phases. In general, however, the evaluation of (126), containing contributions from all three unknown parameters, presents considerable difficulty.

Noting that the observed intensity profiles $I_\parallel$ and $I_\perp$ for amorphous Si have the same shape, Alben *et al.* (1973) have concluded that $\alpha' \simeq 0$ for this material. With this assignment for α', and with the assumption ${r_{ij}}^{-1}\gamma = \frac{3}{8}\gamma'$, they have evaluated Raman scattering intensities for Si, their results being shown in Fig. 8. Here Fig. 8a gives the calculated frequency spectrum, $I_\parallel$ is depicted in Fig. 8b, and the measured Raman scattering intensity is shown

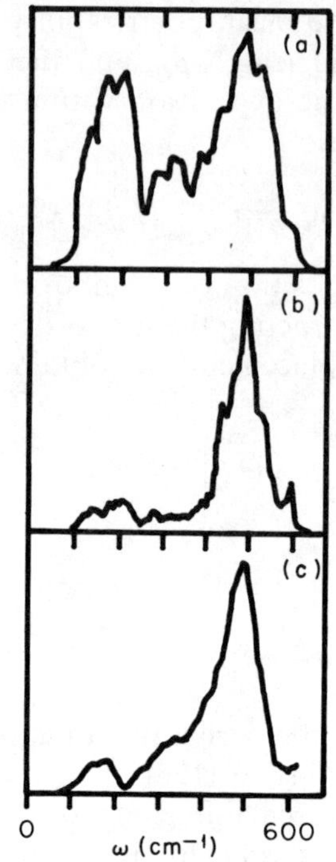

FIG. 8. Raman scattering in amorphous Si. (a) Calculated frequency spectrum. (b) Reduced Raman scattering intensity $\omega_l I/(\bar{n}_l + \frac{1}{2} \pm \frac{1}{2})$, calculated in the bond polarizability approximation described in the text. (c) Observed reduced Raman scattering intensity for amorphous Si. After Alben *et al.* (1973). Reprinted with permission.

in Fig. 8c. The theoretical graphs originally derived as line spectra have been smoothed by use of a Lorentzian broadening procedure. As with the infrared absorption calculations by the same authors, the agreement between theory and experiment is extremely good.

R. J. Bell and D. C. Hibbins–Butler (unpub.) used the bond polarizability as the basis of Raman scattering calculations for SiO_2-type glasses. Here, no definitive information on the relative magnitudes of the polarizability parameters is available, so comparison between calculated and observed results is not straightforward. Figure 9b shows computed values of $g(\omega)[45A^2]$ with α' taken to be 1.0, and values of $g(\omega)[G^2]$ for the two cases $\gamma' = 1, r_{ij}^{-1}\gamma = 0$ and $\gamma' = 0, r_{ij}^{-1}\gamma = 1$. The curves have been produced from line spectra by the use of a Lorentz broadening procedure with halfwidth 30 cm^{-1}. Figure 9a shows the calculated frequency spectrum $g(\omega)$, and 9c gives the measured Raman spectrum for vitreous SiO_2 (after Leadbetter and Stringfellow, 1974). Comparison of Fig. 9a and b shows clearly that allowance for the response

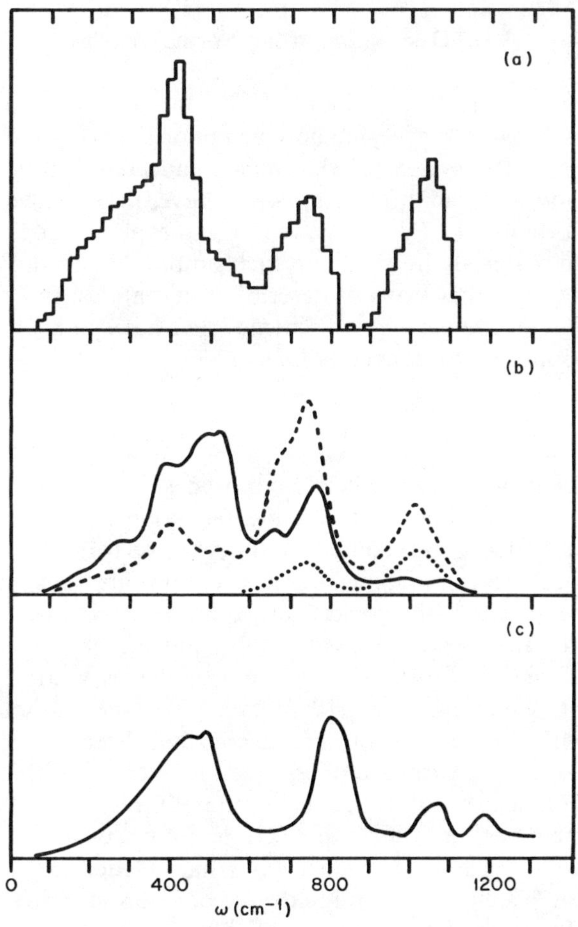

FIG. 9. Raman scattering in vitreous SiO_2. (a) Computed frequency spectrum (Bell *et al.*, 1968). (b) Contributions to the reduced Raman scattering intensity $\omega_l I/(\bar{n}_l + \frac{1}{2} \pm \frac{1}{2})$, calculated in the bond polarizability approximation described in the text: solid line, $g(\omega)[45A^2]$ for $\alpha' = 1$; dashed line, $g(\omega)[G^2]$ for $\gamma' = 1, r_{ij}{}^{-1}\gamma = 0$; dotted line, $g(\omega)[G^2]$ for $\gamma' = 0, r_{ij}{}^{-1}\gamma = 1$. (c) Observed reduced Raman scattering intensity for vitreous SiO_2 (after Leadbetter and Stringfellow, 1974). Reprinted with permission.

of the modes to the exciting radiation modifies the spectral profile considerably. A direct comparison between calculated intensities and experimental results is difficult, for the reasons given above, but it seems unclear that agreement between the two can be achieved by any reasonable choice of α', γ', and $r_{ij}{}^{-1}\gamma$. Clearly, in this case, exploration of alternative models of the dependence of polarizability on atomic displacement is desirable.

VII. Description of the Normal Modes

In the case of a crystalline solid, the **k** vector and point group symmetry considerations provide a ready-made scheme for the convenient labeling of normal modes of the system at the outset. Indeed, many features of the response of modes to radiation, discussed above, may be deduced directly for crystals on the basis of the wave vector and symmetry characterization. For disordered materials, however, no such simple classification is available, and one must seek other ways of describing the modes. In this section we sketch out some of the descriptive schemes that have so far been proposed for vibrational motion in noncrystalline solids.

A. Fourier Analysis

One significant property of vibrational modes in a number of systems is their Fourier profile. The most suitable basis for Fourier analyzing the modes of a disordered lattice consists of the set of vectors of a corresponding regular lattice subject to the same boundary conditions. This type of analysis indicates to what extent the perfect lattice modes have been disturbed by the presence of the various imperfections, and enables one to draw up something resembling a dispersion relation (ω versus **k** curve) for the disordered system. Champier *et al.* (1966) have given such an analysis for the linear chain with isolated mass and force constant defects.

Expansion of the perturbed modes in terms of those of the unperturbed system is, of course, central to the conventional Green's function approach and, for a system with a point imperfection, an expression for the Fourier components of the perturbed eigenvector **v** may be deduced formally from Eq. (53a,b). For systems with a finite degree of disorder, however, ensemble averaging methods used in conjunction with the Green's function method can result in misleading information about the spatial form of modes.

In numerical investigations of disordered systems, P. Dean (unpublished) dealing with the 2-component mass-disordered chain, and R. J. Bell (unpublished) dealing with topologically disordered chains, have performed Fourier analysis of selected modes. In both cases, modes were found to be wavelike, possessing a small number of significant Fourier components, only in the region near $\omega = 0$. At moderate and high frequencies, a wide spread of components was needed to represent the eigenvectors. These results confirm that, for systems with a finite degree of disorder, the concept of a dispersion relation is valid only near the low frequency limit.

While a Fourier analysis of modes can provide interesting information for disordered systems that are based on geometrically regular chains and lattices, the situation for spatially disordered materials such as glasses is altogether different. In such cases there is generally no direct correspondence between individual atomic sites in the ordered and disordered states, and the choice of a plane wave basis set for the Fourier analysis is not straightforward. It is possible, at least in principle, to generate a set of linear combinations of perfect lattice waves that are orthonormal with respect to summation over a given network of irregularly spaced points; however, it is not clear whether analysis of the modes of the disordered system in terms of such a set can yield anything of useful significance.

B. Normal Mode Assignments

All the information relevant to a given normal mode vibration is contained in the appropriate displacement eigenvector, and it may sometimes be practical, as with highly localized vibrations in 1-dimensional systems, to ascertain the nature of the mode by a superficial inspection of the eigenvector elements. For more extended modes, however, particularly those in higher dimensional systems, the form of the vectors may often be so complex as to yield little useful information at first sight. In such a case, some processing of the information contained in the normal mode vectors is necessary before the essential features of the atomic motion can be understood.

Bell *et al.* (1971) have devised such a scheme for processing data on the vibrational modes of SiO_2-type glasses, using *local* symmetry axes based on local atomic arrangements to separate out different aspects of atomic motion for individual analysis. The **B**, **R**, **S** axis system used by these authors is shown in Fig. 10. Briefly, for each fully coordinated O atom in the system, one defines a **B** axis parallel to the bisector of the local Si–O–Si angle, an **S** axis perpendicular to the bisector and lying within the Si–O–Si plane, and an **R** axis normal to the Si–O–Si plane. By resolving the normal mode motion of each fully coordinated, or bridging, O atom along such symmetry axes, one may divide the vibrational energy of these atoms into contributions from B (bending), R (rocking), and S (stretching) types of motion. If singly coordinated, or nonbridging, O atoms are also free to vibrate in the structure, their energy can be resolved into a NS (nonbridging stretching) contribution arising from motion parallel to their respective Si–O bonds, and a NB (nonbridging bending) contribution arising from motion perpendicular to the bonds. The remaining energy of modes, that of the Si atoms, is designated as a C (cation) contribution.

The results of analyzing vibrational modes of vitreous SiO_2 in this way are shown in Fig. 11. Here the mode assignment diagrams (11c) and (11d) are shown under the appropriate vibrational spectra (11a) and (11b). In each case, the left-hand diagram refers to a calculation with the fixed end boundary condition imposed (nonbridging O atoms at the model boundary

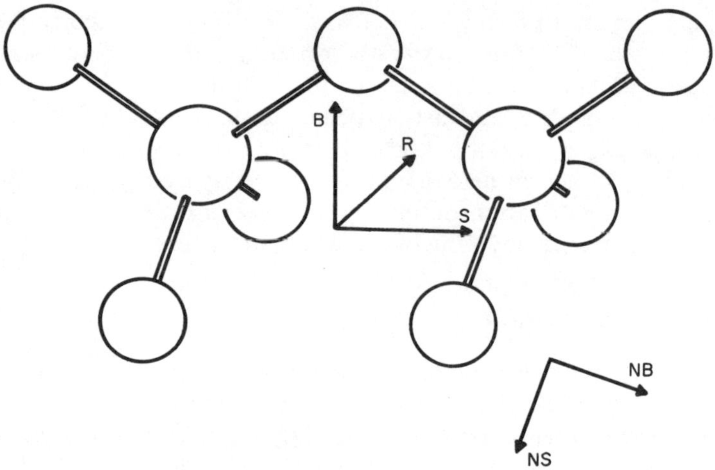

FIG. 10. The **B**, **R**, **S** (bond bending, bond rocking, bond stretching) axis system associated with a typical doubly coordinated O atom in SiO_2. Also shown are the **NB** and **NS** (nonbridging bending and nonbridging stretching) directions for a nonbridging atom of the structure.

constrained to be stationary) while the right-hand diagram refers to the free end condition (all atoms free to vibrate). In the assignment diagrams, it is the *distance between* plotted curves, rather than the positions of the curves themselves, that indicates the proportion of energy arising from the various types of motion.

It is possible, on the basis of the assignment analysis, to associate each major spectral band with a characteristic type of atomic motion. Thus the spectral features in the regions around 1050 cm^{-1}, 400 cm^{-1}, and below 350 cm^{-1} in Fig. 11a and b may be labeled S, R, and C, because of their appreciable content of stretching, rocking, and cation motion, respectively. Those at 550 cm^{-1} and 750 cm^{-1} contain a significant proportion of bending motion, modified in the former case by bond stretching and in the latter by cation motion and are therefore designated by B + S and B + C, respectively. Figure 11d confirms clearly that the additional features present at around 300 cm^{-1} and 850 cm^{-1} in the free-end spectrum stem directly from NB (bending) and NS (stretching) motions, respectively, of the nonbridging O atoms in the structure, as indicated in Section V.

As we mentioned earlier (Section VI, A), information of the kind derived in the normal mode assignments analysis described above is of interest in connection with neutron scattering studies. Bell *et al.* (1971) have already given similar assignments for vitreous GeO_2 and BeF_2, and there is no reason in principle why the method should not be extended to other simple glass formers such as $ZnCl_2$, B_2O_3, As_2O_3, and P_2O_5.

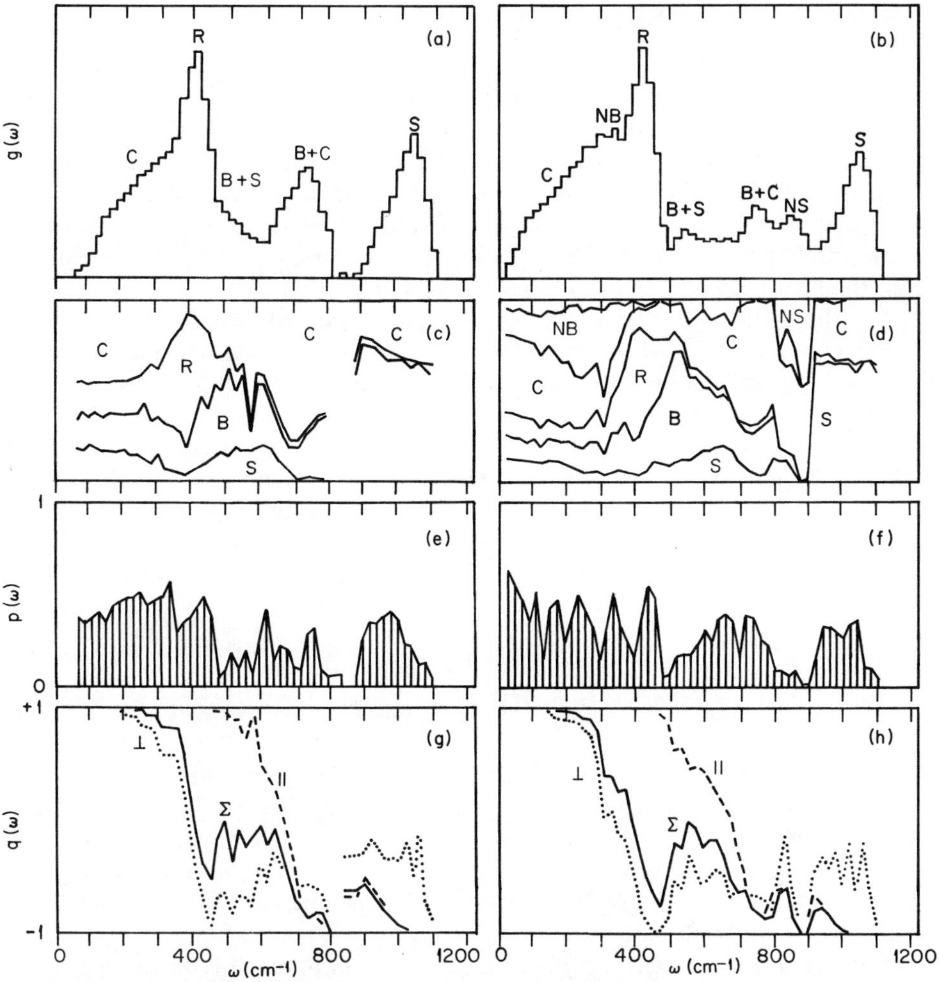

FIG. 11. Descriptive parameters for phonons in vitreous SiO_2 (Bell *et al.*, 1968, 1970, 1971; Bell and Hibbins–Butler, 1975); left-hand diagrams refer to a calculation with the fixed end boundary condition imposed, while those on the right refer to a free end calculation: (a) and (b) computed frequency spectra. (c) and (d) Analysis of the mode energies into contributions from the different types of atomic motion described in the text; in these diagrams, it is the *distance between* curves, rather than the positions of the curves themselves, that indicates the proportion of vibrational energy arising from the various types of motion. (e) and (f) Participation ratios, based on analysis of the vibrational motion of Si atoms in the network. (g) and (h) Phase quotients for modes in vitreous SiO_2; here the curves labeled || and ⊥ refer to projected motion parallel and perpendicular to the Si–O bonds, while Σ gives a phase quotient for the unprojected motion.

C. The Spatial Extent of Modes

The spatial extent of vibrational modes is of interest in connection with heat transfer in a material. In the harmonic approximation, the modes can be regarded as individual channels carrying energy across the solid. If a mode is extended in space, that is, if its amplitude is reasonably large throughout the solid, it acts as an efficient conductor of heat. Any tendency for a mode to be localized, on the other hand, inhibits its ability to transport (although not to contain) energy. In this way, localized modes can contribute to the heat capacity of a solid, but not its conductivity.

In the earlier numerical investigations of disordered lattices, the localization characteristics of modes were generally assessed on a rather ad hoc basis, by graphical representation and visual inspection of the displacement vectors. In recent years, however, more qualitative criteria for discussing the spatial extent of modes have been evolved, the most commonly used being the participation ratio $p(\omega)$. This quantity is defined simply in terms of the mass-dependent amplitudes $\mathbf{u}_{i,l}$ for a mode with frequency ω_l by

$$p(\omega_l) = (M_1)^2/(M_0 M_2) \tag{127}$$

where

$$M_n = M_n(\omega_l) = \sum_i |\mathbf{u}_{i,l}|^{2n} \tag{128}$$

is the nth energy moment of the mode. The participation ratio $p(\omega_l)$ gives a measure of the proportion of atoms which participate effectively in the motion of a mode with frequency ω_l. Taking the case of a monatomic system as an example, a vibration in which all atoms are displaced with equal amplitude (as in a simple translation of the system) has $p(\omega_l) = 1$, while a mode in which only one atom moves has $p(\omega_l) = 1/N$, N being the number of atoms in the system; if Np atoms participate appreciably in the motion, with roughly equal amplitudes, then $p(\omega_l) \simeq p$. For the sinusoidal modes characteristic of crystals, one expects a participation ratio of around $p(\omega_l) = \frac{2}{3}$. In a polyatomic system, it may often prove convenient to define a participation ratio $p_t(\omega_l)$ for atoms of a particular species t: the definition of $p_t(\omega_l)$ is identical to that of $p(\omega_l)$, except that the summation (128) is restricted to t-type atoms $i(t)$.

A number of authors, particularly Bell *et al.* (1970), Visscher (1972), Edwards and Thouless (1972), and Economou and Papatriantafillou (1975) have used the participation ratio in discussions of the spatial extent of vibrational (or electron) states in disordered solids. The results of Bell *et al.* (1970) for SiO_2 glass are shown in Fig. 11e and f below the corresponding spectra and normal mode assignment diagrams. As noted by Bell *et al.*, the vibrations are on the whole less extended in space than the plane wavelike modes that characterize regular crystals. Localization tends to be greatest at high frequencies and, especially, near band edges. Calculations performed with a free-end boundary condition (right-hand diagrams), indicate that modes arising from NS vibrations of the nonbridging atoms are very localized indeed, with the atomic motion being effectively restricted to no more than a few molecules in the structure. The NB vibrations, on the other hand, do not appear to be severely localized in this way, and are probably similar in kind to the modes associated with in-band resonances discussed in Section III.

R. J. Bell (unpublished) has studied the localization of modes in a topologically disordered chain (cf. Section V), in both real space and in reciprocal space. Since the Fourier transform of a plane wave is simply a δ function, and vice versa, one would expect to find a strong inverse correlation between localization in the two representations. This is borne out by Fig. 12

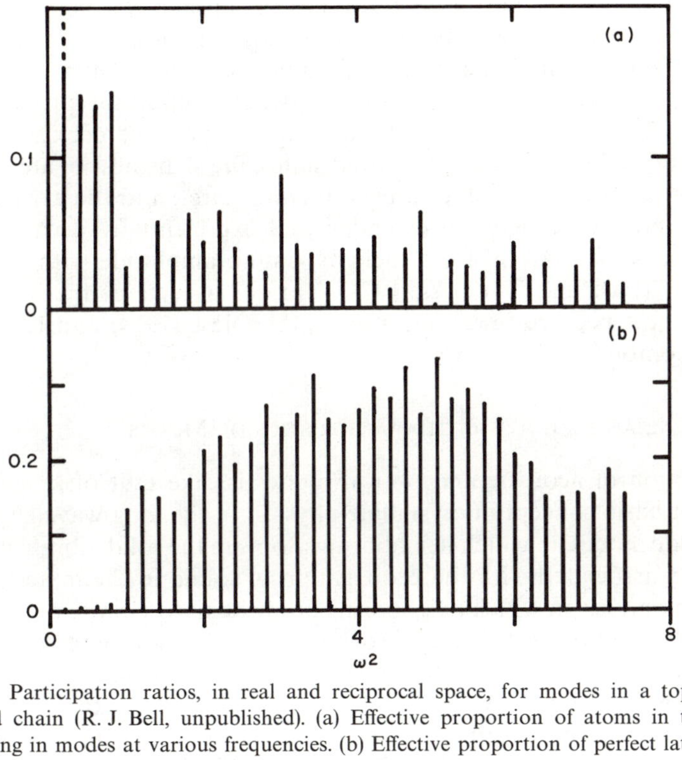

FIG. 12. Participation ratios, in real and reciprocal space, for modes in a topologically disordered chain (R. J. Bell, unpublished). (a) Effective proportion of atoms in the system participating in modes at various frequencies. (b) Effective proportion of perfect lattice waves contributing significantly to the Fourier decomposition of these modes.

(R. J. Bell, unpublished) which gives participation ratios for the modes of the topologically disordered chain in conventional and Fourier transform representations.

In their numerical studies of phonons in glasses R. J. Bell, P. Dean, and D. C. Hibbins–Butler (unpublished), have studied the distribution of vibrational energy in selected modes, as a function of distance from the point of maximum mode amplitude. Dealing with the equivalent problem of electrons in disordered systems, Thouless (1972) and Edwards and Thouless (1972) have used the sensitivity of eigenvalues to boundary condition as a criterion for gauging the spatial localization of eigenstates.

While the numerical investigations have yielded a considerable amount of background information about the spatial distribution of modes in 1-dimensional chains, they have not generally provided any cogent evidence on the ultimate form of vibrational states in 3-dimensional systems: it is simply not feasible to make deductions about the ultimate decay of vibrational amplitudes (with the possible exception of the case of highly localized out-of-band modes) on the basis of sample model lattices whose linear dimension is of the order of several times the atomic spacing. If any truly rigorous conclusions are to be reached in this area, it seems it must occur through the medium of the analytical methods, with the numerical results perhaps providing useful supporting evidence. Important problems that remain to be resolved concern the question of possible coexistence (at a given energy) of extended and localized states, the sharpness of any mobility edge between conducting and nonconducting states, and the existence of a regime intermediate between extended and exponentially decaying states where modes may fall off from their maximum amplitude with an inverse power law (Last and Thouless, 1974). Various aspects of the localization problem have been reviewed recently by Thouless (1974) and Licciardello and Economou (1975).

D. Optical and Acoustic Characteristics of Modes

The notion of acoustic and optical modes is, like that of **k** vectors, an important basic concept in crystalline physics. In the long wavelength limit the acoustic modes, in which neighboring atoms in the solid vibrate in-phase, are closely associated with the propagation of sound in the material, while the optical modes, in which neighboring atoms tend to vibrate out-of-phase, can give rise to first-order light absorption and emission processes.

There are no *precise* counterparts of acoustic and optical modes in glasses, just as there is no precise demarcation between optically active and inactive vibrations. It can be useful, however, to gauge the extent to which atomic vibrations tend to be largely in-phase or out-of-phase over the solid as a whole, since this gives some guidance as to the likely optical activity of the modes. Bell and Hibbins–Butler (1975) have proposed a scheme for assessing the phase relationship between neighboring atoms throughout the glass. They define a *phase quotient*

$$q_\Sigma(\omega_l) = \sum_{ij}^{\text{bonds}} \mathbf{u}_{i,l} \cdot \mathbf{u}_{j,l} \Big/ \sum_{ij}^{\text{bonds}} |\mathbf{u}_{i,l} \cdot \mathbf{u}_{j,l}| \tag{129}$$

the summation being over all nearest-neighbor atomic pairs in the solid. If the motion of the two atoms in a nearest-neighbor pair is roughly parallel (i.e., if $\mathbf{u}_{i,l} \cdot \mathbf{u}_{j,l}$ is positive) for each nearest-neighbor pair in the system, $q_\Sigma(\omega_l)$ has the value $+1$. If, on the other hand, the motion is roughly antiparallel (i.e., if $\mathbf{u}_{i,l} \cdot \mathbf{u}_{j,l}$ is negative) for each pair, then $q_\Sigma(\omega_l)$ is -1. Clearly $q_\Sigma(\omega_l)$ gives some measure of the phase relationship between neighboring atoms in a vibrating solid, ranging from $+1$ for modes which are largely acoustic in character to -1 for those which are largely optical.

Phase quotients for vitreous SiO_2, GeO_2, and BeF_2 have been computed by Bell and Hibbins–Butler, their results for SiO_2 being shown in Fig. 11g, (fixed end boundary condition) and Fig. 11h (free end boundary condition). As might be expected on the basis of experience of vibrations in crystals, the motion of modes in the glass becomes increasingly optical, that is, out-of-phase, with increasing frequency.

It is interesting to consider, too, phase quotients $q_\parallel(\omega_l)$ and $q_\perp(\omega_l)$ associated with atomic motion parallel and perpendicular to the interatomic Si–O bonds. Specifically, these are defined by

$$\begin{aligned} q_\parallel(\omega_l) &= \sum_{ij}^{\text{bonds}} \mathbf{u}_{i,l} \cdot (\hat{\mathbf{r}}_{ij}\hat{\mathbf{r}}_{ij}) \cdot \mathbf{u}_{j,l} \Big/ \sum_{ij}^{\text{bonds}} |\mathbf{u}_{i,l} \cdot (\hat{\mathbf{r}}_{ij}\hat{\mathbf{r}}_{ij}) \cdot \mathbf{u}_{j,l}| \\ q_\perp(\omega_l) &= \sum_{ij}^{\text{bonds}} \mathbf{u}_{i,l} \cdot (\mathbf{1} - \hat{\mathbf{r}}_{ij}\hat{\mathbf{r}}_{ij}) \cdot \mathbf{u}_{j,l} \Big/ \sum_{ij}^{\text{bonds}} |\mathbf{u}_{i,l} \cdot (\mathbf{1} - \hat{\mathbf{r}}_{ij}\hat{\mathbf{r}}_{ij}) \cdot \mathbf{u}_{j,l}| \end{aligned} \tag{130}$$

The phase quotients for parallel and perpendicular motion calculated by Bell and Hibbins–Butler are represented in Fig. 11g and h, by dashed and dotted lines, respectively. It is possible, by correlating $q_\parallel(\omega_l)$ and $q_\perp(\omega_l)$ with data from the normal mode assignment analysis, to make tentative predictions about likely infrared activity of modes. Thus Bell and Hibbins–Butler conclude that all modes in vitreous SiO_2, with the exception of the low frequency C region, should be strongly active. Surprisingly, this relatively crude procedure gives a rather similar prediction of the activity of bands in vitreous SiO_2 to that from the point charge model calculation mentioned in Section VI, B.

VIII. Concluding Remarks

It is clear that numerical calculations based on the negative eigenvalue theorem have given, and are likely to continue to give, useful and accurate information on vibrational frequency spectra and the nature of normal mode displacements in real disordered solids. The introduction of descriptive schemes, such as those concerned with normal mode assignments, mode localization, and phase relationship, has contributed significantly to the understanding of normal modes in glasses. The general task of providing suitable descriptive parameters for phonons in amorphous solids remains an extremely important one, and considerable further effort needs to be devoted to this problem. In the area of response to radiation, while inelastic neutron scattering processes seem to be satisfactorily dealt with by existing theory, it is clear that the point-charge and bond-polarizability models do not always give an adequate account of infrared and Raman activities in glasses, and a systematic investigation of other feasible models of the charge distribution is called for.

The derivation of results of a rigorous formal nature on phonon properties in amorphous materials ought, in principle, to proceed through the medium of analytical investigations. Up until now, however, the analytical studies have been largely restricted to idealized 1-dimensional systems exhibiting very simple kinds of disorder. Until it becomes feasible to extend such studies to more realistic models of the amorphous state, the numerical methods are likely to continue to provide the main source of useful information on real disordered solids.

Acknowledgments

The author wishes to thank Dr. P. Dean for the benefit of many interesting and fruitful discussions on lattice dynamics over the past ten years. This article is written by the author in a personal capacity and not as a staff member of the National Physical Laboratory.

References

Abramowitz, A., and Stegun, I., eds. (1964). "Handbook of Mathematical Functions," Appl. Math. Ser. No. 55. Nat. Bur. Stand., Washington, D.C.

Agacy, R. L. (1964). *Proc. Phys. Soc.* **83**, 591–596.

Alben, R., Smith, J. E., Jr., Brodsky, M. H., and Weaire, D. (1973). *Phys. Rev. Lett.* **30**, 1141–1144.

Alben, R., Weaire, D., Smith, J. E., Jr., and Brodsky, M. H. (1974). *Proc. Int. Conf. Amorphous Liquid Semicond., 5th, 1973* pp. 1231–1238.

Anderson, P. W. (1958). *Phys. Rev.* **109**, 1492–1505.

Anderson, P. W. (1970). *Comments Solid State Phys.* **2**, 193–198.
Bell, R. J. (1972a). *Rep. Progr. Phys.* **35**, 1315–1409.
Bell, R. J. (1972b). *J. Phys. C* **5**, L315–L318.
Bell, R. J., and Dean, P. (1972). *Phil. Mag.* [8] **25**, 1381–1398.
Bell, R. J., and Hibbins–Butler, D. C. (1975). *J. Phys. C* **8**, 787–792.
Bell, R. J., Bird, N. F., and Dean, P. (1968). *J. Phys. C* **1**, 299–303.
Bell, R. J., Dean, P., and Hibbins–Butler, D. C. (1970). *J. Phys. C.* **3**, 2111–2118.
Bell, R. J., Dean, P., and Hibbins–Butler, D. C. (1971). *J. Phys. C* **4**, 1214–1220.
Bell, R. J., Bird, N. F., and Dean, P. (1974). *J. Phys. C* **7**, 2457–2466.
Bermudez, V. M. (1971). *J. Chem. Phys.* **54**, 4150–4159.
Bermudez, V. M. (1972a). *J. Chem. Phys.* **56**, 681–682.
Bermudez, V. M. (1972b). *J. Chem. Phys.* **57**, 2793–2799.
Bobovich, Ya. S., and Tulub, T. P. (1958). *Usp. Fiz. Nauk* **66**, 3–41.
Borland, R. E. (1963). *Proc. Roy. Soc., Ser. A* **274**, 529–545.
Bottger, H. (1974). *Phys. Status Solidi B* **62**, 9–42.
Brouers, F., and Deltour, J. (1967). *Physica (Utrecht)* **37**, 139–144.
Champier, G., Janot, C., and Deviot, B. (1966). *Phys. Status Solidi* **15**, 277–290.
Dean, P. (1960). *Proc. Roy. Soc., Ser. A* **254**, 507–521.
Dean, P. (1961). *Proc. Roy. Soc., Ser. A* **260**, 263–272.
Dean, P. (1964). *Proc. Phys. Soc., London* **84**, 727–744.
Dean, P. (1967). *Proc. Phys. Soc., London* **90**, 479–485.
Dean, P. (1972). *Rev. Mod. Phys.* **44**, 127–168.
Dean, P. (1974). *In* "Transfer and Storage of Energy by Molecules" (North, Burnett, and Sherwood, eds.), Vol. 4, pp. 329–403. Wiley, New York.
Dean, P., and Bacon, M. D. (1962). *Nature (London)* **194**, 541–542.
Dean, P., and Bacon, M. D. (1963). *Proc. Phys. Soc., London* **81**, 642–647.
Dean, P., and Bacon, M. D. (1965). *Proc. Roy. Soc., Ser. A* **283**, 64–82.
Dean, P., and Martin, J. L. (1960). *Proc. Roy. Soc., Ser. A* **259**, 409–418.
Dyson, F. J. (1953). *Phys. Rev.* **92**, 1331–1338.
Economou, E. N., and Cohen, M. H. (1971). *Phys. Rev. B* **4**, 396–404.
Economou, E. N., and Papatriantafillou, C. (1972). *Solid State Commun.* **11**, 197–201.
Economou, E. N., and Papatriantafillou, C. (1975). To be published.
Edwards, J. T., and Thouless, D. J. (1972). *J. Phys. C* **5**, 807–820.
Elliott, R. J., and Taylor, D. W. (1967). *Proc. Roy. Soc., Ser. A* **296**, 161–188.
Elliott, R. J., Krumhansl, J. A., and Leath, P. L. (1974). *Rev. Mod. Phys.* **46**, 465–543.
Florinskaya, V. A., and Pechenkina, R. S. (1952). *Dokl. Akad. Nauk SSSR* **85**, 1265.
Friedman, B. (1961). *Proc. Cambridge Phil. Soc.* **57**, 37–49.
Fujita, T., and Hori, J. (1972). *J. Phys. C* **5**, 1059–1066.
Fujita, T., and Hori, J. (1973). *J. Phys. C* **6**, 51–56.
Gaskell, P. H., and Johnson, D. W. (1975). To be published.
Gubernatis, J. E., and Taylor, P. L. (1971). *J. Phys. C* **4**, L94–L96.
Gubernatis, J. E., and Taylor, P. L. (1973). *J. Phys. C* **6**, 1889–1895.
Hanna, R. (1965). *J. Amer. Ceram. Soc.* **48**, 595–599.
Harrand, M. (1954). *C. R. Acad. Sci.* **238**, 784–786.
Hass, M. (1970). *J. Phys. Chem. Solids* **31**, 415–422.
Hass, M., Rosenstock, H. B., and McGill, R. E. (1969). *Solid State Commun.* **7**, 1–5.
Hori, J. (1968). "Spectral Properties of Disordered Chains and Lattices." Pergamon, Oxford.
Katsura, S., Morita, T., Inawashiro, S., Horiguchi, T., and Abe, Y. (1971). *J. Math. Phys.* **12**, 892–895.
Kirkpatrick, S. (1973). *Rev. Mod. Phys.* **45**, 574–588.

Krishnan, R. S. (1953). *Proc. Indian Acad. Sci.* **37**, 377–384.
Last, B. J., and Thouless, D. J. (1974). *J. Phys. C* **7**, 699–715.
Leadbetter, A. J., and Stringfellow, M. W. (1974). *In* "Neutron Inelastic Scattering," *Proc. Grenoble Conf., 1972* pp. 501–514. I.A.E.A., Vienna.
Licciardello, D. C., and Economou, E. N. (1975). To be published.
Lifschitz, E. M. (1956). *Nuovo Cimento* **3**, Suppl. 19, 716.
Lifschitz, E. M. (1964). *Advan. Phys.* **13**, 483–536.
Lippincott, E. R., Valkenberg, A. V., Weir, C. E., and Bunting, E. N. (1958). *J. Res. Nat. Bur. Stand.* **61**, 61–70.
Lomer, W. M., and Low, G. G. (1965). *In* "Thermal Neutron Scattering" (P. A. Egelstaff, ed.), pp. 2–52. Academic Press, New York.
Ludwig, W. (1967). "Recent Developments in Lattice Theory." Springer–Verlag, Berlin and New York.
Maradudin, A. A. (1963). *In* "Astrophysics and the Many Body Problem," *1962 Brandeis Lectures in Theoretical Physics* (K. W. Ford, ed.), Vol. 2, pp. 109–320. Benjamin, New York.
Maradudin, A. A. (1965). *Rep. Progr. Phys.* **28**, 331–380.
Maradudin, A. A., Montroll, E. W., and Weiss, G. H. (1963). *Solid State Phys., Suppl.* **3**.
Martin, J. L. (1961). *Proc. Roy. Soc., Ser. A* **260**, 139–146.
Mazur, P., Montroll, E. W., and Potts, R. B. (1956). *In* "3rd Berkeley Symposium on Mathematical Statistics and Probability," Vol. 3. Univ. of California Press, Berkeley.
Miler, M. (1968). *Czech. J. Phys. B* **18**, 354–362.
Montroll, E. W. (1956). *In* "3rd Berkeley Symposium on Mathematical Statistics and Probability," Vol. 3. Univ. of California Press, Berkeley.
Montroll, E. W., and Potts, R. B. (1955). *Phys. Rev.* **100**, 525–543.
Montroll, E. W., and Potts, R. B. (1956). *Phys. Rev.* **102**, 72–84.
Payton, D. N., III. (1966). "Dynamics of Disordered Harmonic Lattices," Report LA-3510. Los Alamos Sci. Lab., New Mexico.
Payton, D. N., III, and Visscher, W. M. (1966). "Dynamics of Disordered Harmonic Lattices in One, Two and Three Dimensions," Report LA-3471-MS. Los Alamos Sci. Lab., New Mexico.
Payton, D. N., III, and Visscher, W. M. (1967a). *Phys. Rev.* **154**, 802–811.
Payton, D. N., III, and Visscher, W. M. (1967b). *Phys. Rev.* **156**, 1032–1038.
Payton, D. N., III, and Visscher, W. M. (1968). *Phys. Rev.* **175**, 1201–1207.
Piseri, L., and Zerbi, G. (1968). *Chem. Phys. Lett.* **2**, 127–131.
Sah, P., and Srivastava, K. P. (1970). *Physica (Utrecht)* **45**, 537–545.
Schmidt, H. (1957). *Phys. Rev.* **105**, 425–441.
Shawyer, R. E., and Dean, P. (1972). *J. Phys. C* **5**, 1028–1037.
Simon, I. (1957). *J. Amer. Ceram. Soc.* **40**, 150–153.
Taylor, D. W. (1967). *Phys. Rev.* **156**, 1017–1029.
Thouless, D. J. (1972). *J. Non-Cryst. Solids* **8–10**, 461–469.
Thouless, D. J. (1974). *Phys. Rep.* **13***C*, 94–142.
Visscher, W. M. (1972). *J. Non-Cryst. Solids* **8–10**, 477–484.
Weaire, D., and Alben, R. (1974). *J. Phys. C* **7**, L189–L191.
Wilkinson, J. H. (1965). "The Algebraic Eigenvalue Problem." Oxford Univ. Press (Clarendon), London and New York.
Wong, J., and Angell, C. A. (1971). *Appl. Spectrosc. Rev.* **4**, 155–232.
Wong, J., and Angell, C. A. (1974). "Vitreous State Spectroscopy." Dekker, New York.
Zerbi, G., Piseri, L., and Cabasi, F. (1971). *Mol. Phys.* **22**, 241–256.

Lattice Dynamics of Quantum Crystals

T. R. KOEHLER

IBM RESEARCH LABORATORY
SAN JOSE, CALIFORNIA

I. Introduction: Unique Aspects of Quantum Solids

HISTORICALLY, THE TERM QUANTUM solids was first applied to the solid isotopes of helium, but current usage is to consider quantum solids as differentiated from ordinary solids by the importance of zero-point energy, so that, at 0°K when all thermal excitations are frozen out, there is an appreciable motion of the particles in the solid due to quantum-mechanical effects. This requires some combination of light molecular mass and weak intermolecular binding which effectively reduces the possible candidates for quantum solids to the simple Van der Waals solids—the rare gas solids and the solid isotopes of hydrogen and helium.

The above considerations may be systematized if one characterizes each substance by its de Boer quantum parameter $\lambda = [\hbar/\sqrt{(m\varepsilon)}]/\sigma$, which is the ratio of the de Broglie wavelength of a particle of mass m and typical intermolecular binding energy ε to a typical intermolecular spacing σ in the solid. It is also a measure of the kinetic energy required to localize a molecule in the space available in the solid. The precise numerical values of λ depend on the empirical parameters ε and σ and should not be taken too seriously, but their general trend provides valuable clues to the real behavior of the simple Van der Waals solids. One expects that solid argon ($\lambda \sim 0.03$) should be quite classical, solid neon ($\lambda \sim 0.09$) marginal, and solid D_2, HD, H_2, ^{4}He, and ^{3}He (λ from 0.2 to 0.5) increasingly quantum mechanical. The ease of solidification, for example, follows this trend; in fact, helium remains a liquid down to 0°K unless approximately 25 atmospheres pressure is applied. More examples can be found in de Boer (1957) and additional comments relevant to quantum crystals in a recent review (Koehler, 1975b).

Most important from the point of view of lattice dynamics is the fact that the standard quasiharmonic approximation (QHA) breaks down for the substances solid D_2 through ^{3}He and the theory predicts that all the vibrational modes for these should be imaginary—i.e., phonons would not exist as well-defined excitations. Thus, developing a theory which predicted real phonons was the first challenge one faced in treating the lattice dynamics of quantum crystals, and an outline of such a development will be given in Section II, B.

Other challenges arose because of the fascinating, specialized nature of the problem: Of all the physical effects that must be considered in the general theory of lattice dynamics, some must be treated much more carefully for the quantum solids than for other substances and some may be more justifiably neglected altogether.

As an example of the latter, one should be able to model the simple Van der Waals solids as a collection of particles which interact through a pairwise-additive, spherically symmetric intermolecular potential $v(r)$, assume that all electronic effects are contained in this potential, and ignore atomic polarizabilities, long range coulomb interactions, electron–phonon interactions and band structure. This approximation should be more justifiable for helium than for any other substance since helium has only two tightly bound (24.5 eV) 1s electrons per atom. This model of solid helium will be the sole concern of this article, although the model has been fruitfully applied to a solid of rotationally symmetric ($J = 0$) hydrogen molecules and to other simple Van der Waals solids.

The phase diagrams for solid ^{3}He and ^{4}He are shown in Fig. 1a and b. Features of special relevance to lattice dynamics are the existence of bcc, fcc, and hcp phases for each substance and the possibility of changing the molar volume by a factor of two by the application of experimentally realizable pressures. Thus the density dependence of a calculation can be tested over a wide range.

An additional simplifying feature for solid helium is that at all pressures, it melts at a temperature low compared to the Debye temperature. Thus the

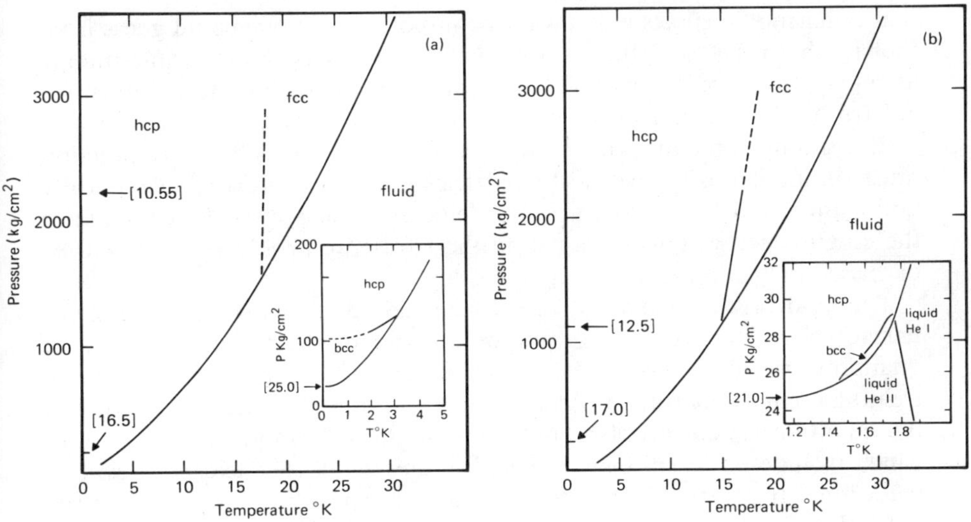

FIG. 1. (a) Phase diagram for ^{3}He. (b) Phase diagram for ^{4}He. Both diagrams are after Dugdale (1965). The numbers in square brackets indicate approximate molar volumes in cubic centimeters per mole at 0°K from data in Wilks (1967).

lattice dynamics at 0°K for the appropriate specific volume should be adequate and this assumption will be implicit throughout the remainder of this paper.

However, the very feature that makes a substance a quantum solid—appreciable zero-point motion—introduces theoretical and computational complexities. The fundamental assumption of classical lattice dynamics is that particles undergo small amplitude vibrations about their equilibrium positions and the strict harmonic approximation (HA) has the additional assumption that the equilibrium positions are close to their classical values at $T = 0$°K. The first assumption automatically leads to a picture, at the lowest level of approximation, of a set of noninteracting vibrational modes or phonons. Then at a higher level of approximation, the introduction of weak interactions between the phonons (anharmonicity) still leads to well-defined, but slightly damped, phonon modes. Around the Debye temperature, the thermally induced vibrational amplitude is large and interactions are important, but that is not of concern here. In solid helium the situation is quite different. Here the zero-point energy is such that the lattice spacing has increased to about 130% of its classically optimal value, which completely invalidates HA, and the mean square displacement of an atom from its lattice site is about 30% of the lattice spacing. Ideas based on the concept of small amplitude displacements are obviously unreliable and one anticipates that anharmonic effects will always be important. As one could guess from the fact that it barely solidifies, solid helium at 0°K is about as anharmonic as most solids are at melting. The basic expressions for the treatment of anharmonic interactions will be reviewed in Section II, C.

Still another difficulty is introduced by zero-point effects. The wavefunction must simultaneously localize the particles enough to prevent energetically unfavorable hard-core overlap and delocalize them enough to minimize the kinetic energy. This gives rise to the hard-core problem, which will be discussed in Section II, D.

The hard-core problem can actually be viewed as an extreme case of anharmonicity; however, the review of theories of the lattice dynamics of quantum crystals to be given in Section III will make clear the logic of treating them as separate problems. Since both are of central importance in the quantum crystal problem, only computationally implemented theories which include both will be considered in any detail. The exception will be variational treatments of the ground-state energy, which will be discussed in Section III, A.

The results of calculations will be compared with experiment in Section IV and a perspective on the current status of the theory of the lattice dynamics of quantum crystals and some recommendations for future work will be given in Section V.

In this article, I will review virtually all of the expressions that are significant in the computational implementation of the theories that meet the criteria given above. The expressions will be explained, but derivations will be avoided. Familiarity with the general concepts and terminology of lattice dynamics is assumed. Programming and computer time considerations for various calculations will be given. Within these bounds, the coverage of topics will be thorough, but somewhat superficial, and will be primarily theoretical. More details can be found in the original works cited here and a broader perspective on quantum crystals is available in recent primarily theoretical reviews by Glyde (1974), Koehler (1975b), and Guyer (1969b) and primarily experimental reviews by Trickey *et al.* (1972a), Wilks (1967), and Dugdale (1965). The older review of Domb and Dugdale (1957) is an excellent survey of the entire field at that time.

II. Necessary Theoretical Tools

A. Formulation of the Problem

The Hamiltonian for the model described in the introduction in units of ε and σ is

$$\mathscr{H} = -\tfrac{1}{2}\lambda^2 \sum \nabla_\alpha(i)\nabla_\alpha(i) + \tfrac{1}{2}\sum_{i \neq j} v(ij), \tag{1}$$

where a fairly standard form of lattice dynamics notation and a few additional conventions have been used: $v(ij) = v[r(ij)]$ with $r(ij) = |\mathbf{r}(i) - \mathbf{r}(j)|$, where $\mathbf{r}(i)$ is the vector denoting the position of the ith molecule and its α-Cartesian component is $r_\alpha(i)$, gradients are designated by $\nabla_\alpha(i) = \partial/\partial r_\alpha(i)$, a summation sign which exhibits the summation indices implies summation over the entire range of the indices and a summation sign which does not exhibit summation indices is used as a modified form of the Einstein summation convention and indicates that all repeated indices are to be summed over their entire range. These conventions will be followed throughout this text. The notation introduced above is specialized to the case of one particle per unit cell, the necessary modifications for the treatment of hcp solid helium are discussed in Gillis *et al.* (1968).

A useful universal potential is the well-known Lennard–Jones potential

$$v(r) = 4\varepsilon[(\sigma/r)^{12} - (\sigma/r)^6], \tag{2}$$

which is illustrated in Fig. 2a, where it is clear that the depth of the potential well is ε and the hard core radius—the distance at which the potential is

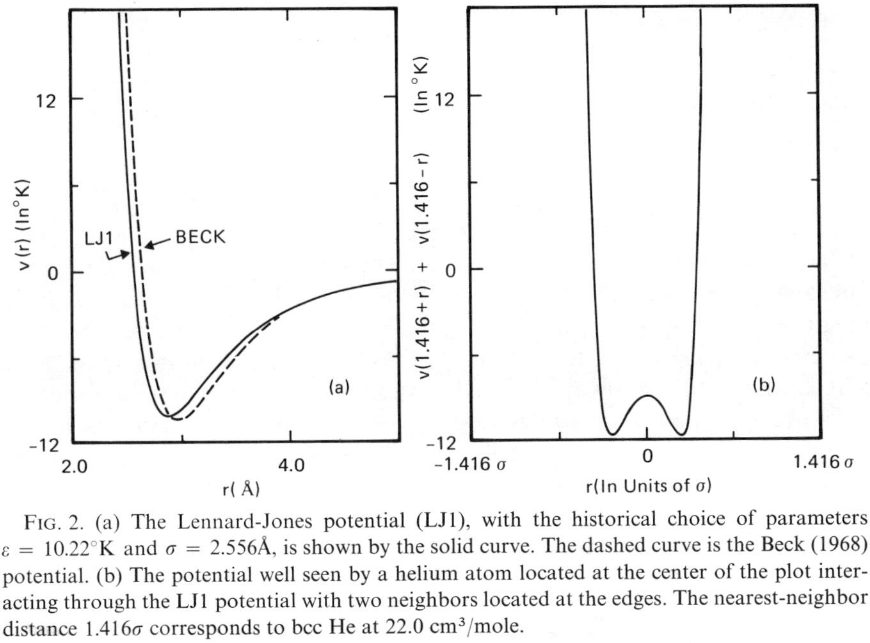

FIG. 2. (a) The Lennard-Jones potential (LJ1), with the historical choice of parameters $\varepsilon = 10.22°$K and $\sigma = 2.556$Å, is shown by the solid curve. The dashed curve is the Beck (1968) potential. (b) The potential well seen by a helium atom located at the center of the plot interacting through the LJ1 potential with two neighbors located at the edges. The nearest-neighbor distance 1.416σ corresponds to bcc He at 22.0 cm^3/mole.

zero and below which it rapidly becomes strongly repulsive—is σ. The spirit in which one uses such a potential is to assign values to ε and σ for each substance so that selected calculated properties agree with experimental data. A customary procedure is to fit gas phase virial coefficients. One then hopes to calculate the remaining properties of the substance, for example the lattice dynamics, using this potential and the appropriate calculational technique. The historical choice of Lennard–Jones parameters for helium is $\varepsilon = 10.2°\text{K}$ and $\sigma = 2.556\text{Å}$, which leads to $\lambda(^3\text{He}) = 0.492$ and $\lambda(^4\text{He}) = 0.426$.

The use of the Lennard–Jones potential will be tacitly assumed for most of this article, although a few remarks on better potentials for helium, including a seemingly better choice of Lennard–Jones parameters, will be made in Section IV, A. The philosophy is that the same theoretical and computational methods which can reliably evaluate the properties of a Lennard–Jones quantum solid will work for any other reasonable intermolecular potential.

It is clear from Eq. (1) that the importance of quantum mechanics is controlled by λ; as $\lambda \to 0$, the kinetic energy term in the Hamiltonian is turned off and the system becomes increasingly classical. Another feature of quantum mechanics is ignored here—the symmetry properties of the wavefunction. These will be unimportant here because the exchange energy is small compared to energies of interest in lattice dynamics.

B. Self-Consistent Phonon Theory

As a start toward understanding the lattice dynamics of quantum crystals, one would like to have some kind of model harmonic Hamiltonian

$$H = -\tfrac{1}{2}\lambda^2 \sum \nabla(i)^2 + V_0 + \tfrac{1}{2}\sum u_\alpha(i)\Phi_{\alpha\beta}(ij)u_\beta(j) \tag{3}$$

as an approximation to the Hamiltonian (1) in order to have some estimate of the phonon properties of the crystal and to obtain a basis set of noninteracting phonons. This Hamiltonian has a kinetic energy term in which the mass is represented by λ, a static term V_0, and a term quadratic in the displacements $\mathbf{u}(i) = \mathbf{r}(i) - \mathbf{R}(i)$ of the particles from their equilibrium positions $\mathbf{R}(i)$. An italic H will always be used to designate a harmonic Hamiltonian, as in (3), and a script $\mathscr{H}$ to designate a true crystal Hamiltonian, as in (1). The $\Phi_{\alpha\beta}(ij)$ are commonly called force constants and they are elements of the force constant matrix Φ.

The phonon modes associated with (3) may be obtained by a standard procedure. A normal coordinate transformation

$$Q_j(\mathbf{q}) = \frac{1}{\sqrt{N}} \sum e_\alpha \begin{pmatrix} \mathbf{q} \\ j \end{pmatrix} u_\alpha(i) \exp[-i\mathbf{q} \cdot \mathbf{R}(i)] \tag{4}$$

is made on the $\mathbf{u}(i)$ and similarly on the $\mathbf{V}(i)$. The exponential on the right-hand side of (4) is dictated by the translational properties of the crystal lattice and this part of the transformation alone is sufficient to partially diagonalize the force-constant matrix into a set of smaller (3×3 for the case of one atom per unit cell) matrices, one for each $\mathbf{q}$ value,

$$\mathbf{D}(q) = (\lambda^2/N)\sum \mathbf{\Phi}(ij) \exp[i\mathbf{q} \cdot \mathbf{R}(ij)] \tag{5}$$

which are called the dynamical matrices. The wave vectors $\mathbf{q}$ are members of the lattice which is reciprocal to the space lattice defined by the $\mathbf{R}(i)$.

If one now chooses the $\mathbf{e}\binom{\mathbf{q}}{j}$ to be eigenvectors of $\mathbf{D}(\mathbf{q})$ so that

$$\sum_{\beta} D_{\alpha\beta}(\mathbf{q}) e_{\beta}\begin{pmatrix}\mathbf{q}\\ j\end{pmatrix} = e_{\alpha}\begin{pmatrix}\mathbf{q}\\ j\end{pmatrix} \omega_j(\mathbf{q})^2 \tag{6}$$

is satisfied, (3) completely decouples into a set of independent harmonic oscillator equations in the $Q_j(\mathbf{q})$, with three branches (corresponding to $j = 1, 2, 3$) for each $\mathbf{q}$ value. The phonon frequencies are the $\omega_j(\mathbf{q})$ and their polarization vectors are the $\mathbf{e}\binom{\mathbf{q}}{j}$. Thus, the squares of the phonon frequencies are proportional to the roots of the $\mathbf{\Phi}$ matrix.

The QHA prescription for obtaining the model is to expand the potential energy term $V = \frac{1}{2}\sum_{i \neq j} v(ij)$ of (1) in a Taylor series in the $\mathbf{u}(i)$ and truncate the expansion at the quadratic term. The term linear in the displacements vanishes because the expansion is about an equilibrium point. This leads to the well-known results that V_0 in (3) is the equilibrium value of V and the force constants are the equilibrium value of the second derivatives of V.

When applied to the quantum crystals, this prescription immediately gets one into trouble. In these crystals, the zero-point motion leads to an equilibrium position in which nearest neighbors are separated by a distance greater than the inflection point in the potential. The situation is illustrated in Fig. 2b where the potential felt by a helium atom along the line of centers between two diametrically opposed nearest neighbors is shown.

The equilibrium position of the atom is the center of the drawing. It is clear that the equilibrium value of second derivative of the potential and therefore the force constant is negative. The frequency, which is proportional to the square root of the force constant, is imaginary. This illustration and the subsequent discussion give an accurate guide to the situation in three dimensions. It was shown by de Wette and Nijboer (1965) that every frequency in the entire Brillouin zone was imaginary for solid He. They used a Lennard–Jones potential, but the result would hold for any physically realistic potential.

The way out of this difficulty was first suggested by Born (1951), and applications to a simple, one-dimensional model were made by Hooton (1955a, b). However, the method was not used by others then and was virtually forgotten—probably because its computational implementation in the case of a real, three-dimensional crystal could not have been achieved at that time.

Some time later, the original method and several variants were independently developed by Ranninger (1965), Boccara and Sarma (1965), Choquard (1967), in a series of papers by Plakida and Siklós which are referenced in Siklós (1972), Fredkin and Werthamer (1965), Koehler (1966b), and Horner (1967). The latter three were immediately computationally implemented and applied to real three-dimensional Van der Waals solids. By now there are many additional developments of lattice dynamics which depart from the QHA and these will not be cited individually. For terminology here, I will use self-consistent phonon theories (SPT) to designate any or all of these and will use self-consistent harmonic approximation (SHA) for a particular approach whose relationship to the QHA will become apparent shortly.

The logic which leads to the SHA is as follows: Because one is interested in a theory that contains phonons, one should not abandon the idea of modeling (1) by a harmonic Hamiltonian (3). Rather one should find an alternative way for relating the force constant matrix to the potential V. The simplest approach is variational. One starts with the ground-state eigenfunction of the now assumed arbitrary harmonic Hamiltonian (3), which, when written in coordinate space, has the form of a correlated Gaussian (CG) wavefunction

$$|0\rangle \propto \exp[-\tfrac{1}{2}\textstyle\sum u_\alpha(i)G_{\alpha\beta}(ij)u_\beta(j)], \tag{7}$$

where the matrix relationship $G^2 = \Phi$ holds. The bra and ket symbols $|0\rangle$ and $\langle 0|$ will frequently and exclusively be used to denote the normalized ground-state eigenfunction of (3).

One can then obtain a variational estimate for the ground-state energy as

$$\mathscr{E}_0(G) = \tfrac{1}{4}\lambda^2 \operatorname{Tr} G + \langle 0|V|0\rangle. \tag{8}$$

The algebra necessary to satisfy the variational condition $\partial\mathscr{E}_0(G)/\partial G_{\alpha\beta}(ij) = 0$ can readily be carried out and one finds that the variationally optimal harmonic model for (1) is (3) with

$$V_0 = \langle 0|V|0\rangle \tag{9}$$

and

$$\Phi_{\alpha\beta}(ij) = \langle 0|\nabla_\alpha(i)\nabla_\beta(j)V|0\rangle. \tag{10}$$

The latter is the self-consistent equation. The self-consistency arises because the force constant matrix is involved explicitly on the left-hand side of (10) and implicitly on the right-hand side through (7) and the relationship between G and Φ.

The appearance of ground-state average rather than equilibrium values is eminently sensible from the point of view of quantum mechanics. One should bear in mind that quantum mechanics does not enter into the usual derivations of QHA at all, it only enters later through the quantum statistics of the phonons. It has also been noted by Koehler (1968) that (9) and (10) result if one expands the potential in a hermite polynomial (the natural set of functions for the harmonic oscillator problem) rather than a Taylor series basis set. In the classical limit $\lambda \to 0$, $|0\rangle$ becomes an array of delta functions centered at the equilibrium positions and (9) and (10) become the equilibrium values of the quantities involved. The prescription for obtaining phonon frequencies implied by Eqs. (5), (6), and (10) is specific to the SHA; however, similar expressions are encountered in other modern formulations of lattice dynamics.

The numerical implementation of the SHA is one or two orders of magnitude more difficult than for the QHA, despite the surface similarity of the two. As such, the implementation requires the use of electronic computers. Since V is a sum of two body terms, the evaluation of (9) and (10) involves the use of the normalized harmonic pair function, which is defined as

$$g_h(12) = \int \Psi_h(1, 2, 3, \ldots, N)^2 \, d\mathbf{r}_3 \cdots d\mathbf{r}_N \tag{11}$$

where $\Psi_h = |0\rangle$, and 1 and 2 instead of i and j have been used to designate two arbitrary particles. This convention will be used whenever it is convenient. It has been pointed out (Koehler, 1966a; Choquard, 1967) that g_h is a CG function involving the difference coordinates $u(ij)$ whose explicit form is

$$g_h(il) = [\pi^3|\mathbf{\Lambda}(il)|]^{-1/2} \exp\{-\mathbf{u}(il) \cdot [\mathbf{\Lambda}(il)^{-1}] \cdot \mathbf{u}(il)\} \tag{12}$$

where

$$\mathbf{\Lambda}(il) = \frac{2\lambda^2}{N} \sum_{\mathbf{q}j} \mathbf{e}\begin{pmatrix}\mathbf{q}\\ j\end{pmatrix} \mathbf{e}\begin{pmatrix}\mathbf{q}\\ j\end{pmatrix} [\omega_j(\mathbf{q})]^{-1}\{1 - \cos[\mathbf{q} \cdot \mathbf{R}(il)]\} \tag{13}$$

is a 3×3 matrix whose inverse is $\mathbf{\Lambda}(ij)^{-1}$.

A SHA calculation involves an iterative cycle: One starts with a guess for a set of force constants, for example, the QHA set. Then one finds the phonon frequencies and eigenvectors at a mesh of points in $\mathbf{q}$ space throughout the irreducible Brillouin zone in order to obtain the $\mathbf{\Lambda}(ij)$. Thus one cannot get by with frequencies along symmetry directions only as is usually the case in QHA calculations. It is possible that approximation schemes for evaluating the $\mathbf{\Lambda}(ij)$ could be devised, but the original SHA development took place in the machine computation era and such schemes were never pursued.

The pair functions are then constructed from the $\mathbf{\Lambda}(ij)$ and the integrals implied in (9) and (10) are evaluated numerically. Any standard method for performing numerical integrations can be used or a specialized method (this has been described in Koehler, 1975a) which involves Taylor series expanding the potential and evaluating the CG average of the resulting polynomials analytically. Along some symmetry directions the $\mathbf{\Lambda}(ij)$ are cylindrically symmetric and the integrals simplify. In general, however, a three-dimensional integral must be performed.

Each iteration cycle starts with one set of force constants and results in a set of average values of $\nabla\nabla V$. Some linear combination of these can be used as a new choice of force constants. The cycle converges when the new and the old force constants agree well enough. Usually around five cycles are sufficient. On a modern fast computer like the IBM 360/195, the entire calculation takes only a few seconds for a simple, nonquantum solid like solid argon.

The SHA does lead to positive definite frequencies for classically unstable potentials like in Fig. 2b. However, it is only a step along the way to a full theory of quantum solids. Other considerations, which will be discussed in the remainder of this section, make the SHA an incomplete theory for quantum solids.

C. Anharmonicity

The true crystal Hamiltonian (1) consists of the model harmonic Hamiltonian (3) plus the difference between the two ($\mathscr{H} - H$). This difference gives rise to anharmonic interactions between the otherwise undamped, harmonic phonons and to perturbative corrections to the ground-state energy. The total anharmonic term is usually broken up into terms introducing interactions between a particular number n of phonons which provides an n-phonon vertex in a diagrammatic perturbation expansion.

In QHA or SHA the terms are obtained by continuing the Taylor series or hermite polynomial expansion of the potential. The vertices are proportional to λ^n while, as can be seen from (5) and (6), the frequencies, which will enter into energy denominators, are proportional to λ. Thus, perturbative corrections are ordered both with respect to the number of phonons involved and to their order in perturbation theory.

The leading corrections are expected to be four-phonon interactions to first order and three-phonon interactions to second order. These are usually called the quartic and cubic contributions, respectively. The quartic term will not be considered here; it does not appear in the SHA, where it is incorporated into the lowest order terms, or in any theory of the lattice dynamics of quantum crystals.

In SHA, the cubic term makes a second-order contribution to the ground-state energy of

$$\delta\mathscr{E}_0^{(3)} = -\frac{1}{6}\sum \frac{|\langle 0|\mathscr{H}|123\rangle|^2}{\omega(1)+\omega(2)+\omega(3)} \tag{14}$$

where an obvious notation has been used to indicate matrix elements of $\mathscr{H}$ between normalized phonon eigenstates, 1 is used to indicate $\mathbf{q}_1$, j_1, etc. The matrix elements involve the cubic force constants

$$\Phi(123) = \langle 0|\nabla(1)\nabla(2)\nabla(3)V|0\rangle \tag{15}$$

in the following way:

$$\langle 0|\mathscr{H}|123\rangle = \sum_{i_1 i_2 i_3} \prod_{l=1,3} \left[\left(\frac{\lambda^2}{2N\omega(l)}\right)^{1/2} \exp[i\mathbf{q}_l\cdot\mathbf{R}(il)]\mathbf{e}(l)\right]\cdot\mathbf{\Phi}(i_1 i_2 i_3), \tag{16}$$

where centered dots (·) between boldface quantities indicate summation over cartesian indices. In (16) one of the summations over lattice vectors gives a factor $N\Delta(\mathbf{q}_1+\mathbf{q}_2+\mathbf{q}_3)$, where $\Delta(\mathbf{q}) = 0$ unless $\mathbf{q} = 0$ or a reciprocal lattice vector. This will not be exhibited explicitly in any of the anharmonic expressions.

Of greater importance to lattice dynamics, the three-phonon terms make a contribution to the phonon self energy

$$M_{jl}(\mathbf{q},\Omega) = -\Delta_{jl}(\mathbf{q},\Omega) + i\Gamma_{jl}(\mathbf{q},\Omega) \tag{17}$$

of

$$\begin{Bmatrix}\Delta_{jl}(\mathbf{q},\Omega)\\ \Gamma_{jl}(\mathbf{q},\Omega)\end{Bmatrix} = -\frac{1}{2}N\sum\left\langle \begin{matrix}-\mathbf{q}\\ j\end{matrix}\middle|\mathscr{H}\middle|12\right\rangle\left\langle 12\middle|\mathscr{H}\middle|\begin{matrix}\mathbf{q}\\ j\end{matrix}\right\rangle \times\begin{Bmatrix}P[\omega(1)+\omega(2)+\Omega]^{-1} + P[\omega(1)+\omega(2)-\Omega]^{-1}\\ \pi\,\delta[\omega(1)+\omega(2)+\Omega] - \pi\,\delta[\omega(1)+\omega(2)-\Omega]\end{Bmatrix} \tag{18}$$

where P means that the principal value is to be taken and the matrix elements in (18) are obtained from (16) and the relationship

$$\left\langle \begin{matrix} -\mathbf{q} \\ j \end{matrix} \middle| \mathscr{H} \middle| 12 \right\rangle = \left\langle 0 \middle| \mathscr{H} \middle| \begin{matrix} \mathbf{q} \\ j \end{matrix}\, 12 \right\rangle.$$

The self-energy enters into the phonon Green's function $G(\mathbf{q}, \Omega)$ in the usual way:

$$\sum_k \left\{ \left[\omega_j(\mathbf{q})^2 - \Omega^2 \right] \delta_{jk} - 2\omega_j(\mathbf{q}) M_{jk}(\mathbf{q}, \Omega) \right\} G_{kl}(\mathbf{q}, \Omega) = 2\omega_j(\mathbf{q})\, \delta_{jl}. \quad (19)$$

For vanishing anharmonicity, the Green's function is a δ function at the bare frequency. When the anharmonicity is small, $M_{jl}(\mathbf{q}, \Omega) \ll \omega_j(\mathbf{q})$, the frequency dependence of M may be ignored. The real and imaginary parts, evaluated at the bare frequency, provide, respectively, a shift in the energy and a linewidth or lifetime. In the case of the quantum crystals where the anharmonicity is large, one must evaluate M as a function of Ω and then plot the spectral function $A(\mathbf{q}, \Omega) \propto \operatorname{Im} G(\mathbf{q}, \Omega)$, which is closely related to what one measures in an inelastic neutron scattering experiment and therefore provides a convenient meeting ground for the confrontation of theory and experiment. This will be discussed in more detail in Section V.

Strictly speaking, the concept of anharmonicity only has validity in a perturbative theory based on a model Hamiltonian. However, the general form of Eqs. (14)–(19) are appropriate to any theory of interacting phonons that is dominated by three-phonon processes. This is likely to be the case in any system that has a well-defined phonon like excitation. Thus one should anticipate that the framework provided by these equations will be valid for quantum crystals but the interpretation of the physical quantities should be liberalized: bare phonons will not necessarily arise from a harmonic Hamiltonian and may have lifetimes already and anharmonicity will simply refer to interactions between phonons.

One reason for trying to remain within the formalistic framework provided by the QHA with anharmonic corrections is that it is conceptually clear. Another is that, at the time when interest in quantum crystals was increasing, the computational implementation of anharmonic calculations was becoming more commonplace. Thus a theory that had the right general form could be plugged into existing computer programs.

Evaluation of the anharmonic expressions is by far the most time consuming part of any complete lattice dynamics calculation. One is typically interested in evaluating either the energy correction or plotting the spectral function vs. Ω for several $\mathbf{q}$ vectors along symmetry directions. Either takes two to three orders of magnitude more computer time than a SHA calculation. Compared to this, the time required to evaluate the matrix elements according to the prescriptions of the various theories of Section III is not significant although the programming effort may be large.

D. Hard-Core Problems

The SHA, even with anharmonic corrections, is not a useful, complete theory for quantum crystals. Consider evaluating the ground-state average of the Lennard–Jones potential or its derivatives. The potential diverges while the ground-state wavefunction is a Gaussian and never vanishes completely where the potential is infinite. Thus all the matrix elements that enter the theory are divergent and the theory does not work.

In reality, the divergence is only important in quantum solids. The width of the ground-state wavefunction is determined from the balance of kinetic and potential energies that minimized the total energy. In even a moderately quantum-mechanical substance like solid neon, the width is narrow enough so that the divergence occurs in a quite small region in which the squared wavefunction is many orders of magnitude below its maximum. The divergence can be eliminated by simply flattening out the potential for, say, $r \lesssim 0.1\sigma$. This is not mathematically rigorous, but can have no observable effect on the physical properties of the system.

The escape clause based on the grounds of physical reasonableness does not hold for the quantum crystals where an unreasonable expenditure of kinetic energy is required to avoid appreciable hard-core overlap. The problem is unavoidable, since the large zero-point energy gives the particle a somewhat gaslike behavior and hard-core collisions will occur frequently.

The theories that will be discussed in the next section are the results of attempts to blend the more traditional ideas of Sections II, B and C with some mechanism for treating the hard core. These theories actually provide for the first time a rigorously convergent theory of lattice dynamics for the Lennard–Jones solid. The QHA is also formally divergent because the Taylor series expansion of a singular potential diverges if carried out to enough terms. This fact can normally be overlooked. It is only for the quantum solids that divergences associated with harmonic theories are encountered immediately.

A treatment of the hard-core problem that has not yet led to anharmonic calculations and therefore will not be discussed in Section III is the application of point transform theory by Trickey *et al.* (1972b, 1973). Here one makes a coordinate transformation that removes the hard core at the expense of complicating the potential. The results of some moderately successful numerical applications of the method are presented in the references.

III. Theories

A. Jastrow Function Variational

The theory of Nosanow (1964, 1966) was a landmark in the computation of properties of the quantum crystals: it led to the first quantitively reliable numerical evaluation of the ground-state energy of solid helium. Furthermore, it was a factor in the first successful calculations of the lattice dynamics of quantum crystals, because its development coincided in time with derivations of generalized theories of lattice dynamics and it provided a convenient vehicle for their application to solid helium. This will be discussed in Section III, B. Here, the conceptual basis of the theory and its application to ground-state energy calculations will be treated.

The theory is based on the use of a variational wavefunction

$$\Psi_0 = \Psi_J \Psi_h \tag{20}$$

which is the product of a Jastrow function* whose square is

$$\Psi_J^2 = \prod_{i<j} f(ij), \qquad f(r \to 0) \to 0, \qquad f(r \to \infty) \to 1 \tag{21}$$

and another function which will be restricted to be a harmonic function here. The original theory of Nosanow used a noncorrelated Gaussian (NCG) wavefunction which can be obtained by setting $G_{\alpha\beta}(ij) = A\,\delta_{\alpha\beta}\,\delta_{ij}$ in (7). This is a Hartree-type wavefunction and is the product of independent Gaussians in each particle coordinate. The treatment in this section will assume a NCG with A chosen variationally; the switch to a CG will be made in Section III, B. In (20), the harmonic part provides the lattice structure and the property $f(r \to 0) \to 0$ introduces short-range correlation and prevents hard-core overlap.

If one uses the hermiticity of $-i\nabla_\alpha(i)$, the variational estimate for the ground-state energy $\mathscr{E}_0 = \langle\Psi_0|\mathscr{H}|\Psi_0\rangle/\langle\Psi_0|\Psi_0\rangle$ may be manipulated into the form

$$\mathscr{E}_0 = \tfrac{1}{4}\lambda^2 \operatorname{Tr} G + \frac{\langle 0|V_{\text{eff}}|0\rangle}{\langle 0|\Psi_J^2|0\rangle}, \tag{22}$$

where

$$V_{\text{eff}} = \Psi_J^2[V - \tfrac{1}{4}\lambda^2 \textstyle\sum \nabla(i)\cdot\nabla(i)\ln\Psi_J]. \tag{23}$$

* This function was originally (Jastrow, 1955) used in a nuclear physics context. A discussion of its properties appropriate to the context of quantum solids or quantum liquids can be found in Nosanow (1966) or Feenberg (1969), respectively.

Since $f(ij)$ cuts off the potential as $r(ij) \to 0$, the ground-state energy involves a harmonic ground-state average of an effective potential V_{eff}, which involves not only a softened potential $f(r)v(r)$, but also an additional term which contributes positively and represents the price in kinetic energy one must pay for softening the potential.

Two methods have been used to evaluate Eq. (22). (1) The integrals may be done exactly by Monte Carlo (MC) integration. This approach will be treated in Section IV, A. (2) A cluster expansion may be used. The working expressions for the lowest order cluster expansion will be reviewed in the following because they have been used extensively in numerical work on solid helium. The development of the cluster expansion and many attendant technical details have been discussed by Nosanow (1966) and Guyer (1969b). In addition, some higher order cluster expansion calculations have been reported by Hetherington *et al.* (1967).

Since the term in square brackets in (23) is a sum of pair terms, the contribution from V_{eff} to (22) may be evaluated from a knowledge of the pair function $g_J(ij)$ for Ψ_J, where g_J is defined by an obvious expression analogous to (11). The cluster expansion may be viewed as a systematic means for obtaining g_J without performing the full many-body integration. Nosanow's expansion is based on one by Van Kampen (1961) and, in lowest order, is simply

$$g_J(ij) = f(ij)g_h(ij). \tag{24}$$

The rationale for this approximation is that g_h is the pair function for Ψ_h and it will not be affected too much by integrating over Ψ_J too, since the Jastrow function is unity for most of the range of the integrations. However, $f(ij)$ must dominate the short-range behavior and this is inserted explicitly. The above is overly simplified, but it does contain the basic idea. In practice it is found that the cluster expansion sometimes works and sometimes does not, a point which will be amplified in Section IV, A.

B. Correlated Basis Function

The first merger of the SHA concepts with the Jastrow-variational formalism resulted in the theories of Koehler (1967, 1968) and Horner (1967). The extension to include anharmonic effects was made by Koehler and Werthamer (1971) who noted the parallel between their treatment of solid helium and the correlated basis function (CBF) method that had been extensively developed and applied to liquid helium by Feenberg and coworkers. A comprehensive review of the latter may be found in Feenberg (1969).

For solids, the basic idea in the CBF method is that one can generate a complete set of states for use in a matrix approach to quantum mechanics by orthogonalizing functions that are products of polynomials in the normal coordinates $Q_j(\mathbf{q})$ and a function that should be a reasonable approximation to the ground-state wavefunction such as (20). One hopes that the polynomial of degree n will be a good approximation to an n-phonon state. For the case $f(ij) = 1$, the identification is exact. If Ψ_h is the appropriate CG, Ψ_0 becomes the true harmonic ground-state wavefunction and the polynomials are Hermite polynomials in the $Q_j(\mathbf{q})$.

Having a prescription for a complete set of states, one can evaluate matrix elements of $\mathscr{H}$ and, in principle, obtain all the properties of the system. Thus the CBF procedure, if carried to high enough order, is rigorous, but practical considerations limit one to the computational framework described in Section II, C. Then the CBF approach is justified if the ansatz for the ground-state wavefunction and the method used for constructing approximate excited states are good enough. The tests of these criteria are agreement of calculations with experiment; internal consistencies, such as corrections to lowest order quantities being small; and possible estimates to indicate the smallness of higher order terms. As will be seen in Section IV, B, the agreement between the CBF calculations and experiment is satisfactory so it should be considered a valid approach. The key equations of the theory will be written and discussed here, but details of the derivation will be omitted.

It is obvious that one would like to variationally optimize Ψ_0 in some way. As will be discussed in Section IV, A, the choice of f is somewhat restricted, so let us consider optimizing with respect to G. The result

$$\mathbf{\Phi}(ij) = \frac{\langle 0|\nabla(i)\nabla(j)V_{\text{eff}}|0\rangle}{\langle 0|\Psi_J{}^2|0\rangle} - \frac{\langle 0|\nabla(i)\nabla(j)\Psi_J{}^2|0\rangle\langle 0|V_{\text{eff}}|0\rangle}{\langle 0|\Psi_J{}^2|0\rangle^2} \tag{25}$$

is similar to (10). The major difference is the replacement of V by V_{eff}, since the second term on the right-hand side of (25) proves to be small. The CBF recipe for a normalized one-phonon state is

$$\left|\begin{matrix}\mathbf{q}\\ j\end{matrix}\right\rangle = Q_j(q)\Psi_J|0\rangle$$

and the phonon energy to lowest order is

$$\nu_j(\mathbf{q}) = \left\langle\begin{matrix}\mathbf{q}\\ j\end{matrix}\right|\mathscr{H}\left|\begin{matrix}\mathbf{q}\\ j\end{matrix}\right\rangle \Big/ \left\langle\begin{matrix}\mathbf{q}\\ j\end{matrix}\middle|\begin{matrix}\mathbf{q}\\ j\end{matrix}\right\rangle - \mathscr{E}_0. \tag{26}$$

This leads to the result that the $\nu_j(\mathbf{q})$ are the roots of the inverse of the matrix whose elements are

$$\langle \Psi_0 | u_\alpha(i) u_\beta(j) | \Psi_0 \rangle / \langle \Psi_0 | \Psi_0 \rangle \tag{27}$$

or, equivalently,

$$\mathbf{G}^{-1}(ij) + \tfrac{1}{2}\lambda^2 \sum_{1,2} \mathbf{G}^{-1}(i1) \frac{\langle 0 | \boldsymbol{\nabla}(1)\boldsymbol{\nabla}(2)\Psi_J{}^2 | 0 \rangle}{\langle \Psi_0 | \Psi_0 \rangle} G^{-1}(2j), \tag{28}$$

where $\mathbf{G}^{-1}(ij)$ is the (ij) element of the matrix G^{-1}. Equation (27) can be recognized as the static expression for the one-phonon Green's function so its form is not unexpected. However, its equivalent, Eq. (28) is not so satisfying as it says that the phonon frequencies $\nu_j(\mathbf{q})$ are not the same as the $\omega_j(\mathbf{q})$ that enter into the harmonic part of the wavefunction through (7) and (25). This is a troublesome point to which we will return in Section III, 4.

The CBF theory at this level can be used to calculate the ground-state energy and dispersion curves for noninteracting phonons. Anharmonic corrections to these are obtained by constructing the restricted manifold of one, two, and three phonon states and using these to work out the matrix elements that are to be inserted in the formulas of Section II, C. Various technical details of this process are explained in Koehler and Werthamer (1972) and arguments justifying the use of the anharmonic expressions can be found there and in Gillessen and Biem (1971).

The final expressions in the lowest order cluster expansion for anharmonic corrections to the CBF theory are those of Section II, C with two modifications: (1) The third derivative of $v(ij)$ is replaced by the third derivative of

$$\tilde{v}(ij) = v_{\text{eff}}(ij) - \langle 0 | v_{\text{eff}}(ij) | 0 \rangle f(ij), \tag{29}$$

where

$$v_{\text{eff}}(ij) = f(ij)[v(ij) - \tfrac{1}{8}\lambda^2 \boldsymbol{\nabla}(i)^2 \ln f(ij)]. \tag{30}$$

(2) The $\omega_j(\mathbf{q})$ in the energy denominators are replaced by the $\nu_j(\mathbf{q})$. The $\omega_j(\mathbf{q})$ associated with nondiagonal matrix elements remain the roots of G.

C. *t*-Matrix Theories

The *t*-matrix approach, which was devised to cope with the hard-core problem in nuclear physics, has by now been applied to quantum solids by Iwamoto and Namaizawa (1966, 1971), Guyer (1969a,b), Horner (1970a,b), Ebner and Sung (1971), Mullin (1971), Østgaard (1971), Glyde and Khanna

(1971, 1972a,b), Namaizawa (1972), Brandow (1972), Homma *et al.* (1973), and Kurihara *et al.* (1974).

The reference by Brandow is a lengthy review that compares technical differences in the various t-matrix approaches. Because of this, we shall not delve too deeply into the t-matrix formalism but will discuss a simple model that is a blend of the derivations by Glyde and Khanna and by Guyer. In fact, the term "t-matrix theory" is used somewhat loosely here to denote methods that attempt to derive an effective potential by focusing attention on the hard-core interaction between individual pairs of particles, and the discussion will be one which is appropriate in the context of quantum crystals.

If one starts with a model harmonic Harmiltonian, the perturbation that turns the model into the true crystal Hamiltonian is $\frac{1}{2}\sum v_p(ij)$, where

$$v_p(ij) = v(ij) - \mathbf{u}(ij) \cdot \mathbf{\Phi}(ij) \cdot \mathbf{u}(ij). \tag{31}$$

The well-known level shift expression

$$\mathscr{E}_0 - E_0 = \left\langle 0 \middle| \frac{1}{2} \sum_{i \neq j} v_p(ij) \middle| \Psi_0 \right\rangle \Big/ \langle 0 | \Psi_0 \rangle \tag{32}$$

expresses the true ground-state energy $\mathscr{E}_0$ in terms of its shift from the harmonic ground-state energy E_0. The t matrix for the pair (ij) is now defined as

$$\langle 0 | t(ij) | 0 \rangle = \langle 0 | v_p(ij) | \Psi_0 \rangle / \langle 0 | \Psi_0 \rangle \tag{33}$$

and enters into the theory similarly to $v_{\text{eff}}(ij)$ in Eq. (30).

One sees that the t matrix is an operator that acts on the unperturbed ground-state wavefunction $|0\rangle$ to produce a function whose projection on $|0\rangle$ is the same as that of the product of the perturbation (which contains the hard core) and the true ground-state wavefunction Ψ_0 (which must cut it off). If the operator is replaced by a function, the t-matrix method can be viewed as a way of obtaining a softened, effective potential that can be used in place of the real potential so that one can work with the convenient harmonic wavefunctions and expressions.

The t matrix should be obtained from the summation of the set of diagrams in perturbation theory which account for the hard-core pairwise interaction. However, a computationally convenient approximation can be obtained from a more intuitive derivation. Suppose one starts with a harmonic Hamiltonian and replaces the harmonic potential between two atoms by the true potential obtaining the Hamiltonian

$$H'(ij) = H + v_p(ij). \tag{34}$$

Since the major effect of $v_p(ij)$ is the introduction of short-range repulsion between i and j, the ground-state eigenfunction of Eq. (34) could be approximated by

$$\Psi'_0(ij) = f(ij) \mid 0\rangle, \tag{35}$$

where $f(ij)$, which is assumed to be normalized, can be determined variationally from H'.

Returning to the original problem, one can obtain a first-order expression for the energy shift caused by replacing the harmonic by the true potential in the entire crystal as

$$\mathscr{E}_0 - E_0 \approx \tfrac{1}{2} \sum_{i \neq j} \langle 0|v^t_{\text{eff}}(ij)|0\rangle \equiv \tfrac{1}{2} \sum_{i \neq j} \langle 0|t(ij)|0\rangle, \tag{36}$$

if one assumes that the individual contributions are additive. The quantity

$$v^t_{\text{eff}}(ij) = f(ij) v_p(ij) \tag{37}$$

can be seen from Eqs. (22) and (36) to be the effective potential appropriate to this t-matrix theory.

The above provides a method for the evaluation of the ground-state energy for a hard-core system. One can extend the theory and perform lattice dynamics calculations for quantum solids by simply replacing v by v_{eff} in the expressions of Sections II, B and C. Although this approach is not rigorously correct, the anharmonic lattice dynamics of solid helium has proved to be sufficiently difficult that intuitively appealing, numerically tractable theories are certainly worth trying. The results of such attempts will be presented in Section IV, B.

The t-matrix derivation outlined here starts from a bare phonon Hamiltonian. Most of those referred to in the beginning of this section start from a single-particle picture, which is the usual case for nuclear theories. This has the disadvantage of making the introduction of phonons more difficult and the advantage of allowing ground-state energy calculations without the use of lattice dynamics programs.

D. Fully Consistent Theories

The term "fully consistent theory" is used here to refer to theories that use renormalized phonon propagators throughout. Such theories have been developed by Ranninger (1965), Horner (1967, 1971), Choquard (1967), and Werthamer (1969, 1970a). Since the method of Horner is the only one that has been computationally implemented and applied to quantum crystals, it is the only one that will be considered here. This theory is rather complex; however, it features several judiciously introduced, physically appealing, and mathematically justified approximations which make the theory computationally tractable. Since these are important and may have wider application, our discussion will center on them and their formal underpinning will only be treated briefly.

The formal development uses the method of De Dominicis and Martin (1964). A force term $\sum_{\alpha i} U_\alpha(i, t) r_\alpha(i, t)$ is added to $\mathscr{H}$ to obtain a time-dependent Hamiltonian $\mathscr{H}(t)$. By repeatedly differentiating a generalized partition function

$$Z = \mathrm{Tr}\left\{T_\tau \exp\left[-i\int_0^{-i\beta} \mathscr{H}(t)\,dt\right]\right\} \tag{38}$$

with respect to the $U_\alpha(i, t)$, one can generate a series of n-point functions, the first two of which are the one-point function

$$d_\alpha(i, t) = -i\langle r_\alpha(i, t)\rangle \tag{39}$$

and the two-point function

$$d_{\alpha\beta}(ij, tt') = -\langle T_\tau r_\alpha(i, t) r_\beta(j, t')\rangle + \langle r_\alpha(i, t)\rangle\langle r_\beta(j, t')\rangle \tag{40}$$

where $\beta = 1/k_B T$, the average of an operator is defined in the usual way and T_τ denotes time averaging along the imaginary time axis.

We first consider the derivation of the pair function which will ultimately be used in the evaluation of matrix elements. A static approximation is eventually made for this quantity and it is simplest to temporarily suspend the time dependence of the point functions and to use the averaging symbol to mean the ground-state average.

The pair function is given by

$$g(ij) = \langle\delta[\mathbf{r}'(i) - \mathbf{r}'(j) - \mathbf{r}(ij)]\rangle \tag{41a}$$

$$= \left\langle \int \frac{d\mathbf{k}}{(2\pi)^3} \exp\{-i\mathbf{k}\cdot[\mathbf{r}'(i) - \mathbf{r}'(j) - \mathbf{r}(ij)]\}\right\rangle, \tag{41b}$$

where the averaging is with respect to r. The right-hand side of (41b) can be expressed in terms of cumulants to give

$$g(ij) = \int \frac{d\mathbf{k}}{(2\pi)^3} \exp\{-i\mathbf{k}\cdot[i\mathbf{d}(i) - \mathbf{r}(i) - i\mathbf{d}(j) + \mathbf{r}(j)]\}$$
$$\times \exp\{-\tfrac{1}{2}\mathbf{k}\cdot[\mathbf{d}(ii) - 2\mathbf{d}(ij) + \mathbf{d}(jj)]\cdot\mathbf{k}\} \times \phi(\mathbf{k}) \tag{42}$$

where $\phi(\mathbf{k})$ involves products of $\mathbf{k}$ with higher point functions. This term can be moved in front of the integral sign by replacing $\mathbf{k}$ by $i\nabla(ij)$ in the term. The integral is now a quadratic in $\mathbf{k}$ and may be performed exactly to give

$$g(ij) = \phi[i\nabla(ij)]g_h(ij) \tag{43a}$$

$$= f(ij)g_h(ij) \tag{43b}$$

where f is defined by these equations and $g_h(ij)$ is identical to (12) with the identifications

$$\mathbf{R}(ij) = i[\mathbf{d}(i) - \mathbf{d}(j)] \tag{44}$$

and

$$\Lambda(ij) = -\tfrac{1}{2}[\mathbf{d}(ii) - 2\mathbf{d}(ij) + \mathbf{d}(jj)]. \tag{45}$$

The two-point functions are the phonon propagators, so as in the SHA, the $\Lambda(ij)$ are related to the phonons. The short-range behavior of the pair function is then contained in $f(ij)$, but it is a short-range function that is uniquely separated from the harmonic part in the following way. The averages in (39) and (40) can be obtained from $g(ij)$; however, these equations become identities if g_h rather than g is used. This leads to three consistency conditions involving the zeroth, which affects only the normalization, the first

$$\int d\mathbf{r}(ij)f(ij)\mathbf{u}(ij)g_h(ij) = 0 \tag{46}$$

and the second

$$\int d\mathbf{r}(ij)\mathbf{u}(ij)\mathbf{u}(ij)f(ij)g_h(ij) = \tfrac{1}{2}\Lambda(ij) \tag{47}$$

moments of $\mathbf{u}(ij)$. These relationships are important; they essentially prohibit the short-range correlations from altering the average distance between a pair of particles or the phonon-caused, mean square width of their distribution. They should probably be imposed in the t-matrix and CBF theories. In the former, they would provide one with a unique way of finding the starting Hamiltonian in the derivation given; in the latter, Werthamer (1973) has shown that the consistency relationships are satisfied if a variationally optimized Jastrow function is used. This also removes the term in Eq. (28) that leads to the difference between ω and ν.

The power of the consistency relationships is that they allow one to determine a good approximate form for $f(ij)$ in a simple way. For example, Horner

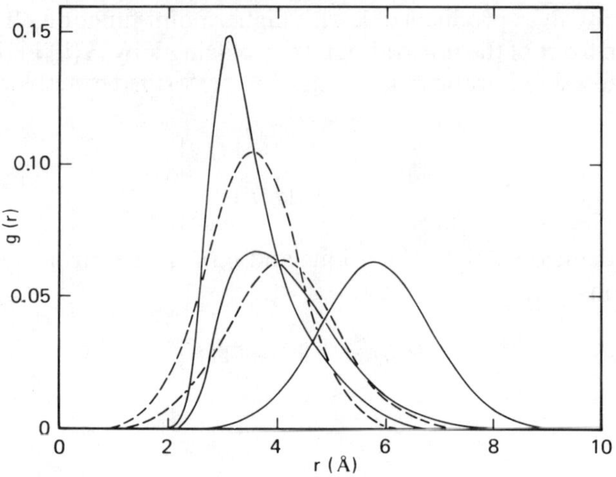

FIG. 3. The solid lines show the pair function for the first three shells of neighbors in bcc ^4He at 20.9 cm^3/mole as determined by Horner (1972a). The dashed lines, which cannot be distinguished for the third shell, give the Gaussian part alone.

used an s-wave scattering solution multiplied by a polynomial which could be adjusted to fulfill the longitudinal (the most important) components of the consistency equations. The difference between g and g_h is shown in Fig. 3 for the first three shells of bcc ^4He. This clearly shows the interference between the short-range and the harmonic part of the pair function.

In order to obtain phonon propagators, it is necessary to return to the time-dependent equations. A quantity

$$\mathbf{K}(i, t) = \sum_{j \neq i} \iint d\mathbf{r}(i)\, d\mathbf{r}(j) g(ij, t) \nabla(j) v(ij), \tag{48}$$

which is identified as the internal force on particle i, is extracted from the equation of motion for the one-point function obtained from $[\mathscr{H}(t), [\mathscr{H}(t), r(i, t)]]$. The equation of motion for the two-point function leads to an expression for the phonon self-energy

$$M_{\alpha\beta}(ij, tt') = -i \frac{\delta K_\alpha(i, t)}{\delta d_\beta(j, t')} \tag{49}$$

which plays the role of a generalized, time-dependent force constant matrix. It is interpreted as the change in the average force on particle i when the average position of particle j has been altered by the external force. A detailed analysis of M is made in Horner (1971). He first notes that $\mathbf{M}(ij, tt')$ is determined from the dependence of $g(ij, t)$ on $\mathbf{d}(j, t')$ and then finds that, if g is separated into a short-range and a harmonic part, three contributions to **M** can be distinguished:

1. An instantaneous contribution

$$\mathbf{M}^{(1)}(ij, tt) = -\int d\mathbf{r}(ij) g_h(ij, t) \boldsymbol{\nabla}(ij)[f(ij, t)\boldsymbol{\nabla}(ij)v(ij)] \tag{50}$$

from the explicit dependence of $g_h(ij, t)$ on $\mathbf{R}(ij, t)$. This reduces to $\langle \boldsymbol{\nabla}\boldsymbol{\nabla} V\rangle$ in the limit $f \to 1$.

2. A dispersive part $M^{(2)}(ij, tt')$ due to the dependence of $g_h(ij, t)$ on $\boldsymbol{\Lambda}(ij, t)$. This is a complicated term because it contains all the anharmonic corrections to the SHA theory in the limit $f \to 1$. The analogous approximation to retaining only three-phonon vertices leads to expressions resembling those of Section II, B with (15) replaced by

$$\boldsymbol{\Phi}(123) = \frac{1}{2} \sum_{i \neq j} \int d\mathbf{r}(ij) \boldsymbol{\nabla}(1) \boldsymbol{\nabla}(2)[f(ij)\boldsymbol{\nabla}(3)v(ij)] \tag{51}$$

and the full phonon propagators instead of the energy denominators of (18). A computational technique for using the propagators is described in Horner (1972a).

3. A term arising from the functional dependence of $f(ij, t)$ on $\mathbf{R}(ij, t)$ which can be reduced by certain approximations to

$$M^{(3)}_{\alpha\beta}(ij, tt) = \sum_{l=i} \int d\mathbf{r}(il) g_h(il) \frac{\delta f(il)}{\delta R_\beta(il)} \nabla_\alpha(ij) v(il), \tag{52}$$

an instantaneous contribution which has no equivalent in the SHA theory and is simply added to the force-constant analog (51) in a computation. An expression for the total instantaneous contribution that is related to the sum of (50) and (52) has also been obtained by Meissner (1968a,b).

Thus we see that even this rather complicated theory can be fitted into the general computational procedure described in Section II.

IV. Comparison with Experiment

A. Ground-State Energy Results

The first reliable quantitative theoretical results for the ground-state energy of solid helium were those obtained by Nosanow (1964, 1966) from a Jastrow-variational calculation. This basic approach was refined in subsequent calculations. Nosanow used a NCG function for Ψ_h in (20); Koehler (1965) had shown that a variationally optimal CG always gave a lower value for the ground-state energy than a NCG and this was born out by the CG–Jastrow calculation of Koehler (1967). Finally, the CBF theory with anharmonic corrections was applied by Koehler and Werthamer (1972).

All of these calculations used the lowest-order cluster expansion and $f(r) = \exp[-cv(r)]$, where v is the potential and c is chosen variationally. This is not the optimum Jastrow factor, but it has certain essential properties. Studies of the cluster expansion approximation by Hetherington *et al.* (1967), Guyer (1969b), Trickey and Nuttall (1969), and Kuebbing and Trickey (1972) have shown that it converges fairly well for this form of Jastrow function, but not for all forms. In particular, not for $f = \exp(-c/r^5)$, which is especially desirable because it is the short range solution to the two-body Schrödinger equation for a Lennard-Jones system. Thus the Jastrow factor cannot be freely varied in a cluster expansion calculation, which imposes a severe limitation on the method.

The results of the various Jastrow-cluster calculations mentioned above for a Lennard-Jones model of bcc ^{3}He are shown in Fig. 4 together with the experimental results of Pandorf and Edwards (1968) and the Monte Carlo (MC) results of Hansen (1970). It is now widely felt that the latter are accurate for the Lennard-Jones system to better than 1°K and can be considered to be the best available "experimental" data for that system. From this perspective, the noteworthy features of Fig. 4 are as follows. (1) All of the cluster results are too high, probably due to the restriction on the choice of Jastrow factor [Hansen used $\exp(-c/r^5)$]. (2) The lowering in energy is achieved by using a CG instead of a NCG. (3) There is an incorrect density dependence of the CG calculations. This is a strong indication that the consistency condition (46) should be imposed on the cluster approximated pair function.

The MC technique is sufficiently powerful that it has been used by Hansen and Pollock (1972) to test the suitability of various potentials by evaluating variational ground-state energies over wide density ranges for fcc ^{3}He and fcc ^{4}He. Their results for the Beck (1968), Bruch-McGee (1970), usual Lennard-Jones (LJ1) and new Lennard-Jones (LJ2) (with $\sigma = 2.62$Å) potentials are shown in Fig. 5 together with the experimental results of Dugdale and Franck (1964) and Straty and Adams (1968). Difference values are also plotted to show the good density dependence of the two best potentials—the Beck and the LJ2. The quality of the latter is somewhat surprising because the LJ form cannot be correct. The explanation is probably that the zero-point motion is such that the first few moments of the potential are what really matter and the LJ2 has the correct moments. The Beck potential appears to be the best and detailed arguments in support of this have been given by Glyde (1971).

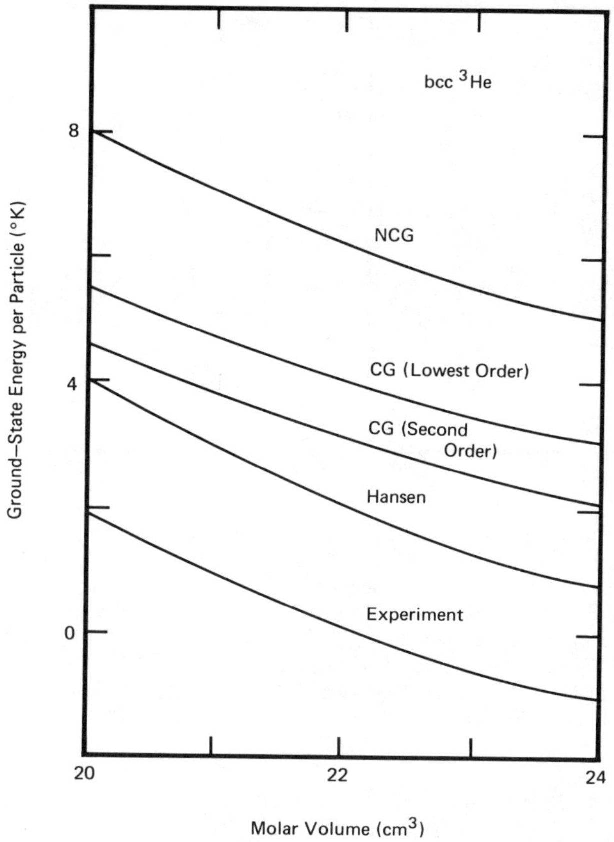

FIG. 4. Theoretical volume dependence of the ground-state energy per particle for bcc ^{3}He from various calculations using a Jastrow–NCG or CG wavefunction. The top three curves use the cluster expansion. The NCG curve is the result of the Nosanow (1966) method and the two CG curves are results of the CBF approach in lowest order and with cubic anharmonic corrections. All of the calculations used the Lennard-Jones (LJ1) potential.

The results of various t-matrix calculations for bcc ^{3}He are shown in Fig. 6. The single-particle basis and the LJ1 potential were used by Horner (1970b), Guyer (1969a), and Ebner and Sung (1971). The MC results of Hansen (1970) are shown for comparison with these. Possible causes for the variation in results have been discussed by Brandow (1972). Glyde and Khanna (1972a) used the Beck potential and both the single-particle and collective basis sets. The energy difference between the two is similar to that shown in Fig. 4 for the NCG and CG calculations. Again, the bad density dependence of their lower curve indicates that a condition like (46) should be imposed on the pair function. Results similar to the above have been obtained by Iwamoto and

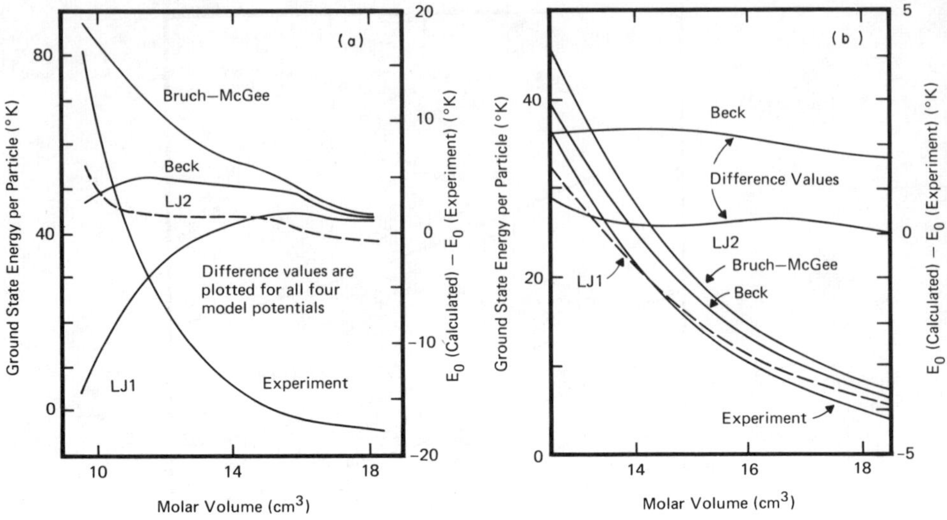

FIG. 5. The results of the Monte Carlo calculations of the ground-state energy of (a) fcc ^{4}He and (b) fcc ^{3}He by Hansen and Pollock (1972) for various model potentials, as described in the text, are contrasted with experimental results. For ^{4}He (a) the difference between the calculated values and the experimental results of Dugdale and Franck (1964) is shown. For ^{3}He (b) the calculated results for three potentials and the experimental results for the hcp phase deduced from Straty and Adams (1968) are shown; in addition, the difference values are plotted for the Beck and LJ2 potentials. The difference values in both figures have been somewhat smoothed by hand.

Namaizawa (1966), Østgaard (1971), Canuto *et al.* (1973), Homma *et al.* (1973), and Kurihara *et al.* (1974).

A current perspective on ground state energy calculations is that the problem can be considered solved by the MC method; other techniques are interesting because of their computational convenience or as tests of perturbation theory techniques. It is not now clear whether a correction like (14), which would occur in a CBF–MC combination theory, would make a contribution of the order of the 1°K found by Koehler and Werthamer (1972) or would be very small, having already been accounted for by an accurate lowest order calculation.

B. Phonon Properties

Recalling that the QHA predicts imaginary frequencies for solid helium, it is worth pointing out that two self-consistent phonon theories predicted real phonon energies before neutron scattering experiments confirmed their reality. A single-particle (NCG in practice) based theory developed by

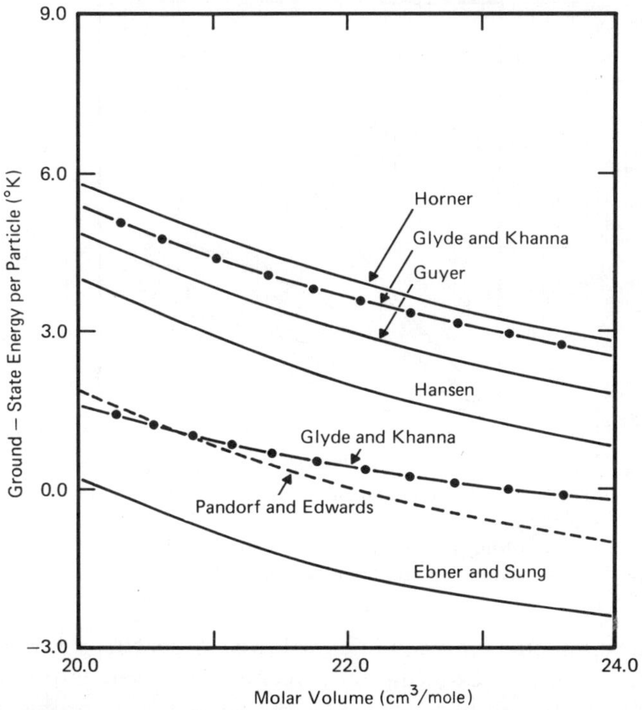

FIG. 6. Volume dependence of the ground-state energy of bcc ^{3}He from various t-matrix calculations together with the Monte Carlo results of Hansen (1970) and the experimental results (dashed curve) of Pandorf and Edwards (1968). The solid and dot–dash curves are from calculations with the LJ1 and Beck potentials, respectively. The four upper curves are t-matrix calculations that start from the single-particle picture, and the lower curve of Glyde and Khanna (1972a) is a t-matrix calculation that starts from the collective picture.

Fredkin and Werthamer (1965) and Gillis and Werthamer (1968) was first used by Nosanow and Werthamer (1965) to calculate sound velocities in bcc ^{3}He and hcp ^{4}He and later by De Wette *et al.* (1967) to evaluate complete phonon dispersion curves in bcc ^{3}He and ^{4}He at a variety of densities. Their working formula was Eq. (25) with the $\langle 0|\nabla\nabla\Psi_J{}^2|0\rangle$ omitted from the right-hand side.

The lowest order CBF theory was first applied by Koehler (1967) to bcc ^{3}He and subsequently by Gillis *et al.* (1968) to hcp ^{4}He. The CBF results are contrasted with the single-particle results for bcc ^{3}He in Fig. 7. The differences are due primarily to the difference between ω and v shown in Eq. (28) and not very much to the presence or absence in (25) of the $\nabla\nabla\Psi_J{}^2$ term, which is small, or to the CG vs. NCG average of $\nabla\nabla V$.

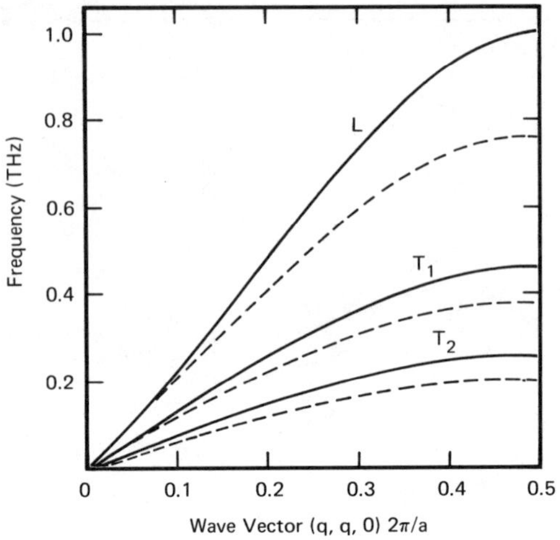

FIG. 7. Lowest-order phonon dispersion curves in bcc ^{3}He at 21.5 cm^3/mole as calculated from two Jastrow-cluster theories. The solid curves are the CBF results from Koehler (1967) and the dashed curves, according to the theory of Nosanow and Werthamer (1965), are obtained from a Jastrow–NCG wavefunction and omit the $\langle 0|\nabla\nabla f|0\rangle$ term.

The greatest impetus for extending the lowest order theories was provided by the inelastic neutron scattering experiments. These were important experiments, but only the results and none of the technical problems will be discussed here. The experiments to date are hcp ^{4}He at 21.1 cm^3/mole by Minkiewicz *et al.* (1968), hcp ^{4}He at 16.0 cm^3/mole by Reese *et al.* (1971), polycrystalline hcp ^{4}He at 20.9 cm^3/mole by Bitter *et al.* (1967), bcc ^{4}He at 21.0 cm^3/mole by Osgood *et al.* (1972), and fcc ^{4}He at 11.7 cm^3/mole by Traylor *et al.* (1972). Technical reasons would make experiments on ^{3}He very difficult and none have been reported yet.

The experiments on the hcp phase were the first to be performed since it is the most accessible phase. The results obtained by Reese *et al.* (1971) together with some points borrowed from Minkiewicz *et al.* (1968) by a scaling procedure described in the previous reference are compared to the theoretical results of Gillis *et al.* (1968) in Fig. 8. The agreement between theory and experiment was quite satisfactory for an application of a no adjustable parameter theory to a difficult substance. The major discrepancies were the incorrect degree of anisotropy in the theoretical curves and an incorrect density dependence when the theory was compared to both sets of experimental data.

The eventual necessity of using a higher order theory was also indicated by the experiments. Substantial linewidths and indications of asymmetries were found in some directions and some lines, shown cross-hatched in Fig. 8, were

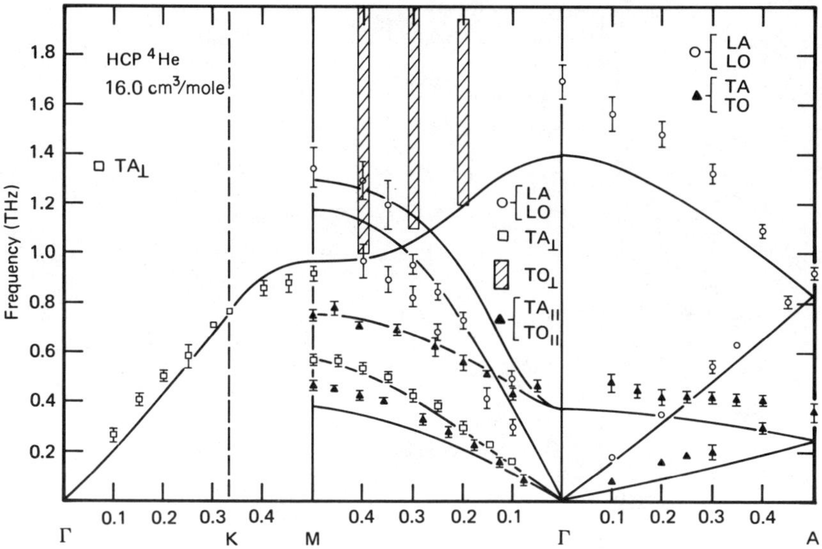

FIG. 8. Theoretical phonon dispersion curves for hcp ^{4}He at 16.0 cm^3/mole obtained by Gillis *et al.* (1968) together with the experimental results of Reese *et al.* (1971).

ill-defined and were found at what appeared to be too high an energy. The anharmonic theories of Section III have not been applied to this phase because it is computationally considerably more difficult to treat, although the feasibility of such computations has been demonstrated by R. D. Grey and T. R. Koehler (unpublished, 1972).

The anharmonic theories have all been applied to the cubic phases. For brevity, only the bcc phase will be considered here; the agreement between theory and experiment is similar for both phases.

The agreement involves details that would have been ignored before the current generation of computations. The experiment measures a line which, after adjustments to eliminate noise and compensate for instrumental considerations, indicates by its position and shape the energy and wave vector dependence of the response of the crystal to the neutron probe. The theorist calculates what this response should be. To a first approximation it is the spectral function $A(\Omega)$, which is obtained from the one-phonon Green's function (19). Results at this level will be discussed in this section. In helium, it has proved necessary to go further and this treatment will be considered in Section IV, C.

Some of the neutron groups of Osgood *et al.* (1972) are shown in Fig. 9. These are actually atypical and are primarily of importance in the next section. Plots of $A(\Omega)$ obtained by Koehler and Werthamer (1972) from the

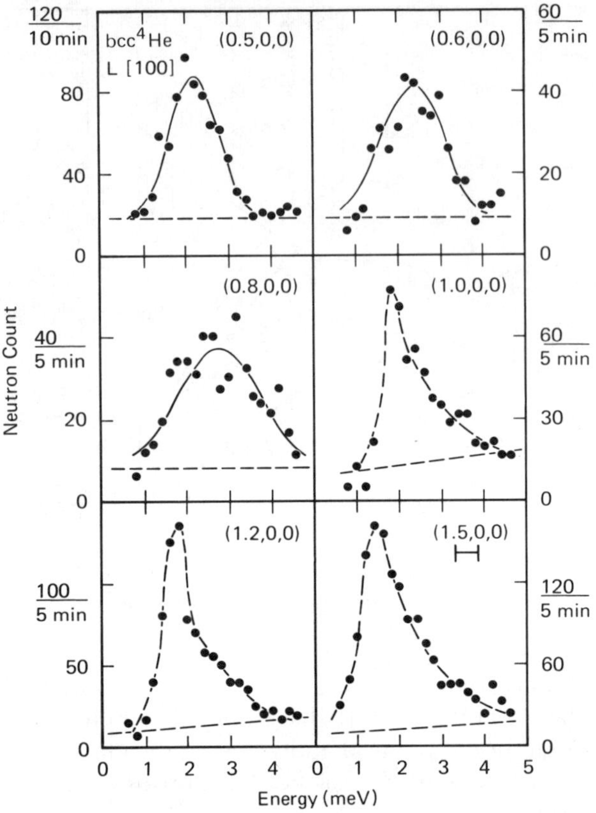

FIG. 9. Phonon profiles for the $L(100)$ branch from Osgood *et al.* (1972), where the value of the reduced momentum transfer $\mathbf{Q}a/2\pi$ is indicated. The solid lines are fits to the data and the straight dashed lines are estimates of the background. The instrumental resolution is indicated in the (1.5, 0, 0) figure.

CBF theory for bcc ^{4}He at $\mathbf{q} = (0.4, 0.4, 0)2\pi/a$, where a is the lattice constant, are shown in Fig. 10 for all three branches together with $\Delta(\Omega)$ and $\Gamma(\Omega)$. These are typical results for solid helium—low lying modes are quite narrow, medium energy modes are somewhat broadened, but symmetrical, and more energetic modes are both broad and unsymmetrical. Note that the lowest mode is renormalized downward considerably even though it is quite narrow.

The renormalized dispersion curves for the CBF theory with anharmonic corrections obtained by Koehler and Werthamer (1972) are compared to the experimental results of Osgood *et al.* (1972) in Fig. 11. The most obvious disagreement is near the zone boundary along (1, 0, 0). Elsewhere, the agreement is not too bad, although one could hardly say that the renormalized theory provided any better agreement than the lowest order theory. A detailed

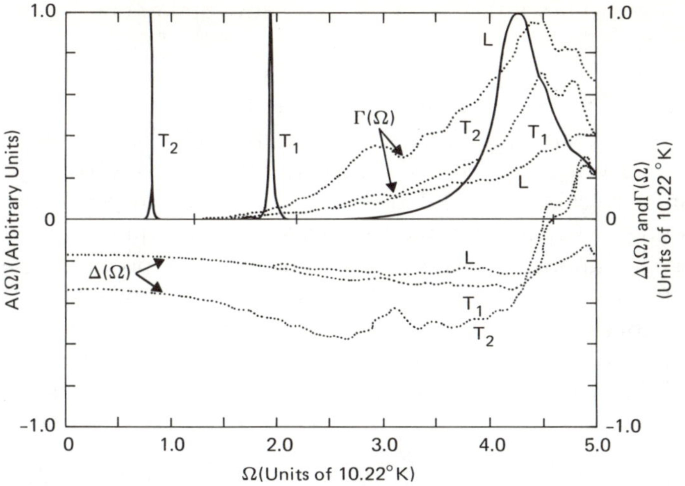

FIG. 10. The solid lines show the theoretical phonon spectral functions for the wave vector $\mathbf{q} = (0.4, 0.4, 0)2\pi/a$ in bcc ^{3}He at 21.5 cm^3/mole according to the calculation of Koehler and Werthamer (1972) with the CBF theory. The contributions $\Delta(\Omega)$ and $\Gamma(\Omega)$ to the phonon self-energy are indicated by the dotted lines. The short vertical lines show the positions of the bare phonon energies.

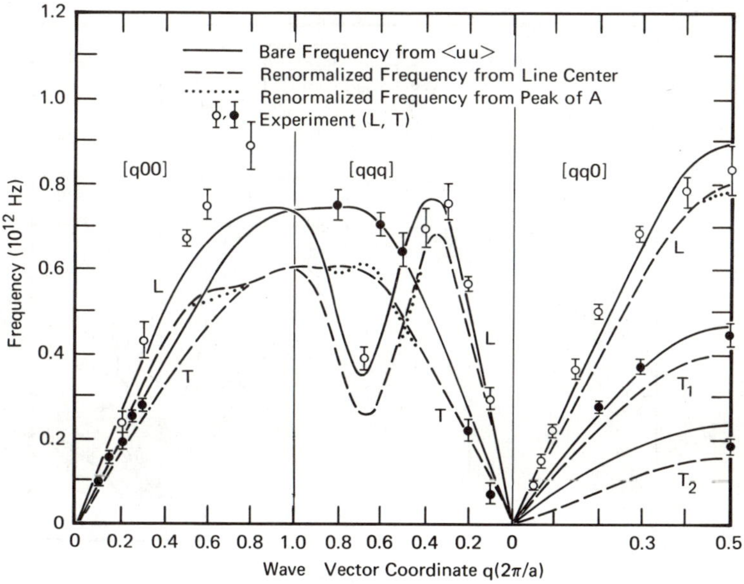

FIG. 11. Theoretical phonon dispersion curves in bcc ^{4}He at 21.0 cm^3/mole from Koehler and Werthamer (1972). The solid curves are the lowest order CBF frequencies and the dotted and dashed curves include cubic anharmonic corrections and are extracted from the spectral function in different ways as described in the text. The experimental points are those of Osgood *et al.* (1972).

comparison of linewidths, however, did show quite good agreement with experiment.

The linewidths obtained in the *t*-matrix calculations of Glyde and Khanna (1972b) were about half those found in the CBF calculations and so agree less well with experiment. The overall agreement for the complete dispersion curves was about the same and these will not be shown. By happenstance, a first-order calculation by Glyde (1970) gave better dispersion curves than either of the above anharmonic treatments. This calculation used the Beck potential and a nonrigorous theory in which the true potential was replaced by the effective potential (23) in all lattice dynamics expressions.

It is obvious from Figs. 9 and 10 that, for a broad unsymmetrical line, it is difficult to choose a unique value as *the* phonon energy. For example, the peak and the mean of the lines are not coincident. Since this is important in comparing theory and experiment, Koehler and Werthamer (1972) demonstrated the effect of different definitions of the phonon energy in comparing their results to experiment. This is shown in Fig. 11 where the definition dependence is not really too great; it was considerably more so for bcc ^{3}He, which is not shown here.

Finally, the results for bcc ^{4}He of the fully consistent theory obtained by Horner (1972a) are shown in Fig. 12. Curves for the peak and mean of the

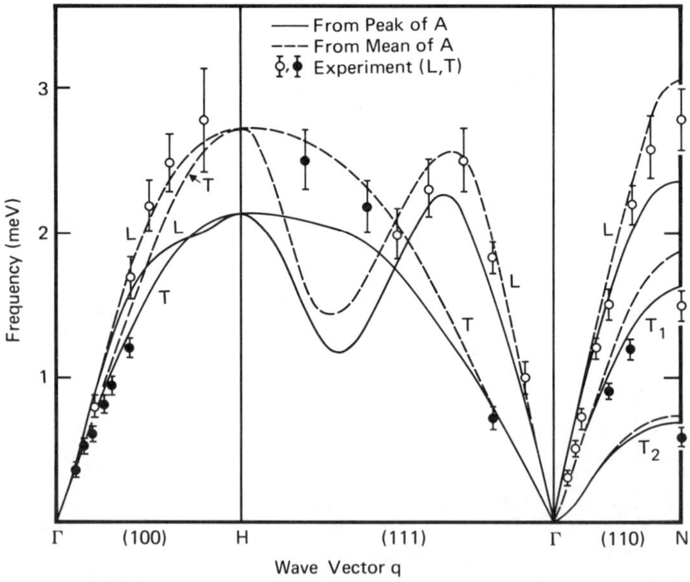

FIG. 12. Theoretical phonon dispersion curves for bcc ^{4}He at 20.92 cm^3/mole from Horner (1972a). The explanation of what is meant by the mean of *A* may be found in the reference. The experimental points are those of Osgood *et al.* (1972).

spectral function are shown. These are in quite satisfactory agreement with experiment. Equally good agreement is found in his fcc calculations, which indicates a correct density dependence for the method. This is substantiated by calculations of the density dependence of the velocity of sound presented in the same paper; good agreement with the experiments of Greywall (1971) and Wanner (1971) was found for all branches except the lowest in the (1,1,0) direction. However, anharmonic calculations (Koehler and Werthamer, 1972) and interpretation of sound velocity experiments (Wanner, 1971) are particularly prone to error for this branch.

The success of the calculations based on the fully consistent theory is probably due to a combination of all the ingredients: the use of the Beck potential, the use of renormalized propagators, the consistency conditions for the pair function, and the identification of the three individual contributions to the self-energy. One can hope that the relative importance of each of these might be singled out in future work.

Although the results of all the anharmonic theories discussed here are in satisfactory agreement with experiment, there are significant discrepancies between certain details. These are considered in an illuminating paper by McMahan and Beck (1973), which clarifies the inner workings of the various methods.

C. Interference Effects

Rather than the spectral function, the correct expression for neutron scattering is proportional to

$$S(\mathbf{Q}, \Omega) = \frac{1}{2\pi} \int_{-\infty}^{\infty} dt\, e^{i\Omega t} \sum_{j} \exp[-i\mathbf{Q} \cdot \mathbf{R}(ij)] \times \langle \exp[-i\mathbf{Q} \cdot \mathbf{u}(i, t)] \exp[i\mathbf{Q} \cdot \mathbf{u}(j, 0)] \rangle, \tag{53}$$

where the brackets denote ensemble averages and $\mathbf{Q}$ is the momentum transfer in the scattering. This is the Fourier transform of the density–density correlation function.

For a purely harmonic crystal, the analysis of (53) is straightforward (e.g., March *et al.*, 1967). For a quantum crystal, it is necessary to refer to the more sophisticated treatments of Ambegaokar *et al.* (1965), Werthamer (1970b), or Beck and Meier (1972). The essential elements of these for the purposes of this section are described as qualitatively as possible in the following.

$S(\mathbf{Q}, \Omega)$ can be expressed as the sum of three separate contributions: (1) an elastic scattering term which is unimportant here; (2) a multiphonon background term which is slowly varying in Ω and will be called $\mathbf{S}_2(\mathbf{Q}, \Omega)$

here, and (3) a term which contains the most important features of the structure

$$S_1(\mathbf{Q}, \Omega) = e^{-2W(\mathbf{Q})} \sum_j \Delta(\mathbf{Q} - \mathbf{q}) \times \operatorname{Im}\left[G_{jj}(\mathbf{q}, \Omega) \left\{ \left[\mathbf{Q} \cdot \mathbf{e}\begin{pmatrix} \mathbf{q} \\ j \end{pmatrix} \right]^2 + L_{jj}(\mathbf{q}, \mathbf{Q}, \Omega) \right\} \right], \tag{54}$$

where G is the one-phonon Green's function of (19),

$$e^{-2W(\mathbf{Q})} \equiv |\langle \exp[i\mathbf{Q} \cdot \mathbf{u}(i)] \rangle_{\mathrm{L}}|^2, \tag{55}$$

the Debye–Waller factor, is basically an intensity factor and $L_{jj}(\mathbf{q}, \mathbf{Q}, \Omega)$ is a is a complex quantity, which arises from interference effects between one- and multiphonon processes, has a fairly smooth frequency dependence, and is proportional to $\mathbf{Q}^3$ at small Q. A linked cluster expansion of the enclosed quantity is indicated by the $\langle\ \ \rangle_{\mathrm{L}}$ in (55).

S_1 is written in a form appropriate to symmetry directions where G and L are diagonal. It can be separated into two contributions: S_1^* [Eq. (54) with $L = 0$] which has the shape of the spectral function, and S_{int}, which contains all the interference effects. Either S_1 or S_1^* obey the ACB (Ambegaokar *et al.*, 1965) sum rule

$$\int \frac{d\Omega}{2\pi} \Omega S_1(\mathbf{Q}, \Omega) = e^{-2W} Q^2/2m, \tag{56}$$

which is important in the analysis of neutron scattering experiments.

This analysis usually contains the tacit assumption that S_1 is S_1^*. When this approach was applied to the bcc ^{4}He data of Osgood *et al.* (1972), two apparent anomalies were found. First, $S_1^*(\mathbf{Q}, \Omega)$ is invariant under translations of $\mathbf{Q}$ by a reciprocal lattice vector. This invariance is clearly violated for the equivalent pairs (0.5, 0, 0), (1.5, 0, 0) and (0.8, 0, 0), (1.2, 0, 0) of $\mathbf{q}$ vectors shown in Fig. 9. Second, the Debye–Waller factor for harmonic crystals is essentially a simple Gaussian in Q^2. Even for quantum crystals, (55) contains the ground-state average of a rather simple quantity and independent evaluations of this average by Sears and Khanna (1972), McMahan and Guyer (1974), and Horner (1972b) show that deviations from the Gaussian should be small. However, when the Debye–Waller factor was extracted from the experimental response function using (56), anomalously large scattering in terms of departure from the expected Gaussian behavior was found for $Q \simeq (1.6)2\pi/a$.

The first convincing explanation of these anomalies that seemed to fit all the facts and to be accompanied by supportive computations was that advanced by Horner (1972b, 1974) although important pieces can be found in the other works cited in the previous paragraph and in Werthamer (1972). The key is the realization that L produces two effects that influence the interpretation of experimental data: (1) it destroys the invariance in reciprocal space and (2) it mixes the dispersive real part of G with the absorptive imaginary part and the mixing coefficients change phase as a zone boundary is crossed.

Computations by Horner (1972b) that illustrate the situation are shown in Fig. 13. Note that S_1 is the same for each **Q** value but that S_{int} is different

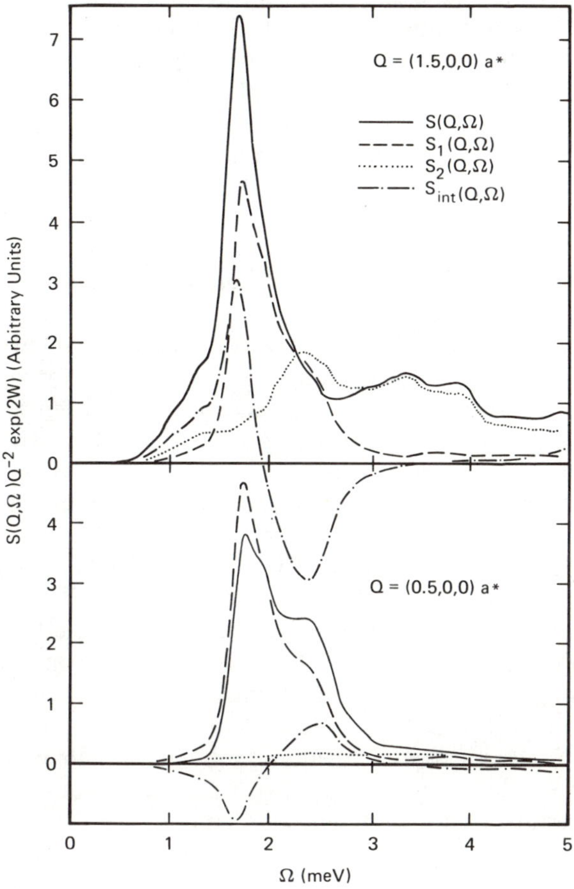

FIG. 13. The scattering function $S(\mathbf{Q}, \Omega)$ in bcc ^{4}He as obtained by Horner (1972b) for two scattering vectors $\mathbf{Q} = (0.5, 0, 0)2\pi/a$ and $\mathbf{Q} = (1.5, 0, 0)2\pi/a$, which are equivalent in reciprocal space, is shown by the solid curves. The separate contributions are (1) the one-phonon spectral function $S_1(Q, \Omega)$ (dashed), (2) the one, two-phonon interference terms $S_{\text{int}}(Q, \Omega)$ (dot-dash), and (3) the multiphonon background (dotted).

in magnitude and sign, but would apparently integrate to zero in (56). The relationship between the total scattering function S is in qualitative agreement with that seen in Fig. 9. He also evaluated the multiphonon background S_2 for each case. Realizing that this and the experimental noise must be subtracted out in using (56), one sees that the contribution from the tail of the line, at somewhat higher frequency than the main peak, is likely to get eliminated in the subtraction. However, at $\mathbf{Q} = (1.5, 0.0)2\pi/a$, a substantial part of the tail has been shifted to the main peak where it will stand out above the background. At $\mathbf{Q} = (0.5, 0.0)2\pi/a$ the opposite happens and the estimated intensity will be too large in the former case and too small in the latter.

Thus, the shape and intensity anomalies are not true anomalies, but are in agreement with the predictions of the inelastic neutron scattering intensity expression (53) when it is correctly applied to extremely anharmonic substances. The demonstration of this, however, required a calculation using all the sophisticated computational techniques encountered in the quantum crystal problem.

V. Conclusions and Future Prospects

We have seen that rapid progress was made in the theory of the lattice dynamics of quantum solids from about 1964 to 1972. Prior to this period, there was no direct experimental evidence for the existence of phonons of finite wave vector and no theory that predicted phonons whose frequencies were not imaginary. Now, dispersion curves have been measured in several inelastic neutron scattering experiments and a variety of theoretical results agree not only qualitatively with these experiments but also quantitatively in such fine details as line shapes. Even anomalous line shapes and intensities appear to be well understood.

The theoretical progress would have been impossible before the advent of the machine computation era. The use of some form of self-consistent phonon theory and the inclusion of anharmonically induced interactions between phonons are essential ingredients in all of the rival theories of quantum crystals; their implementation by hand calculations would be virtually impossible. The interplay between theory, computationally obtained theoretical results, and experiment was vital to the development of good theories.

The necessity of using these specialized computational tools is implicit in the nature of the quantum solids because they are differentiated from ordinary solids by the presence of large atomic motion at 0°K. Even though there are only a few quantum solids, the fruits of studying them are likely to be a much better understanding of the lattice dynamics of all solids at higher temperatures, where the amplitude of thermal motion is large. In fact, this is already proving to be so.

The state of the art in the theory of the lattice dynamics of quantum crystals as presented here is a satisfying one, although many of the ideas are new and should be explored in considerably more detail. Two obvious areas for future research, which will involve less than an order of magnitude additional computational complexity, are available. The first is the study of anharmonic effects in the hcp phase and the second is the study of phonon transport properties. In addition, other topics will undoubtedly arise in due course.

Acknowledgment

The figures in this article have been reprinted with permission from the North-Holland Publishing Company and originally appeared in Koehler (1975b).

References

Ambegaokar, V., Conway, J. M., and Baym, G. (1965). *In* "Lattice Dynamics" (R. F. Wallis, ed.), p. 261. Pergamon, Oxford.
Beck, D. E. (1968). *Mol. Phys.* **14**, 311.
Beck, H., and Meier, P. F. (1972). *Phys. Kondens. Mater.* **14**, 336.
Bitter, M., Gissler, W., and Springer, T. (1967). *Phys. Status Solidi* **23**, K155.
Boccara, N., and Sarma, G. (1965). *Physics* **1**, 219.
Born, M. (1951). *Goetting. Akad. Festschr.* p. 1.
Brandow, B. H. (1972). *Ann. Phys.* (New York) **74**, 112.
Bruch, L. W., and McGee, I. J. (1970). *J. Chem. Phys.* **52**, 5884.
Canuto, V., Lodenquai, J., and Chitre, S. M. (1973). *Phys. Rev. A* **8**, 949.
Choquard, P. (1967). "The Anharmonic Crystal." Benjamin, New York.
de Boer, J. (1957). *Progr. Low Temp. Phys.* **2**, 1.
De Dominicis, C., and Martin, P. C. (1964). *J. Math. Phys.* **5**, 14.
De Wette, F. W., and Nijboer, B. R. A. (1965). *Phys. Lett.* **18**, 19.
De Wette, F. W., Nosanow, L. H., and Werthamer, N. R. (1967). *Phys. Rev.* **162**, 824.
Domb, C., and Dugdale, J. S. (1957). *Progr. Low Temp. Phys.* **2**, 338.
Dugdale, J. S. (1965). *In* "Physics of High Pressure and the Condensed Phase" (A. Van Itterbeek, ed.), p. 382. North-Holland Publ., Amsterdam.
Dugdale, J. S., and Franck, J. P. (1964). *Phil. Trans. Roy. Soc. London* **257**, 1.
Ebner, C., and Sung, C. C. (1971). *Phys. Rev. A* **4**, 269.
Feenberg, E. (1969). "Quantum Theory of Fluids." Academic Press, New York.
Fredkin, D. R., and Werthamer, N. R. (1965). *Phys. Rev.* **138**, A1527.
Gillessen, P., and Biem, W. (1971). *Z. Phys.* **242**, 250.
Gillis, N. S., and Werthamer, N. R. (1968). *Phys. Rev.* **167**, 607.
Gillis, N. S., Koehler, T. R., and Werthamer, N. R. (1968). *Phys. Rev.* **175**, 1110.
Glyde, H. R. (1970). *J. Low Temp. Phys.* **3**, 559.
Glyde, H. R. (1971). *Can. J. Phys.* **49**, 761.
Glyde, H. R. (1974). *In* "Rare Gas Solids" (M. L. Klein and J. A. Venables, eds.). Academic Press, New York (to be published).

Glyde, H. R., and Khanna, F. C. (1971). *Can. J. Phys.* **49**, 2997.
Glyde, H. R., and Khanna, F. C. (1972a). *Can. J. Phys.* **50**, 1143.
Glyde, H. R., and Khanna, F. C. (1972b). *Can. J. Phys.* **50**, 1152.
Greywall, D. S. (1971). *Phys. Rev. A* **3**, 2106.
Guyer, R. A. (1969a). *Solid State Commun.* **7**, 315.
Guyer, R. A. (1969b). *Solid State Phys.* **23**, 413.
Hansen, J. P. (1970). *J. Phys. (Paris)* **31**, Suppl. C3, 67.
Hansen, J. P., and Pollock, E. L. (1972). *Phys. Rev. A* **5**, 2651.
Hetherington, J. H., Mullin, W. J., and Nosanow, L. H. (1967). *Phys. Rev.* **154**, 175.
Homma, S., Nagai, K., and Namaizawa, H. (1973). *Progr. Theor. Phys.* **49**, 1779.
Hooton, D. J. (1955a). *Phil. Mag.* [7] **46**, 422 and 433.
Hooton, D. J. (1955b). *Z. Phys.* **142**, 42.
Horner, H. (1967). *Z. Phys.* **205**, 72.
Horner, H. (1970a). *Phys. Rev. A* **1**, 1712.
Horner, H. (1970b). *Phys. Rev. A* **1**, 1722.
Horner, H. (1971). *Z. Phys.* **242**, 432.
Horner, H. (1972a). *J. Low Temp. Phys.* **8**, 511.
Horner, H. (1972b). *Phys. Rev. Lett.* **29**, 556.
Horner, H. (1974). *Proc. Int. Conf. Low Temp. Phys. 13th, 1972* Vol. 2, p. 3.
Iwamoto, F., and Namaizawa, H. (1966). *Progr. Theor. Phys., Suppl.* **37–38**, 234.
Iwamoto, F., and Namaizawa, H. (1971). *Progr. Theor. Phys.* **45**, 682.
Jastrow, R. (1955). *Phys. Rev.* **98**, 1479.
Koehler, T. R. (1965). *Phys. Rev.* **139**, A1097.
Koehler, T. R. (1966a). *Phys. Rev.* **144**, 789.
Koehler, T. R. (1966b). *Phys. Rev. Lett.* **17**, 89.
Koehler, T. R. (1967). *Phys. Rev. Lett.* **18**, 654.
Koehler, T. R. (1968). *Phys. Rev.* **165**, 942.
Koehler, T. R. (1975a). *In* "Proceedings of the International School of Physics Enrico Fermi, Course 55" (S. Califano, ed.). Academic Press, New York.
Koehler, T. R. (1975b). *In* "Dynamical Properties of Solids" (A. A. Maradudin and G. K. Horton, eds.). North-Holland Publ., Amsterdam.
Koehler, T. R., and Werthamer, N. R. (1971). *Phys. Rev. A* **3**, 2074.
Koehler, T. R., and Werthamer, N. R. (1972). *Phys. Rev. A* **5**, 2230.
Kuebbing, S. J. M., and Trickey, S. B. (1972). *J. Low Temp. Phys.* **8**, 499.
Kurihara, Y., Kuroda, Y., and Ishimura, N. (1974). *Progr. Theor. Phys.* **51**, 959.
McMahan, A. K., and Beck, H. (1973). *Phys. Rev. A* **8**, 3247.
McMahan, A. K., and Guyer, R. A. (1974). *Proc. Int. Conf. Low Temp. Phys., 13th, 1972* Vol. 2, p. 110.
March, N. H., Young, W. H., and Sampanthar, S. (1967). "The Many-Body Problem in Quantum Mechanics." Cambridge Univ. Press, London and New York.
Meissner, G. (1968a). *Phys. Rev. Lett.* **21**, 435.
Meissner, G. (1968b). *Phys. Lett. A* **27**, 261.
Minkiewicz, V. J., Kitchens, T. A., Lipschultz, F. P., Nathans, R., and Shirane, G. (1968). *Phys. Rev.* **174**, 267.
Mullin, W. J. (1971). *J. Low Temp. Phys.* **4**, 135.
Namaizawa, H. (1972). *Progr. Theor. Phys.* **48**, 709.
Nosanow, L. H. (1964). *Phys. Rev. Lett.* **13**, 270.
Nosanow, L. H. (1966). *Phys. Rev.* **146**, 120.
Nosanow, L. H., and Werthamer, N. R. (1965). *Phys. Rev. Lett.* **15**, 618.
Osgood, E. B., Minkiewicz, V. J., Kitchens, T. A., and Shirane, G. (1972). *Phys. Rev. A* **5**, 1537.

Østgaard, E. (1971). *J. Low Temp. Phys.* **5**, 237.
Pandorf, R. C., and Edwards, D. O. (1968). *Phys. Rev.* **169**, 222.
Ranninger, J. (1965). *Phys. Rev.* **140**, A2031.
Reese, R. A., Sinha, S. K., Brun, T. O., and Tilford, C. R. (1971). *Phys. Rev. A* **3**, 1688.
Sears, V. F., and Khanna, F. C. (1972). *Phys. Rev. Lett.* **29**, 549.
Siklós, T. (1972). *Acta Phys.* **34**, 327.
Straty, G. C., and Adams, E. D. (1968). *Phys. Rev.* **169**, 232.
Traylor, J. G., Stassis, C., Reese, R. A., and Sinha, S. K. (1972). *In* "Inelastic Scattering of Neutrons." IAEA, Vienna. To be published.
Trickey, S. B., and Nuttall, J. (1969). *J. Low Temp. Phys.* **1**, 109.
Trickey, S. B., Kirk, W. P., and Adams, E. D. (1972a). *Rev. Mod. Phys.* **44**, 668.
Trickey, S. B., Witriol, N. M., and Morley, G. L. (1972b). *Solid State Commun.* **11**, 139.
Trickey, S. B., Witriol, N. M., and Morley, G. L. (1973). *Phys. Rev. A* **7**, 1662.
Van Kampen, N. G. (1961). *Physica (Utrecht)* **27**, 783.
Wanner, R. (1971). *Phys. Rev. A* **3**, 448.
Werthamer, N. R. (1969). *Amer. J. Phys.* **37**, 763.
Werthamer, N. R. (1970a). *Phys. Rev. B* **1**, 572.
Werthamer, N. R. (1970b). *Phys. Rev. A* **2**, 2050.
Werthamer, N. R. (1972). *Phys. Rev. Lett.* **28**, 1102.
Werthamer, N. R. (1973). *Phys. Rev. A* **7**, 254.
Wilks, J. (1967). "The Properties of Liquid and Solid Helium," pp. 560–664. Oxford Univ. Press, London and New York.

Methods of Brillouin Zone Integration

G. GILAT

DEPARTMENT OF PHYSICS
TECHNION—ISRAEL INSTITUTE OF TECHNOLOGY,
HAIFA, ISRAEL

I. Introduction

A. The Basic Problem

THE CALCULATION OF MANY properties in crystalline solids is closely associated with summations over the first Brillouin zone. In particular we refer here to properties that can be conveniently named "spectral properties of solids," the definition of which will be given. All these properties, although encompassing a broad area in solid state physics, have one feature in common.

They can all be described in a functional form where an "intensity-like" property is given as a function of an "energy-like" variable. For example, in an experiment of inelastic incoherent scattering of neutrons the number of scattered neutrons is given as a function of the energy (or wavelength) of the neutrons. In another experiment of superconducting tunneling between a superconductor and a normal metal the current intensity is usually measured through an insulating gap as a function of a potential difference across this gap. In the case of infrared absorption in solids the intensity absorbed in the solid is measured as a function of the wavelength. These are examples of spectral properties of solids. The property mentioned first in each case is intensity-like, whereas the second one is energy-like. Other examples of spectral properties are second-order Raman spectroscopy in solids, vibronic sidebands of electronic transitions of impurities in solids, optical transitions in solids, dynamical susceptibilities, etc. Other important properties, that are also simpler in nature, are the densities of states of different excitations in solids, such as of phonons, electrons, magnons, etc. The analogy between all these properties is not very impressive, perhaps, if only the experimental aspect is considered. It becomes more justifiable if we look also at the method of derivation of these properties, where the analogy is more convincing. All these properties are obtainable as similar summations over the first Brillouin zone of the (crystalline) solid under observation. To formulate this we note that all these properties can be mathematically described as either the imaginary or the real part of a function $G(\omega)$, which is a slight generalization of a Green's function, and ω is the energy-like variable. $G(\omega)$ is defined by

$$G(\omega) = \lim_{\varepsilon = 0+} (1/N) \sum_{k} F(\mathbf{k})(\omega - \omega(\mathbf{k}) - i\varepsilon)^{-1}. \tag{1}$$

We denote the imaginary and real (principal value) parts of $G(\omega)$ by $I(\omega)$ and $R(\omega)$, respectively, which are given by

$$I(\omega) = (\pi/N) \sum_{k} F(\mathbf{k})\, \delta(\omega - \omega(\mathbf{k})) \tag{2}$$

and

$$R(\omega) = (1/N) \sum_{k} F(\mathbf{k})(\omega - \omega(\mathbf{k}))^{-1} \tag{3}$$

Equations (1)–(3) describe a variety of phenomena which is appreciably richer than that closely relevant to solid state physics alone. For this reason no specific meaning has been assigned to the variables of these equations.

Next we limit our discussion to the solid state, so that $I(\omega)$ and $R(\omega)$ represent, in fact, what we termed spectral properties of solids. For practical reasons it is convenient to write Eqs. (1) and (2) in their integral equivalents, and this is done by using the transformation

$$\frac{1}{N}\sum_{k} \rightarrow \frac{1}{(2\pi)^3 V_{BZ}} \int d^3k, \tag{4}$$

where N is the number of different $\mathbf{k}$ vectors for a given crystal, and V_{BZ} is the volume of its first Brillouin zone. The expressions for $I(\omega)$ and $R(\omega)$ are now given by

$$I(\omega) = \frac{1}{8\pi^2 V_{BZ}} \int F(\mathbf{k})\,\delta(\omega - \omega(\mathbf{k}))\,d^3k \tag{5}$$

and

$$R(\omega) = \frac{1}{(2\pi)^3 V_{BZ}} \int F(\mathbf{k})(\omega - \omega(\mathbf{k}))^{-1}. \tag{6}$$

The interpretation of the various symbols of Eqs. (5)–(6) is as follows: $\omega(\mathbf{k})$ is the energy of an elementary excitation in the solid (i.e., phonons, electrons, magnons) which is usually derivable as an eigenvalue of a certain appropriate Hamiltonian that contains the interaction that gives rise to this excitation. For the case of phonons $\omega(\mathbf{k})$ is known as phonon dispersion relations. The meaning of $F(\mathbf{k})$ can be understood by realizing that in order to observe $I(\omega)$, one usually requires an external probe into the solid (such as a photon, an electron, or a neutron), which interacts with the solid and thus gives rise to $I(\omega)$. $F(\mathbf{k})$ is the transition probability of the excitation of the solid under the interaction with the external probe. $F(\mathbf{k})$ has different names for different branches of the field and is known also as matrix element, coupling constant, oscillator strength, etc.

The functions $I(\omega)$ and $R(\omega)$ are not independent of each other, and they obey the Hilbert transform (Kramers–Krönig relations) given by

$$R(\omega) = \frac{1}{\pi} \int_{-\infty}^{+\infty} \frac{I(\alpha)}{\alpha - \omega}\,d\alpha, \tag{7}$$

and reciprocally

$$I(\omega) = -\frac{1}{\pi} \int_{-\infty}^{\infty} \frac{R(\alpha)}{\alpha - \omega}\,d\alpha. \tag{8}$$

Our main concern in this article is the numerical evaluation of integrals which are represented by Eqs. (5) and (6). A characteristic feature of these integrals is their highly singular nature. As a result, ordinary procedures of numerical integration are inefficient for this purpose, and therefore methods based on new and different ideas have been developed in recent years. In the present article we discuss only computational methods and do not include other methods known as analytical methods. Those methods are based on the general assumption that the integrands are expandable in some appropriate analytical series that admits exact integration. The high singularity of the integrands usually makes the convergence of these expansions very slow, a feature that limits the practicality of these methods. Analytical methods for zone integrations have been extensively discussed by Maradudin *et al.* (1971).

B. Relationship between Computation and Observation

The numerical evaluation of Eqs. (5) and (6) constitutes an important part of many calculations, the end product of which is a certain spectral property of a solid. The numerical computations of these properties may be quite awkward and lengthy, a fact that can seriously affect the quality of the computations. Indeed, prior to the development of high speed electronic computers, most of these calculations were prohibitively long and complex. There are at least two reasons that contribute to the difficulty of making these computations. One reason is associated with the evaluation of the eigenvalues $\omega(\mathbf{k})$. In many cases these are derivable from a Hamiltonian that is given by a sizable matrix. The diagonalization of a matrix of the order of a few tens is a lengthy and costly procedure. Another reason is connected with the fineness of the mesh that approximates the volume of integration. A computation performed for a crude mesh is considerably shorter, but its quality is also correspondingly lower. One must, therefore, adjust one's mesh to the requirement of the computation. Another difficulty of more fundamental nature is associated with $F(\mathbf{k})$. $F(\mathbf{k})$ is better known as matrix element and can be written as $|(i|M|j)|^2$, which represents the transition probability of the excitation under observation in the solid between the $|i\rangle$ and the $|j\rangle$ states. In some cases, such as superconducting tunneling, the theoretical estimates for $F(\mathbf{k})$ are still rather crude, for lack of better knowledge. Even crude estimates for $F(\mathbf{k})$ may require lengthy computation, in particular because they require the knowledge of the eigenstates $|i\rangle$ and $|j\rangle$.

Most computations relate certain theoretical models to experimental measurements. In the case of spectral property one relates a multivalued function $I_{\text{obs}}(\omega)$ obtained by experimental observation to another multivalued function $I_{\text{calc}}(\omega)$, which is a result of a calculation. Now, in contrast to analytical expressions that can be evaluated at a practically unlimited multitude of values, these functions can be given only for a limited number of points. In the experimental case this number is limited by the resolution of the apparatus that measures the data. In computation, the number is limited by the mesh size that we can afford, or more generally by the computational effort that is required. Also in the case of $I_{calc}(\omega)$ it is possible to define a computational resolution (Gilat, 1972) that is determined by the size of the mesh. In both cases it is convenient to estimate the resolution by a pure number N_ω that is related to the energy resolution by

$$N_\omega = (\omega_{max} - \omega_{min})/\Delta\omega, \tag{9}$$

where $\omega_{max} - \omega_{min}$ is the energy range accessible to the measurement or calculation, and $\Delta\omega$ is the resolution. The question that is naturally asked is: How large must N_ω be for a given calculation? Instead of attempting a direct answer to this, it is more convenient to bring in a few arguments that relate to this question.

1. In many experimental fields the technology of measurement has become fairly advanced and is still developing at a brisk pace. An important outcome of this is the improvement of resolution in these fields. In some fields the experimental resolution is already of the order of $N_\omega = 10^2$–10^3. One example is the field of superconducting tunneling between a superconductor and a normal metal. In these measurements one observes the current through an insulating thin gap as a function of the voltage across this gap. In practice one measures the current $I(V)$, as a function of the voltage V, and also dI/dV, the gap conductance, using modulation technique. In some cases one also requires d^2I/d^2V. From these data it is possible to deduce a function $\alpha^2(\omega)F(\omega)$, where α^2 is called an effective electron-phonon coupling and $F(\omega)$ is the ordinary phonon density of states. In Fig. 1 we represent a typical measurement of dI/dV for lead between the normal and the superconducting states (McMillan and Rowell, 1969).

Another example of a high resolution measurement of spectral property in solids is the phonon sidebands observed for narrow electronic transitions of impurity ions in a host lattice. Because of the lack of translation symmetry at the impurity ion, the phonon spectrum of the host crystal can show up as sidebands of the electronic (optical) transition of the impurity ion under certain conditions (Pryce, 1966). The experimental resolution for this observation is of the order of 1 cm^{-1}, which amounts to several times 10^2 for N_ω. In Fig. 2 a typical measurement of phonon sidebands (also called vibronic transitions) is shown (Cohen and Guggenheim, 1968).

It is self-evident that if a computed spectrum is to be presented and compared with experiment, it must be required that its resolution be at least

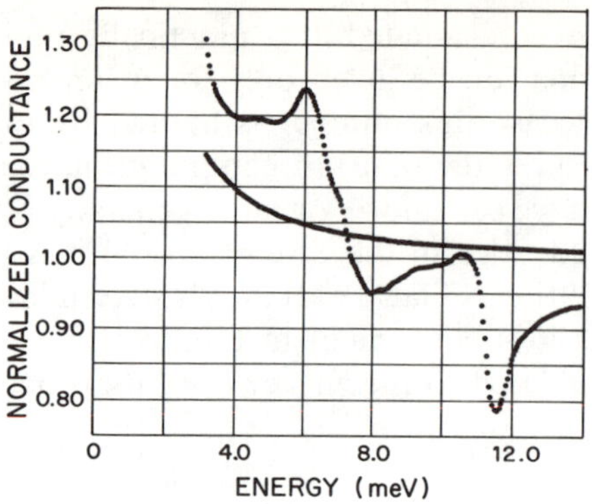

FIG. 1. The conductance dI/dV of a Pb–I–Pb junction in the superconducting state normalized by the conductance in the normal state versus voltage. Also shown is the calculated conductance of the BCS density of state that contains no phonon structure. The data are from McMillan and Rowell (1969).

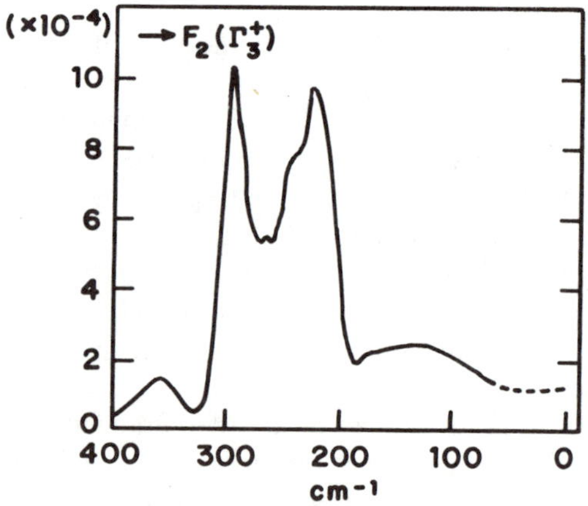

FIG. 2. Vibronic transitions of Sm^{2+} in SrF_2 associated with the electronic transition $^5D_0(\Gamma_1{}^+) \rightarrow {}^7F_2(\Gamma_3{}^+)$. The relative emission intensity is plotted versus the phonon energy. The data are from Cohen and Guggenheim (1968).

as good as the experimental resolution. As a matter of fact it is desirable to have a computed spectrum that is even one step ahead of the experimental spectrum as far as resolution and accuracy are concerned. It is not rare that computational resolutions of the order of $N_\omega = 10^3$ or more should be required in various calculations.

2. In many cases a theoretical model is ultimately described by a certain spectrum $I(\omega)$ derived from this model. An inadequate method of computation usually adds a considerable amount of statistical noise to $I(\omega)$. This

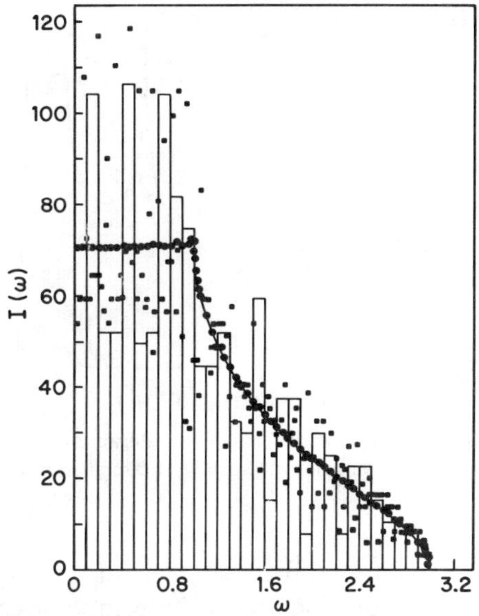

FIG. 3. Model calculation of $I(\omega) = \Sigma_k \delta(\omega - \Sigma_i \cos \frac{1}{2}k_i)$, $(i = x, y, z)$ for simple cubic lattice. Only the positive half of ω is shown. The histograms and square dots represent low resolution calculations using root sampling method (see Section II). The circular points represent high resolution calculations.

can make the computed $I(\omega)$ unreliable in some cases. More important, in some cases it is interesting to study the effects of fine adjustments in the model on $I(\omega)$. This is almost impossible to do unless the "noise level" of the computation is very low. For this reason it is very important in some cases to choose a method of computation of $I(\omega)$ that is capable of representing a given model with high accuracy. To illustrate this point a model calculation of $I(\omega)$ is shown in Fig. 3. $\omega(\mathbf{k})$ is given by the expression $\omega(\mathbf{k}) = \sum_i \cos(\frac{1}{2}k_i)$, where $i = x, y, z$ and $I(\omega) = \sum_{\mathbf{k}} \delta(\omega - \omega_{\mathbf{k}})$, and it is calculated for a simple cubic lattice. The integral of $I(\omega)$ can be worked out exactly (Jelitto, 1969) in this case, and it is represented by the solid curve. The histograms and the square dots represent numerical computations of $I(\omega)$ made by a method that gives poor resolution and accuracy. The circular dots are computed by a high resolution method. The main objective of Fig. 3 is to demonstrate the huge disagreement between numerical computations of the same spectrum $I(\omega)$ that can occur when performed by different computational methods.

3. A typical feature of most of the calculations of $I(\omega)$ is the existence of the singularities which were predicted by Van-Hove (1953). These singularities, or critical points, are ordinarily finite and continuous for three-dimensional lattices, but the derivatives $dI/d\omega$ are discontinuous and can be

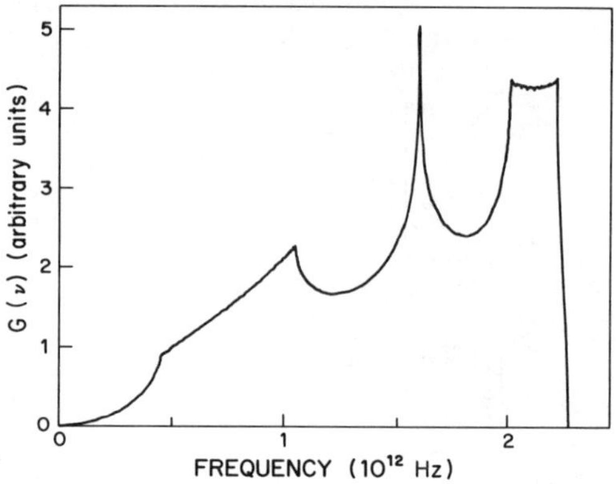

FIG. 4. Model calculation of $g(\nu)$ for first-neighbor interactions in bcc crystals. The resolution here is $N_\omega = 2500$ and it is sufficient to resolve the logarithmic singularity at 1.60 THz.

infinite at certain values of ω. Obviously, it is of particular interest to observe these points in a computation of $g(\omega)$. Now in some cases the singularity of these points can be rather weak, and good resolution is required to expose it. Moreover, in some special cases a theoretical model might predict outstanding features, such as weak infinities in $I(\omega)$. Naturally there is much interest in such theoretical predictions, but unless very fine and accurate calculations of $I(\omega)$ are performed, it is very likely that such features will be obliterated and disappear altogether as a result of too crude a computational resolution. In Fig. 4 a calculation of $g(\omega)$ based on only a first-neighbor interaction of bcc crystals is shown. The singularity in this $g(\omega)$ has a logarithmic infinity, the origin of which is well understood (Maradudin *et al.*, 1971). The resolution required to sufficiently expose this singularity is $N_\omega \simeq 2500$.

4. It is possible in some cases to discover measurable effects by performing accurate calculations. An interesting example of such a case is the lattice specific heat of aluminum which displays an anomaly at low temperature. This anomaly shows up as a small maximum in the Debye temperature $\theta_D(T)$ at $T = 14°K$. This anomaly was discovered first by computation and later confirmed experimentally (Berg, 1968). In this example the anomaly could be predicted only because of the high quality of the computation (see Fig. 5) (Gilat and Nicklow, 1966).

5. In certain calculations the theoretical model can be very complex, and extremely lengthy calculations are required to obtain its eigenvalues. Another case that requires lengthy calculations occurs when double or triple summations over the Brillouin zone are needed. For such cases an efficient method of calculating $I(\omega)$ is greatly needed. High resolution methods are also highly efficient in obtaining $I(\omega)$ rapidly, but here one usually sacrifices resolution

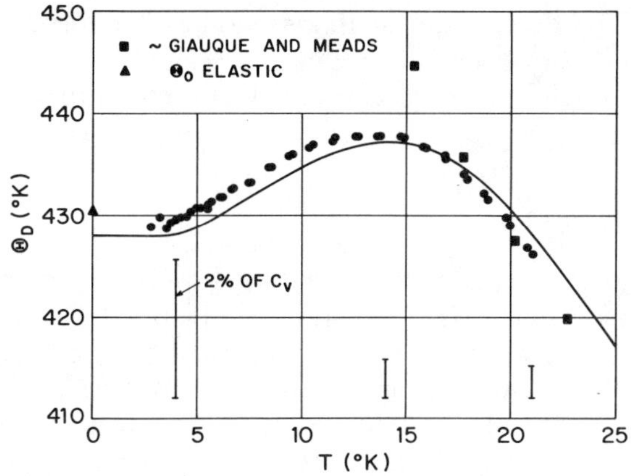

FIG. 5. Comparison between measured and calculated specific heat of aluminum at low temperatures. The experimental data is of Berg (1968) and the calculation is based on a model by Gilat and Nicklow (1966). Figure is from Berg (1968).

in favor of computer time. Resolution is therefore only one option that can be exercised; a complementary option is a very substantial saving in computer time, that can reach several orders of magnitude in some cases. Moreover, efficient computing methods are the *only* means to obtain $I(\omega)$ within reasonable expenditure of computing time in some extreme cases.

6. In certain cases it is desirable to obtain phenomenological models that are fitted to measured $I(\omega)$. These models are usually obtained by iterative procedure of optimizing parameters during the course of which a computed $I(\omega)$ is fitted to an experimental $I(\omega)$. Such a procedure can be excessive in computing time and therefore requires a highly efficient computational scheme, which can be provided by the high resolution methods of zone integration.

These are a few examples that are brought forward to convince the reader of the necessity for employing computational procedures which are somewhat more sophisticated, but at the same time combine high computational precision with very substantial saving in computing expenditure. The characteristics of the high resolution methods are described in the following section.

II. Methods of Zone Integration

In this section a detailed description of the various methods of zone integration is presented. The various sections are arranged according to the development of the different methods which are logically linked with one another. The description in this section is confined, however, to the calculation of $g(\omega)$ rather than to that of the more general $I(\omega)$, and this is so for the purpose of simplicity. The introduction of $F(\mathbf{k})$, the matrix element, and its effect on the calculation of $I(\omega)$ is deferred to Section II, G.

A. Root Sampling

The first numerical method for evaluating Eq. (2) was proposed and implemented by Blackman (1937). Blackman's motivation for this calculation was to show that departures from the Debye model for the lattice specific heat of solids could be significant and measurable. Kellerman (1940) also used this method to evaluate the frequency distribution function $g(\omega)$ of NaCl from his rigid ion model. The idea behind this so-called root sampling (RS) method is simple and straightforward. Since it is practically impossible to consider all the phonons in the solid, one takes a sample of a few **k** vectors situated on a regular mesh in the irreducible part of the Brillouin zone (IBZ). Then the dynamical matrix of the lattice for these vectors is solved and the eigenfrequencies are sorted out into a number of frequency channels $(\omega, \omega + \Delta\omega)$ in the range of frequencies that exists between ω_1 and ω_2 for the excitation spectrum (for lattice vibrations $\omega_1 = 0$). A typical computation of $g(\omega)$ for iron based on the inelastic neutron experiment of Minkiewicz *et al.* (1967) is shown in Fig. 6. A characteristic feature of root sampling calculations that are usually seen in publications is the considerable amount of statistical fluctuations in the histograms. The reason for this can be understood on the basis of the analysis of the various methods of zone integration.

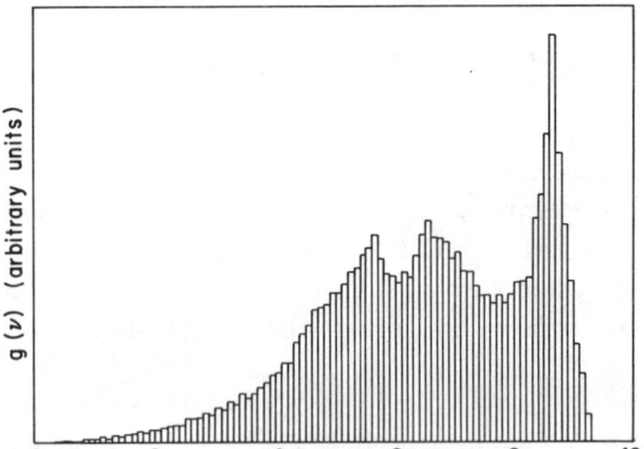

FIG. 6. Root-sampling calculation of $g(\nu)$ for iron. The model is based on neutron scattering data of Minkiewicz *et al.* (1967). The number of mesh points required for this $g(\nu)$ is $N_k = 5200$, and the resolution is $N_\nu = 93$.

Although this matter will be further discussed in Section II, F, it ought to be mentioned here that in many cases the computational resolution N_ω is much smaller than the actual number of histograms employed, and this causes the statistical fluctuations.

The RS method is the simplest of all zone integration techniques. For the root sampling method only the values of the various eigenvalues of the dynamical matrix for each **k** vector are required. This is in contrast to the more elaborate methods that require additional information, such as first and sometimes even second derivatives of eigenvalues at each **k** vector. Its simplicity makes the RS technique very useful whenever only crude features of $I(\omega)$ are required. In cases where high resolution is required, or when the computational effort of obtaining each eigenvalue is considerable, or when double summations are wanted, the root sampling method is not to be recommended and more efficient methods ought to be employed.

Probably the most elaborate and extensive calculations using the RS method were performed by Karo and Hardy (1966) who calculated the vibrational properties and second-order Raman effect in NaCl. In Fig. 7 a typical computation of $g(\omega)$ of NaCl is presented. The number of **k** vectors in the IBZ for this particular case is 64,000 and $\Delta\omega = 0.06 \times 10^{13}$ rad/sec or $N_\omega = 85$. Although the location of the major Van-Hove singularities is well resolved, the weaker singularities are somewhat less apparent owing to

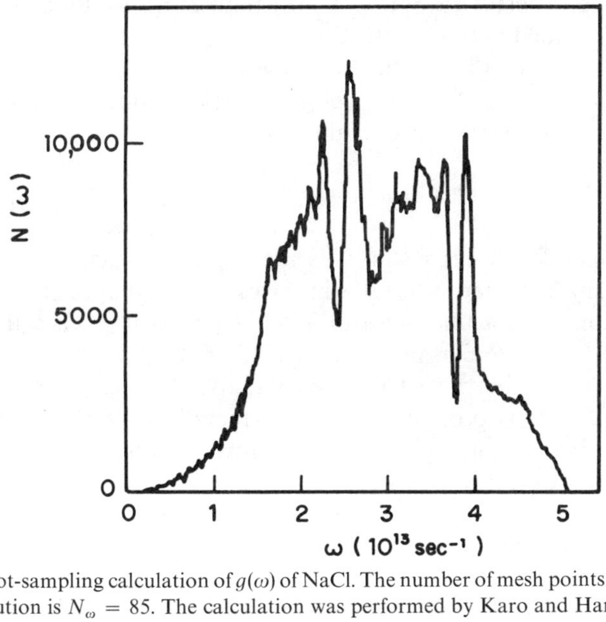

FIG. 7. Root-sampling calculation of $g(\omega)$ of NaCl. The number of mesh points is $N_k = 64{,}000$ and the resolution is $N_\omega = 85$. The calculation was performed by Karo and Hardy (1966).

the large amount of statistical "background" noise. This noise shows up especially in the lower part of $g(\omega)$, which looks quite wriggly instead of being smooth as required by Debye theory. In another computation that involves a mesh of 512,000 **k** vectors the resolution is somewhat better and more, although not all, of the Van-Hove singularities are resolved.

In an effort to reduce the statistical noise, Karo and Hardy (1969) applied a procedure of Gaussian smoothing, by which the height of each Gaussian was given by the histogram height, and the half-width was adjusted to obtain the best smoothing consistent with retaining the structure of $g(\omega)$. It is obvious that a Gaussian smoothing cannot improve the resolution of a given mesh of points. All it does, in fact, is to optimize the resolution, as is explained in Section II, F.

B. Discrete Methods

A first significant improvement over the root sampling method was proposed by Gilat and Dolling (1964). Their idea was to employ a crude mesh of a few hundreds or thousands of **k** vectors, then to form cubic cells about each of these points and to interpolate for many additional eigenvalues throughout each of these cells. In order to be able to do so one is in need of not only the values of $\omega(\mathbf{k})$ but of $\nabla\omega(\mathbf{k})$ as well. The evaluation of $\nabla\omega(\mathbf{k})$ requires additional calculations for each **k** vector, but in most cases it is much easier and faster to obtain $\nabla\omega(\mathbf{k})$ than $\omega(\mathbf{k})$. Methods of obtaining $\nabla\omega(\mathbf{k})$ are described in Section II, E.

Let $\mathbf{k}_c$ be the **k** vector on the crude mesh, located at a center of a cubic cell. If $\mathbf{k}_i$ is any vector inside the cell, then $\omega(\mathbf{k}_i)$ is obtainable by using the linear term of a Taylor expansion about $\mathbf{k}_c$, i.e.,

$$\omega(\mathbf{k}_i) = \omega(\mathbf{k}_c) + \nabla\omega \cdot (\mathbf{k}_i - \mathbf{k}_c). \tag{10}$$

Clearly, once $\nabla\omega(\mathbf{k}_c)$ is given, it is incomparably easier to derive $\omega(\mathbf{k}_i)$ than $\omega(\mathbf{k}_c)$, since the latter usually involves matrix diagonalization. Resultantly one can form a fine mesh of, say, M points in each cell where $i = 1, \ldots, M$. If the crude mesh is of size M_c, then the final mesh is of the size of $M_f = MM_c$. In this way it is an easy matter to increase the size of the initial mesh M_c by two or three orders of magnitude without considerably lengthening the computation time and effort. In Fig. 8 a computation of phonon density of states of sodium computed by this method is shown. This calculation is based on a fifth-nearest-neighbor model of Woods *et al.* (1962). In this calculation $M_c = 440$, $M = 729$. The number Z of frequencies for a given mesh represents the sampling size of this mesh. Z depends on the number of eigenvalues per mesh point and to some extent on weighing

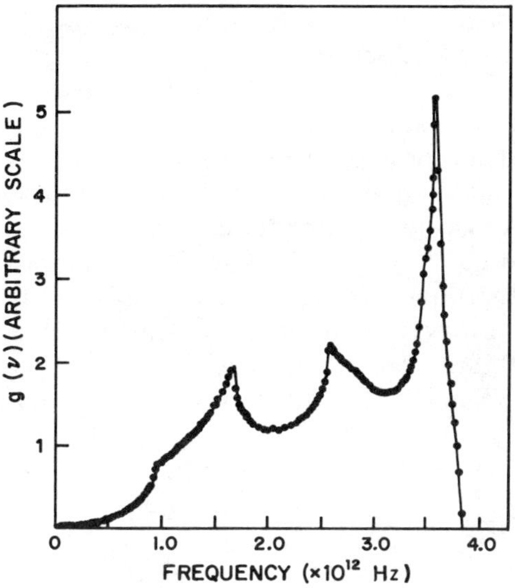

FIG. 8. Comparison between calculations of $g(\nu)$ of sodium using the RS and the LD methods. The solid curve represents the RS calculation based on $N_k = 180{,}441$, and the dots are due to the LD calculation with $N_k = 440$ and $M = 729$. The RS calculation was performed by Dixon *et al.* (1963).

factors associated for each mesh point. In the computation shown in Fig. 8 $Z = 34{,}992{,}000$. For comparison, on the same graph a RS method calculation (Dixon *et al.*, 1963) that employs $N_k = 180{,}441$ is shown. For this case $Z = 24{,}576{,}000$. The computer time required for the second calculation was about 250 times longer than for the first one.

One method of obtaining $\nabla\omega(\mathbf{k}_c)$ requires first-order perturbation approximations which can be somewhat awkward whenever there are degenerate eigenvalues $\omega(\mathbf{k}_c)$. Degeneracy usually occurs at high symmetry **k** vectors, and it became necessary to avoid these points. In early computations, prior to the extensive use of electronic computers, it was customary and convenient to take advantage of the symmetry properties of these **k** vectors to facilitate significantly the evaluation of eigenvalues at these points. The use of high speed computers makes in many cases the "brute force method" of direct diagonalization almost as speedy as the sophisticated group theoretical techniques. Moreover, the statistical weight of high symmetry points is considerably smaller than for general **k** vectors. For these reasons it is found convenient in many cases to shift the crude mesh away from high symmetry points.

The name of linear discrete (LD) is chosen for this method for the reason that only linear terms are included in Eq. (10) and because the final mesh of points, although much denser, is still discrete and finite. In this sense the root sampling method is also discrete.

In some cases, usually encountered in electronic band structure or in lattice dynamical calculations with many atoms per unit cell, large matrices are involved, and the procedure of obtaining eigenvalues is lengthy. For these cases it becomes important to save as much as possible on the size of the crude mesh. A plausible way of doing this is to extend the Taylor expansion in Eq. (10) to quadratic terms. This was first suggested by Brust (1965) who calculated the photoelectric effect in silicon by the pseudopotential method in band theory. In order to carry out quadratic interpolation for points $\mathbf{k}_i$ in the vicinity of the crude mesh points $\mathbf{k}_c$, it is required to obtain the full tensor of $\nabla\nabla\omega(\mathbf{k}_c)$. The values for $\omega(\mathbf{k}_i)$ are given by

$$\omega(\mathbf{k}_i) = \omega(\mathbf{k}_c) + \nabla\omega \cdot (\mathbf{k}_i - \mathbf{k}_c) + \tfrac{1}{2}(\mathbf{k}_i - \mathbf{k}_c) \cdot \nabla\nabla\omega \cdot (\mathbf{k}_i - \mathbf{k}_c). \quad (11)$$

The quadratic discrete (QD) method goes a step beyond the linear discrete (LD) method in the sense that it improves statistics of $I(\omega)$ for a given mesh size. The QD method was described in a detailed manner by Mueller *et al.* (1971; also see Mueller, 1971) who employed it to compute the electronic density of states of platinum. In Fig. 9 the S-band part of $g(E)$ for platinum is shown. The number of crude mesh points used for this calculation was 1876. The total number of mesh points was about 10^6, and the computational resolution was $N_\omega = 850$. The calculation reveals all the major critical points, as well as the less pronounced ones. Nevertheless, there is still a certain amount of noise that shows up as irregular wriggles in the spectrum.

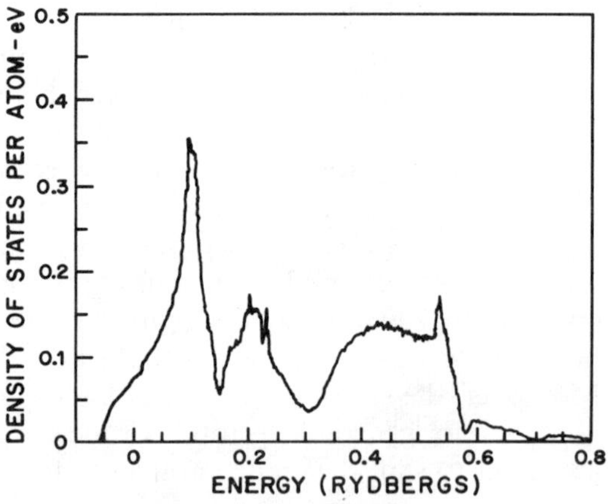

FIG. 9. S-band part of the electronic density of states for platinum. In this graph $N_k = 1876$ points and the total (fine) mesh points is 10^6. Resolution is $N_\omega = 850$. The calculation was performed by Mueller *et al.* (1971; Mueller, 1971) who used the QD method.

C. Analytical (Continuous) Method

The discrete methods could in principle be further developed by using higher and higher terms in the Taylor expansion. This procedure is not to be recommended, however, since the marginal gain in statistics by applying a next step in the Taylor series is becoming progressively smaller, while the computational effort in obtaining it increases. As a matter of fact, the discrete linear and quadratic methods are in many cases adequately efficient for obtaining the calculated $I(\omega)$ in fair resolution.

The linear analytic (LA) method was first introduced by Gilat and Raubenheimer (1966) who used $\nabla\omega(\mathbf{k}_c)$ for obtaining analytical expressions for the contributions coming from each cell around $\mathbf{k}_c$. In the first version all the cells were assumed to be cubic, but later this restriction was lifted, and orthorhombic cells were used as well (Raubenheimer and Gilat, 1967).

The idea behind the linear analytic methods starts from the well-known expression for $g(\omega)$

$$g(\omega) = \sum_j \int \frac{dS}{|\nabla\omega|}, \tag{12}$$

where integration takes place over constant frequency surfaces $\omega_j(\mathbf{k}) = \omega$, and the summation is over the different bands j. The next step is to approximate these surfaces inside every small orthorhombic cell by a set of parallel planes. The direction cosines of these planes are given by the direction cosines l_i of $\nabla\omega(\mathbf{k}_c)$. If an eigenvalue $\omega(\mathbf{k}_c)$ is known, then its variation throughout the cell can be approximated by Eq. (10). Let w be the distance of a particular frequency plane from the center $\mathbf{k}_c$ of the cell. By virtue of the linear approximation all the jth eigenvalues for all the points $\mathbf{k}$ that lie on the plane are equal. Next we choose a frequency interval $d\omega$ and form a partial density of states function $g_j(\mathbf{k}_c;\,\omega)$, that gives the contribution of the jth eigenfrequency coming from the cell centered at $\mathbf{k}_c$. Under the linear assumption it is easy to evaluate $g_j(\mathbf{k}_c;\,\omega)d\omega$, and in fact it is proportional to the volume confined within the two parallel constant frequency planes of ω and $\omega + d\omega$ and the cell boundaries. To find this volume we must find the area of intersection between these planes and the cell and multiply it by their separation dw. For cubic cells we have

$$\begin{aligned} g_j(\mathbf{k}_c;\,\omega)\cdot d\omega &= CW(\mathbf{k}_c)S_j(l_1 l_2 l_3;\,w)\cdot dw = CW(\mathbf{k}_c)S_j(l_i;\,w)\cdot(\mathbf{d}\omega/|\nabla\omega_j(\mathbf{k}_c)|) \\ &\quad \text{for} \quad \omega_j(\mathbf{k}_c) - w_{\max}|\nabla\omega_j(\mathbf{k}_c)| \leqslant \omega \leqslant \omega_j(\mathbf{k}_c) + w_{\max}|\nabla\omega_j(\mathbf{k}_c)| \\ &= 0, \qquad \text{otherwise.} \end{aligned} \tag{13}$$

Where w is related to $\omega - \omega_j(\mathbf{k}_c)$ by

$$w = [\omega - \omega_j(\mathbf{k}_c)]/|\nabla\omega_j(\mathbf{k}_c)|, \tag{14}$$

C is a suitable constant of normalization, and $W(\mathbf{k}_c)$, the statistical weight of the point $\mathbf{k}_c$, is inversely proportional to the number of IBZ sharing the cubic cell centered at $\mathbf{k}_c$. w_{max} is the maximum length w confined within each cell. It is important to point out that the use of $W(\mathbf{k}_c)$ is not obligatory and it is possible to do without it if the IBZ is filled exactly and in an exhaustive manner by the cells of $\mathbf{k}_c$. For cubic lattices, it was found convenient to overfill the IBZ to a certain extent and then correct this with the aid of $W(\mathbf{k}_c)$. In contrast, for hexagonal lattices no use was made of $W(\mathbf{k}_c)$, but part of the IBZ had to be filled with triangular prisms. The question of the most efficient shapes for the cells will be discussed again in Section IV.

The analytical expressions for $S_j(l_j; w)$ were worked out by Gilat and Raubenheimer (1966). The span of w in the cubic cell is divided into four ranges given by the four distances $w_i (i = 1, 2, 3, 4)$ of the corners of the cube from its center, i.e.,

$$w_1 = b|l_1 - l_2 - l_3|,$$

$$w_2 = b(l_1 - l_2 + l_3),$$

$$w_3 = b(l_1 + l_2 - l_3),$$

and

$$w_4 = b(l_1 + l_2 + l_3), \tag{15}$$

where $2b$ denotes the side of the cubic cell, and the direction cosines l_1 are ordered in a decreasing sequence $l_1 \geqslant l_2 \geqslant l_3 \geqslant 0$. $S_j(l_1; w)$ themselves are given for the various ranges of w by (w_i, w_{i+1}). For the first range $0 \leqslant w \leqslant w_1$ there are two possibilities, depending on whether $l_1 \geqslant l_2 + l_3$ or $l_1 < l_2 + l_3$. For the former case the cross section of the cubic cell is a parallelogram of area

$$S_j(l_1 l_2 l_3; w) = 4b^2/l_1 \quad \text{for} \quad l_1 \geqslant l_2 + l_3 \quad \text{and} \quad 0 \leqslant w \leqslant w_1, \tag{16}$$

while for $l_1 < l_2 + l_3$ the cross section is hexagonal, of area

$$S_j(l_1 l_2 l_3; w) = [2b^2(l_1 l_2 + l_2 l_3 + l_3 l_1) - (w^2 + b^2)]/(l_1 l_2 l_3) \quad \text{for} \quad l_1 < l_2 + l_3 \quad \text{and} \quad 0 \leqslant w \leqslant w_1. \tag{17}$$

For the second range (w_1, w_2) of w the cross section is a pentagon of area

$$S_j(l_1l_2l_3; w) = [b^2(3l_2l_3 + l_1l_2 + l_3l_1) + bw(l_1 - l_2 - l_3) - \tfrac{1}{2}(w^2 + wb^2)]/(l_1l_2l_3) \quad \text{for} \quad w_1 \leqslant w \leqslant w_2. \tag{18}$$

In the next range (w_2, w_3) the shape of the cross section is quadrilateral of area

$$S_j(l_1l_2l_3; w) = 2b[b(l_1 + l_2) - w]/(l_1l_2) \quad \text{for} \quad w_2 \leqslant w \leqslant w_3; \tag{19}$$

and finally we obtain for (w_3, w_4) a triangle of area

$$S_j(l_1l_2l_3; w) = [b(l_1 + l_2 + l_3) - w]^2/(2l_1l_2l_3) \quad \text{for} \quad w_3 \leqslant w \leqslant w_4. \tag{20}$$

All these expressions, as well as their first derivatives, are continuous at their respective range boundaries. These expressions are determined for the range of $w < 0$ by the relation $S_j(l_i; -w) = S_j(l_i; w)$. Typical and schematic graphs of $S_j(l_i; w)$ are shown in Fig. 10 for the two possibilities of Eq. (16) and (17). The area underneath the curves is proportional to the volume of the cubic integration cell.

The extension of the LA method to orthorhombic cells is readily accomplished by setting $\lambda_i = l_i a_i$, where $2a_1$, $2a_2$, $2a_3$ are the side lengths of the orthorhombic cell. The cross-section areas are given by the following expressions in analogy to Eqs. (16–20):

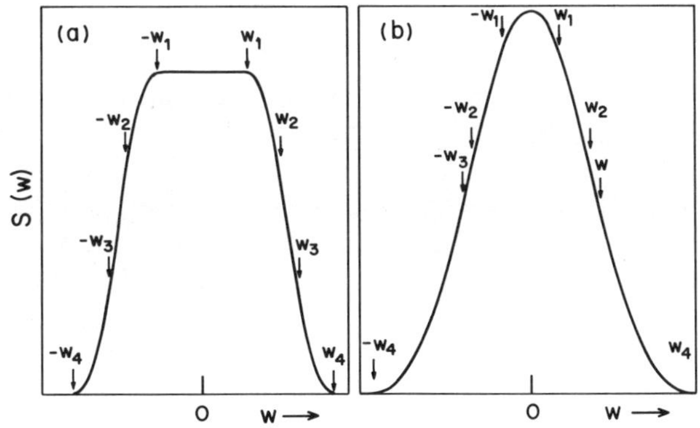

FIG. 10. Typical dependence of $S(l_i; w)$ on w. The locations of the various corners are indicated by w_i. (a) The case of $l_1 \geqslant l_2 + l_3$ and (b) the case of $l_1 < l_2 + l_3$.

$$S_j(\lambda_1\lambda_2\lambda_3; w) = V/2\lambda_1 \qquad \text{for} \qquad \lambda_1 \geqslant \lambda_2 + \lambda_3, \qquad 0 \leqslant w \leqslant w_1, \tag{21}$$

$$\begin{aligned} S_j(\lambda_1\lambda_2\lambda_3; w) = {} & (V/8\lambda_1\lambda_2\lambda_3)[2(\lambda_1\lambda_2 + \lambda_2\lambda_3 + \lambda_3\lambda_1) \\ & - (w^2 + {\lambda_1}^2 + {\lambda_2}^2 + {\lambda_3}^2)] \\ & \text{for} \quad \lambda_1 < \lambda_2 + \lambda_3, \quad 0 \leqslant w \leqslant w_1, \end{aligned} \tag{22}$$

$$\begin{aligned} S_j(\lambda_1\lambda_2\lambda_3; w) = {} & (V/8\lambda_1\lambda_2\lambda_3)[3\lambda_2\lambda_3 + \lambda_1\lambda_2 + \lambda_3\lambda_1 + w(\lambda_1 - \lambda_2 - \lambda_3) \\ & - \tfrac{1}{2}(w^2 + {\lambda_1}^2 + {\lambda_2}^2 + {\lambda_3}^2)] \\ & \text{for } w_1 \leqslant w \leqslant w_2, \end{aligned} \tag{23}$$

$$S_j(\lambda_1\lambda_2\lambda_3; w) = (V/4\lambda_1\lambda_2)(\lambda_1 + \lambda_2 - w) \qquad \text{for} \quad w_2 \leqslant w \leqslant w_3, \tag{24}$$

and

$$\begin{aligned} S_j(\lambda_1\lambda_2\lambda_3; w) = {} & (V/16\lambda_1\lambda_2\lambda_3)(\lambda_1 + \lambda_2 + \lambda_3 - w)^2 \\ & \text{for} \quad w_3 < w < w_4. \end{aligned} \tag{25}$$

In these expressions $V = 8a_1a_2a_3$ is the volume of the orthorhombic cell. The various ranges (w_{i-1}, w_i) are determined by

$$\begin{aligned} w_1 &= |\lambda_1 - \lambda_2 - \lambda_3|, \qquad & w_2 &= \lambda_1 - \lambda_2 + \lambda_3, \\ w_3 &= \lambda_1 + \lambda_2 - \lambda_3, \qquad & w_4 &= \lambda_1 + \lambda_2 + \lambda_3. \end{aligned} \tag{26}$$

Once $S_j(l_i; w)$ or $S_j(\lambda_i; w)$ is known, it is possible to obtain $g_j(\mathbf{k}_c; \omega)$ by Eq. (13). The total $g(\omega)$ is obtainable by summing over all $\mathbf{k}_c$ and j, namely

$$g(\omega) = \sum_{\mathbf{k}_c, j} g_j(\mathbf{k}_c; \omega). \tag{27}$$

The LA method has already been applied to numerous crystals and has proved its efficiency. Here only a few examples are brought. In Fig. 11 the $g(\omega)$ for iron is shown. In fact the same density of states for iron has already been shown in Fig. 6, but the method of computation is different here and based on the LA method. The number of mesh points employed for the calculation of $g(\omega)$ in both figures was $N_k = 5200$, but the resolution in Fig. 11 is $N_\omega = 6200$, whereas in Fig. 6 $N_\omega = 93$. This comparison is meant to demonstrate that by using the LA rather than the RS method, gain in N_ω becomes almost two orders of magnitude for the same mesh size. This is so while the increase in the computational effort is only marginal.

In Fig. 12 the $g(\omega)$ for sodium is shown. The calculation is based on the data of Woods *et al.* (1962). In fact, this $g(\omega)$ is based on the same model

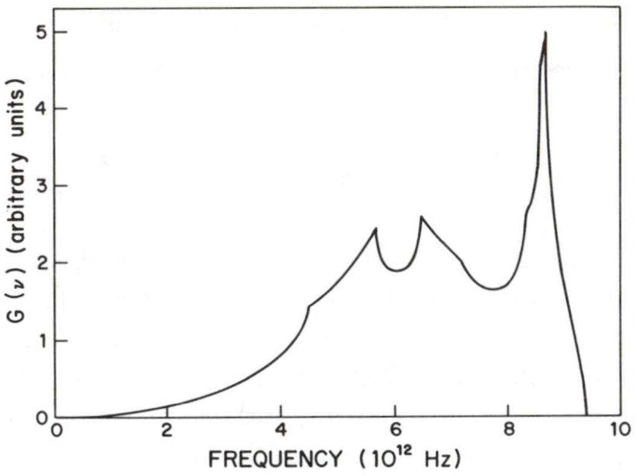

FIG. 11. LA method calculation of $g(\nu)$ of iron. This calculation uses the same model and N_k as in Fig. 6, but the resolution here is $N_\nu = 6200$.

as that of Fig. 8, but the mesh size of Fig. 12 is larger and is $N_k = 3080$, and the resolution $N_\omega = 3800$. The interesting feature about $g(\omega)$ of sodium is its extremely sharp singularity at $\nu = 3.591$ THz (1 THz $= 10^{12}$Hz). The shape of this singularity may indicate that although very weak it could be a genuine infinity. It must be stressed here that it is not a trivial matter to detect a weak infinity numerically. The reason for this is that in numerical calculations the product $g(\omega)\,\Delta\omega$ is computed and not just $g(\omega)$. Therefore on a histogram plot, large as N_ω may be, $g(\omega)\,\Delta\omega$ is always finite. On the other hand, if an infinity in $g(\omega)$ does exist at, say $\omega = \omega_0$, then it is possible

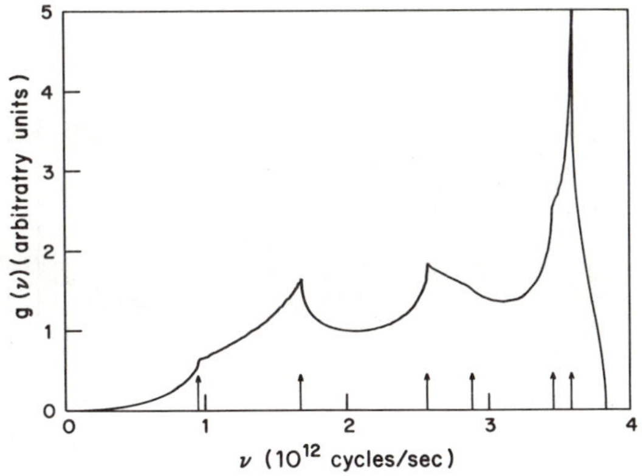

FIG. 12. LA method calculation of $g(\nu)$ for sodium. This calculation uses the same model as in Fig. 8, but here $N_k = 3080$ and $N_\nu = 3800$. The location of Van–Hove singularities are shown by arrows. At $\nu = 3.59$ THz there is a strong singularity that may be a weak infinity.

to perform several computations with various resolutions $\Delta\omega$ and to see if $g(\omega)$ is bounded at ω_0 where $g(\omega)$ is given by

$$g(\omega) = \lim_{\Delta\omega = 0} \frac{1}{2\,\Delta\omega} \int_{\omega_0 - \Delta\omega}^{\omega + \Delta\omega} g(\omega)\, d\omega \tag{28}$$

In the case of sodium it was found that $g(\nu)$ did not reach a limit when $\Delta\nu$ was made as small as $\Delta\nu = 0.0005$ THz. As a matter of fact, comparing the relative heights of the peaks at $\nu = 3.591$ THz in Figs. 8 and 12, it can be observed how these heights are effected by resolution. Although the evidence is still inconclusive, there is a strong indication that the singularity at $\nu = 3.591$ THz is a weak infinity. Unlike the case of Fig. 3, where the origin for the infinity is simple and well understood, the reason for the present infinity is less clear. It is the only known case of an infinity in $g(\omega)$ that occurs for a quite general force model, and it is important to understand its origin. It is clear that this singularity is associated with a triple degeneracy that exists at $\mathbf{k} = (1, 1, 1)$, but an attempt by Gilat (1967) to explain it was incorrect. Another feature of the computation shown in Fig. 12 is that it resolves for the first time a new critical point at $\nu = 2.86$ THz. This is attributed to the efficiency of the LA method.

In Fig. 13 the $g(\nu)$ for zinc is presented. Zinc has a hexagonal symmetry, and as such it was at first somewhat more difficult to obtain for it a highly resolved $g(\nu)$. The calculation is based on a modified axially symmetric model by DeWames *et al.* (1965). The mesh size in this calculation was $N_k = 3600$, and the resolution $N_\omega = 2500$. The critical points are all well resolved, and it was interesting to find out that at least one pronounced

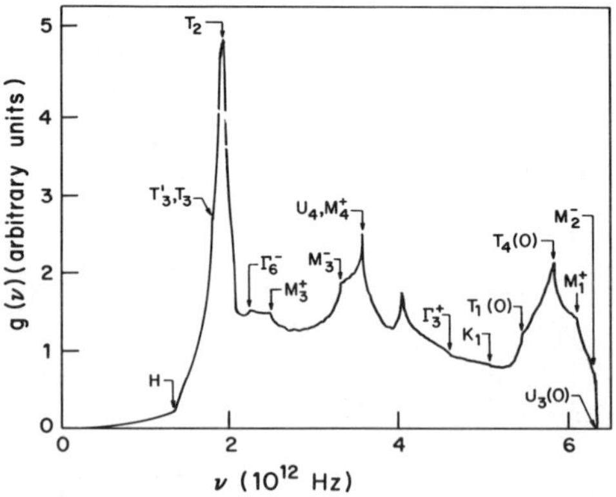

FIG. 13. LA method calculation of $g(\nu)$ for zinc. The lattice dynamical model is of DeWames *et al.* (1965). In this calculation $N_k = 3600$ and $N_\nu = 2500$. The assignment of the critical points is according to Raubenheimer and Gilat (1967).

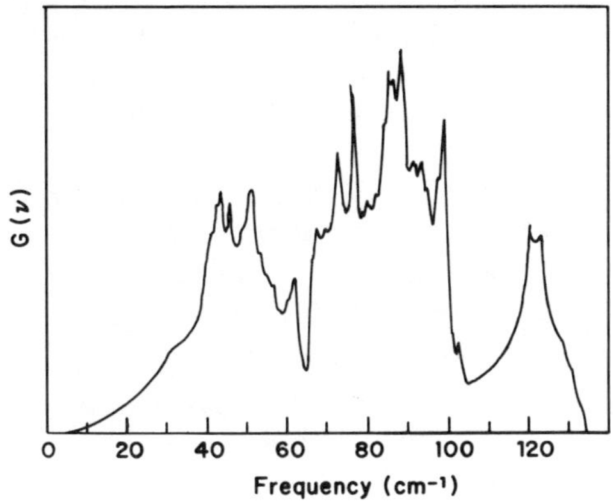

FIG. 14. LA method calculation of $g(\nu)$ of crystalline benzene. The crystal symmetry is orthorhombic, and the calculation was performed by Bonadeo and Taddei (1973). The mesh number is $N_k = 1000$ and $N_\nu = 1300$.

singularity originated from an off-symmetry **k** vector. As a matter of fact, this finding also showed up in the computation of $g(\omega)$ of magnesium, beryllium, and white-tin (Kam and Gilat, 1968). It is therefore important to realize that critical points can originate from off-symmetry **k** points, and the way to find them is by improving resolution in the calculation of $g(\omega)$. Incidentally, for symmetries lower than cubic we find at the same time fewer high symmetry points and more Van-Hove singularities, and therefore it is not surprising that at least some of them come from general **k** vectors.

The LA method is not confined to cubic or hexagonal lattices only. In principle it can be extended to any symmetry, but each symmetry may need appropriate modifications as in the case of hexagonal and trigonal crystals that require triangular integration cells. As yet the LA method has been extended to tetragonal and orthorhombic systems. In Fig. 14 the $g(\omega)$ of crystalline (orthorhombic) benzene is shown. The calculations of the phonon dispersion relations as well as of $g(\omega)$ are based on an empirical atom–atom potential and were performed by Bonadeo and Taddei (1973). The number of **k** vectors employed was $N_k = 1000$, and the resolution is $N_\omega = 1300$. Owing to the large number (24) of branches, the structure of $g(\omega)$ is very rich in critical points, most of which are well resolved by the computation. In an earlier calculation of $g(\omega)$ of benzene by Harada and Shimanouchi (1967) the RS method was employed, and the resolution was only $N_\omega = 25$, so that many critical points in $g(\omega)$ remained unresolved.

D. COMBINED (HYBRID) METHOD

The considerable success of the LA method may encourage the extension of the analytic method to include terms higher than linear. Unfortunately this task is prohibitively difficult, and even if possible it is likely to be inefficient. Moreover, it is impossible to extend the extrapolation method of obtaining $\nabla\omega$ beyond the linear term. The reason for this is the use of perturbation theory in deriving $\nabla\omega$. The interpolation method, however, can readily be extended to second-order terms. Alternatively, it is much simpler to give up higher order terms and to compensate for them by increasing N_k. There is, however, another possibility of improvement on the ordinary LA method by combining it with the QD method. This idea was first employed by Janak *et al.* (1970, Janak, 1969) and independently by Cooke and Wood (1972). The underlying idea of the combined method is to evaluate $\omega_j(\mathbf{k}_c)$ at a "gross" mesh point $\mathbf{k}_c$, then to form a fine regular mesh $\mathbf{k}_i$ in the vicinity of $\mathbf{k}_c$ exactly in the same manner as used for the discrete methods. At each fine mesh point $\mathbf{k}_i$ one proceeds to evaluate $\omega_j(\mathbf{k}_i)$ as well as the three components of $\nabla\omega_j(\mathbf{k}_i)$. The values for $\omega_j(\mathbf{k}_i)$ are obtainable from Eq. (11), and $\nabla\omega_j(\mathbf{k}_i)$ is interpolated from

$$\nabla\omega_j(\mathbf{k}_i) = \nabla\omega_j(\mathbf{k}_i) + (\mathbf{k}_i - \mathbf{k}_c) \cdot \nabla\nabla\omega_j(\mathbf{k}_c). \tag{29}$$

The next step is to subdivide the gross integration cell into many cells, each one centered at a respective $\mathbf{k}_i$ point, and then to apply the LA method to each of these cells. In this manner it becomes possible to further reduce the number N_k of gross mesh points. This saving can be quite important in cases where the computation of $\omega_j(\mathbf{k}_c)$ is a lengthy procedure, as it is for large dynamical matrices or for many electronic band structure calculations. Like the QD method, the combined method requires ten parameters [one for $\omega_j(\mathbf{k}_c)$, three for $\nabla\omega_j(\mathbf{k}_c)$ and six for $\nabla\nabla\omega_j(\mathbf{k}_c)$] for each branch j and each $\mathbf{k}_c$, a fact that adds to the complexity of the method.

The combined method may be regarded as a final step in the present sequence of methods for calculating densities of states and spectral properties in solids. It seems unlikely that an attempt to extend the analytical method beyond the linear term will prove successful. It is possible to extend the discrete methods to third-order terms and beyond, but the marginal gain in doing so is discouragingly small. The five methods discussed here are graphically summarized in Fig. 15. On the second column the various methods of integrating at $\mathbf{k}_c$ are shown schematically. The resultant $g(\omega)$ or $I(\omega)$ are shown on the third column. The discrete methods show up in $I(\omega)$ as discrete contributions and this is in contrast to the analytical methods, the contributions of which are continuous second order functions in ω.

The combined method has already been applied to a few spectral calculations. In Fig. 16 a comparative calculation of $g(\omega)$ for copper is shown. The comparison is between the QD, LA, and the combined methods. The calculations were performed by Cooke and Wood (1972) and are based on a

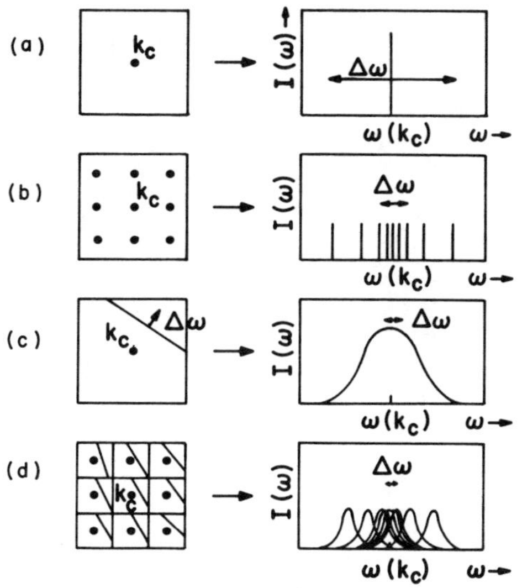

FIG. 15. Schematic description of the various zone integration methods. In the left column the different integration procedures in **k** space are symbolized. In the right column the resultant $g(\mathbf{k}_c; \omega)$ for every integration cell is shown. The discrete methods [RS (a), LD and QD (b)] yield discrete values for $g(\mathbf{k}_c; \omega)$ whereas analytic methods [LA (c) and combined (d)] yield continuous spectra. The error $\Delta\omega$ typical of each method is also indicated.

lattice dynamical model of Svensson *et al.* (1967). The energy resolution in all three graphs of $g(\omega)$ is $N_\omega = 1800$, but the smoothness of the curves is not the same. The LA method for which $N_k = 4218$ mesh points gave the smoothest $g(\omega)$ of the three. The QD method used $N_k = 1000$ gross mesh points and $M = 6000$ points randomly situated inside each small cube. The QD method gave the least smooth curve of the three and also required the longest computing time. The numerical details concerning the combined method do not give a specific value for N_k, but it is estimated to be close to 100. The fine mesh for each cell is $M = 1000$. As expected the combined method is the fastest of the three, but its accuracy for the example given is intermediate between the LA and the QD methods. It should also be mentioned that both the QD and the combined methods use interpolation, whereas the LA method uses extrapolation, and a certain systematic error associated with interpolation is caused. This point is further discussed in Section II, E.

Another comparative calculation is shown in Fig. 17, and it concerns the electronic band structure of ferromagnetic nickel. The band structure itself was computed by the KKR method. The electronic density of states $g(E)$ for 12 bands was calculated by Cooke and Wood (1972), and the comparison is between the QD and the combined methods. The computed $g(E)$ looks

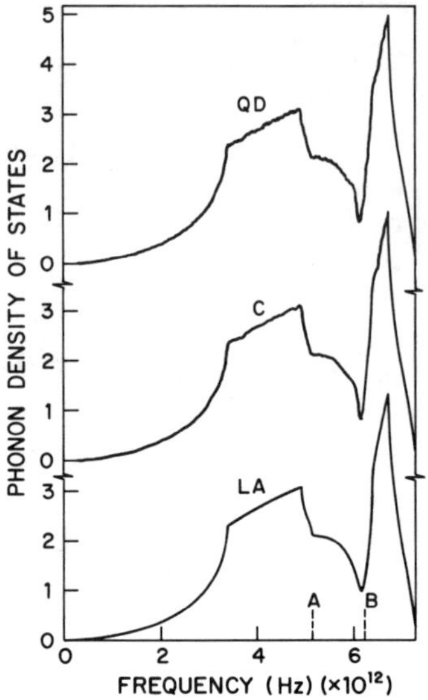

FIG. 16. Comparative calculation of $g(\nu)$ of copper using the QD, LA, and the combined methods. The dynamical model is of Svensson *et al.* (1967). The resolution is $N_\nu = 1800$ for all calculations. For the LA method $N_k = 4218$, for the QD method $N_k = 1000$, and in each cell $M = 6000$. The combined method uses about $N_k = 100$, and $M = 1000$. The calculations were performed by Cooke and Wood (1972).

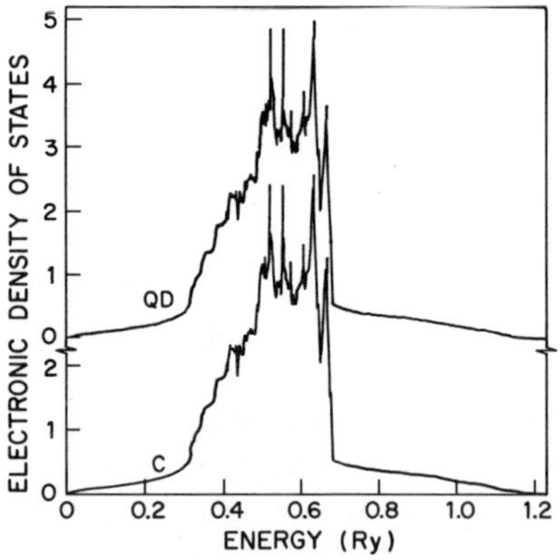

FIG. 17. Comparison between calculations of electronic $g(E)$ for nickel using the QD and the combined methods. The band structure was calculated by the KKR method (Faulkner *et al.*, 1967). The mesh number in both calculations is $N_k = 916$ and resolution is $N_E = 1800$. The calculation was performed by Cooke and Wood (1972).

much the same for both calculations with very minor differences. Both calculations employ $N_k = 916$ gross mesh points and energy resolution of $N_\omega = 1800$, but the combined method was quicker by about a factor of 11 in computing time.

In summary, the combined method which couples together the best features of the linear–analytic and the quadratic–discrete methods is the best available method for lengthy and complex calculations of densities of states and spectral properties of solids. However, by its very virtue of high efficiency, direct (and lengthy) calculations are substituted to a large extent by approximations, and one should be wary of the danger of overstretching these approximations at the cost of accuracy.

E. Interpolation and Extrapolation Methods

A problem very closely linked with the various methods discussed in Sections II, B, C, and D concerns the ways of obtaining the various parameters of $\nabla\omega(\mathbf{k}_c)$ and $\nabla\nabla\omega(\mathbf{k}_c)$ which are essential for the implementation of these methods.

In all the calculations of $g(\omega)$ and $I(\omega)$ it is necessary to obtain many values of eigenvalues $\omega_j(\mathbf{k}_c)$ at different $\mathbf{k}$ vectors, which can be a lengthy procedure owing to large matrices or because of complex force models. Apart from $\omega_j(\mathbf{k})$, it is also necessary for all methods except for the RS method to evaluate $\nabla\omega_j(\mathbf{k})$, and in some cases even $\nabla\nabla\omega_j(\mathbf{k})$ is required. However, $\nabla\omega$ and $\nabla\nabla\omega$ are only auxiliary for the computation in contrast to ω_j, which is essential. For this reason the values of $\nabla\omega_j$ and $\nabla\nabla\omega_j$ are required at a lower accuracy than ω_j. The error in $\nabla\omega_j$ to be permitted is of the order of the variations of $\nabla\omega_j$ over an integration cell. The error permitted in $\nabla\nabla\omega_j$ is even larger. In view of this assumption, it is convenient to apply relatively crude but quick methods of evaluating $\nabla\omega$ and $\nabla\nabla\omega$ at each $\mathbf{k}$ vector.

In simple cases where it is possible to obtain analytical expressions for $\omega_j(\mathbf{k})$ the evaluation of $\nabla\omega_j$ and $\nabla\nabla\omega_j$ is direct and simple, such as the case of a single band Hamiltonian of spin waves. Unfortunately, such cases occur very seldom. In most practical cases it is required to apply approximate methods for this purpose.

Two methods for evaluating $\nabla\omega$ and $\nabla\nabla\omega$ have emerged in recent years, and they are known under the names of extrapolation and interpolation methods. The extrapolation method was first introduced by Gilat and Dolling (1964) in connection with the LD method, and it is closely tied to first-order perturbation technique. The eigenvalues of the dynamical matrix $\omega_j^2(\mathbf{k})$ are evaluated at a point $\mathbf{k}$. Then small increments δk_x, δk_y, and δk_z are added to $\mathbf{k}$, and the respective changes $\varepsilon_j(\mathbf{k})$ in the eigenvalues are found as first-order perturbations. Let $D_{ij}(\mathbf{k})$ be an element of the dynamical matrix and $\Delta_{ij}{}^{\alpha}(\mathbf{k})$ the change in D_{ij} incurred by adding $\delta\mathbf{k}$ to $\mathbf{k}$, where $\alpha = x, y, z$. Then

$$\Delta_{ij}{}^{\alpha}(\mathbf{k}) = D_{ij}(\mathbf{k} + \mathbf{e}_{\alpha}\,\delta k_{\alpha}) - D_{ij}(\mathbf{k}), \tag{30}$$

and $\mathbf{e}_{\alpha}$ is a unit vector along the α axis. The change $\varepsilon_j(\mathbf{k})$ in the jth eigenvalue is given by

$$\varepsilon_j(\mathbf{k}) = \Delta'^{\alpha}_{jj}(\mathbf{k}) + \sum_{i \neq j} \frac{(\Delta'^{\alpha}_{ij})^2}{\omega_{jj}{}^2 - \omega_{ii}{}^2} + \cdots, \tag{31}$$

where $\Delta'^{\alpha}_{ij}(\mathbf{k})$ is given by

$$\Delta'^{\alpha}_{ij}(\mathbf{k}) = \sum_{st} u^{*}_{is}(\mathbf{k})\,\Delta_{st}{}^{\alpha}(\mathbf{k}) u_{tj}(\mathbf{k}), \tag{32}$$

and $u_{is}(\mathbf{k})$ is the sth element of the ith eigenvector of the dynamical matrix at $\mathbf{k}$. In order to simplify the computation of $\varepsilon_j(\mathbf{k})$, it is desirable to neglect the second term in Eq. (31) in comparison with Δ'^{α}_{jj}. This is justifiable if the increments δk_{α} are sufficiently small. A value of $\delta k_{\alpha} \cong 10^{-4}$ was found to be sufficiently small. Another requirement that concerns Eq. (31) is that there be no degenerate eigenvalues of D at $\mathbf{k}$. Degeneracies usually occur at high symmetry values of $\mathbf{k}$, and it is therefore advisable to choose a mesh of $\mathbf{k}$ vectors with as few highly symmetric $\mathbf{k}$ vectors as possible. Gilat and Dolling (1964) used a cubic mesh shifted away from the origin along the (1, 1, 1) direction. In this manner it was possible to avoid all $\mathbf{k}$ vectors with zero components, and, consequently, the degeneracies associated with them. On the other hand, the degeneracies associated with $\mathbf{k}$ vectors having all three components equal (Λ points) were still present. The nature of symmetry of Λ points has a certain effect on the transverse modes along the (1, 1, 1) direction (Gilat, 1969). Its effect on the calculation of $g(\omega)$ is negligible, however.

Having obtained an estimate for $\varepsilon_j(\mathbf{k})$ it is straightforward to obtain $\boldsymbol{\nabla}\omega_j(\mathbf{k})$ by

$$\nabla_{\alpha}\omega_j(\mathbf{k}) = [1/2\omega_j(\mathbf{k})]\,[\Delta'^{\alpha}_{jj}(\mathbf{k})/\delta k_{\alpha}]. \tag{33}$$

It is important to point out that because of the use of first-order perturbation technique in the calculation of $\boldsymbol{\nabla}\omega_j(\mathbf{k})$, extrapolation is unsuited for the calculation of higher order derivatives of $\omega_j(\mathbf{k})$.

The method of interpolation is based on solving Eq. (11) for the unknown quantities of $\omega_j(\mathbf{k})$, $\nabla\omega_j(\mathbf{k})$, and $\nabla\nabla\omega_j(\mathbf{k})$. If $\mathbf{k}_c = \mathbf{k}$ is a given mesh point and $\mathbf{k}_i$ are neighboring points, then for a three-dimensional cubic mesh there are usually 27 neighboring points. In Eq. (11) there are 10 unknowns that can be evaluated from the 27 equations for the given eigenvalues at $\mathbf{k}_i$ and $\mathbf{k}_c$ by some optimization procedure (Mueller *et al.*, 1971; Mueller, 1971).

In applying any interpolation scheme which involves several eigenvalues per **k** vector, it is required to relate in a systematic way the various eigenvalues of the different points. The usual procedure adopted is to form a sequence of increasing order of eigenvalues that satisfy

$$\omega_{j+1}(\mathbf{k}) \geqslant \omega_j(\mathbf{k}) \geqslant \omega_{j-1}(\mathbf{k}). \tag{34}$$

This procedure conforms with the usual assignment of bands. There is, however, a certain problem associated with this method of identifying bands, which is liable to occur at regions where two or more bands cross, or sharply approach each other. The case of the actual band crossing is limited only to bands that belong to different representations of the symmetry group of a given **k** vector. For general **k** vectors the dynamical matrix, or the Hamiltonian, are usually irreducible, so that all the bands belong to the same representation and therefore cannot cross each other. The case of sharp approaching of two bands is generally permitted and actually occurs in many cases. At a point of maximal approach of two bands the gradients of these bands are zero. Now, the region required for the interpolation process is not negligible in comparison to the total IBZ, and therefore the event of mutual approach of two bands is likely to happen within this range. As a result, interpolation may predict too small values for $|\nabla\omega|$ over a region of mutual approach of two bands. A direct outcome of this can be the wrong prediction of extra singularities in $g(\omega)$. A close examination of Fig. 16 actually reveals such a case. In the range of frequencies marked by the points A and B we can spot a weak singularity for the two upper curves that were computed with the aid of interpolation. In contrast to these, the lower graph which uses extrapolation does not show this singularity. The origin of this fortuitous critical point is well known, and it comes from the crossing of two phonon bands along the [110] direction (see also Gilat, 1973).

Very recently, Cooke *et al.* (1975) performed a comparative computation of the electronic density of states of niobium, using interpolation and extrapolation procedures. In comparing the results, Cooke *et al.* noticed that an additional and fortuitous critical point was present for the case when interpolation was employed. Moreover, this extra critical point is associated with an actual band crossing that occurs in the electronic band structure of Nb.

F. Analysis of Resolution, Accuracy, and Computing Effort

It is important to be able to make a quantitative comparison between the various methods described in this article, with a special emphasis on such properties as resolution, accuracy, and computing effort. Such an analysis was recently made by Gilat (1972); the main conclusions are briefly brought here, and the same notation is used here. The parameter that mostly affects resolution, accuracy, and computing effort is N_k, the number of gross mesh points in the IBZ. However, the same N_k can yield very different results for these properties when different zone integration methods are used. For discrete methods there is another parameter that affects these merits, and it is the number of fine mesh points M in the close vicinity of each gross mesh point $\mathbf{k}_c$. It can be observed from Eq. (11) that if $\omega_j(\mathbf{k})$ is expanded in power series about $\mathbf{k}_c$, then the error in $\omega_j(\mathbf{k})$ made by neglecting all powers of $(\mathbf{k} - \mathbf{k}_c)$ beyond a certain power, can be estimated by the first term that is neglected. Following this idea, the RS method, being of zeroth order in $\mathbf{k} - \mathbf{k}_c$, admits an error estimated by $(\mathbf{k} - \mathbf{k}_c) \cdot \boldsymbol{\nabla}\omega$. As a matter of fact all the discrete methods (LD and QD) are of zeroth order because every small cell is represented only by a single value of ω_j. The advantage of the discrete methods over the RS method is in the substantial and easy increase of mesh points from N_k to practically MN_k. It can be shown that the resolution N_ω given by Eq. (9), is related to N_k for the RS method by (Gilat, 1972)

$$N_\omega = \alpha_0 N_k^{1/3}, \tag{35}$$

where α_0 is a number of the order of unity the exact value of which depends to some extent on each specific case. For the discrete methods (LD and QD) Eq. (35) is modified to

$$N_\omega = \alpha_0 (MN_k)^{1/3}. \tag{36}$$

The analytic methods (LA and the combined methods) are linear in $\mathbf{k} - \mathbf{k}_c$, and hence the error for each ω_j is estimated by $\frac{1}{2}(\mathbf{k} - \mathbf{k}_c) \cdot \boldsymbol{\nabla\nabla}\omega \cdot (\mathbf{k} - \mathbf{k}_c)$. For the LA method the relation between N_ω and N_k is given by

$$N_\omega = \alpha_1 N_k^{2/3}, \tag{37}$$

where α_1, like α_0, is a number of the order of unity. For the combined method the same Eq. (37) can be used, but N_k is substituted by MN_k. Expression for α_0 and α_1 are given (Gilat, 1972), but these must be considered only as crude guidelines for estimation. The substantial gain in resolution for a given mesh size N_k can be demonstrated by Table I where we show on the first line how

TABLE I

NUMERICAL ESTIMATES OF RESOLUTION AND ACCURACY PER GIVEN MESH SIZE AND OF COMPUTING EFFORT (N_k) PER GIVEN RESOLUTION FOR THE VARIOUS METHODS DESCRIBED IN THE TEXT[a]

	Root sampling	Linear discrete $M = 100$	Quadratic discrete $M = 100$	Linear analytical	Combined $M = 100$
N_ω for given $N_k = 1000$	30	130	300	1050	4800
$\delta I/I$ for $N_k = 1000$	1%	0.2%	0.1%	0.03%	0.006%
N_k for $N_\omega = 100$	43,500	440	40	30	⟨1

[a] The number of bands is $n = 9$ for these examples.

the resolution N_ω increases for a given $N_k = 1000$ by using different methods of increasing efficiency. On the third line the resolution N_ω is set constant ($N_\omega = 100$), and the number of mesh points N_k is accordingly reduced. It should be pointed out, however, that the relations between N_k and N_ω are valid only for a sufficiently large N_k that permits the neglect of the effects of mesh points located at the surface of the IBZ, N_ω which is determined by Eqs. (35)–(37), is referred to as "optimal resolution."

A measure for the accuracy of a given spectral calculation is given by the relative amplitude error $\delta I/I$ and is closely related to the resolution N_ω. For values of ω that are not too close to critical points $\delta I/I$ is proportional to N_ω^{-1}. In the vicinity of highly peaked critical points, the error becomes somewhat larger. In Table I the numerical estimates for $\delta I/I$ for the various computational methods are given.

In the cases of the LD, QD, and the combined methods, use is made of the fine mesh parameter M. Obviously resolution is improved by increasing M. On the other hand, there must be some limit to the size of M. An important practical limit on M is associated with the computing time and is dealt with at the end of this section. Presently we are interested in defining an optimal value of M. By generating a set of fine mesh points, a certain error in estimating ω_k is caused at each point. If this error is adjusted with the overall error for the gross integration cell, it is possible to define an optimal value M_{op} for M, which is directly related to N_k, but in different ways for different computational methods. For the LD method we obtain

$$M_{\text{op}} = \eta_{\text{LD}} N_k, \tag{38}$$

while for the QD and the combined methods we have, respectively:

$$M_{\text{op}} = \eta_{\text{QD}} N_k^2 \tag{39}$$

and

$$M_{\text{op}} = \eta_{\text{C}} N_k^{\ 1/2}. \tag{40}$$

The numerical coefficients η_{LD}, η_{QD}, and η_{C} can be estimated in a rather crude manner, much along the same lines as for the coefficients α_0 and α_1. The values given for these coefficients are $\eta_{\text{LD}} \simeq 50$, $\eta_{\text{QD}} \simeq 9000$, and $\eta_{\text{C}} \simeq 13$. The reader must be cautioned not to take these values too literally, and they must be used rather as general guidelines for choosing the fine mesh number M. More important than the specific value of M_{op} is its functional dependence on N_k. In all case M_{op} is an increasing function of N_k. The meaning of this is of limiting nature, i.e., a gain in resolution may be achieved by increasing M up to the value of M_{op}, but this gain is limited, and no additional gain in resolution is achieved by increasing M beyond the value suggested by M_{op}. Unfortunately M_{op} can be very large and it is not always easy even to approach M_{op}. This is so because of the functional dependence of M_{op} on N_k. For the QD method M_{op} is of the order of $M_{\text{op}} = 10^{10}$ for $N_k = 1000$. This makes it highly impractical, and M that is actually used is of the order of 10^3. For the LD method M_{op} is of the order of 10^4, much more within the reach of available computers. The best manageable M_{op} is for the combined method, where M_{op} is of the order of 10^2 for $N_k = 1000$.

It is interesting and also clarifying to point out that if M_{op} is used for the LD method for a given N_k, then the optimal resolution N_ω is equal to the same N_ω obtained for the LA method with the same N_k. This means that the LA method is intrinsically optimal and is equivalent to the LD method with its appropriate M_{op}. It is impossible to draw a similar comparison for the QD method because a quadratic–analytic method is still nonexistent.

So far we have made no use in our considerations of the computing effort associated with the various methods. In order to do so we must consider the computing time involved in each of the elements of computation that are required to obtain $g(\omega)$ by each computational method. We denote the computing times required for each computational method by T_i, where i = RS, LD, QD, LA, and C. For all the methods it is required to evaluate a set of eigenvalues $\omega_j(\mathbf{k}_c)$ for every $\mathbf{k}_c$. The time required for this is denoted by T_k, the value of which is independent of the method used. Apart from T_k it takes time T_ω to find $g(\mathbf{k}_c, j; \omega)$ for each j and $\mathbf{k}_c$. There is additional time T_∇ required to evaluate $\nabla\omega(\mathbf{k}_c)$ and $\nabla\nabla\omega(\mathbf{k}_c)$ whenever they are required. Also, for the LD, QD, and the C methods we must generate a fine mesh that requires T_M for each additional mesh point. Whenever different methods require different times for the same element of computation, this is denoted by a superscript i. In the following expressions we bring estimates for the time consumption for each method:

$$T_{\mathrm{RS}} = N_k(T_k + T_\omega), \tag{41}$$

$$T_{\mathrm{LD}} = N_k[T_k + T_\nabla^{\mathrm{LD}} + (M - 1)(T_M + T_\omega)], \tag{42}$$

$$T_{\mathrm{QD}} = N_k[T_k + T_\nabla^{\mathrm{QD}} + (M - 1)(T_M + T_\omega)], \tag{43}$$

$$T_{\mathrm{LA}} = N_k(T_k + T_\nabla^{\mathrm{LD}} + T_\omega^{\mathrm{LA}}), \tag{44}$$

$$T_{\mathrm{C}} = N_k[T_k + T_\nabla^{\mathrm{QD}} + (M - 1)(T_M + T_\omega^{\mathrm{LA}})]. \tag{45}$$

Of all time elements included in the right-hand sides of Eqs. (41)–(45), T_k is in most cases by far the longest. In fact it is exactly for this reason that more efficient methods are needed. There is, however, one exception, where T_ω^{LA} might be comparable to T_k, and this is when very large N_ω are required (of the order of 10^3–10^4). The comparison between different computing times T_i required by different methods must be made for the same resolution N_ω, which is related to N_k in different ways for different methods via Eqs. (35)–(37). Of special interest in our analysis is the effect on computing time of setting up fine meshes. This can be studied by comparing T_{RS} with T_{LD} and T_{QD}, respectively, for the same N_ω. The ratio $T_{\mathrm{RS}}/T_{\mathrm{LD}}$, as well as $T_{\mathrm{RS}}/T_{\mathrm{QD}}$, becomes in the limit of $M \to \infty$:

$$M_{\mathrm{eff}} = \lim_{M\to\infty} (T_{\mathrm{RS}}/T_{\mathrm{LD}}) = (T_k + T_\omega)/(T_M + T_\omega). \tag{46}$$

This ratio limits the efficiency of the LD and the QD methods relative to the RS method. The value of M to be recommended by this analysis is of the order of the ratio given by Eq. (46), which is denoted by M_{eff}. Similar analysis can be made for the combined method against the LA method. In this case M_{eff} has a slightly different expression (Gilat, 1972). In most cases M_{eff} is of the order of 10^2–10^3. For the combined method it is even smaller and practically is equal to M_{op}. Incidentally, when very large N_ω is required, T_ω^{LA} becomes of the order of T_k, then M_{eff} approaches unity. This limits the efficiency of the combined method for large N_ω.

The material concerning resolution, accuracy and computing effort is summarized in Table I (see Gilat, 1972).

G. Transition Probabilities

In the preceding sections it was preferred for reasons of simplicity to avoid the effect of the transition probability $F(\mathbf{k})$ on the calculation of $I(\omega)$. In reality, spectra $I(\omega)$ are usually measured rather than densities of states $g(\omega)$, and $I(\omega)$ includes the transition probability $F(\mathbf{k})$ for the excitations involved (phonons, electrons, magnons, etc.) under the influence of some external probe (neutrons, photons, electrons, etc.) It is therefore very important to understand the effect of $F(\mathbf{k})$ on the calculation of $I(\omega)$. Unfortunately, $F(\mathbf{k})$ involves a large variety of phenomena of different natures, and it is not easy at all to put forward arguments that are sufficiently general to describe every possible case.

For example, in the case of optical transitions in solids the external probe is a photon that causes transitions between electronic bands. The transition probability for this case is given by $|\langle i|\mathsf{O}_p|j\rangle|^2$, which involves the electronic eigenstates $\langle i|$ and $\langle j|$ and the operator O_p that causes the transition. The optical spectrum itself is given by

$$I(\omega) = \sum_i \int |\langle i|\mathsf{O}_p|j\rangle|^2 \, \delta(E_i - E_j \pm h\omega) \, d^3k. \tag{47}$$

The calculation of $I(\omega)$ requires evaluation of $|\langle i|\mathsf{O}_p|j\rangle|^2$ at every $\mathbf{k}$ vector, which may be a difficult task. Apart from this $E_i - E_j$ rather than E_i must be obtained at each $\mathbf{k}$, but this in itself is not a serious complication.

A second example concerns the calculation of superconducting tunneling. The spectra $I(\omega)$ in this case involve the calculation of the electron–phonon interaction that is associated with $F(\mathbf{k})$ as well as the phonon dispersion relations, but the range of summation itself is confined to the volume included within the first Fermi surface. This, for instance, is a complication peculiar to this specific problem. In the calculation of second-order Raman effect we have to perform two-phonon summations so that their net wave vector is zero. Also in the calculation of anharmonic effects, two-phonon summations are required.

The consequence of these examples is that each spectrum $I(\omega)$ may have its own peculiarities that must be studied and understood in each calculation. In Section III a few examples will be studied in more detail.

One feature that is common to all spectra $I(\omega)$ is that they all are related to Eqs. (1)–(3). In particular, Eq. (2) represents the essential feature common to most spectra $I(\omega)$, which is its dependence in the form of the highly singular δ functions on the elementary excitations of the solid. It is generally believed that the $\mathbf{k}$ dependence of $F(\mathbf{k})$ is of much smoother nature than the corresponding dependence of the δ function. The dependence of $F(\mathbf{k})$ on $\mathbf{k}$ can be separated into two parts: (i) its dependence via the eigenvectors which are slowly varying functions of $\mathbf{k}$ except near points of degeneracy; (ii) its dependence on $\mathbf{k}$ through the matrix element, which in many cases is not well understood.

The computational implication of the mild $\mathbf{k}$-dependence of $F(\mathbf{k})$ is that $F(\mathbf{k})$ can be regarded as a constant $F(\mathbf{k}_c)$ over each integration cell centered at $\mathbf{k}_c$. We refer to this approximation as the "unrefined" spectrum $I(\omega)$. Most computations of $I(\omega)$ that are presented in literature are performed by the unrefined approximation. In order to refine the calculations, variations of $F(\mathbf{k})$ in the vicinity of $\mathbf{k}_c$ must be taken into account. For this purpose it is necessary to know the gradient $\nabla F(\mathbf{k}_c)$. In the case of the LA method it is possible to incorporate the exact effect of the linear variation of $F(\mathbf{k})$ in an analytic way. It was shown by Dalton (1970) that this can be done in a general and straightforward way by employing general relationships between integrals that involve spectral calculations. Dalton introduces the function $S(\boldsymbol{\lambda}; \omega)$ defined by

$$S(\boldsymbol{\lambda} : \omega) = \int_v \theta(\omega - \boldsymbol{\lambda} \cdot \mathbf{k})\, d^3k, \tag{48}$$

where $\theta(x)$ is a positive step function, and v is any volume of integration. By noting that $d\theta(x)/dx = \delta(x)$ it can be deduced that

$$I(\boldsymbol{\lambda}; \omega) = \partial S(\boldsymbol{\lambda}; \omega)/\partial \omega, \tag{49}$$

where $I(\boldsymbol{\lambda}; \omega)$ is the spectral function $I(\omega)$ obtained for the volume v of Eq. (48), which in the practical case is the ordinary integration cell. $I(\boldsymbol{\lambda}; \omega)$ is to be associated with $g(\mathbf{k}_c; \omega)$ and is given by

$$I(\boldsymbol{\lambda}; \omega) = \int_v \delta(\omega - \boldsymbol{\lambda} \cdot \mathbf{k})\, d^3k. \tag{50}$$

Next we write the expressions $I_i(\boldsymbol{\lambda}; \omega)$ that are required for evaluating the variations of $F(\mathbf{k})$ near $\mathbf{k}_c$

$$I_i(\boldsymbol{\lambda}; \omega) = \int_v k_i\, \delta(\omega - \boldsymbol{\lambda} \cdot \mathbf{k})\, d^3k, \tag{51}$$

where $i = x, y, z$. Now $I_i(\boldsymbol{\lambda}; \omega)$ can be related to $S(\boldsymbol{\lambda}; \omega)$ in the simple manner:

$$I_i(\boldsymbol{\lambda}; \omega) = -\partial S(\boldsymbol{\lambda}; \omega)/\partial \lambda_i. \tag{52}$$

The way to obtain the explicit expressions for $I_i(\boldsymbol{\lambda}; \omega)$ is by writing the integral relation between $I(\boldsymbol{\lambda}; \omega)$ and $S(\boldsymbol{\lambda}; \omega)$, namely,

$$S(\boldsymbol{\lambda}; \omega) = \int_{-\infty}^{\omega} I(\boldsymbol{\lambda}; \alpha)\, d\alpha, \tag{53}$$

and then by eliminating $S(\boldsymbol{\lambda}; \omega)$, $I_i(\boldsymbol{\lambda}; \omega)$ can be derived directly from $I(\boldsymbol{\lambda}; \omega)$. In the practical case Eqs. (50)–(51) are restricted to an integration cell for which $|\nabla \omega|$ is constant, and $\boldsymbol{\lambda}_i$ are its direction cosines. For orthorhombic cells the expressions for $I(\boldsymbol{\lambda};\, \omega)$ are given by Eqs. (21)–(25), where $w = \omega/|\nabla\omega|$. The complete expressions for computing the refined part of $I(\omega)$ were first derived by Gilat and Kam (1969) for orthorhombic cells. The refined part of $I(\omega)$ adds for each integration cell an asymmetric correction to the symmetric function $S(l_i;\, w)$ shown in Fig. 10.

The significance of refining the spectrum $I(\omega)$ has not been thoroughly tested as yet for realistic calculations. The information gathered from a few model calculations is that refining can affect $I(\omega)$ by 3–5% for N_k of the order of 10^2–10^3. This effect is expected, however, to be much more significant whenever $F(\mathbf{k})$ displays strong variations as a function of $\mathbf{k}$.

In many cases that occur in actual computations very little is known of $F(\mathbf{k})$, and as a result in some cases only its dependence on eigenvectors is considered. Now for this particular case it is known that the rate of variation of eigenvectors with respect to $\mathbf{k}$ is considerably slower in comparison to that of eigenvalues. In the framework of the extrapolation method the $\mathbf{k}$-dependence of eigenvectors is altogether neglected within each integration cell, so that for this special case the refined part of the $I(\omega)$ is zero. It is interesting to note that in the case of band crossing, discussed in Section II, E, the part of $F(\mathbf{k})$ that depends on the eigenvectors varies strongly in the neighborhood of the point of nearest approach of bands. This effect has not been accounted for as yet in the spectrum calculations that include $F(\mathbf{k})$.

H. Real Part Calculations

So far only the imaginary part of the spectrum that is associated with Eq. (5) has been discussed, and very little has been said about the real part $R(\omega)$ of the spectrum. Calculations of $R(\omega)$ are less frequent than $I(\omega)$ in solid state physics, but nevertheless they do occur. For example, they happen in the case of optical absorption in solids in magnetic susceptibilities $\chi(\mathbf{q}, \omega)$, and also in the calculation of the spectral properties of impurities in lattices.

There are two computational ways of obtaining $R(\omega)$ from a model that can yield $\omega_j(\mathbf{k})$. The first way is the direct one of solving Eq. (6). The same computational methods used for the calculation of $I(\omega)$ are also applicable to $R(\omega)$, but in contrast to $I(\omega)$ the use of discrete methods is to be discouraged. The reason for this is associated with the nature of the singularity $[\omega - \omega(\mathbf{k})]^{-1}$ that is typical of $R(\omega)$. In taking a finite (discrete) sample of values of $\omega(\mathbf{k})$ it is possible for the difference $\omega(\mathbf{k}) - \omega$ to attain values arbitrarily close to zero. Moreover, this difference can have both positive or negative values. In computing $R(\omega)$ for a finite sample there is the danger of a very large noise level, because it is very difficult to obtain exact cancellations of positive and negative contributions by using a sampling method. In the case of the LA method, the analytical expressions for the contribution to $R(\omega)$ from each integration cell are by their very nature insensitive to this problem. These expressions were first derived by Gilat and Bohlin (1969) for orthorhombic cells. They are also readily derivable by the way of Dalton (1970). Dalton defines the contribution to $R(\omega)$ coming from an arbitrary cell by $R(\boldsymbol{\lambda}; \omega)$,

$$R(\boldsymbol{\lambda}: \omega) = \int_v \frac{d^3k}{\omega - \boldsymbol{\lambda} \cdot \mathbf{k}}, \tag{54}$$

where all the symbols are explained in Section II, G. The integrand $(\omega - \boldsymbol{\lambda} \cdot \mathbf{k})^{-1}$ can be written in the following way:

$$(\omega - \boldsymbol{\lambda} \cdot \mathbf{k}) = \int_{-\infty}^{+\infty} z^{-1}\, \delta(z - \omega + \boldsymbol{\lambda} \cdot \mathbf{k})\, dz. \tag{55}$$

Inserting back the result of Eq. (55) into (54), and interchanging the order of integration and using Eqs. (49)–(50), we obtain the result

$$R(\boldsymbol{\lambda}; \omega) = \int_{-\infty}^{+\infty} \frac{\partial S(\boldsymbol{\lambda}; \omega - z)}{\partial \omega} z^{-1}\, dz = \int_{-\infty}^{+\infty} I(\boldsymbol{\lambda}; \omega - z) z^{-1}\, dz. \tag{56}$$

A numerical example of the calculation of $R(\omega)$ is shown in Fig. 18. The function $R(\omega)$ is given by $R(\omega) = \sum_{\mathbf{k}} (\omega - \omega_{\mathbf{k}})^{-1}$ in a simple cubic lattice, where $\omega_{\mathbf{k}} = \sum_i \cos(\frac{1}{2}k_i)$ and $i = x, y, z$. This function is the principal value (real part) of the one shown in Fig. 3. The solid line in Fig. 18 was obtained

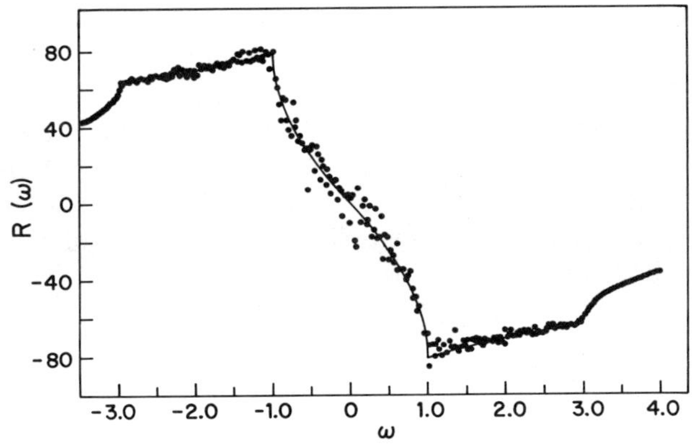

FIG. 18. Comparative calculation of the real part $R(\omega)$ given by $R(\omega) = \Sigma_k(\omega - \Sigma_i \cos \frac{1}{2}k_i)^{-1}$, $(i = x, y, z)$ for simple cubic lattices. The solid curve represents an LA calculation with $N_k = 1540$ and $N_\omega = 3000$, the dots represent an RS calculation with $N_k = 11{,}480$ and $N_\omega = 240$.

by using the LA method, and the dots were obtained by the RS method. As expected, the dots fluctuate considerably in the range $3.0 \geqslant \omega \geqslant -3.0$, where $R(\omega)$ is highly singular, and they fit the theoretical curve very well outside the range of singularity.

Another way of computing $R(\omega)$ is by calculating $I(\omega)$ first and then applying the Kramers–Krönig relations of Eqs. (7)–(8). In fact the right-hand side of Eq. (56) implies these relations for each integration cell separately. In applying these relations care must be taken to integrate out analytically the pole at $\alpha = \omega$. Failing to do so can cause a significant noise level in the computed $R(\omega)$.

The two ways of calculating $R(\omega)$ require approximately the same computer effort, and they give equal resolutions for the same N_k. Therefore they are both to be recommended for practical calculations.

III. Examples of Spectral Properties in Solids

In this section a few specific examples for spectral properties of solids are discussed, and their computational aspects are described. The treatments given here are not meant to represent the theoretical backgrounds for examples given here, but rather their computational aspects. The theory relating to these examples can be found elsewhere in specialized articles. Consequently, the subjects are given here in a compact form, so as to cover their respective computational aspects.

A. Phonon Densities of States

This section deals with the phonon densities of states that can be measured by the scattering of slow neutrons and their comparison with model calculations. Until recently it was believed that the only means of obtaining experimentally the phonon density of states $g(\nu)$ was by incoherent scattering of neutrons (Placzek and Van-Hove, 1954). Bredov *et al.* (1967) proposed that $g(\nu)$ of crystals be measured by coherent inelastic scattering of slow neutrons from a crystalline powder. The single phonon coherent scattering of neutrons yields phonon dispersion relations, since in this process both energy and momentum are conserved, namely

$$E_1 - E_0 = \pm\hbar\omega_j(\mathbf{k}), \tag{57}$$

$$\mathbf{k}_1 - \mathbf{k}_0 = \mathbf{k} + \mathbf{G}, \tag{58}$$

where E_i $(i = 0,1)$ are the energies of the incident and scattered neutrons respectively, and $\hbar\mathbf{k}_i$ are the corresponding neutrons momenta. $\mathbf{k}$ is the phonon wave vector, and $\mathbf{G}$ is a reciprocal lattice vector. These relations are obtainable from the expression of the one-phonon coherent scattering cross section for a single crystal. The process of averaging over all directions for polycrystalline powder is not straightforward, but can be approximated by making a few simplifying assumptions. The most daring assumption is that k_i are spherically symmetric. Bredov *et al.* (1967) obtain the differential cross section for one-phonon coherent scattering of polycrystalline powder

$$\frac{d^2\sigma}{d\Omega\, dE} = \frac{h^2 b^2}{2M} e^{-2W} \frac{k_1}{k_0} (k_1{}^2 + k_0{}^2) \frac{g(\omega)}{\hbar\omega} \frac{1}{\exp(\hbar\omega/k_B T) - 1}. \tag{59}$$

In this expression b is the nuclear coherent scattering length, e^{-2W} is the Debye–Waller factor, and M is the atomic mass of scattering nuclei. Equation (59) is suitable for monatomic cubic crystals only.

Equation (59) provides the justification for the proposal of measuring $g(\omega)$ by inelastic coherent scattering of neutrons. This expression includes a Boltzmann term $[\exp(\hbar\omega/k_B T) - 1]^{-1}$, which is only ω-dependent, and there are no $\mathbf{k}$-dependent terms in Eq. (59). Obviously this expression becomes somewhat questionable whenever k_i deviate considerably from spherical symmetry.

More recently Gompf *et al.* (1972) applied this method to the measurement of $g(\nu)$ of several monatomic cubic metals. In Fig. 19 a comparison between three kinds of $g(\nu)$ for aluminum is shown. The first $g(\nu)$, presented by a solid line, is computed from a force constant model obtained from a best fit to the experimental dispersion relations obtained by Stedman and Nilsson (1966). Points represented by crosses are computed from Eq. (59) for values of scattering angle of neutrons between 10° and 170°. For comparison the same $g(\nu)$ was computed again by the RS method, and is represented by circles. The scatter of both the crosses and the circles about the solid curve reflects the statistical noise produced by the RS method of computation. Otherwise Fig. 19 shows that "measured" $g(\nu)$ fits the computed one.

In Fig. 20 a comparison between actual measurement of neutron scattering and model calculation is shown. The solid curve is produced from the curve in Fig. 19 to match the experimental resolution, which is energy dependent and much cruder.

The technique of direct measurement of $g(\omega)$ for phonons is still rather crude and depends on strong assumptions. It is hoped it will become a more useful technique when the experimental resolution is improved. Also the generalization to polyatomic crystals and to crystals of lower symmetries will make it more useful and versatile.

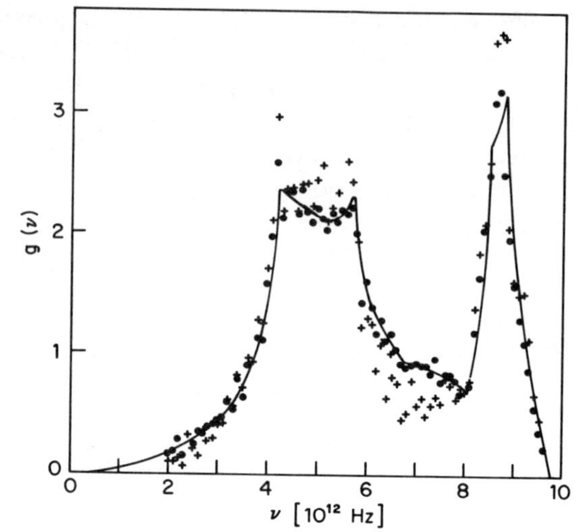

FIG. 19. Computer simulation of coherent inelastic scattering of slow neutrons from aluminum. The solid line represents $g(\nu)$ for aluminum, computed at high resolution. The crosses represent the energy spectrum of the neutrons scattered by aluminum, and the circles show the phonon density of states, both computed by the root-sampling method. The calculations were performed by Gompf *et al.* (1972).

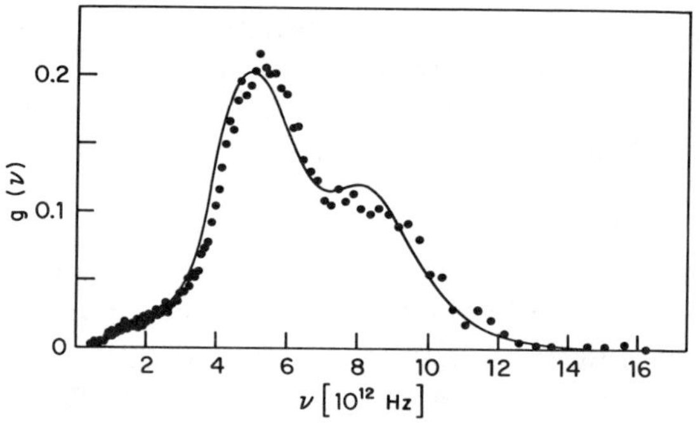

FIG. 20. Comparison between coherent inelastic scattering of neutrons by crystalline powder of aluminum and calculated $g(\nu)$ of aluminum. The resolution of the calculated $g(\nu)$ was artificially lowered to match experimental resolution. The experiment was performed by Gompf *et al.* (1972).

Incidentally, the **k**-dependent part of the one-phonon scattering from a single crystal can, in principle, be included exactly in the averaging integral over all the directions of the powder. In this case one obtains a function $I(\omega)$ which is not identical with $g(\omega)$, but can be computed by employing the **k**-dependence of $F(\mathbf{k})$. Likewise, the $I(\omega)$ for polyatomic crystals and crystals of lower symmetry can be accounted for in the same manner.

B. TUNNELING

There are a few examples for tunneling in solid state physics, but here we confine the discussion to tunneling between a superconductor and a normal metal. The reason for this is mainly that the energy relevant to this tunneling is the phonon energy.

From 1960 on, it became evident, especially in strong coupling superconductors, that the BCS model (Bardeen *et al.*, 1957) could not explain the fine structure of energy in tunneling experiments. In many experiments, where the current $I(V)$, and the conductivity dI/dV and even d^2I/dV^2 are measured as a function of potential difference V across the insulating gap, it is possible to observe a certain structure, as in Fig. 1, which is due to the phonon spectrum in the normal metal, and which cannot be explained by the BCS model. A generalized version of the BCS theory that includes electron–phonon interaction was developed by Eliashberg (1960). Schrieffer *et al.* (1963) used the equations developed by Eliashberg to explain the deviations from the BCS model in strongly coupled superconductors. McMillan and Rowell (1965) developed a method to solve Eliashberg's gap equations, which enabled them to write the tunneling energy spectrum in the form of

$$I(\omega) = \alpha^2(\omega)F(\omega), \tag{60}$$

where $\alpha^2(\omega)$ represents the energy spectrum of the electron–phonon interaction, and $F(\omega)$ is identical with $g(\omega)$. Combining their method for computing $\alpha^2(\omega)F(\omega)$ together with many tunneling measurements, McMillan and Rowell (1969) produced the tunneling spectra of numerous superconducting systems.

As might be expected, the function $I(\omega)$ in Eq. (60) can be directly calculated by performing an appropriate Brillouin zone integration. In order to do so, it is required to have a theoretical expression for the electron–phonon interaction. Carbotte and Dynes (1968) used the pseudopotential approach to calculate the electron–phonon interaction and obtained a closed form for $\alpha^2(\omega)F(\omega)$

$$\begin{aligned}\alpha^2(\omega)F(\omega) = \frac{\pi N(0)}{4h^2 k_F{}^2 MN} \sum_j \int_{<2k_F} \frac{|\mathbf{q}\cdot\mathbf{u}(\mathbf{q};j)|^2}{q\omega_j(\mathbf{q})} \\ \times |\langle \mathbf{k}_F + \mathbf{q}|w|\mathbf{k}_F\rangle|^2\, \delta(\omega - \omega_j(\mathbf{q}))\, d^3q.\end{aligned} \tag{61}$$

In this expression $N(0)$ is the electronic density of states at the Fermi energy E_F. M is the atomic mass, N is ion number density, $\mathbf{u}(\mathbf{q};j)$ is the polarization vector for the j-th phonon band and the wavevector $|\langle \mathbf{k}_F + \mathbf{q}|w|\mathbf{k}_F\rangle|$ is the pseudopotential form factor (Harrison, 1966). Equation (61) is only for

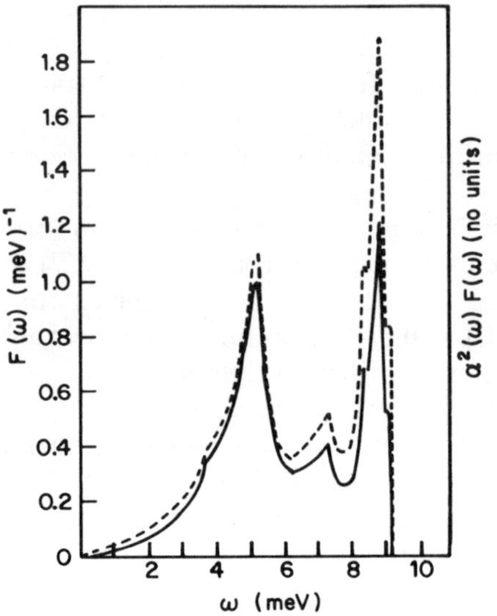

FIG. 21. Comparison between $\alpha^2(\omega)F(\omega)$ and the phonon density $F(\omega)$ of lead. The calculations were carried out by Dynes *et al.* (1969).

monatomic cubic crystals. The electron–phonon coupling constant is proportional to the integrand of Eq. (61) except for the δ-function. In fact, this constant is exactly $F(\mathbf{k})$ in Eq. (5) that is suitable to this case. A property peculiar to Eq. (61) concerns the range of integration for q, which is $0 \leqslant q \leqslant 2k_F$. The reason for this is that an electron can be scattered by a phonon only between two states inside the Fermi sphere. For this reason $F(\mathbf{q}) = 0$ for $q > 2k_F$.

An example of a computation of $\alpha^2(\omega)F(\omega)$ from Eq. (61) is shown in Fig. 21. The calculation was performed by Dynes *et al.* (1969), who investigated tunneling in Tl–Pb–Bi alloys. In this figure comparison between $\alpha^2(\omega)F(\omega)$ and $F(\omega)$ is shown, the latter being, in fact, $g(\omega)$, for Pb. In their computation Dynes *et al.* (1969) used the LA method but they made no use of the gradient of $F(\mathbf{q})$, so that the spectrum shown in Fig. 21 is unrefined according to Section II, G.

C. Phonon Sidebands

The present discussion concerns phonon sidebands to electronic transitions of an impurity in host crystal. This phenomenon is also known as vibronic transitions and it is discussed here in a qualitative manner only. In experiment optical transitions of impurity ions are accompanied by phonon, or other excitation, sidebands, usually of weaker intensity. The transition probability of vibronics depends on the initial and final electronic states of the impurity, on the modes of vibrations of the host crystal, and on the coupling between the impurity ion and the crystal. The theoretical situation is well described by Wagner (1968), who also brings a comprehensive list of references to earlier works.

The experimental resolution of vibronic transitions is rather high and is of the order of $N_\omega = 10^2$. For this reason it is a favorable case for making comparisons between observation and calculation. On the other hand, being an impurity problem, it may include undesirable side effects that can obscure the interpretation of the data. To minimize these, it is required to choose impurities the electronic states of which are well localized. Good candidates for this are rare earth ions, the 4f shell of which is effectively screened off from the crystalline field. Under this condition it can be assumed that the crystal field, being a small perturbation of the interionic interactions, is modulated by the genuine lattice vibrations of the host crystal.

Next, in order to analyze the situation properly, it is helpful to choose for a host a crystal the lattice dynamics of which is well understood theoretically and experimentally, and that accepts readily the rare earth ion as a substitutional impurity. It is also desirable to find impurities that have the same mass and force constants as the host crystal, in order to unravel the meaning of the vibronic spectrum. These conditions are approximately satisfied for the system $BaF_2:Sm^{2+}$, which was studied theoretically by Hurrell *et al.* (1972). The theory is based on the weak electron–lattice coupling discussed by Wagner (1968). The term in the interaction Hamiltonian that gives rise to the vibronic transitions is treated by second-order perturbation theory. The electron–lattice interaction consists of an electronic part that depends only on the configuration of the ion and can be separated out, and of a lattice dynamical part that is responsible for the $I(\omega)$ of the vibronic sidebands. In the case of $BAF_2:Sm^{2+}$ the lattice dynamics of BaF_2 is treated by the shell model (Cowley *et al.*, 1963) developed for ionic crystals. The interionic potential can be developed into multipole series, the contribution of each to $I(\omega)$ being taken separately into account.

The vibronic transitions, being associated with a point defect in the crystal, do not obey the $\mathbf{k} = 0$ conservation law, as does the Raman scattering. Therefore, vibronic transitions are actually associated with one-phonon transitions. In the case of several electronic transitions, as it is for Sm^{2+} ions, each one can be accompanied by phonon sidebands if the lifetime of the excited level is sufficiently long. In Fig. 22 we present a computation of vibronic transition accompanying the optical transition of $^5D_0 \to {}^7F_0$, as well as experimental results. The calculation of $I(\omega)$ involves an expression

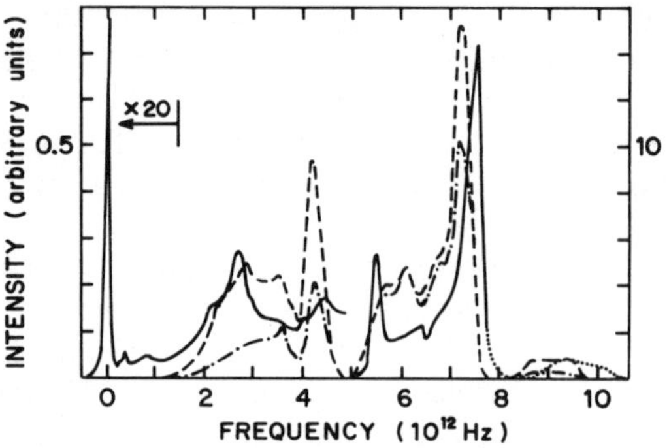

FIG. 22. Vibronic spectrum accompanying $^5D_0$ to $^7F_0(\Gamma_{1g})$ transition. Solid curve represents experimental data, dot-dashed curve is rigid-ion contribution and dashed curve ion plus shell contribution. Data is taken from Hurrell *et al.* (1972).

identical with Eq. (5), which makes the methods described in this article suitable for the evaluation of $I(\omega)$ for the vibronic transitions. $I(\omega)$ shown in Fig. 22 is computed for different models, such as the rigid-ion and the shell models. Although experiments are of high resolution, the theoretical side of the problem of vibronic transitions is still inadequate to obtain spectacular agreement with observation.

D. SECOND-ORDER RAMAN EFFECT

The calculation of second-order Raman (SOR) effect differs from the previous examples in one important aspect. SOR effect involves two-phonon transitions in contrast to previous examples that are single-phonon excitations. It is therefore expected that SOR computations will be considerably more complex than single-excitation spectra. However, the $\mathbf{k} = 0$ rule simplifies matters considerably in the case of SOR effect.

Experimentally, due to recent developments of laser light sources, SOR effect gradually has become measurable at relatively high resolution, which makes it more accessible to theoretical interpretations.

The basic phenomenological theory for SOR effect was given by Born and Bradburn (1947) in terms of an expansion of the polarizability tensor elements in powers of the atomic displacements from their equilibrium positions. Cowley (1964) applied the shell model to the SOR effect and obtained direct expressions that relate elements of the polarizability tensor to the normal modes of vibrations. Cowley's theory is therefore limited to ionic crystals. The expression for the SOR scattered intensity $I(\omega)$ for a given polarization of the incident and the scattered photon is given by

$$I_{\alpha\beta\gamma\delta}(\omega) = \int d^3k \sum_{j_1 j_2} P^{(2)}_{\alpha\beta}(\mathbf{k}; j_1 j_2) P^{*(2)}_{\gamma\delta}(\mathbf{k}; j_1 j_2) I_T(\omega, \mathbf{k}; j_1 j_2), \tag{62}$$

where Greek indices represent different polarizations and run over x, y, z. $P^{(2)}_{\alpha\beta}(\mathbf{k}; j_1 j_2)$ are the Fourier coefficients of the polarizability tensor that involves two phonons specified by $(\mathbf{k}; j_1)$ and $(-\mathbf{k}; j_2)$. $I_T(\omega, \mathbf{k}; j_1 j_2)$ is the integrand of the thermally weighted two-phonon density of states. For the Stokes part of the spectrum

$$\begin{aligned} I_T(\omega, \mathbf{k}; j_1 j_2) = {} & [n(\mathbf{k}; j_1) + 1][n(-\mathbf{k}; j_2) + 1] \\ & \times \delta[\omega - \omega(\mathbf{k}; j_1) - \omega(-\mathbf{k}; j_2)] \\ & + [n(\mathbf{k}; j_1) + 1][n(-\mathbf{k}; j_2)] \\ & \times \delta[\omega - \omega(\mathbf{k}; j_1) + \omega(-\mathbf{k}; j_2)], \end{aligned} \tag{63}$$

where $\omega(\mathbf{k}; j_1) \geqslant \omega(-\mathbf{k}; j_2)$ and $n(\mathbf{k}; j)$ is the phonon occupation number.

The evaluation of the elements of the polarizability tensor depends on far reaching theoretical assumptions that concern the nature of the interactions between electrons and ions. In most cases these elements are not known. The product of the elements of $P^{(2)}$ in Eq. (62) are to be identified as $F(\mathbf{k})$ in Eq. (5). The evaluation of $I_{\alpha\beta\gamma\delta}(\omega)$ in Eq. (62) is, apart from the two-phonon transitions, very similar to earlier evaluations of $I(\omega)$. In the SOR scattering computations sums and differences of phonon energies must be taken into account, as well as their appropriate gradients. Resultantly, there are n^2 combinations of n eigenfrequencies at each $\mathbf{k}_c$. This makes the "bookkeeping" of the calculation slightly more complex, and adds an amount of computation time in comparison to single-phonon cases. An SOR calculation for KBr performed by Pasternak *et al.* (1974) is presented in Fig. 23. The curves in the upper parts of this figure represent observed and calculated SOR spectra for the I_{xxxx} and I_{xyxy} light polarizations. The joint density of states is also shown. The method applied in these calculations was the LA method, and the resolution is $N_\omega = 100$. It is beyond the scope of this article to go into details of these spectra, and the interested reader is referred to the original paper.

In concluding this section, it is important to mention that the examples brought here by no means exhaust the domain of applicability of the Brillouin zone integration methods described in this article. They can be readily applied to other problems such as of impurity modes, of infrared absorption, of anharmonic effects and others, to mention only lattice dynamical effects. There are more examples in the domains of electronic, optical, and magnetic properties of solids.

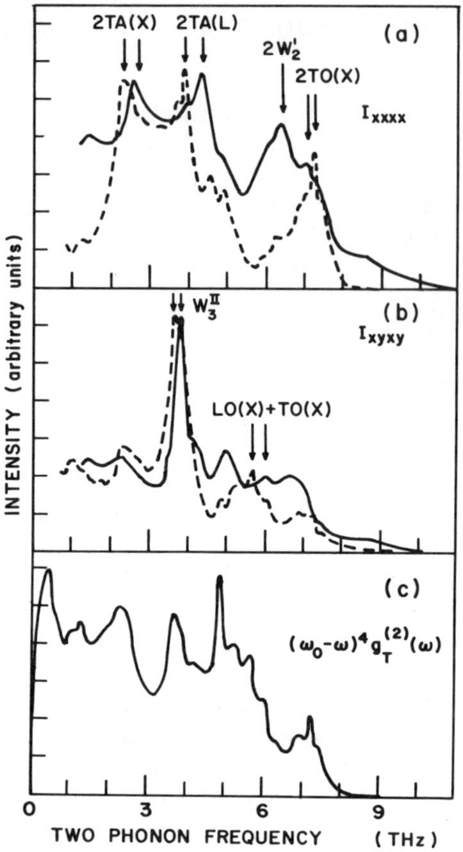

FIG. 23. Comparison between calculated and observed second-order Raman scattering of KBr. The calculated data are taken from Pasternak *et al.* (1974). Experimental data are by Krauzman (1967). In (c) the joint density of states of KBr is shown.

IV. Other Problems Related to Zone Integration

A. LIFETIME EFFECTS

In all preceding discussions the elementary excitations giving rise to spectra $I(\omega)$ and $R(\omega)$ via Eqs. (5)–(6) were considered to be infinitely sharp, i.e., with infinitely long lifetimes. In physical cases this assumption is only approximately correct, and for phonons it can occasionally be strongly violated. In order to correct for finite lifetimes, Eq. (1) must be replaced by

$$G(\omega) = \frac{1}{N} \sum_{\mathbf{k}} \frac{F(\mathbf{k})}{\omega - \omega_{\mathbf{k}} - \Delta_{\mathbf{k}}(\omega) - i\Gamma_{\mathbf{k}}(\omega)} \equiv R(\omega) + iI(\omega), \tag{64}$$

where $\Delta_{\mathbf{k}}(\omega)$ and $\Gamma_{\mathbf{k}}(\omega)$ are the real and imaginary parts of $\Sigma_{\mathbf{k}}(\omega)$, the self-energy of the excitation $\omega_{\mathbf{k}}$. In the case where $|\Sigma_{\mathbf{k}}| \ll \omega_{\mathbf{k}}$ it is possible to describe excitations as quasi-particles, where $\Delta_{\mathbf{k}}$ causes a shift in the energy $\hbar\omega_{\mathbf{k}}$ and $\Gamma_{\mathbf{k}}$ broadens the δ-type excitation into a Lorentzian-like shape of width $\Gamma_{\mathbf{k}}$.

In order to compute spectral functions $G(\omega)$ from Eq. (64), it is required to evaluate this integral for the IBZ. We are therefore faced again with the same difficulty that necessitated the development of efficient methods for zone integration. There is, however, one difference between the case of Eqs. (5)–(6) and the present case. For finite $\Sigma_{\mathbf{k}}(\omega)$ the integrand of Eq. (64) is no longer singular, so that critical points become less pronounced in both $R(\omega)$ and $I(\omega)$.

The computational problem of Eq. (64) was solved by Dalton and Gilat (1972) within the general framework of the LA method. The first step is to form a mesh of $\mathbf{k}_c$ points throughout the IBZ. Let $G_c(\omega)$ be the contribution from a single cell centered at $\mathbf{k}_c$; we then have

$$G(\omega) = [1/(2\pi)^3 V_{\mathrm{BZ}}]\Sigma_{\mathbf{k}_c} G_c(\omega), \tag{65}$$

which is the usual transformation from sums into integrals over the IBZ. Next we develop the integrand inside each cell into power series, retaining only the linear terms of the expansion. To simplify the expressions let us define $\Omega_{\mathbf{k}}(\omega)$ as

$$\Omega_{\mathbf{k}}(\omega) = \omega_{\mathbf{k}} + \Delta_{\mathbf{k}}(\omega) + i\Gamma_{\mathbf{k}}(\omega). \tag{66}$$

The expression for $G_c(\omega)$ is given by

$$G_c(\omega) = \frac{1}{|\nabla\Omega|} \int \frac{F(\mathbf{k}_c) + \mathbf{q} \cdot \nabla F}{\alpha - \mathbf{l} \cdot \mathbf{q}} \, d^3q, \tag{67}$$

l_i are the direction cosines of $\nabla\Omega$ at $\mathbf{k} = \mathbf{k}_c$, where $\mathbf{q} = \mathbf{k} - \mathbf{k}_c$ and α is given by

$$\alpha = [\omega - \Omega_{\mathbf{k}}(\omega)]/|\nabla\omega| \tag{68}$$

The integrand of Eq. (67) includes the unrefined as well as the refined contributions to the spectrum, and these are denoted by $G_c{}^0(\omega)$ and $G_c{}^i(\omega)$, respectively, where $i = x, y, z$, so that

$$G_c(\omega) = G_c^{\,0}(\omega) + \textstyle\sum_i G_c^{\,i}(\omega). \tag{69}$$

The explicit expressions for $G_c^{\,0}$ and $G_c^{\,i}$ are

$$G_c^{\,0} = \frac{F(\mathbf{k}_c)}{2\Lambda} \sum_{P_j = \pm 1} P_1 P_2 P_3 W_{P_j}^{\,2} \log W_{P_j} \tag{70}$$

$$G_c^{\,i} = \frac{\partial F/\partial k_i}{6\Lambda l_i}\left[\sum_{P_j = \pm 1} P_1 P_2 P_3 W_{P_j}^{\,3} \log W_{P_j} - 3\lambda_i P_i \sum_{P_j = \pm 1} P_1 P_2 P_3 W_{P_j}^{\,2} \log W_{P_j}\right] - \frac{V_c\, \partial F/\partial k_i}{3 l_i}, \tag{71}$$

where P_j ($j = 1, 2, 3$) runs over all the possible values of $P_j = \pm 1$, $\Lambda = l_1 l_2 l_3 |\nabla\Omega|$, $V_c = 8a_1 a_2 a_3$ is the volume of the integration cell, $\lambda_i = a_i l_i$ and W_{P_j} is given by

$$W_{P_1 P_2 P_3} = \alpha + \textstyle\sum_i P_i \lambda_i. \tag{72}$$

A useful relationship that may serve as a safeguard against computational errors, misprints and other mistakes is

$$\textstyle\sum_i l_i [G_c^{\,i}/(\partial F/\partial k_i)] = G_c^{\,0}/F - V_0/|\nabla\Omega|. \tag{73}$$

It is straightforward to obtain separate expressions for the real and imaginary parts of Eqs. (70)–(71), but they are quite lengthy and therefore are not reproduced here. They are given in the original article (Dalton and Gilat, 1972).

An example for a calculation of $G(\omega)$ that includes lifetime effects is shown in Fig. 24. This is a model calculation (D. Falik and G. Gilat, unpublished, 1974) of the real part of $G(\omega)$ evaluated for a bcc lattice, where $\omega_{\mathbf{k}} = \cos k_x \cos k_y \cos k_z$. $\Gamma_{\mathbf{k}}(\omega)$ is taken as an imaginary constant that is varied against the computational resolution in order to study its effect on the Van-Hove singularities. The case of $\Gamma = 0$ is represented by the solid curve, and it has an infinite singularity at $\omega = 0$. By making Γ finite the infinity disappears and $R(\omega)$ becomes gradually flatter as Γ increases.

The LA method that includes lifetime effects has not yet been applied to actual cases. One important candidate for application is the calculation of anharmonic effects in lattice dynamics (Cowley, 1963). In these calculations

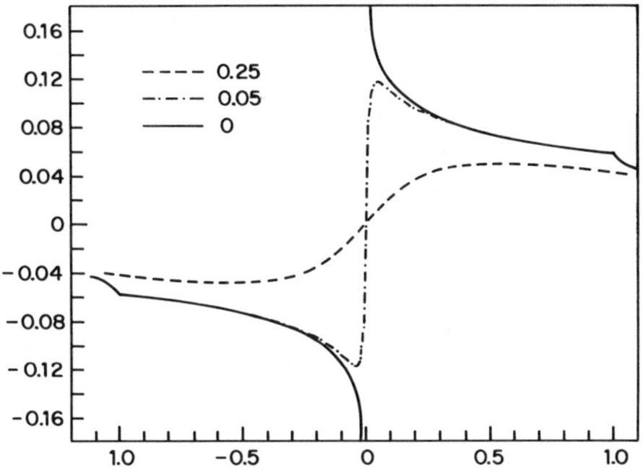

FIG. 24. Comparison between curves of different lifetimes for the real part function $R(\omega) = \Sigma_k(\omega - \omega_k)^{-1}$, $\omega_k = \cos k_x \cos k_y \cos k_z$ for bcc lattices. The different curves are specified by different values for Γ. Calculation was performed by D. Falik and G. Gilat (unpublished, 1974).

it is possible to obtain explicit expressions for $\Sigma_{\mathbf{k}}(\omega)$ and then apply the method described here in order to obtain the required $I(\omega)$ or $g(\omega)$.

B. TWO-EXCITATION TRANSITIONS (E.G., PHOTOEMISSION)

In some special cases it is important to calculate two-excitation transitions. The appropriate expression for such cases is given by

$$I(E_1, E_2) = \int_{\mathrm{BZ}} F(\mathbf{k})\, \delta(E_1 - E_1(\mathbf{k}))\, \delta(E_2 - E_2(\mathbf{k}))\, d^3k. \tag{74}$$

The best known example for two-excitation transitions is the photoemission of electrons from surfaces of solids, following irradiation by ultraviolet or soft X rays. In these experiments it is possible in some cases to vary the energy of the photons and correlate it to the energy of the emitted electrons. In order to compute $I(E_1, E_2)$, it is required to know the detailed electronic band structure of the solid. Since electronic properties of solids are beyond the scope of this article, only the computational part of the problem is sketched here, and even this is done very briefly.

In the framework of the LA method, expressions such as Eq. (74) can be computed by finding the intersection of the two constant energy surfaces $E_1 = E_1(\mathbf{k})$ and $E_2 = E_2(\mathbf{k})$. Within each integration cell the intersection of two surfaces (planes) is a line segment, the length of which is proportional to the contribution of the cell to $I(E_1, E_2)$. Based on this approach, Janak (1971) developed a computer algorithm to calculate $I(E_1, E_2)$. More recently,

Dalton and Gilat (1973) derived the analytical expressions for the lengths of the segments confined in an orthorhombic integration cell. Numerical calculations based on this approach were performed by Janak *et al.* (1970) who analyzed the photoemission of Pd.

Another problem, somewhat similar in its computational nature to the problem of photoemission, is the problem of computing the orbits of electrons in the de Haas–van Alphen effect. In this problem it is required to find the curves of intersection between the Fermi surface and a magnetic field plane. Janak (1972) treated this problem by evaluating line integrals within the framework of the LA method.

C. Derivatives of Spectral Functions

A powerful technique for measuring spectral properties of solids has been developed recently, and it is the modulation spectroscopy (Cardona, 1969). In this technique, the energy of excitation is modulated by some appropriate means (e.g., external field, pressure, temperature, frequently modulation) and the differential spectrum $dI(\omega)/d\omega$ is obtained rather than the more familiar and direct spectrum $I(\omega)$. Recently the differential second-order Raman spectrum of $KTaO_3$ was measured by Yacoby and Linz (1974), using modulation techniques. In this method singularities of the critical points become much more pronounced, because of the infinite slopes at these points.

The computational problem of $dI(\omega)/d\omega$ is not simple, since the direct way of obtaining $I(\omega)$ and then performing numerical differentiation on it is not to be recommended. The reason for this is that regardless of how good the calculation of $I(\omega)$ is, its numerical differentiation is bound to produce an uncontrollable amount of statistical noise that may render meaningless results for $dI(\omega)/d\omega$. The correct way of calculating $dI(\omega)/d\omega$ is first to obtain an analytical expression for $dI/d\omega$, and then to apply a reliable computational method to this expression.

The analytical expression for $dI/d\omega$ was first obtained by Gilat (1974). The derivation of this expression begins with the well-known formula for $I(\omega)$

$$I(\omega) = \int \frac{F(\mathbf{k})\, dS}{|\nabla\omega|}, \tag{75}$$

where integration is performed over the constant frequency surface $\omega_j(\mathbf{k}) = \omega$. It is possible to use Gauss' theorem to transform this integral into a volume integral, to perform the differentiation with respect to ω, and then to obtain the surface integral

$$\frac{dI(\omega)}{d\omega} = \int \frac{\operatorname{div} Z}{|\nabla\omega|}\, dS, \tag{76}$$

where Z is given by

$$\mathbf{Z} = F(\mathbf{k})\nabla\omega/|\nabla\omega|^2, \tag{77}$$

and

$$\operatorname{div} Z = |\nabla\omega|^{-2}(\nabla^2\omega - 2\mathbf{l}\cdot\nabla\nabla\omega\cdot\mathbf{l})F(\mathbf{k}) + |\nabla\omega|^{-1}(\mathbf{l}\cdot\nabla F(\mathbf{k})), \tag{78}$$

where $\mathbf{l}$ is the direction-cosines vector of $\nabla\omega$, and $\nabla\nabla\omega$ is the full tensor of $\partial^2\omega/\partial k_i\,\partial k_j$. For $F(\mathbf{k}) = 1$, $g(\omega)$ rather than $I(\omega)$ is obtained, and Eq. (76) simplifies to

$$\frac{dg(\omega)}{d\omega} = \int \frac{\nabla^2\omega - 2\mathbf{l}\cdot\nabla\nabla\omega\cdot\mathbf{l}}{|\nabla\omega|^3}\,dS. \tag{79}$$

The method of obtaining $dI/d\omega$ is not limited to first derivatives only and can be readily extended to any derivative of $I(\omega)$.

Lifetime effects play a very important role in modulation spectroscopy. The reason for this is that Van-Hove singularities become the major part of $dI/d\omega$. In real observations infinite singularities do not exist due to the finite lifetimes of the excitations. Therefore it is very important to take the self-energy into account right from the beginning. The appropriate expression for $G(\omega)$ is given by Eq. (64) and since this integral does not include any singularities, the differentiation with respect to ω is direct. We obtain for $I(\omega)$ and $R(\omega)$

$$\frac{dI}{d\omega} = \int_{\mathrm{BZ}} \frac{F[\Gamma'_\omega(\Omega^2 - \Gamma^2) - 2\Gamma\Omega(1 - \Delta'_\omega)]}{\Omega^2 + \Gamma^2}\,d^3k, \tag{80}$$

and

$$\frac{dR}{d\omega} = \int_{\mathrm{BZ}} \frac{F[(\Delta'_\omega - 1)(\Omega^2 - \Gamma^2) - 2\Omega\Gamma\Gamma'_\omega]}{\Omega^2 + \Gamma^2}\,d^3k \tag{81}$$

where primes denote differentiation with respect to ω, and $\Omega = \omega - \omega_{\mathbf{k}} - \Delta_{\mathbf{k}}(\omega)$. If $\Sigma_{\mathbf{k}}(\omega)$ is independent of ω, we obtain $\Gamma'_\omega = \Delta'_\omega = 0$ and Eqs. (80)–(81) simplify considerably.

No numerical calculations of derivatives of spectral functions $dI/d\omega$ have been performed as yet.

D. Meshes and Shapes of Integration Cells

All the methods described in this article use one kind of mesh or another to obtain the required spectral function $I(\omega)$. It may, therefore, be of some importance to indulge a little on different ways of constructing meshes of points. The simplest and the most widely used mesh is certainly the regular mesh, which is, in fact, a periodic lattice of points of orthorhombic structure. In many cases, especially for cubic crystals, this mesh is cubic. In some cases the contribution coming from different parts of the Brillouin zone may have different effects on $I(\omega)$. A well-known example is associated with the low phonon density of states for small ω. If a good calculation of $g(\omega)$ is required at low ω, then it may be necessary to increase the concentration of mesh points at small values of $\mathbf{k}$. This practice was adopted by a few authors (Gilat and Dolling, 1964; Gilat and Raubenheimer, 1966; Hardy *et al.*, 1973), and it is recommended for the calculation of thermodynamical properties of solids at low temperatures.

In most cases the origin for the mesh is chosen for simplicity at $\mathbf{k} = (0, 0, 0)$. This choice may not be very convenient whenever $\nabla\omega$ is required, especially for cubic symmetry. The reason for this is that whenever phonon degeneracies exist, there may be complications in the calculation of $\nabla\omega$ due to the use of perturbation technique. The remedy for this is to shift the whole mesh by a constant vector as described in Section II, E.

Meshes other than regular ones are also used occasionally. In particular, random mesh points are employed for the RS method. The reason for using random samples of points is to avoid systematical errors. Random mesh points were also used for fine mesh (Mueller *et al.*, 1971; Mueller, 1971; Brust, 1968) for the QD method, in order to prevent systematic oscillations in the density of states.

Another problem which is of interest for the LA method concerns the shapes of integration cells. These shapes are determined mainly by practical considerations of convenience, simplicity, and efficiency. There is a considerable amount of freedom in forming these shapes, but symmetry properties of different Brillouin zones impose certain constraints on these shapes.

In the case of cubic crystals it was preferred to employ cubic integration cells. Cubic cells are very simple and have a high degree of symmetry that allows for considerable simplification. On the other hand, the cross-section area $S(l_i; w)$ consists of four different expressions for each cell [Eqs. (16)–(20)]. It was recognized that tetrahedral cells may be competitive with cubic cells since they require only three different expressions for $S(l_i; w)$ per cell. Tetrahedral cells were actually implemented by Jepsen and Andersen (1971) and by Lehmann and Taut (1972) who also made comparative numerical calculations by using cubic and tetrahedral cells. Their findings confirmed that cubic and tetrahedral cells give about the same computational performance, and if there is any advantage in tetrahedral cells it is quite marginal. Since three tetrahedra exactly fill a cube, or a rectangular prism, they can be used for symmetries other than cubic ones. Tetrahedra are also useful whenever integrations over Fermi surfaces are performed.

It should be stressed that the number of expressions of $S(l_i; w)$ is no serious reason for concern. The difficulty of these expressions is, perhaps, only in their derivation, and once they are obtained they can be suitably coded and used without any further difficulty.

Very recently Rath and Freeman (1975) and Lindgård (1975) applied successfully the tetrahedron method to the calculation of magnetic susceptibility in hexagonal metals (Sc, Gd, Tb, and Dy). Gilat and Bharatiya (1975) expanded this method to include also matrix elements.

Another kind of simplification in the shape of the integration cell emerged recently (Johnson and Dresselhaus, 1973; Fehlner and Loly 1974), and it consists of the use of spheres rather than cubes or other polyhedra. It is true that spheres are the simplest shapes and have the highest degree of symmetry of all solid bodies. The only snag about spheres is that it is impossible to fill space exactly with spheres. As a result, the price of simplicity is paid by a sacrifice in accuracy. Moreover, the improvement in simplicity is not very crucial, since once the cross-section areas for cubic or tetrahedral cells are coded, the gain in simplicity becomes insignificant.

The source of error is obvious. If a cube of a side length of $2a$ is replaced by a sphere of radius R of equal volume, then $R/a \simeq 1.24$. This means that about 4.5% of the volume of the Brillouin zone is duplicated, whereas an equal volume at the corners of each (cubic cell) is neglected. Unfortunately, the contribution to $I(\omega)$ of the overlapping volumes does not compensate precisely for the neglect of the contributions coming from the corners. In the absence of comparative numerical calculations it is hard to give a good estimate of this error, but it can be of the order of a few percents. In view of the systematic error caused by spherical integration cells, their use for the calculation of $I(\omega)$ is to be discouraged.

V. Summary and Conclusions

The main objective of the present article was to give an adequate account of existing methods of integration over **k** space. All the methods described in the text are numerical ones, and almost no mention is made of analytical methods. In particular, five methods are described in greater detail: the root-sampling (RS) method, the discrete (LD and QD) methods, the linear–analytic (LA) method, and the combined linear–quadratic (C) method. All these methods are related to one another, but their computational efficiency rises roughly with their order of listing. Many examples for the use of these methods were brought in the text, but there are many more areas in the realm of solid state physics and chemistry to which they can be successfully applied. It becomes gradually evident that no serious progress in the computation of spectral properties of solids can be made without the use of efficient methods designed for this purpose.

A few fields relating to the dynamical properties of crystals are included in the text, and their relation to the computational problem of the relevant spectral property is emphasized. In another article in this volume, various ramifications of the subject, such as lifetime effects, derivatives of spectral functions, etc., are considered. These include new developments that to a large extent have not been applied to actual calculations as yet.

In conclusion, the subject of Brillouin zone integration has reached the stage where efficient methods for performing integrations do exist, and further developments in this field are not expected to involve major changes. What requires further investigation is the subject of the statistical errors incurred by the use of each of the various techniques. The range of applicability of these methods, being already quite large, will probably increase considerably in the future.

Acknowledgment

It is a pleasure to thank Miss Mary Kroiter, who typed the manuscript and whose help in completing the article was indispensable.

References

Bardeen, J., Cooper, L. N., and Schrieffer, J. R. (1957). *Phys. Rev.* **108**, 1175.
Berg, W. T. (1968). *Phys. Rev.* **167**, 583.
Blackman, M. (1937). *Proc. Roy. Soc., Ser. A* **159**, 416.
Bonadeo, H., and Taddei, G. (1973). *J. Chem. Phys.* **58**, 979.
Born, M., and Bradburn, M. (1947). *Proc. Roy. Soc., Ser. A* **188**, 161.
Bredov, M. M., Kotov, B. A., Okuneva, N. M., Oskotskii, V. S., and Shakh–Budagov, A. L. (1967). *Fiz. Tverd. Tela* **9**, 287; *Sov. Phys.—Solid State* **9**, 214 (1967).
Brust, D. (1965). *Phys. Rev.* **139**, A489.
Brust, D. (1968). *Methods Comput. Phys.* **8**, 33.
Carbotte, J. P., and Dynes, R. C. (1968). *Phys. Rev.* **172**, 476.
Cardona, M. (1969). "Modulation Spectroscopy," Solid State Phys., Suppl. 11. Academic Press, New York.
Cohen, E., and Guggenheim, H. J. (1968). *Phys. Rev.* **175**, 354.
Cooke, J. F., and Wood, R. F. (1972). *Phys. Rev. B* **5**, 1276.
Cooke, J. F., Davis, H. L., and Mostoller, M. (1975). *Phys. Rev. B* **11**, 706.
Cowley, R. A. (1963). *Advan. Phys.* **12**, 421.
Cowley, R. A. (1964). *Proc. Phys. Soc., London* **84**, 281.
Cowley, R. A., Cochran, W., Brockhouse, B. N., and Woods, A. D. B. (1963). *Phys. Rev.* **131**, 1020.

Dalton, N. W. (1970). *Solid State Commun.* **8**, 2047.
Dalton, N. W., and Gilat, G. (1972). *Solid State Commun.* **10**, 287. There is a mistake in Eqs. 16, 17, 22–25, the last term on each must be divided by $|\nabla\Omega|$.
Dalton, N. W., and Gilat, G. (1973). *Solid State Commun.* **12**, 211.
DeWames, R. E., Wolfram, T., and Lehman, G. W. (1965). *Phys. Rev.* **138**, A717.
Dixon, A. E., Woods, A. D. B., and Brockhouse, B. N. (1963). *Proc. Phys. Soc., London* **81**, 973.
Dynes, R. C., Carbotte, J. P., Taylor, D. W., and Campbell, C. K. (1969). *Phys. Rev.* **178**, 713.
Eliashberg, G. M. (1960). *Zh. Eksp. Teor. Fiz.* **38**, 966; *Sov. Phys.—JETP* **11**, 696 (1960).
Faulkner, J. S., Davis, H. L., and Joy, H. W. (1967). *Phys. Rev.* **161**, 656.
Fehlner, W. R., and Loly, P. D. (1974). *Solid State Commun.* **15**, 69.
Gilat, G. (1967). *Phys. Rev.* **157**, 540.
Gilat, G. (1969). *Solid State Commun.* **7**, 55.
Gilat, G. (1972). *J. Comput. Phys.* **10**, 432.
Gilat, G. (1973). *Phys. Rev. B* **7**, 891.
Gilat, G. (1974). *Solid State Commun.* **14**, 263. A factor of 2 must be placed before the term $\mathbf{l}\cdot\nabla\nabla\omega\cdot\mathbf{l}$ in Eqs. (10) and (12).
Gilat, G. and Bharatiya, N. R. (1975). To be published.
Gilat, G., and Bohlin, L. (1969). *Solid State Commun.* **7**, 1727.
Gilat, G., and Kam, Z. (1969). *Phys. Rev. Lett.* **22**, 715.
Gilat, G., and Dolling, G. (1964). *Phys. Lett.* **8**, 304.
Gilat, G., and Nicklow, R. M. (1966). *Phys. Rev.* **143**, 487.
Gilat, G., and Raubenheimer, L. J. (1966). *Phys. Rev.* **144**, 390.
Gompf, F., Lu, H., Reichardt, W., and Salgado, J. (1972). *In* "Neutron Inelastic Scattering 1972," p. 137. IAEA, Vienna.
Harada, J., and Shimanouchi, T. (1967). *J. Chem. Phys.* **46**, 2708.
Hardy, R. J., Morrison, I. W., and Bijanki, S. (1973). *J. Comput. Phys.* **13**, 591.
Harrison, W. A. (1966). "Pseudopotentials in the Theory of Metals." Benjamin, New York.
Hurrell, J. P., Kam, Z., and Cohen, E. (1972). *Phys. Rev. B* **6**, 1999.
Janak, J. F. (1969). *Phys. Lett. A* **28**, 570.
Janak, J. F. (1971). *In* "Computational Methods in Band Theory" (P. M. Marcus, J. F. Janak, and A. R. Williams, eds.), p. 323. Plenum, New York.
Janak, J. F. (1972). *Solid State Commun.* **10**, 833.
Janak, J. F., Eastman, D. E., and Williams, A. R. (1970). *Solid State Commun.* **8**, 271.
Jelitto, R. J. (1969). *J. Phys. Chem. Solids* **30**, 609.
Jepsen, O., and Andersen, O. K. (1971). *Solid State Commun.* **9**, 1763.
Johnson, R. L., and Dresselhaus, G. (1973). *Phys. Rev. B* **7**, 2275.
Kam, Z., and Gilat, G. (1968). *Phys. Rev.* **175**, 1156.
Karo, A. M., and Hardy, J. R. (1966). *Phys. Rev.* **141**, 696.
Karo, A. M., and Hardy, J. R. (1969). *Phys. Rev.* **181**, 1272.
Kellerman, E. W. (1940). *Phil. Trans. Roy. Soc. London, Ser. A* **238**, 513.
Krauzman, M. (1967). *C.R. Acad. Sci., Ser. B* **265**, 1029.
Lehmann, G., and Taut, M. (1972). *Phys. Status Solidi B* **54**, 469.
Lindgård, P. A. (1975). *Solid State Commun.* **16**, 481.
McMillan, W. L., and Rowell, J. M. (1965). *Phys. Rev. Lett.* **14**, 108.
McMillan, W. L., and Rowell, J. M. (1969). *In* "Superconductivity" (R. D. Parks, ed.), p. 561. Dekker, New York.
Maradudin, A. A., Ipatova, I. P., Montroll, E. W., and Weiss, G. H. (1971). "Theory of Lattice Dynamics in the Harmonic Approximation," 2nd ed., Solid State Phys., Suppl. 3. Academic Press, New York.
Minkiewicz, V. J., Shirane, G., and Nathans, R. (1967). *Phys. Rev.* **162**, 528.

Mueller, F. M. (1971). *In* "Computational Methods in Band Theory" (P. M. Marcus, J. F. Janak, and A. R. Williams, eds.), p. 305. Plenum, New York.
Mueller, F. M., Garland, J. W., Cohen, M. H., and Bennemann, K. H. (1971). *Ann. Phys. (New York)* **67**, 19.
Pasternak, A., Cohen, E., and Gilat, G. (1974). *Phys. Rev. B* **9**, 4584.
Placzek, G., and Van–Hove, L. (1954). *Phys. Rev.* **93**, 325.
Pryce, M. H. L. (1966). *In* "Phonons in Perfect Lattices and in Lattices with Point Imperfections" (R. W. H. Stevenson, ed.), p. 403. Plenum, New York.
Rath, J., and Freeman, A. J. (1975). *Phys. Rev. B* **11**, 2109.
Raubenheimer, L. J., and Gilat, G. (1967). *Phys. Rev.* **157**, 586.
Schrieffer, J. R., Scalapino, D. J., and Wilkins, J. W. (1963). *Phys. Rev. Lett.* **10**, 333.
Stedman, R., and Nilsson, G. (1966). *Phys. Rev.* **145**, 492.
Svensson, E. C., Brockhouse, B. N., and Rowe, J. M. (1967). *Phys. Rev.* **155**, 619.
Van–Hove, L. (1953). *Phys. Rev.* **89**, 1189.
Wagner, M. (1968). *Z. Phys.* **214**, 78.
Woods, A. D. B., Brockhouse, B. N., March, R. H., Stewart, A. T., and Bowers, R. (1962). *Phys. Rev.* **128**, 1112.
Yacoby, Y., and Linz, A. (1974). *Phys. Rev. B* **9**, 2723.

Computer Studies of Transport Properties in Simple Models of Solids*

WILLIAM M. VISSCHER

THEORETICAL DIVISION
LOS ALAMOS SCIENTIFIC LABORATORY, UNIVERSITY OF CALIFORNIA
LOS ALAMOS, NEW MEXICO

I. History and Introduction

THE HISTORY OF THE development of the theory of transport in solids is long and interesting, but it has been thoroughly covered in a number of recent reviews (see, for example, Kubo, 1973), so no detailed recitation will be given here, only a brief resume.

A. THE BOLTZMANN EQUATION

Until relatively recently, the only reasonably well-founded calculations of conductivity (both thermal and electrical) in solids (or any material) were

* Work performed under the auspices of the United States Atomic Energy Commission.

based on the Boltzmann equation, or variations of it. The Boltzmann equation was originally derived in the context of the kinetic theory, and it is most nearly credible in dilute gases. It is simple and transparent, following from equations of motion with only one necessary assumption which is undesirably less than rigorous. (The *Stosszahlansatz*; a physically reasonable postulate about the randomness of the distribution of velocities of the particles prior to each collision.) The transparency and usefulness of the Boltzmann equation in kinetic theory and fluid physics in general is such that most physicists have considerable faith in its soundness, even though it has so far not been possible to rigorously justify it. That the limits to the validity of the Boltzmann equation are not remote is clear from the now well-established fact (Alder and Wainwright, 1970) that velocity autocorrelation functions in molecular dynamical computer experiments persist at infinite times, in contradiction to the consequences of the *Stosszahlansatz*.

Early in this century, the Boltzmann equation was adapted to apply to an electron gas in a metal by Lorentz (1905). This was its first application to solid state physics and required little stretching of credulity to accept, even though it was clear that electrons in a lattice (they were first taken not to interact with each other) were far from the freely moving gas molecules for which the Boltzmann equation was first derived. In large part Lorentz's theory contradicted experiments, though, because quantum mechanics is essential to describe solids, except for nonmetals at quite high temperatures.

With the invention of quantum mechanics the Boltzmann equation was immediately generalized. Peierls (1929) extended it to apply to phonons in solids, greatly decreasing the intuitive transparency of the original Boltzmann equation, because phonons are not localizable with a definite velocity like the atoms in a classical gas. But the Boltzmann–Peierls equation, although it is still harder to justify than the original Boltzmann equation, was and is useful in understanding lattice thermal conductivity.

Our present understanding of electric conductivity in solids had its genesis with Bloch (1928), also just after quantum mechanics was invented. Bloch adapted the Boltzmann–Lorentz equation for classical electrons to quantum mechanical electrons in a solid which possesses internal excitations (phonons). The dependent variable in the Boltzmann–Bloch equation is the distribution function for the electrons; the phonons are assumed to be in thermal equilibrium.

All of the applications of the Boltzmann equation to calculation of transport coefficients in solids are, to a greater or lesser extent, subject to the criticism that they have sacrificed rigor for transparency. This is all to the good as long as they give correct results. Whether or not they do can be established in several ways. One is to calculate their predictions of results of physical experiments. Another is to see if the formal expressions they give for observables can be obtained from other theories. The former has the drawback that it is only a necessary condition; sufficiency would require predictions and results to be compared in *all* physical situations. The latter is more satisfactory only if the other theories are more firmly based than the Boltzmann equation.

B. Green–Kubo Formulas

Such is the case for the theories of the Green–Kubo expressions for the transport coefficients, and in some cases a correspondence can be established between Green–Kubo expressions and Boltzmann equation results. The Green–Kubo formulas for transport coefficients can be derived relatively rigorously and quite easily from first principles (for a review, see Chester, 1963; Zwanzig, 1965; see also Visscher, 1974), as opposed to the Boltzmann expressions, for which some derivations exist, but not rigorous ones.

Since the Boltzmann equation is a hydrodynamic-type equation which is manifestly physically intuitive in form, it has some advantage over the Green–Kubo expressions for transport coefficients. The Boltzmann–Peierls equation (for phonons), for example, explicitly exhibits on the right-hand side the collision terms between electrons and phonons whereby the energy the electron gains from an applied electric field is dissipated.

C. Computer Models

Similarly, it is physical transparency which is the strong point of the simple models of transport processes which we will review in this article. Because of the uncertainty of the foundations of the Boltzmann equation and because of the relative opacity of the Green–Kubo formulas, it would be desirable to model systems complex enough to be physically nontrivial but yet simple enough to be calculationally tractable (either analytically or computationally). Then one would be in a position to intuitively understand (as one is with the Boltzmann equation) exactly what physical processes are at the root of whatever effect one is studying. The ideal in inventing a heuristic computer model of a physical system is always simplicity, ease and speed of computation, with the constraint that the model not be so trivial as to be uninteresting physically.

Dimensionality is usually the first parameter to be set. Clearly there are some problems for which one could not invent a meaningful one dimensional (1d) model—turbulence is an obvious one in fluid physics. Some features of thermal and electric conductivity cannot be studied in a 1d system—interaction of transverse and longitudinal phonons is an example. But these are examples of phenomena which are intrinsically multidimensional—the higher dimensional systems are phenomenologically richer than lower dimensional ones. Physical phenomena which exist in 3d (transverse phonons, turbulence) do not necessarily have analogs in 1d. As far as we know, though, the converse (that 1d phenomena have 3d analogs) has no exceptions, so that if the physics of a 1d or 2d system is interesting, then it is profitable to study it. It is certainly true that in most cases by so doing one ignores some essential physics; but one also isolates and clarifies some other essential physics and may simplify the problem enough to render it tractable computationally. So, although it is only seldom that the simple heuristic models we will discuss in this article have sufficient physical realism that results derived from them can be compared with experiment, the clarity with which they can isolate and illustrate some of the physical principles involved in transport in solids partly compensates for that drawback.

II. Thermal Conductivity

Lattice thermal conductivity is in principle the simplest transport process in a solid. This is because it involves only the motion of the bound lattice atoms themselves, and for understanding it one does not need to consider any mobile particles in the lattice (impurities diffusing or conducting electrons). Conceptually the question is very simple; wiggle the atoms at one end of a lattice by putting them in contact with a heat reservoir of some sort, and ask how fast the lattice delivers heat to a colder reservoir at the other end of the lattice. The lattice can be imagined to have all degrees of complexity. The simplest models (periodic structures with harmonic forces and certain 1d anharmonic ones) can be attacked analytically. We will mention some of them in the following when it is prerequisite to developing computer models or helpful in understanding computer results.

The next simplest models are disordered lattices but still with harmonic forces. Quite a large body of exact analytic knowledge about these systems in one dimension has by now been accumulated; we will refer to some of it when discussing certain limits of more realistic computer models. As far as we know, no exact analytic results exist for thermal properties of 2d disordered harmonic systems, but the results of computer experiments indicate that they have many common features with their 1d analogs.

The jump from harmonic to anharmonic forces is a big one, which up to now has not been accomplished by any analytic method for calculating thermal transport. Some particular solutions for some very interesting 1d systems have been obtained (see, for example, Toda, 1970), but the tools for finding the general solutions that are needed in the heat conductivity problem are not in sight. This is a domain that is not yet accessible to any theoretical tools but the computer experiment; although analytic perturbative methods for periodic anharmonic lattices have been successful, perturbation theory fails completely for the disordered anharmonic case. The reason for this and its physical clarification will be explained in the following for the case of the 1d disordered anharmonic chain.

Two-dimensional periodic and disordered anharmonic systems have been subjected to computer experiment, too, because of the chance that the extra dimension would introduce some essential new physics (for example, the transverse phonon branches). The results are qualitatively the same as for the 1d system. Any further extensions of computer measurements, for instance to 3d anharmonic systems, or to systems with conduction electrons, would strain the memory capacity and speed of our present computers, but they should not be put beyond contemplation. These more complicated, more physically realistic models must at present be left to analytical perturbative methods, and to adaptations of other approximate methods, such as some varieties of the coherent potential method, which have in certain cases succeeded in obtaining semiquantitative agreement with exact numerical results of computer experiments. [A complete and up-to-date bibliography of coherent potential approximation (CPA) is given by Yonezawa and Morigaki (1973).]

There are at least three classes of theoretical tools available to the theorist seeking to calculate thermal conductivity. First, exact analytic methods. If these were available for physically realistic systems the other tools would be superfluous, but unfortunately the domain in which they apply is limited to the simplest systems, far from the real world. Next, approximate analytic methods, mostly perturbation theories and coherent potential methods. These can be applied, sometimes with considerable difficulty, to realistic systems, but little is known about their accuracy or their convergence. Third are computer experiments. They are intermediate in domain of application between the exact (for very idealized models) and approximate (for more or less realistic models) analytic methods; a perturbation theory can be devised that will apply to some systems that can be numerically experimented upon, and a numerical experiment can probably be made on any system that will yield to exact analytical treatment. Thus, the exact analytical calculation can be a check on numerical experiments in their common domain, and a proper role of numerical experiments is to keep the analytic perturbation methods honest by providing benchmarks otherwise unavailable in systems which are tractable by both methods.

A. Harmonic One-Dimensional Chains

The simplest imaginable system that might have some bearing on heat conduction in solids is the monatomic harmonic chain. It can be exactly treated by analytic methods. Because we will use a similar approach in a

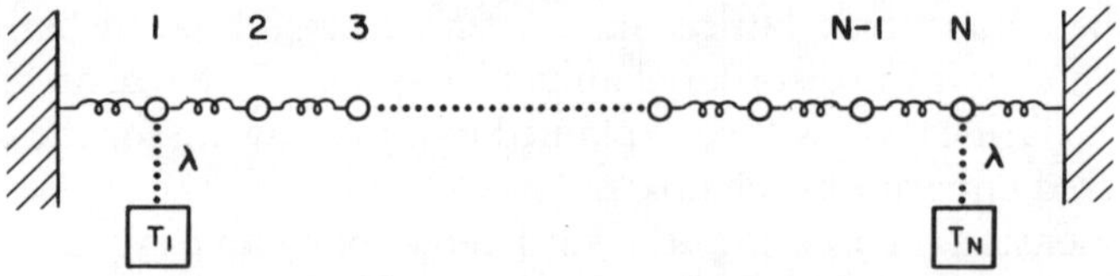

FIG. 1. The harmonic chain of Rieder *et al.* (1967). The boundaries (particles 0 and $N + 1$) are fixed; particles 1 and N interact with thermal reservoirs at temperatures T_1 and T_N via Fokker–Planck forces of strength λ.

computer calculation to be described later, we will describe a model which was invented and exactly solved by Rieder *et al.* (1967). It is illustrated in Fig. 1. N particles, each connected by Hooke's law springs to its nearest neighbors, form a 1d harmonic chain. Particles 0 and $N + 1$ are rigidly fixed; particles 1 and N are in stochastic interaction via a Fokker–Planck force of strength λ with heat reservoirs of temperature T_1 and T_N, respectively. In the monatomic case all masses and force constants are the same; more complicated cases can be considered by letting the masses be a binary or more complex mixture, randomly disordered or otherwise arranged. Rieder *et al.* (1967) solved for the average values (in the steady-state conduction situation) of all the bilinear products of momenta and position coordinates; they found that in the part of the chain far from the ends there was no temperature gradient and the temperature was $(T_1 + T_N)/2$, and the heat current from the hot end to the cold end was, for $N \gg 1$,

$$J = C(\lambda)(T_1 - T_N), \tag{1}$$

where $C(\lambda)$ depends on the spring constant and particle mass as well as on λ but not on the length N of the chain. So the monatomic harmonic chain has an abnormal thermal conductivity (does not obey Fourier's law, which would say the heat current is proportional to the temperature *gradient*, not the temperature difference), as was pointed out more than 60 years ago by Debye (1914) in a more general context. Rieder, Lebowitz, and Lieb solved this steady-state nonequilibrium problem by setting up matrix equations for averages of bilinear forms in positions and momenta (a covariance matrix) and they find the exact solutions analytically. The same matrix equations (with appropriately modified matrices) apply in the more general case where the masses are no longer all the same, but now they can no longer be solved exactly analytically. Ishii and Matsuda (1970) undertook the task of working out formal expressions for the temperature and heat currents in this disordered harmonic chain, in the limit of small coupling to the reservoirs (λ). Their results are written in terms of sums over the normal modes of the disordered systems: for the heat current, for example, they give, to first order in λ,

$$J = k\lambda(T_1 - T_N) \sum_{\alpha=1}^{N} [1/(x_{\alpha 1}{}^{-2} + x_{\alpha N}{}^{-2})], \tag{2}$$

where $u_{\alpha n} = x_{\alpha n}/(m_n)^{1/2}$ is the displacement of the nth atom in the αth normal mode and $\sum_n x_{\alpha n} x_{\beta n} = \delta_{\alpha\beta}$. Equation (2) is valid in systems for which $|\omega_\alpha - \omega_\beta| \gg \lambda$ if $\alpha \neq \beta$; i.e., no degeneracy within the width induced by the reservoir interactions, at least for the modes whose amplitudes are large at the ends of the chain. [See Visscher (1971) for a discussion of the limits of validity of Eq. (2).]

One can easily see that for the ordered monatomic case Eq. (2) agrees with Eq. (1), because then

$$x_{\alpha n} = (2/N)^{1/2} \sin[\pi \alpha n/(N + 1)] \tag{3}$$

for fixed boundary conditions, the same expression with a cosine if the springs at the ends are cut. Each of the modes [Eq. (3)] which has nonzero amplitude at the ends contributes $0(1/N)$ to the sum; since there are $0(N)$ of them, $J = 0(N^0)$ in agreement with Eq. (1).

But now if the chain is disordered it is well known that nearly all of the normal modes are localized [see Bell (1972) for a recent review of dynamics of disordered lattices]; only those of very low frequency (long effective wavelength) can have appreciable amplitude at both ends of the chain, which is necessary for them to contribute to Eq. (2). Ishii and Matsuda showed that these ($\omega \to 0$) least localized modes extended over of the order of

$$l = [\langle m \rangle^2/(\langle m^2 \rangle - \langle m \rangle^2)]^{1/2} \omega^{-2} \tag{4}$$

atoms, if the assumption was made that l^{-1} could be identified with the average blowup rate (logarithmic decrement) of the successive amplitudes in the semiinfinite disordered chain. (This they call the IF-assumption—that $\langle |u_{i+i}/u_i| \rangle = e^{1/l}$.) Equation (4) implies that only a limited number of modes can contribute to Eq. (2), namely approximately

$$n_d = (12/\pi)N^{1/2}. \tag{5}$$

If a factor n_d replaces the sum in Eq. (2) and an average amplitude for the n_d extended modes is used, i.e., $|x_{\alpha 1}| \sim N^{-1}$ for fixed ends, $|x_{\alpha 1}| \sim N^{-1/2}$ for free ends, then one sees that

$$\begin{aligned} J &\propto \lambda(T_1 - T_N)N^{-3/2} \quad \text{(fixed)} \\ J &\propto \lambda(T_1 - T_N)N^{-1/2} \quad \text{(free)} \end{aligned} \tag{6}$$

for the indicated boundary conditions. Equation (6), if true, shows that, contrary to what many people had expected, disorder in a harmonic chain is not enough to render the heat conduction normal; in fact the thermal conductivity, defined as $\lim_{N\to\infty} J/[(T_1 - T_N)/N]$, vanishes for fixed ends, diverges for free ends. Some support is lent to this conclusion by a computer verification of Eq. (5) (Visscher, 1971) and by some recent work on localization in disordered chains by Painter and Hartmann (1974).

This discussion has been admittedly quite intuitive. For a rigorous presentation of the same problem, which unfortunately leads to no such definite conclusion as Eq. (6), see O'Connor and Lebowitz (1974). They rightfully raise the objection that Eq. (2) is only the first term in a power series expansion in λ, an operation which may well not commute with taking the thermodynamic limit ($N \to \infty$).

That the disordered harmonic chain seemingly has abnormal thermal transport properties is an important enough fact that it should be verified by independent means. Because of its simplicity it has been much studied both by analytic and numerical methods; perhaps out of familiarity we will cite some of our own work as direct numerical confirmation.

One important part of the argument leading to the conclusion that the thermal conductivity of the disordered harmonic chain is abnormal was the invocation (Ishii and Matsuda, 1970; Ishii, 1973) of the IF-assumption which led to the estimate [Eq. (5)] for n_d, the number of normal modes contributing to the sum of Eq. (2). Visscher (1971) took several isotopically disordered chains of lengths from 100 to 1000 atoms, consisting of random mixtures of atoms of two species with mass ratio 2. By finding eigenvalues and eigenvectors of the tridiagonal dynamical matrix of the equations of motion

$$-m_i\omega_\alpha{}^2 u_{i\alpha} = \gamma(u_{i+1\alpha} - 2u_{i\alpha} + u_{I-1\alpha}) \tag{7}$$

he numerically performed the sum in Eq. (2). The results for two typical cases are shown in Fig. 2. One sees that contributions to the sum cease completely after a certain mode number is exceeded. If v_c is defined more or less arbitrarily to be the mode number at which the sum attains 90% of its ultimate value then Fig. 3 shows v_c as a function of chain length N, for several randomly disordered chains. The straight line is

$$v_c = 5.5N^{1/2}, \tag{8}$$

which agrees unexpectedly well with Eq. (5) which is dependent on the IF-assumption, thus lending some support to the IF-assumption. This, we believe, demonstrates the thermal abnormality of the disordered harmonic chain, at least to lowest order in λ. The question of what happens for finite λ

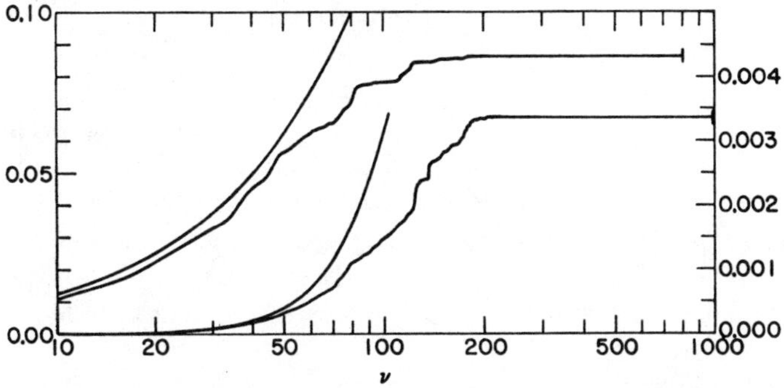

FIG. 2. Cumulative contributions to the sum in Eq. (2), as a function of mode number. The upper curve (left ordinate scale) is for a random chain with free ends of length 800 composed of equal numbers of two species of mass ratio 2; the lower curve is for a similar chain of length 1000 with fixed end boundary conditions. The smooth curves are the corresponding cumulative contributions for the normal modes of a perfect monatomic chain. From Visscher (1971).

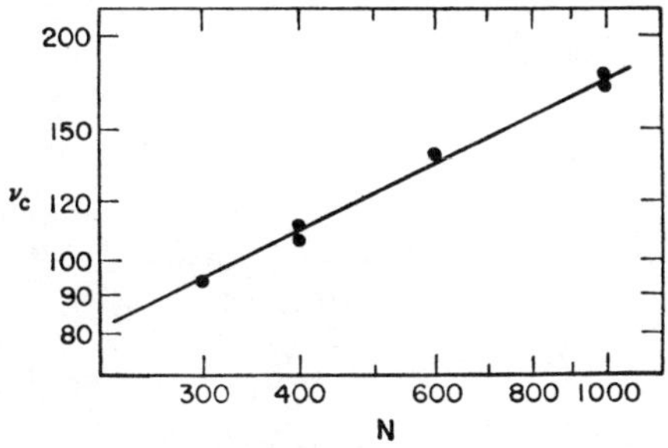

FIG. 3. Critical mode number ν_c as a function of chain length N for the random chain composed equally of two species with mass ratio 2 and fixed end boundary conditions. From Visscher (1971).

remains an open one. We have done some numerical experiments bearing on this question (with stochastic impulsive interactions with end reservoirs of particles with Maxwellian velocity distributions) for disordered harmonic chains up to $N = 1000$ with results that indicate abnormal thermal conduction, but statistical errors were such that definite conclusions could not be drawn. [See also Rubin and Greer (1971) for some numerical results on a different, but closely related, system.]

B. Anharmonic Ordered One-Dimensional Chains

The ineffectiveness of disorder in rendering the thermal conductivity of the harmonic chain normal being seemingly apparent, one next asks whether anharmonicity will do it. Anharmonicity means any deviation from Hooke's law forces between the atoms, and there are many different kinds one might consider.

One might try a Lennard-Jones potential $(r_0/r)^{12} - 2(r_0/r)^6$ or a polynomial approximation to it. A fair amount of computer experimentation along these lines has been done, some of which we will mention in the next section. Cubic and quartic corrections to the harmonic potential are probably good approximations to the effective potential in real solids at moderate temperatures, but in a 1d model it has been shown by Thompson (1961) that in low-order perturbation theory, at least, they do not render the thermal conductivity normal.

A more blatantly anharmonic force acts in the chain of hard rods; an infinite repulsion at $r = 0$, no force at all for $r > 0$. It is intuitively obvious and can be easily verified analytically that this system has an abnormal thermal conductivity of the same kind as the harmonic chain, namely, the heat current is proportional to the temperature difference between the ends and is independent of N.

A slightly more general case is the infinite square well. This, too, will clearly have abnormal conductivity because the velocity distribution is not changed when particles collide, as was the case with hard rods.

1. *The Super-Anharmonic Chain*

Not completely obvious is the result of putting a step in the bottom of the infinite square well, i.e., the interaction potential is V_1 for $0 < r \leqslant a$, V_2 for $a < r < b$, and infinity otherwise, as shown in Fig. 4. The equilibrium properties of this system have been worked out by Mazur and Rubin (1963). As far as thermal conductivity is concerned, it is clear that in the limits of

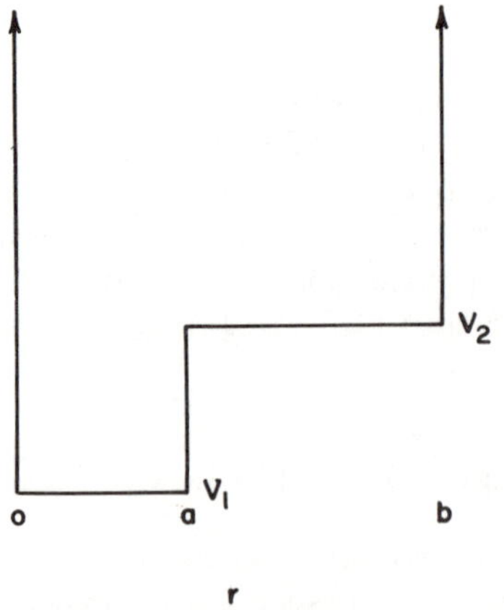

FIG. 4. Interatomic potential for the super-anharmonic chain.

both high and low temperatures the system will behave like the infinite square-well system, because in the former case $kT \gg V_2 - V_1$ and in the latter case the nearest neighbor distances will always be less than a unless the chain is under tension.

From a numerical experimental point of view, one attractive feature of this "super-anharmonic chain" is that the computer does not need to integrate any equations of motion—it only keeps books. The collisions between particles are either elastic, with a reversal of relative velocity, or passing a potential barrier, which, to find the new velocity, only requires the computer to take a square root. The time-development proceeds speedily with only roundoff errors to degrade the accuracy.

A computer experiment on this system has been performed; some typical results are shown in Fig. 5. The chain was confined between two walls a distance $L = 1.2Na$ apart; a stochastic interaction with reservoirs at the ends was imposed such that at regular time intervals a reservoir particle of

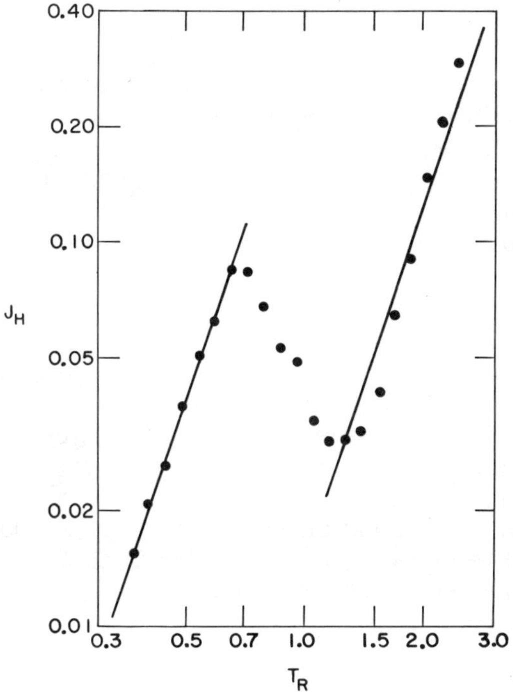

FIG. 5. Heat current in a super-anharmonic chain as a function of right reservoir temperature. The left reservoir temperature is always two-thirds as large. This chain had $N = 100$, $V_2 - V_1 = 1$, and $b = 2a$. The straight lines have slope 3, reflecting the fact that for $V_2 = V_1$, $J \propto T^3$. Each point is the result of an average over 10^6 collisions; statistical errors are about 5%.

mass equal to that of the chain particles, chosen from a Maxwellian distribution characterized by a temperature T_R on the right, $T_L = \frac{2}{3} T_R$ on the left, collided elastically with the end particles. The net flow of energy from the hot end to the cold end is shown in Fig. 5. The asymptotic behavior is as it should be for $|\delta V| \gg kT$ and $|\delta V| \ll kT$, but there is little hope of demonstrating whether the thermal conductivity behavior is normal or not by this kind of experiment. The reason for this is that both asymptotes are N-independent; the only part of the curve which might depend on N is therefore the part in the middle with negative slope. It is, however, in Fig. 5 already smeared over a temperature difference of a factor of $\frac{3}{2}$ (a large temperature difference is necessary in these experiments to reduce statistical errors to a tolerable level); for an infinitesimal temperature difference the corners presumably would be sharper. One can conclude from this computation that if the super-anharmonic chain exhibits normal behavior as far as thermal conductivity is concerned, it is only for a very narrow temperature range.

2. *Harmonic Chain with Hard Cores*

A step up in realism from the super-anharmonic chain is the harmonic monatomic chain with hard cores. The equilibrium properties for this system were studied by Northcote and Potts (1964), who also did some numerical experimentation to elucidate the approach to equilibrium. The interatomic force for this model is shown in Fig. 6.

In the harmonic chain (the limit of this one as $d \to \infty$) and in the chain of hard rods (the limit of this one as $d \to 0$—particles jammed together, so all energy is kinetic) the temperature is merely a scaling factor on the displacements and/or velocities. Here, though, if d^2 is neither much greater than nor much less than the temperature, changes in its ratio to the temperature change the physics of the problem. So if the harmonic chain and the hard rods are zero—parameter systems as far as their extensive thermal transport processes are concerned—the harmonic-hard rod chain (HHR) is a one-parameter system.

Like the super-anharmonic chain, the HHR lattice needs the computer mostly in a bookkeeping capacity. The reason for this is that the equations of motion for the regular harmonic chain can be solved exactly and the solution written down explicitly for the velocities and relative displacements at time t in terms of their initial values. If this solution shows that a collision between neighbors occurs at a certain time, one determines that time (by a modification of Newton's method not requiring calculation of derivatives—the secant method) and restarts the time development with reversed relative velocity for the colliding pair at that time. No numerical integration pro-

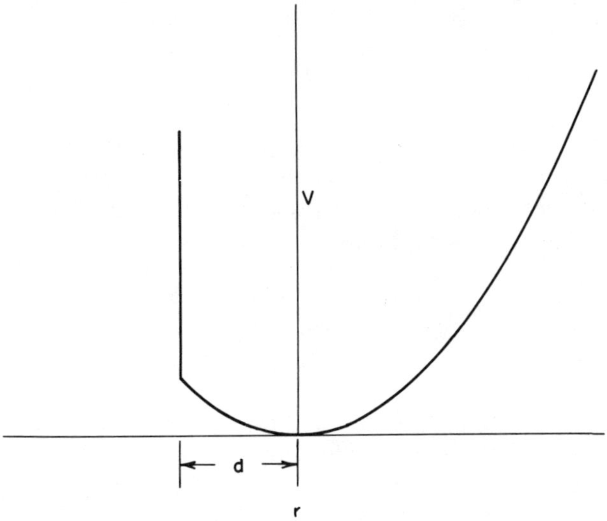

FIG. 6. The interatomic force in the harmonic chain with hard cores [harmonic hard rod (HHR) system]: r is the relative displacement from equilibrium of adjacent atoms; d is the distance of the hard cores from the equilibrium positions. If $d = \infty$, the chain is harmonic, if $d \to 0$, there can be no potential energy if the ends of the chain are fixed, and the system becomes a chain of hard rods.

cedures are needed; the most time-consuming part of the computation is finding the collision times, and that is easy to do quickly nearly to machine accuracy.

The method we use in this computation is Schrödinger's method, which exploits the fact that the equations of motion for the regular harmonic chain, when expressed in certain dependent variables, are identical to a recursion formula for Bessel functions.

Let

$$\begin{aligned} \xi_{2n} &= \sqrt{m}\dot{u}_n \\ \xi_{2n+1} &= \sqrt{\gamma}(u_n - u_{n+1}) \\ \omega_0 &= 2(\gamma/m)^{1/2} \\ \tau &= \omega_0 t. \end{aligned} \tag{9}$$

Then Newton's equations of motion for the harmonic chain are

$$\frac{2}{\omega_0}\dot{\xi}_{2n} = 2\xi'_{2n} = \xi_{2n-1} - \xi_{2n+1} \tag{10}$$

and the relative velocities are

$$\frac{2}{\omega_0}\dot{\xi}_{2n+1} = 2\xi'_{2n+1} = \xi_{2n} - \xi_{2n+2} \tag{11}$$

where the dots denote d/dt, the primes $d/d\tau$. Equations (10) and (11) are satisfied by Bessel functions; in particular if $\xi_k(0)$ are values of velocities and displacements at $\tau = 0$, then

$$\xi_n(\tau) = \sum_{k=-\infty}^{\infty} J_{n-k}(\tau)\xi_k(0) \tag{12}$$

are their values at τ. Any standard kind of boundary conditions can be handled with Eq. (12); for example if particles 0 and $N + 1$ are fixed (N movable particles), then one puts $\xi_{-k}(0) = (-1)^{k+1}\xi_k(0)$ and $\xi_{2N+2+k}(0) = (-1)^{k+1}\xi_{2n+2-k}(0)$.

The calculation proceeds by discretizing the τ axis in increments of h. Then $\xi_n(\tau + h)$ is computed in terms of $\xi_n(\tau)$ by Eq. (12) for $h \sim 0.25$ and is examined for evidences of collisions. [The calculation of the Bessel functions to CDC machine accuracy (10^{-14}) is quick and easy using $J_{n-1} = (2n/z)J_n - J_{n+1}$ starting with $J_{18} = 0$, $J_{17} = 1, \ldots$ and normalizing with $J_0 + 2J_2 + 2J_4 + \cdots = 1$.] $\xi_{2i+1} \leqslant d$ is the hard core constraint which separates particles i and $i + 1$; if $\xi_{2i+1} - d$ changes sign between τ and $\tau + h$ then a collision has occurred. Even if it does not change sign but $\xi'_{2i+1}(\tau) > 0$ and $\xi'_{2i+1}(\tau + h) < 0$, then a collision may or may not have taken place. Exactly when the collision happens, if it does, can be determined by the secant method. The earliest collision is found in this way; its effects are accounted for by propagating the effects of the momentum exchange in the collision backward in time to time τ. This requires resetting the velocities and displacements of first, second, and third neighbors of the colliding pair. This is as far as effects of a collision will propagate, to machine accuracy, in a time $h \leqslant 0.25$. Figure 7 shows the procedure.

This scheme for keeping track of the collisions has the advantage that one needs to compute all coordinates only every h time units; as the successive collisions in each h interval are found, their effects are accumulated by resetting neighboring coordinates at τ.

We have done a considerable amount of numerical experimentation with the HHR system. Like the super-anharmonic system, both limits of variation of the one parameter d are well known. For $d \to \infty$, the system becomes the regular harmonic chain. (The total length can always be taken equal to zero, without loss of generality, because stretching the chain will only shift the

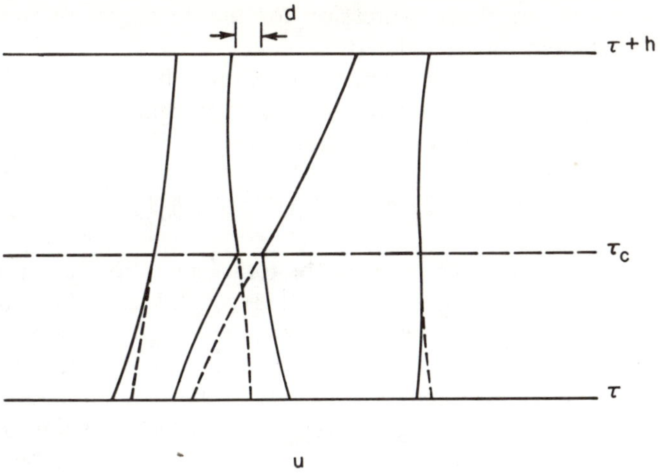

FIG. 7. This illustrates the method by which collisions are taken into account in the HHR calculation. The dashed lines are the orbits for particles without hard core interactions which will give the same orbits as the model system with hard cores after the collision occurs at τ_c.

equilibrium positions and change the static potential energy.) For $d \to 0$, one can neglect the potential energy changes, and we have the one-dimensional hard rod system.

In both limits the thermal conduction properties are abnormal, giving a heat current which is independent of chain length. Only for a small range of variation of d between the two limits is there any hope of a normal conductivity. Changing the hard core radius d is wholly equivalent to changing the temperature because the only quantity with dimensions of energy one can form from the HHR parameters is $\frac{1}{2}d^2$, the potential energy associated with a touching pair.

A computer simulation calculation of steady-state thermal conductivity requires that one end of the chain be in stochastic interaction with a reservoir at one temperature and the other end with a reservoir at another temperature. The two temperatures must be appreciably different (as in Section II, B, 1) in order that the steady-state heat current one seeks not be overwhelmed by noise. But if the HHR system has normal thermal properties, they will obtain only for a limited range of temperature (or d) such that d^2 is of the order of the temperature. One must, therefore, keep the fractional change in temperature over the chain small in order to resolve rapid variations in j_H as was the case for the super-anharmonic chain; this is inconsistent with the need for a large $\Delta T/T$ to insure good statistics.

For this reason the computer simulation of steady-state conduction was not pursued in the HHR system. Instead, a Green–Kubo type approach was adopted, in which an exact expression for the heat current in a definite initial-value problem was calculated.

It is shown in the Appendix that the integrated heat current that passes the center point of a chain in which initially the right and left halves were prepared at different uniform temperatures T_R and $T_L = T_R + \Delta T$ is

$$Q(\tau) = (\Delta T/8)\langle[\Delta E(\tau) - \Delta E(0)]^2\rangle_0 \tag{13}$$

where $\Delta E(\tau)$ is the difference between the total energy in the left half of the chain and that in the right half. Equation (13) is valid to lowest order in ΔT; the brackets denote an equilibrium average. It does not depend on any assumption of normality of thermal properties of the system (validity of Fourier's law), but if the system is normal with thermal conductivity K and heat capacity C, then after a long time the thermal diffusion equation predicts that

$$Q(t) = (K/C)t^{1/2}. \tag{14}$$

It is easy to evaluate Eq. (13) analytically for the harmonic system and for the hard rod system. For both, $Q(t)$ becomes linear in t at long times: for the harmonic ($d \to \infty$) system

$$Q(\tau) = (4/\pi)|\tau - \tau_0| \tag{15}$$

and for hard rods ($d \to 0$),

$$Q(\tau) = (9\pi/8)^{1/2}d^{-1}|\tau - \tau_0|. \tag{16}$$

Equation (13) thus provides a neater way of studying the thermal transport properties than a direct numerical simulation approach, because one needs to consider only one temperature, one does not encounter the resolution troubles mentioned above, and also one need not bother with end reservoirs and the concomitant computational difficulties, time averages, and extra parameters they entail.

$Q(t)$ has been computed for several values of d, for a chain of length 400 with fixed ends. Equation (13) is valid for all times for a finite chain, but Eqs. (15) and (16) are derived under the assumption that the chain is infinite. Equation (13) for a finite chain with large enough or small enough values of d should coincide with Eqs. (15) and (16), however, at least for times up to the sound travel time to the ends from the middle. The results are shown in Fig. 8. The agreement for large d and for small d with the harmonic and hard rod limits is good. The intermediate d's, for which one might hope that

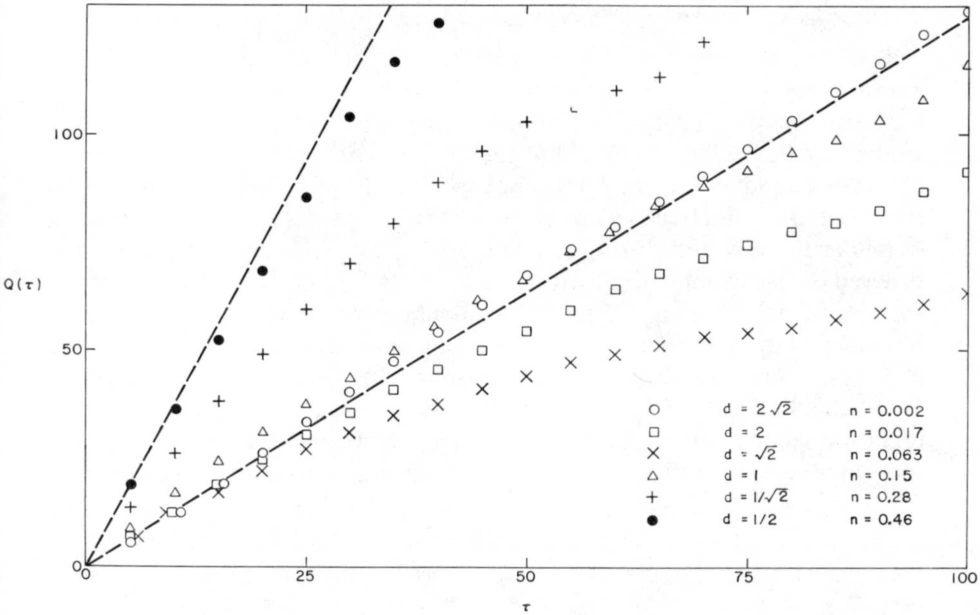

FIG. 8. Integrated heat current past the center of a 400-atom HHR chain with an initially stepped temperature distribution. The lower dashed line is the harmonic ($d \to \infty$) limit; the upper one is the hard rod value for $d = \frac{1}{2}$. The $d = 2^{3/2}$ and $d = \frac{1}{2}$ cases are approaching their respective limits, both of which have $Q(\tau) \propto \tau$. None of the intermediate values of d give curves resembling $Q(\tau) \propto \tau^{1/2}$, so that at least for these short chains no HHR system seems to obey Fourier's law. These data were calculated from Eq. (13), computing orbits from several hundred states of an initial equilibrium ensemble which were generated by the method of Metropolis *et al.* (1953). The various d values have different plotting symbols; n is the average rate of collision of a neighboring pair per τ unit. Temperature is such that $\langle v^2 \rangle = 1$ with unit mass and force constant; the steps in d here correspond to factors of 2 in effective temperature. Sound velocity in the harmonic limit is one half lattice space per τ unit; in the hard rod limit individual velocities propagate with speed $v/2d$ lattice space per τ unit. Since the velocity distribution is Maxwellian, the effects of the boundaries are felt much sooner in the finite d cases than the harmonic sound speed would indicate.

the thermal behavior be normal, do not show $Q \propto t^{1/2}$ [Eq. (14)] as required by Fourier's law. The conclusion is therefore that the HHR system of length up to 400 has abnormal thermal transport properties. One cannot say on the basis of this calculation that longer chains do not obey Fourier's law, but it seems unlikely. Helleman (1972) performed a computer simulation of steady-state conduction on the same system. He used a very large fractional temperature difference between the ends of the chain; for this reason as discussed above the interpretation of his results is difficult.

3. *Disordered Anharmonic Systems*

There is, perhaps, some hope that in spite of the indications that anharmonicity and disorder separately do not make a chain normal, in combination they might. The interaction of anharmonicity with disorder was studied by Payton *et al.* (1967) but the question of normality (independence of thermal conductivity on N) was not specifically addressed. Figure 9 shows the thermal conductivity of chains of length 100 particles for various values of anharmonicity and disorder. The model chosen was an isotopically disordered binary chain with or without Lennard–Jones type anharmonicity in the nearest-neighbor interactions. The calculation was a computer simulation of steady-state conduction with stochastic interactions of the end atoms with reservoirs (impulsive elastic collisions with reservoir particles with a Maxwellian velocity distribution). In the anharmonic cases the numerical integration scheme was a fourth-order Runge–Kutta method; in the harmonic cases an implicit iterative differencing scheme was used which was faster than

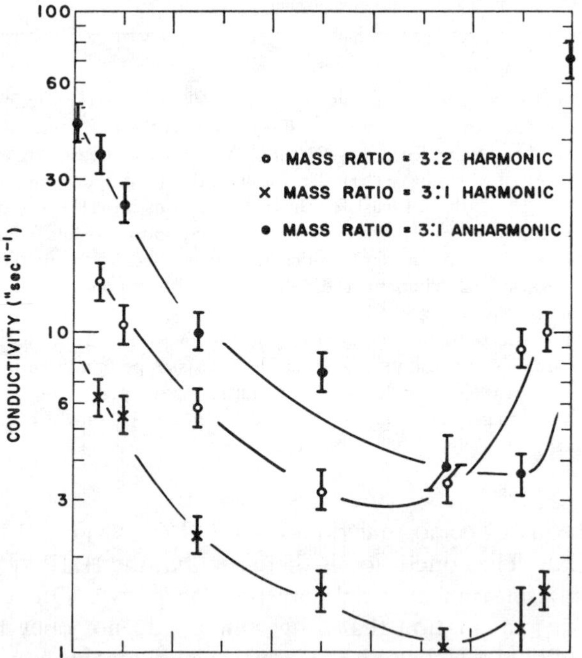

FIG. 9. Thermal conductivity of an isotopically disordered binary chain (mass ratio 2:1) as a function of concentration of light component. Anharmonicity enhances the flow of heat in this system. From Payton *et al.* (1967).

the Runge–Kutta, but which did not reliably converge in the anharmonic case.

The striking feature of the results illustrated in Fig. 9 is that for a disordered system the addition of anharmonicity *increases* the heat current, contrary to what any relaxation time concept would predict. That is, all Boltzmann-equation-type approaches predict that independent scattering mechanisms (e.g., disorder and anharmonicity) will contribute more or less additively to the thermal resistance (at least not subtractively!). But the results of the experiment are that in the system in which both disorder and anharmonicity hinder the heat flow, heat flows better than in one where the only resistive mechanism is disorder!

The solution to the puzzle is easy, when one recalls the nature of the normal modes in the disordered harmonic system. They are all, except for about $\sqrt{N}$ low frequency ones, highly localized, and not able to contribute to heat conduction. But the introduction of anharmonicity enables them to decay into one another; modes localized near a given site can populate other modes with which they overlap, thus providing a mechanism for energy propagation. Therefore in contrast to the situation in the ordered harmonic lattice, wherein the addition of anharmonicity decreases the heat current, in the disordered case it enhances the conductivity. So it is probably futile to attempt to calculate thermal transport properties in a disordered system by a perturbation expansion in a disorder parameter, because the nature of the eigenvectors is very strongly dependent on those parameters near the origin.

4. *Self-Consistent Reservoirs*

The effects of anharmonicities in the interparticle interactions complicate the calculation of heat transport to the extent that, except in special simple cases, no exact analytic work has been done. One way of simulating some of the anharmonic effects while retaining much of the tractability of the harmonic system was invented by Bolsterli *et al.* (1970). Their system is illustrated in Fig. 10. It is a chain of particles connected to each other by harmonic

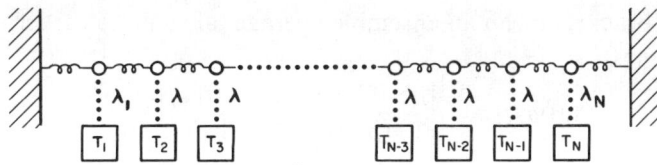

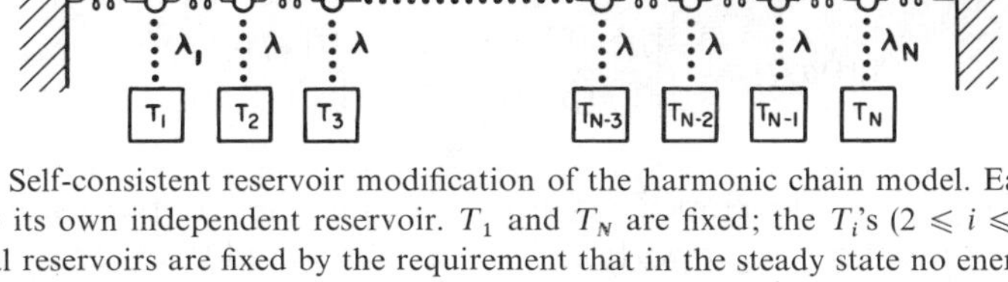

FIG. 10. Self-consistent reservoir modification of the harmonic chain model. Each atom is coupled to its own independent reservoir. T_1 and T_N are fixed; the T_i's $(2 \leqslant i \leqslant N-1)$ of the internal reservoirs are fixed by the requirement that in the steady state no energy flows to or from any reservoir.

springs, but in addition each particle interacts with its own heat reservoir. The function of the reservoirs is to provide a means for the harmonic normal modes to interact with each other. The reservoirs simulate the anharmonic forces which occur in real systems and also simulate interactions of the lattice system with other degrees of freedom which exist in real solids, such as electrons and spin systems. The temperatures of the reservoirs are chosen to be equal to the kinetic temperatures of their respective lattice particles; this self-consistency conditions insures that in the steady state no energy is exchanged between the reservoir and the particle. The reservoir is thus a means for attaining ergodicity, which was manifestly absent in the original harmonic system.

The self-consistent reservoir system can be analyzed by starting with the equations of motion, using Langevin terms for the reservoir interactions.

$$\begin{aligned} \dot{p}_i &= \gamma(x_{i+1} - 2x_i + x_{i-1}) - \lambda_i[p_i/m_i - f_i(t)] \\ m_i\dot{x}_i &= p_i \end{aligned} \tag{17}$$

Appropriate modifications are made for $i = 1$ and $i = N$ depending on the boundary conditions assumed. $f_i(t)$ is a Gaussian Markoffian function of the time; its autocorrelation is proportional to the temperature of the reservoir.

[It might be noted here that Eq. (17), written for the classical case, is true also for the quantum case with p_i, x_i Heisenberg operators. Ford *et al.* (1965) show that reservoir interactions can be expressed as Langevin terms, and the covariance of the random force $f_i(t)$ can be determined. The relation between that covariance and the temperature, however, is different from the classical case, which is given in Eq. (22).]

One defines a $2N \times 2N$ matrix b (a covariance matrix of the harmonic system)

$$b = \begin{bmatrix} \langle p_i p_j \rangle & \langle p_i x_j \rangle \\ \langle x_i p_j \rangle & \langle x_i x_j \rangle \end{bmatrix} \tag{18}$$

where the brackets mean an ensemble average taken at an earlier time; e.g.,

$$\langle x_i(t) p_j(t) \rangle = \frac{\int \rho\{x(0), p(0)\} x_i(t) p_j(t)\, dp(0)\, dx(0)}{\int \rho\{x(0), p(0)\}\, dp(0)\, dx(0)} \tag{19}$$

where $x_i(t)$, $p_i(t)$ are developed via Eq. (17) from their initial values, whose distribution function is proportional to ρ, which could be, for example, an equilibrium Boltzmann factor. If Φ is the matrix of the force constants,

$$\Phi = \gamma \begin{bmatrix} -1 & 1 & 0 & . & . & . \\ 1 & -2 & 1 & 0 & & \\ 0 & 1 & -2 & 1 & & \\ . & & & . & & \\ . & & & & . & \\ . & & & & 1 & -1 \end{bmatrix} \tag{20}$$

(for the fixed boundary condition case), then, for example, from Eq. (17)

$$\begin{aligned}(d/dt)\langle p_i p_j\rangle = \Phi\langle xp\rangle + \langle px\rangle\Phi - \lambda m^{-1}\langle pp\rangle - \langle pp\rangle\lambda m^{-1} \\ + \lambda\langle fp\rangle + \langle pf\rangle\lambda\end{aligned} \tag{21}$$

in an obvious notation. One now makes the usual arguments about $f_i(t)$ being a Markoff process with short memory; using an appropriate limiting process to define the δ function and taking a functional average over Gaussian realizations of $f_i(t)$,

$$\langle f_i(t)\, f_j(t')\rangle = 2kT\,\delta_{ij}\,\delta(t - t') \tag{22}$$

in the classical case [the quantum expressions are developed by Visscher and Rich (1975)]. f is correlated only with p, not with x. By differentiating the other elements of b, one finds

$$\dot{b}(t) = d + b(t)a + a^{\mathrm{T}}b(t) \tag{23}$$

where the superscript T denotes transposition,

$$a = \begin{bmatrix} -\lambda m^{-1} & m^{-1} \\ \Phi & 0 \end{bmatrix} \quad d = \begin{bmatrix} 2\lambda T & 0 \\ 0 & 0 \end{bmatrix} \tag{24}$$

and λ, m, T are diagonal matrices of the λ's, m's, and kT's. Equation (23) was written down and solved for the steady state ($\dot{b} = 0$) by Rieder *et al.* (1967) for the perfect monatomic harmonic chain with reservoir interactions only at the ends. We consider it now for an ordered or disordered classical chain with nonharmonic interactions simulated by self-consistent reservoirs interacting with each atom. The self-consistency condition, that there be no exchange of energy between the reservoirs and the chain (except, of course, at the ends, which act as a source and a sink for energy), can be deduced from some of the diagonal elements of the matrix Eq. (23). Namely, if $\dot{b} = 0$ (steady state),

$$2\lambda T - m^{-1}\lambda\langle pp\rangle - \langle pp\rangle\lambda m^{-1} = -\Phi\langle xp\rangle - \langle px\rangle\Phi \tag{25}$$

the diagonal elements of which are

$$\lambda_i(T_i - m_i\langle p_i p_i\rangle) = \gamma(\langle x_{i-1}p_i\rangle + \langle x_{i+1}p_i\rangle), \tag{26}$$

and the right-hand side is just the energy flow out from site i into the rest of the chain times m_i. The self-consistency condition is then that the kinetic temperature of each particle which does not exchange energy with its reservoir (i.e., all the internal particles) shall be equal to the temperature of its reservoir.

$$kT_i = \langle p_i^2\rangle/m_i. \tag{27}$$

Equation (23) for $\dot{b} = 0$ is a set of linear inhomogeneous equations for the elements of the b matrix. When Eq. (27) is used to eliminate the internal temperatures that occur in d, then the only inhomogeneous terms are the temperatures of the end reservoirs. So in principle it is easy to solve for the self-consistent b matrix using one of the computer linear systems solvers. But in practice a more devious approach is called for, because Eq. (23) is a $4N^2 \times 4N^2$ system of equations, too big too handle for a reasonably large N.

The computer procedure actually used to solve for the steady-state convariance matrix of these systems is the following. One first diagonalizes the $2N \times 2N$ nonsymmetric a matrix:

$$aS = S\hat{a} \tag{28}$$

$\hat{a}$ is a complex diagonal matrix; the columns of S are the eigenvectors of a.

Given Eq. (28), Eq. (23) with $\dot{b} = 0$ reads

$$\hat{d} + \hat{b}\hat{a} + \hat{b}\hat{a}^* = 0, \tag{29}$$

where

$$\hat{b} = S^\dagger bS, \quad \text{etc.} \tag{30}$$

and the dagger denotes Hermitian conjugation. Because $\hat{a}$ is diagonal, it is trivial to solve Eq. (29) for the elements of $\hat{b}$. Then Eq. (30) is inverted to get b. When one has b, Eq. (27) is used to find new reservoir temperatures (the starting ones were guessed) and Eq. (29) is used again to find a new b, from which in turn one gets another set of temperatures, etc. This iterative procedure almost always has converged; the number of iterations needed depends on the goodness of the initial guess. If convergence is defined as no change in successive iterates of reservoir temperatures to five significant figures, then it usually takes about 50 iterations for a disordered chain of 50 atoms.

Some results of self-consistent reservoir calculations for disordered chains are shown in Fig. 11, where we have plotted thermal conductivities, defined as the heat current divided by the internal thermal gradient, as functions of the coupling to the internal reservoirs λ. Both ordered and disordered chains are shown, with free and fixed boundary conditions. The lengths are 25 and 50 atoms for the dotted and dashed lines; the disordered chains are random

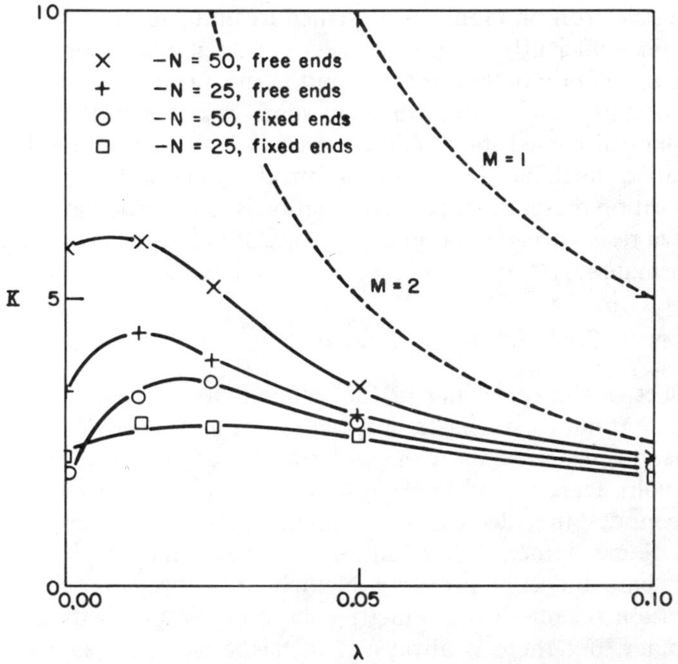

FIG. 11. Thermal conductivity K as a function of coupling to internal reservoirs. $K =$ (steady-state heat current)/(internal temperature gradient); the latter is obtained by a least-squares fit to kinetic temperatures (diagonal elements of b matrix) for a particular realization of a random (equal parts of mass 2 and 1) chain. Statistical errors are large (up to 20%) for $\lambda = 0$, about 10% for $\lambda = 0.0125$, and generally negligible for larger λ. The dashed lines are analytical results for infinite monatomic chains (Bolsterli *et al.*, 1970). For any value of λ, if the chain is long enough, an increase in length should bring the fixed- and free-end results closer together. This is the case for relatively large values of λ in these short chains.

mixtures of approximately equal parts of masses 2 and 1 units. The solid line is the exact monatomic analytic result for $N \to \infty$ (Bolsterli *et al.*, 1970).

As one might expect, the effects of the different treatment of the ends disappears for large λ. And for small λ this system should approach the disordered harmonic system discussed in Section II, A. It seems to do so; K decreases for small λ and large N in the disordered fixed end case as it should, and it increases for free ends.

As mentioned above, the self-consistent reservoir approach can be applied in the quantum-mechanical case. The b matrix elements become expectation values of normal products, and the elements of the d matrix are no longer proportional to the temperatures, but can be written as sums and integrals over Bose functions.

The usefulness of the self-consistent reservoir theory stems from the fact that it is more tractable than the many-body theories (anharmonic and interacting electron-phonon) it is designed to simulate. In the classical case one can (in sufficiently simple models) consider anharmonic lattices by computer simulation of their motion; but in the quantum case one must fall back on perturbation theory which in some cases, as in the classical disordered chain discussed above, fails completely. The reservoir model provides an alternative which has many applications that have not yet been exploited. One application presently under investigation is to the problem of calculating the Kapitza resistance, the thermal resistance at the contact between any two dissimilar materials.

C. Effects of Computational Errors

The effect of the reservoirs in the self-consistent reservoir model is to provide a continuous stochastic interaction for each particle, with the constraint that no energy, on the average, is transferred to or from the reservoirs. The reservoirs act as a passive catalyst, facilitating energy exchange between harmonic modes, thus decreasing or enhancing heat flow, depending on the situation. Some numerical integration schemes, when implemented on a computer, have the same property. Namely, it is always possible to invent an integration scheme which exactly conserves energy, to the accuracy of the computer, but there is always a stochastic element—caused by inaccuracies in the numerical integration or by truncation or roundoff in the computer. One can estimate the importance of this effect in terms of an effective λ. For CDC computers, with a word length of 14 decimal digits, it is completely negligible compared to those used in Section II, B. But in fact a stochastic element even of such a small magnitude is critically important in some situations, like the one studied recently by Bivins *et al.* (1973). They considered a linear chain for which the nearest-neighbor interaction contained a quartic admixture in its potential as well as the harmonic quadratic term, and calculated the time evolution of the amplitudes of the harmonic normal modes. The geometric average of the energies of modes 10 and 12 in a system of 15 particles is shown in Fig. 12, for an initial excitation almost exclusively in mode 11. The admixture of these neighboring modes is seen to grow exponentially, out to a time of order 1500 when they contain energy comparable to the total available. The integration was reversed at $t = 1200$ and at $t = 1500$; if the integration scheme and the computer were infinitely accurate the time development would be exactly retraced, and the system would return to its initial state at $t = 0$. The interpretation of the marked deviation from this expected behavior is that at the time at which the velocities are reversed, there is present a certain roundoff error in the amplitudes of the admixed modes, which grows exponentially in the reversed elapsed time, just as their amplitudes grew in the first place. Confirming this view is the fact that the slopes of the deviant backward-developing lines are the same and equal to the negative of the slope of the forward-developing line; their values at the times of reversal are presumably of the order of magnitude of the roundoff error which started them (these values are consistent with the Maniac II computer word length).

The results shown on Fig. 12 do not show that the system considered is irreversible, mixing, or ergodic—properties that are important in considerations of the approach to equilibrium as well as in the steady-state nonequilibrium statistical mechanics encountered in heat transport calculations. What they do show is that because of errors inherent in the finite computer,

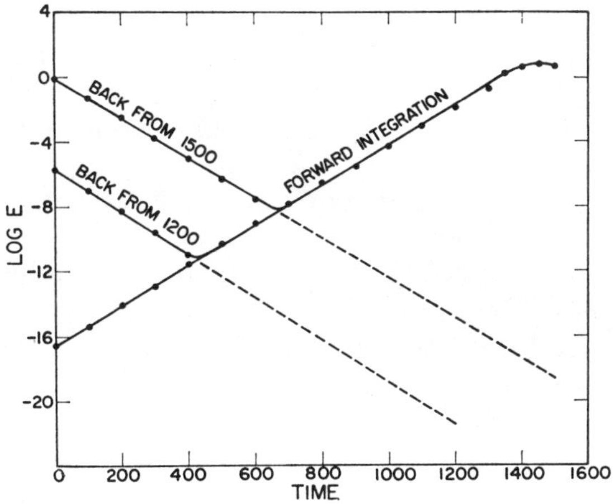

FIG. 12. Geometric mean of energies in modes 10 and 12 of a 15-particle anharmonic chain as a function of time. The velocities are reversed at $t = 1200$, and the equations of motion are integrated backward; the same is done for $t = 1500$. From Bivins *et al.* (1973).

this system might appear, in computed results, to be ergodic, mixing, and irreversible, even if in "reality" (classical reality—a solution of Newton's equations with an infinitely accurate computer) it was not.

The mixing of harmonic normal modes with each other is, as we have seen in Section II, B, 3, mainly responsible for determining the heat current in a disordered anharmonic chain. The calculated mixing can be spurious (initiated by roundoff error or by inaccuracies in the integration scheme) or it can be real, dictated by the equations of motion. To evaluate the reliability of any computer-experimental (molecular dynamical) type of calculation of thermal conductivity it is important to be able to gauge the importance of spurious mixing in a given case.

One way to estimate accuracy, in a given dynamical computer-experimental calculation of nonequilibrium properties, is to check the reversibility of the calculation over times comparable to the relevant physical times involved (equilibration, induction, or ergodicity time). Given displacements and velocities at $t = 0$, the computer integrates the equations of motion to $t = T$; then the velocities are reversed, and at $t = 2T$ the initial displacements should be nearly reproduced, with reversed velocities. What a minimum acceptable T is depends on the system; intuition suggests that it ought to be at least of the order of the time scale of the real ergodicity of the system. For the HHR system that time scale is infinite in both the $d \to 0$ and $d \to \infty$ limits because those systems are not ergodic at all; and according to the results presented in Section II, B, 2 it may be infinite for all d. In the HHR system reversibility of the computer-experimental calculation has been checked for times up to 400. In the system studied by Bivins *et al.* (1973) the maximum T can be read off from Fig. 12, but what the ergodicity time is is not known.

From a practical point of view it seems unlikely that roundoff errors can have any impact on a calculation by computer simulation of transport in solids, because the effective λ due to interactions with other degrees of freedom will be dominant. Apparent exceptions to this statement are low temperature systems in which quantum effects cause all degrees of freedom to be coherent, but they probably cannot be approached by direct computer simulation anyway.

D. Other Numerical Experiments

Jackson *et al.* (1968) have performed a computer experiment in a disordered anharmonic chain which is quite different from the one considered by Payton *et al.* (1967). Their lattice is monatomic with cubic and quartic terms added to the potential of the nearest-neighbor interactions. The coefficients of these nonlinear terms can be chosen randomly, yielding a sort of disordered lattice. They numerically measured the energy flux in the steady state through this lattice when one end is in contact with a heat reservoir at one temperature, the other end in contact with a thermal bath at another temperature. They obtained thermal conductivities in this way. They also calculated the autocorrelation function of the total heat current (the local heat current summed over all the particles) and tried to identify it (via the Green–Kubo formula) with the measured thermal conductivity. Their chain was too short for this to be successful. They did not study the N-dependence of the energy flux; thus whether or not their system has normal thermal conduction properties is unknown.

The Jackson, Pasta, Waters model was also used by Miura (1973), whose interpretation was in the language of solitons. His lattices were also far too short to exhibit normal thermal properties, and he did not study the N-dependence.

Nakazawa (1970) performed a numerical experiment on the same system, only without disorder, as was used by Payton *et al.* (1967), but with a different kind of random force acting on the end atoms. His results were qualitatively similar (a quantitative comparison is difficult because of the different parameters used in the numerical experiments) in that the heat current decreased with increasing anharmonicity. Nakazawa's chains also, were too short ($N = 30$) to enable him to study N-dependence.

Computer experiments designed to measure lattice thermal conductivity have, to a limited extent, been performed on two-dimensional lattices as well (Payton *et al.*, 1967, 1968; Rich *et al.*, 1971). From a computational point of view they differ from the 1d cases mainly in time and memory requirements, which severely limit the sizes of lattices that may be considered. Consequently, no information on N-dependence in 2d lattices is available [except in the ordered harmonic case, which is abnormal (see Helleman, 1972; Nakazawa, 1970)], but the interrelation between anharmonicity and disorder in 2d lattices is qualitatively the same as in 1d; that is, in a disordered lattice anharmonicity enhances the heat current. Zabusky (1973) has interpreted these results in terms of heat conduction by solitons. Figure 13 shows some

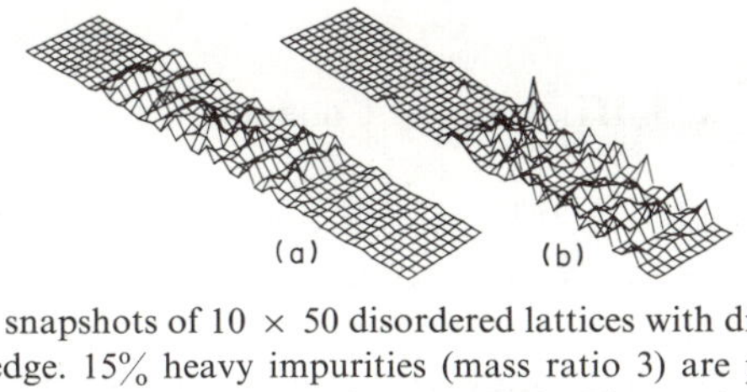

FIG. 13. Equal-time snapshots of 10 × 50 disordered lattices with disturbances propagating from the lower-right edge. 15% heavy impurities (mass ratio 3) are placed at random sites, except in the first third of the lattice, which is monatomic. The vertical coordinate is energy; the lattice in (b) is harmonic, while (a) has Lennard-Jones anharmonicity which materially aids transmission of energy through the disordered lattice. Taken from Payton *et al.* (1968), with permission.

computed results for a disordered 2d lattice in which the atoms along the near edge at $t = 0$ were all given the same initial velocity, while the rest of the lattice atoms were at rest and undisplaced.

The important point here is that the heavy impurities (light impurities have a slightly less spectacular effect) effectively break up the harmonic wavefront, while the anharmonic wavefront (soliton-like) both travels faster than the harmonic one and very tenaciously retains, maybe even reinforces, its coherence. This picture is consistent with the one proposed in Section II, B, 3 for the 1d anharmonic disordered chain, namely, localized modes decaying into others in the direction of flow, but requiring definite phases in the admixed modes to sustain the wavefront. Much of the usefulness of such results comes from the intuitive feeling for the dynamics they impart.

E. Discussion

We are still far from understanding exactly what features in the dynamics generate the right kind of ergodicity to make the thermal conductivity obey Fourier's law. Apparently none of the harmonic systems do, either ordered or disordered, and the anharmonic systems on which numerical experiments have so far been performed do not either. The HHR system was a disappointment in this respect; one had hopes that collisions would mix harmonic modes in an effectively stochastic way and make the systems that are short enough to be computationally tractable ergodic on a sufficiently short time scale to cause the thermal transport to be normal. It may be that one-dimensionality is the stumbling block that no conservative (the self-consistent reservoir model, which explicitly introduces a stochastic element, is thereby excluded) dynamical 1d system can obey Fourier's law—there are too many microscopic conservation laws relative to the number of dimensions. The extra degrees of freedom that come with multidimensionality might be essential, but a computer-experimental verification of this would involve a much bigger calculation than has been done so far.

III. Electric Conductivity

Like the models used to simulate physical systems for thermal conductivity, those that we take to study electric conductivity are chosen as much for their simplicity and tractability as for their fidelity to the real world. Once again it seems to be the case that any models that will yield to exact analytic treatment are too trivial to incorporate most physically interesting features, so again we will look for a simple case which does, and do numerical experiments on it.

A. Model Criteria

It may seem rash to forego quantum mechanics in the name of computational ease simply because the great majority of electric conductivity situations in solids are intrinsically quantum mechanical. Only very high temperature cases are not and may be treated quantitatively classically. But it turns out that even in the classical case there are basic questions still unanswered, and it is amenable to study by computer simulation.

The model we simulate is an ordered 2d array of fixed potentials in which noninteracting electrons move classically in an applied external electric field, with an explicit dissipative mechanism. Two dimensions are chosen because 1d contains little physics in this case (without dissipation or applied external fields an electron in a 1d array will either be trapped or will propagate freely). Noninteracting electrons are chosen because a many-body problem of this kind is too difficult and because the one-body problem already contains much unexplored physics. An explicit dissipative mechanism is needed because without it conductivity is indeterminate, probably infinite; the energy of a drifting electron is a linear function of its position coordinate in the direction of the applied field. A moderately realistic, yet computationally tractable, dissipation mechanism is presented in Section III, C.

It is desirable to obtain exact numerical solutions for conductivity in a simple system like the one described because there is no theory for electric transport in solids which has been derived without inexpiable assumptions (molecular chaos in the Boltzmann equation; analyticity or continuity in linear response theory). If our results show that theoretical results and computer-experimental results do (or do not) agree here, then our confidence in (or skepticism of) theoretical results in more complex and realistic systems will be reinforced.

B. Linear Response Theory

The specific question we address now is whether linear response theory is valid in electric conductivity or not. Van Kampen (1971) has argued that it is, if not wrong, at least ill-founded. His arguments go as follows. Figure 14 shows our conductivity model for a 2d array of hard disks (infinitely repulsive square wells) with a classical electron moving in it. The electron starts at point S with a certain thermal velocity. The solid line is the orbit of the electron with no applied electric field, the dotted line with an applied field of 570 V/m, the dashed line in a field 10 times that great. The point of van Kampen's argument is that linear response theory expresses observables in terms of averages over microscopic orbits; but for physically important magnitudes of fields the orbits obviously cannot be expressed as power series

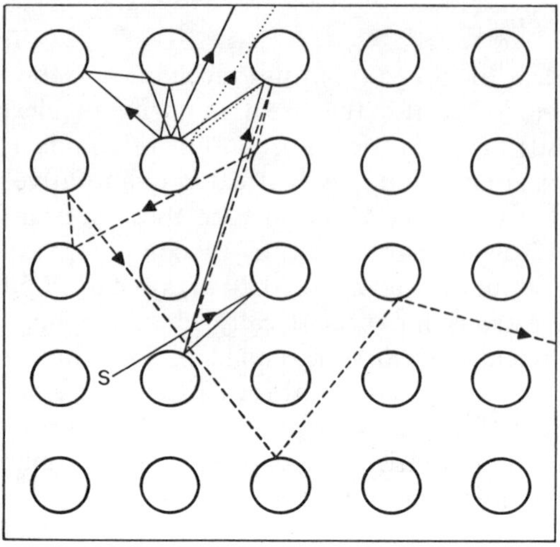

FIG. 14. Orbits of a classical electron in an array of hard disks. The lattice spacing is 10^{-8} cm; the three orbits have $E = 0$ (solid line), $E = 570$ V/cm (dotted line), and $E = 5700$ V/cm (dashed line). From Visscher (1974).

in E, yet that is what linear response theory does, keeping only the first term. So it is important to see in what sense linear response theory is correct, if any.

C. Conductivity Calculation

To do this we (Visscher, 1974) calculated the orbits of classical electrons in the lattice of Fig. 14 averaged over long times and over an ensemble of initial conditions for a variety of potentials, attractive and repulsive. The aim was to find the average drift velocity of the electron in an applied electric field. The applied field was that due to a stepped potential; i.e.,

$$E(x) = aE \sum_{n=-\infty}^{\infty} \delta(x - na). \tag{31}$$

This choice of field insures that the orbits are concatenations of straight lines, which simplifies the computer arithmetic without materially changing the physics.

The only tricky part of this computer experiment was the implementation of the dissipation mechanism. Some stochastic element is clearly required; one wants to have an equilibrium ensemble realized for the electron in the absence of electric fields after long times. One way to do this is to have the circular potentials moving stochastically to simulate the atomic thermal motions in a solid. The potentials would be assigned masses and temperatures and Maxwellian velocity distributions. To actually carry out this simulation properly, meaning in a manner which insures the proper equilibrium velocity distribution for the electrons, would require very complicated analysis of collision times and probabilities, involving stochastic determination not only of the velocity of the potential boundary, but also of its displacement from equilibrium. Any obvious simplification of the collision kinematics does not work, such as using a Gaussian truncated at the electron's radial velocity (because the electron cannot collide with a wall which is receding from it), or using a relative velocity-weighted one-sided Gaussian $v \exp(-\frac{1}{2}\beta v^2)$. One scheme which does work is to assume that the electrons' (they feel no forces except when they cross circular potential boundaries or potential jump lines at $x = na$) motion is governed in some regions of the lattice by a Langevin equation

$$m\dot{v} = -\lambda[v - f(t)], \tag{32}$$

where $f(t)$ is a random function of time. We will take Eq. (32) to hold only in a narrow annulus (see Fig. 15) surrounding each potential, so the straight-line character of the orbits is preserved elsewhere.

If one assumes $f(t)$ to be Gaussian white noise it can be shown (Visscher, 1974) that a particle entering the Brownian medium [in which Eq. (32) holds] at $t = 0$ with velocity $v(0)$ has a velocity distribution after it has been immersed a time t given by (within a normalization factor)

$$P\{v(t)\} = \exp\{-\tfrac{1}{2}\beta(t)m[v(t) - e^{-\eta t}v(0)]^2\} \tag{33}$$

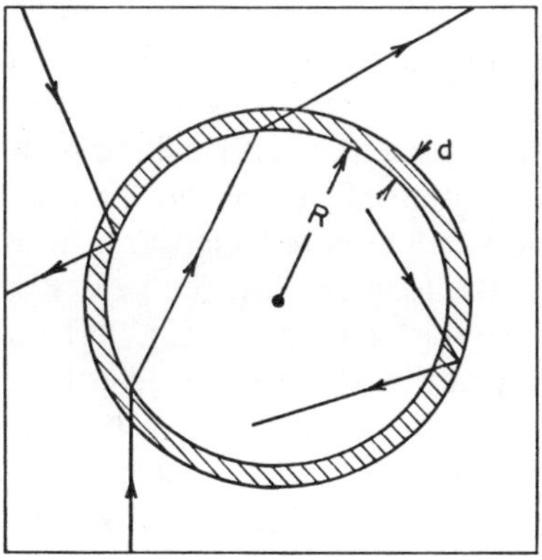

FIG. 15. Ring around each circular potential within which electrons undergo dissipative stochastic interactions with thermal reservoirs. The circular square wells can be attractive or repulsive; the viscous coating is outside them. From Visscher (1974).

where $\eta = \lambda/m$ and

$$\beta^{-1}(t) = kT(1 - e^{-2\eta t}) \tag{34}$$

with T as the reservoir temperature. For short times $P\{v(t)\} \to \delta(v(t) - v(0))$; for long times $P\{v(t)\} \to \exp\{-\frac{1}{2}\beta m v(t)^2\}$. Suppose, given a time of immersion in the Brownian medium t and a random number $Q(0 < Q < 1)$, that one asks what is the velocity $v(t)$ such that $\int_0^{v(t)} P(v)\,dv = Q$. The answer is

$$v(t) = v(0)e^{-\eta t} \pm [2/m\beta(t)]^{1/2} \operatorname{erf}^{-1}(Q), \tag{35}$$

which can be handled by the computer (the inverse error function may be approximated by a piecewise analytic approximation). The computer generates a quasi-random number Q [see Wood (1968) for a discussion of quasi-random number generators and tests of their randomness] uniform on (0, 1) and calculates Eq. (35). The density function for $v(t)$ thus obtained is just $P\{v(t)\}$. Actually it is not exactly Eq. (35) which is computed, but the limit of Eq. (35) for very thin coatings; viz.,

$$v(t) - v(0) = -\eta\, dv(0)/|v^{\mathrm{r}}| \pm 2(\eta d/\beta m|v^{\mathrm{r}}|)^{1/2} \operatorname{erf}^{-1}(Q) \tag{36}$$

because then the position of the particle does not change as it traverses the coating. Here d is the coating thickness and v^{r} is the radial velocity of the particle ($d \to 0$; ηd remains finite). The changes in radial and tangential velocity of the particles encountering one of the sticky circles are calculated by two successive independent applications of Eq. (36). Obviously, particles diffusing freely in a lattice with this kind of reservoir interaction will acquire a Maxwellian velocity distribution. With some other reservoir interactions that have been used in other applications, this is not the case.

The computer, given a starting velocity and position of the charged particle, finds the next encounter with either a circular potential or one of the discontinuities in external potential at $x = na$. In the latter case kinematics determines whether the particle is reflected or transmitted with a changed velocity component. In the former case, the velocity (radial and tangential components independently) is changed by Eq. (36), and if the radial velocity is still inward then kinematics is done to determine whether the particle proceeds to the inside of the potential well. If not, it must penetrate the reservoir coating again to enter the interstitial space. The orbit of each particle is followed through of the order of 2000 potential collisions, and about 1000 different orbits, with initial velocities chosen at random from a Maxwellian distribution, are calculated, and the appropriate averages computed. Some results are shown in Figs. 16 and 17. The quantities shown here

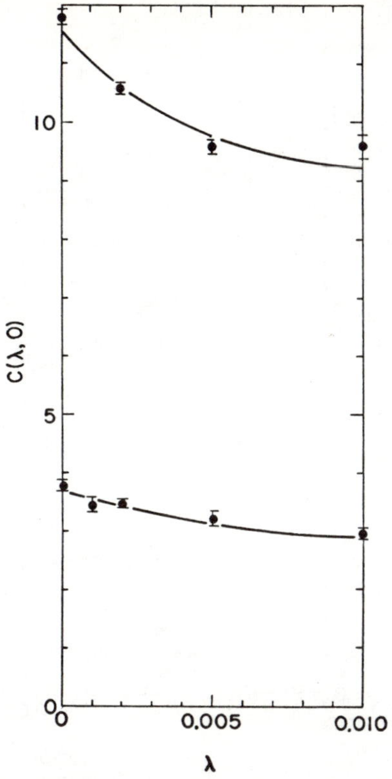

FIG. 16. Diffusion coefficient in $2d$ classical model as a function of dissipation parameter λ. The upper curve is for $V = -40$ (circular square well depth); the lower one for $V = +\infty$. Units: time $= 10^{-15}$ sec, distance $= 10^{-8}$ cm, mass = electron mass, so that energy = 0.057 eV. From Visscher (1974).

are the diffusion coefficient, calculated by averaging $[x(t) - x(t_0)]^2/2(t - t_0)$ for large $t - t_0$ [this is $C(\lambda, 0)$ of Fig. 16], and the drift velocity, the average of $[x(t) - x(t_0)]/(t - t_0)$ for a particular strength of reservoir interaction $\lambda = 0.002$, and various electric field strengths E [this is $C(0.002, E)$ of Fig. 17].

D. DISCUSSION

The prediction of linear response theory is that the conductivity and the diffusivity be numerically equal (for our choices of units and of temperature).

$$C(\lambda, E) \approx C(\lambda, 0). \tag{37}$$

Figure 17 shows that Eq. (37) is reasonably well satisfied for small enough E; thus linear response theory gives the right result with the qualification that

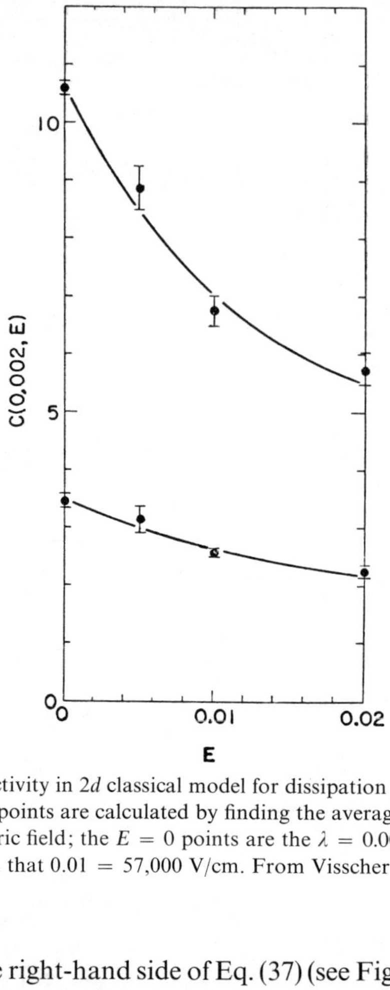

FIG. 17. Electric conductivity in $2d$ classical model for dissipation parameter $\lambda = 0.002$, as a function of E. The $E > 0$ points are calculated by finding the average drift velocity of charged particles in the given electric field; the $E = 0$ points are the $\lambda = 0.002$ diffusion coefficients in Fig. 16. Our units are such that $0.01 = 57{,}000$ V/cm. From Visscher (1974).

$\lambda > 0$. The limit of the right-hand side of Eq. (37) (see Fig. 16) is finite as $\lambda \to 0$, but for $E \neq 0$ the left-hand side is then indeterminate, as we have argued on physical grounds (and it can be numerically made very plausible). Therefore a small amount of dissipation is essential to validate linear response theory. Exactly how much is not important, as long as it is sufficient to keep the electron velocity distribution near equilibrium.

Even for $\lambda > 0$, though, the results shown in Fig. 17 support the conjecture that the conductivity $C(\lambda, E)$ is not analytic at $E = 0$. This is because the numerical results seem to indicate that $C(\lambda, E) \cong a - b|E|$ for small E. The absolute value is necessary because on very general grounds we know that for a system with $x \to -x$ symmetry $C(\lambda, E)$ is an even function of E. But since $|E|$ is singular at $E = 0$, an expansion of $C(\lambda, E)$ in powers of E is illegitimate, and the basic assumption of linear response theory is invalid for this model.

Nevertheless, the numerical results say that Eq. (37) is approximately satisfied for small E, maybe exactly satisfied in the limit $E \to 0$. This is a numerically nontrivial result because the left- and right-hand sides of Eq. (37) are computed in very different ways whose equivalence is alleged by linear response theory. So the prediction of linear response theory (Green–Kubo formula for electric conductivity) is verified in this computer experiment; but at the same time the computer experimental results cast serious doubt on the basic premise of the theory. We have attempted to resolve this seeming paradox by showing (Mjolsness and Visscher, 1972; Visscher, 1974) that Green–Kubo-like expressions for the response can be derived without any assumption of smallness of the driving force or expandability in powers thereof (analyticity). So the question of validity of the linear response result is reduced from asking for analyticity to asking simply for continuity in E (or any other driving force) at the origin. This is, of course, a much less stringent requirement and has been shown numerically to be satisfied for the present model within the accuracy of the computer experiment. A pathological example is exhibited in Visscher (1974) in which the response actually is discontinuous at $E = 0$, and the linear response result does not hold (Ehrenfest's wind-tree model).

Although a numerical proof of a particular analytic or nonanalytic behavior of conductivity is impossible, because one can make an arbitrarily accurate analytic approximation to any reasonable function, the question of whether there is a singularity in response at $E = 0$ is so basic to transport theory that further numerical experiments for more realistic systems ought to be done.

Appendix: Time-Correlation Function Adapted to Computer Experiments

It can be shown (Visscher, 1974) that an exact expression for the response of any dynamical variable A to an applied temperature $\beta^{-1}(x)$ is

$$\langle A(t)\rangle = \langle A(t_0)\rangle + \int_{t_0}^{t} \langle A(t') \textstyle\int dx\, \nabla\beta(x) j_{\mathrm{h}}(x)\rangle\, dt', \tag{38}$$

where j_{h} is the local heat current at time t_0. Similar expressions can be written for response to applied electric fields and density gradients; they are generalized Green–Kubo formulas in that they are exact to all orders in the driving force. In Eq. (38) the "driving force" is $\nabla\beta$, which not only occurs explicitly in a linear fashion but also is contained implictly in the definition of the brackets on the right-hand side, which represent not an equilibrium ensemble average but rather an average in a local equilibrium initial ensemble with an imposed variable temperature $\beta^{-1}(x)$.

If we take A to be a heat current at the center ($x = 0$) of the system (which is the reasonable place to look because there the wall effects are delayed as long as possible and time-correlation functions remain significant), and take $\beta(x)$ to be a step-function with step $\Delta\beta$ at $x = 0$, then Eq. (38) becomes

$$\langle j_0(t)\rangle = \Delta\beta \int_{t_0}^{t} \langle j_0(t')j_0(t_0)\rangle \, dt' \tag{39}$$

where j_0 is the heat current (now specializing to a 1d chain) past the center atom. [If we had taken $\nabla\beta$ to be uniform over the whole length of the chain, then instead of $j_0(t_0)$, the average $N^{-1}\Sigma_n j_n(t_0)$ would have appeared, with $\Delta\beta = N\nabla\beta$.] If we satisfy ourselves with linear response, then the bracket $\langle \ \rangle$ can be replaced with the equilibrium ensemble average $\langle \ \rangle_0$.

Now if $\Delta E(t)$ is defined to be the excess in total energy on the right-hand side of the chain over that on the left side:

$$E(t) = E_{\mathrm{R}}(t) - E_{\mathrm{L}}(t) \tag{40}$$

then

$$\int_{t_0}^{t} j_0(t') \, dt' = \tfrac{1}{2}[\Delta E(t) - \Delta E(t_0)] \equiv Q(t) \tag{41}$$

and Eq. (39) becomes, after a few manipulations utilizing the fact that in an an equilibrium average $\langle A(t)B(t')\rangle_0 = \langle A(t')B(t)\rangle_0$ and after integration over t,

$$Q(t) = (\Delta\beta/8)\langle(\Delta E(t) - \Delta E(t_0))^2\rangle_0. \tag{42}$$

Equation (42) is a form of fluctuation–dissipation theorem for heat flow in this particular problem. In words, it says that the heat unbalance between halves of the chain (which initially were at constant, through different, temperatures) increases in proportion to the mean square heat unbalance in the equilibrium ensemble. It is useful and convenient in a computer experiment because, in addition to the advantages already enumerated, it makes use of the fact that the computer already has integrated the equations of motion to eliminate the time integral in Eq. (39).

Equation (42) can be evaluated for both limits of the HHR chain. For the harmonic chain with $N \to \infty$, $d \to \infty$, it is

$$\tfrac{1}{8}\langle(\Delta E(t) - \Delta E(t_0))^2\rangle_0 = (4/\pi)|t - t_0| \tag{43}$$

and for $d \to 0$, it is

$$\tfrac{1}{8}\langle(\Delta E(t) - \Delta E(t_0))^2\rangle = (9\pi/8)^{1/2}\, d^{-1}|t - t_0|. \tag{44}$$

These are the straight lines plotted on Fig. 8.

The solution of the temperature diffusion equation $d^2T/dx^2 = dT/dx$ for the stepped initial temperature is

$$T(x, t) = \bar{T} - \tfrac{1}{2}\,\Delta T\, \mathrm{sgn}(x)\, \mathrm{erf}(|x|/2\sqrt{t}) \tag{45}$$

implying that the heat current at x is $j = -dT/dx = (\Delta T/2\sqrt{t}) \exp(-x^2/4t)$ and the integrated current past the center ($x = 0$) is

$$Q(t) = \Delta T \sqrt{t}. \tag{46}$$

Comparison of Eqs. (46), (43), and (44) shows that neither the harmonic chain nor the chain of hard rods satisfies Fourier's law, which is no surprise, but Fig. 8 indicates that none of the HHR systems which we have computed does either.

References

Alder, B. J., and Wainwright, T. E. (1970). *Phys. Rev. A* **1**, 18.
Bell, R. J. (1972). *Rep. Progr. Phys.* **35**, 1315–1409.
Bivins, R. L., Metropolis, N., and Pasta, J. R. (1973). *J. Comput. Phys.* **12**, 65–87.
Bloch, F. (1928). *Z. Phys.* **52**, 555.
Bolsterli, M., Rich, M., and Visscher, W. M. (1970). *Phys. Rev. A* **1**, 1086–1088.
Chester, G. V. (1963). *Rep. Progr. Phys.* **26**, 411.
Debye, P. (1914). "Vorträge über die kinetische Theorie der Materie und Electrizetät." Teubner, Berlin.
Ford, G. W., Kac, M., and Mazur, P. (1965). *J. Math. Phys.* **6**, 504–515.
Helleman, R. H. G. (1972). Ph.D. Thesis, Yeshiva University.
Ishii, K. (1973). *Progr. Theor. Phys., Suppl.* **53**, 77–138.
Ishii, K., and Matsuda, H. (1970). *Progr. Theor. Phys., Suppl.* **45**, 56–84.
Jackson, E. A., Pasta, J. R., and Waters, J. F. (1968). *J. Comput. Phys.* **2**, 207.
Kubo, R. (1973). *Acta Phys. Austr., Suppl.* **10**, 301–340.
Lorentz, H. A. (1905). *Proc. Amst. Acad.* **7**, 438.
Mazur, J., and Rubin, R. J. (1963). *Amer. J. Phys.* **31**, 835–836.
Metropolis, N., Rosenbluth, A. N., Rosenbluth, M. N., Teller, A. H., and Teller, E. (1953). *J. Chem. Phys.* **21**, 1087.
Miura, K. (1973). Ph.D. Thesis, University of Illinois, Urbana.
Mjolsness, R. C., and Visscher, W. M. (1972). *Phys. Fluids* **15**, 1854.
Nakazawa, H. (1970). *Progr. Theor. Phys., Suppl.* **45**, 231–262.
Northcote, R. S., and Potts, R. B. (1964). *J. Math. Phys.* **5**, 383–398.
O'Connor, A. J., and Lebowitz, J. L. (1974). *J. Math. Phys.* **15**, 692.

Painter, R. D., and Hartmann, W. M. (1974) *Phys. Rev. B* **10**, 2159–2171.
Payton, D. N., Rich, M., and Visscher, W. M. (1967). *Phys. Rev.* **160**, 706–711.
Payton, D. N., Rich, M., and Visscher, W. M. (1968). *In* "Localized Excitations in Solids" (R. F. Wallis, ed.), pp. 657–664. Plenum, New York.
Peierls, R. E. (1929). *Ann. Phys. (Leipzig)* [5] **3**, 1055.
Rich, M., Visscher, W. M., and Payton, D. N. (1971). *Phys. Rev. A* **4**, 1682.
Rieder, Z., Lebowitz, J. L., and Lieb, E. (1967). *J. Math. Phys.* **8**, 1073.
Rubin, R. J., and Greer, W. L. (1971). *J. Math. Phys.* **12**, 1686.
Thompson, B. V. (1961). M.S. Thesis, University of Birmingham, England.
Toda, M. (1970). *Progr. Theor. Phys., Suppl.* **45**, 174–200.
Van Kampen, N. G. (1971). *Phys. Norv.* **5**, 279.
Visscher, W. M. (1971). *Progr. Theor. Phys.* **46**, 729–736.
Visscher, W. M. (1974). *Phys. Rev. A* **10**, 2461–2472.
Visscher, W. M., and Rich, M. (1975). *Phys. Rev. A* **12**, 675.
Wood, W. W. (1968). *J. Chem. Phys.* **48**, 415.
Yonezawa, F., and Morigaki, K. (1973). *Progr. Theor. Phys., Suppl.* **53**, 1–76.
Zabusky, N. J. (1973). *Comput. Phys. Commun.* **5**, 1–10.
Zwanzig, R. (1965). *Annu. Rev. Phys. Chem.* **16**, 67.

Author Index

Numbers in italics refer to the pages on which the complete references are listed.

Y

Z

Subject Index

D

H

I

J

K

L

S

T

Contents of Previous Volumes

Volume 8: Energy Bands of Solids

Volume 9: Plasma Physics

Volume 10: Atomic and Molecular Scattering

A 6
B 7
C 8
D 9
E 0
F 1
G 2
H 3
I 4
J 5